Rita Singh

Profiling Humans from their Voice

Springer

Rita Singh
Carnegie Mellon University
Pittsburgh, PA, USA

ISBN 978-981-13-8402-8 ISBN 978-981-13-8403-5 (eBook)
https://doi.org/10.1007/978-981-13-8403-5

This Springer imprint is published by the registered company Springer Nature Singapore Pte Ltd.
The registered company address is: 152 Beach Road, #21-01/04 Gateway East, Singapore 189721, Singapore

Profiling Humans from their Voice

"I HAVE SHOWN THAT VOICE QUALITY DEPENDS ABSOLUTELY UPON BONE STRUCTURE, THAT BONE STRUCTURE IS INHERITED, AND THAT THEREFORE VOCAL QUALITY IS INHERITED."

—Walter B. Swift, 1916

The possibility of voice inheritance. In Review of Neurology and Psychiatry, Volume 14, Page 106.

Preface

What is Different

Computational profiling as described in this book differs from prior efforts in one key respect. The fact that voice is correlated to many human parameters has been known to scientists, doctors, philosophers, priests, performers, soothsayers—in fact, to people of myriad vocations for centuries. Many of these correlations have been scientifically investigated, and demonstrated or proven in many ways. The key difference between that earlier science, and what this book represents is simply this: while earlier methods were focused on demonstration and proof of existence (of such relationships), and used observable relations to attempt to predict human parameters from voice, the current science does not require such observables. It is built instead on the hypothesis that *if any factor whatsoever influences the human mind or body, and if that influence can be linked to the human voice production mechanism through any pathway whatsoever, then there **must** exist an effect on voice*. The current science of profiling is then all about discovering those effects. If we hypothesize the existence of some influence, whose effects we cannot model or observe through standard mechanisms known to us, then we must devise artificially intelligent mechanisms that *can* model or observe those. This basically represents a handover of the capability of discovery to intelligent systems designed by us.

This book traverses pathways hewn through information in multiple disciplines, and represents one journey undertaken in search of solutions to such mysteries of the human voice.

About This Book

The ability to shape sound is an amazing gift that most intelligent creatures have. Some do this in response to stimuli, others do this to convey meaning as well. With the evolution of intellectual abilities, it was only natural for this ability to evolve to convey more complex thoughts and deeper meaning. It was also natural, through the process of natural selection, for those kinds of information-embedding mechanisms that were more supportive of survival to remain, propagate, and be refined. Sound is no exception. Deliberately or inadvertently, partly for evolutionary reasons and partly due to our specific biological construction, information about each person is embedded in their voice.

At this point in time, we don't have a good idea of just how much information is embedded in the human voice. Reasoning about speech as a biomechanical, social, and cognitive process leads us to believe that there is a tremendous amount of information in it—more than we are capable of assimilating through our limited capabilities of auditory observation and perception.

This book captures some of my thoughts, ideas, and research on discovering or even guessing the range and extent of information embedded in the human voice, on deriving it quantitatively from voice signals, and using it to infer bio-relevant facts about the speaker and their environment. In my experience, this endeavor is so rife with challenges, and human voice is of such tremendous importance, that it deserves to be assigned the status of a subfield of acoustic intelligence in its own right—so I call it *Profiling Humans from their Voice*, which is also the title of this book.

The book has two parts. Part I takes a sweeping look at the landscape of scientific explorations into the human voice, which has been the subject of an astounding volume of research and observation, literally over centuries. Voice, its acoustics, its content, the effect of various factors on these, and conversely *its* effect on them, and on other humans, its perceptions, and its manipulations are all discussed in this part. This part also dives into the voice *signal* from a signal processing perspective. It (very) briefly elucidates the concepts that might be relevant or foundational to profiling, attempting to link some subjective observations of the quality of voice, based on which most human judgments are made, to explainable or quantifiable signal characteristics.

Part II deals with computational profiling: the computationalization of human judgment (and beyond). Predicated largely on concepts in machine learning and artificial intelligence, this part discusses mechanisms for information discovery, feature engineering, and the deduction of profile parameters from them. It discusses the subject of reconstructing the human persona from voice, and its reverse—the reconstruction of voice from information about the human persona. It ends with a discussion of the applications and future outlook for the science of profiling, and also ethical issues associated with its use. Issues of reliability, confidence estimation, statistical verification, etc., that are extremely important in practical applications, are all discussed.

There are multiple chapters under each part, divided up to make the overall theme of the part easier to navigate. Each part is provided with a summary of the chapters it includes.

Although meant to be a technical exposition, in my mind this book is a canvas with dabs of paint from many fields. Together they form a complete picture—but it is still an underdrawing. There is just so much that one can write in a book of normal dimensions. As I type the last full stop in this book, it feels incomplete in many respects. I hope the studies referenced in this book can fill in the missing details to some extent.

More than anything, I hope this book is both enjoyable and informative to all readers.

Pittsburgh, PA, USA
March 2019

Rita Singh

Acknowledgements

I thank the United States Coast Guard for initiating this research in early December 2014, and for their strong and unfaltering support of it since then. They continue to drive it forward through constructive feedback from actual field tests of this technology.

Numerous colleagues across the world have given their opinion, enhanced my knowledge, and helped me refine my thoughts and perspective. Of those, some who have had an especially transformative influence on this work are Bhiksha Raj from Carnegie Mellon University, and Joseph Keshet from Bar-Ilan University. John McDonough was a source of inspiration to me as a scientist—and continues to be, though he is no more with us. I thank all from the bottom of my heart.

I thank my family, especially my mother Susheela Singh, my aunt Namita Singh who passed away since I began writing this book, my sister Sapna a.k.a Jugnu Singh, and our niece Neha Singh for accomplishing the impossible task of creating Time. This book was written through times of great turmoil. It was written in negative time—in the time I didn't have. For this book to exist, time had to be manufactured somehow. They solved the problem by making every distraction in my life vanish for the duration they were in my physical vicinity. It came at the cost of *their* time. My friend Satya Venneti did so as well by vanishing for brief periods and magically reappearing with things that represented significant time savings for me. I thank Alka Patel, my bravest and ever-smiling friend, for her friendship and support through all this.

I have been fortunate enough to have spent my entire career working alongside intellectuals and scholars, who are giants in their own fields. If those, I would like to especially thank Richard Stern for his invaluable lessons in signal processing and psychoacoustics, and James Karl Baker for sharing his phenomenal insights into the subjects of automatic speech recognition, human language, speech and voice in general. Their scientific perspective, and that of all my colleagues around the world who I interact with has helped shape my thinking and thereby this book.

Two important milestones were crossed while this book was being written. The first was crossed when we demonstrated voice profiling technology as described in this book for the first time, at the World Economic Forum in Tianjin, China, in September 2018. It was tested by nearly a thousand people within a span of 3 days. The systems demonstrated made multiple profile deductions from voice, and also recreated the speaker's face in virtual reality. The second milestone was crossed in February 2019, when we used the reverse of this technology to recreate the voice of one of the world's greatest painters—Rembrandt van Rijn (1606–1669)—from his self-portraits and other information. This was done in collaboration with Rijksmuseum of Holland (and their art experts and historians), J. Walter Thompson Inc. in Amsterdam, and Linguists from Leiden University in Holland with the sponsorship of ING Bank in Europe. My students Mahmoud Al Ismail, Richard Tucker, Yandong Wen, Wayne Zhao, Yolanda Gao, Daanish Ali Khan, Shahan Ali Memon, Hira Yasin, Alex Litzenberger, and Jerry Ding were instrumental in the creation of both landmarks. In addition to the students who directly contributed to these, I thank Abelino Jiminez, Anurag Kumar (a.k.a Anurag Last Name Unknown), Ahmad Shah, Rajat Kulshreshtha, Prakhar Naval, Raymond Xia and Tyler Vuong for their incredible creativity and continuing contributions to this work. I wouldn't have been able to take this research forward without the contribution of all of these students.

Contents

List of Figures

List of Tables

Part I
Profiling and the Human Voice

Part I of this book connects findings from scientific explorations in different fields to define and outline an emergent area of research—that of profiling humans from their voice. It is not a comprehensive review of these fields, but rather a cursory attempt to build a perspective that may be central to future developments in the area. There are six chapters in this part. Chapter 1 defines profiling and segregates its different aspects. The rest of the chapters provide key insights into each aspect, drawn from existing literature that often spans multiple fields. Chapter 2 is a brief overview of the key concepts in voice production and audition that have a bearing on profiling. Chapter 3 describes the relationship of voice to different categories of profile parameters. Chapters 4 and 5 deal with the information content in the voice signal, much of which is based on core signal processing concepts. These are explained where necessary within the chapters. Chapter 6 deals with voice quality, a very subjective and descriptive entity that has been so extensively correlated with human parameters that it warrants a separate chapter in itself.

Chapter 1
Profiling and Its Facets

1.1 Profiling

The term *profiling from voice* refers to the deduction of personal characteristics, and information about the circumstances and environment of a speaker from their voice. At the outset, we note the distinction between the terms *voice* and *speech*. "Voice" refers to sound produced in the human vocal tract. "Speech" is the signal produced by modulating voice into meaningful patterns.

We will use the term "*profiling*" to refer to the process of deduction of these entities, and the term "*parameters*" to refer to the entities deduced. Later in this book we will often use the term "*features*," which will exclusively refer to mathematical representations derived from the voice signal for the purpose of profiling. Other terms such as "*characteristic*," "*entity*" etc. will be used in their normal linguistic sense, depending on context. For example, properties of the voice signal, such as voice quality, speaking rate, loudness etc. will be referred to as "*signal characteristics*." In later chapters, we will also see that features are a subset of signal characteristics, but not vice-versa. As long as we clearly adhere to this specific usage of terms, the exposition in this book can be followed with precision.

We begin with expanding the definition of some of these terms, to establish the specific organization of topics and the vocabulary that we will use in this book.

1.1.1 Parameters

Profiling comprises the deduction of *all* kinds of parameters from voice. This space is vast, and an initial categorization of parameters based on some broad-level commonalities is needed in order to proceed. Although it is obvious that many categorizations are possible, in this book we will follow the specific broad-level subcategorization of parameters given in the list below. This list only includes a few samples of parameters from each category, and is not an exhaustive one.

R. Singh, *Profiling Humans from their Voice*,
https://doi.org/10.1007/978-981-13-8403-5_1

1. **Behavioral parameters**: Dominance, leadership, public and private behavior.
2. **Demographic parameters**: Race, geographical origins, level of education.
3. **Environmental parameters**: Location of the speaker at the time of speaking, objects surrounding the person at the time of speaking, devices and communication channels used for voice capture and transmission.
4. **Medical parameters**: Presence or absence of specific diseases, medications and other substances (drugs, food, intoxicants etc.) in the body, state of physical health, state of mental health, effects of trauma, effects of medical procedures, presence or absence of physical abnormalities or disabilities.
5. **Physical parameters**: Height, weight, body-shape, facial structure.
6. **Physiological parameters**: Age, hormone levels, heart rate, blood pressure.
7. **Psychological parameters**: Personality, emotions.
8. **Sociological parameters**: Social status, income, profession.

In addition to the four key terms (profiling, parameters, features, characteristics) mentioned above, the following terms may sometimes be used in this book:

- **Factors**: A generic term that refers to elements, phenomena, processes, influencing entities etc. For example: "*Atmospheric pressure is a factor that influences vocal parameters.*" The word "*factor*" may substitute for the word "*parameter*" in some contexts that do not immediately involve deduction, e.g. "humidity is also a factor that affects voice characteristics."
- **Voice-print**: Any segment of speech, regardless of its lexical or semantic content, or completeness.
- **Bio-relevant parameters** (or *bio-parameters*): All parameter categories mentioned above except environmental parameters. Note that this is a superset of parameters that have strictly biological relevance.

1.1.2 Features and Signal Characteristics

The set of features includes informative representations of the voice signal. The set of signal characteristics refers to various properties of the voice signal that have traditionally been derived and studied in the context of voice. Some of these are listed below, based on how they are measured (described in detail in later chapters). Again, this list is not exhaustive:

1. **Temporal domain features**: Properties measurable in the time domain, such as signal energy, zero crossing rate, loudness, energy, speaking rate, phonation rate etc.
2. **Spectral domain features**: Properties measurable in the frequency domain, such as pitch or fundamental frequency, harmonics, spectral flux, spectral density, spectral roll-off etc.

3. **Composite domain features**: Require measurements in both frequency and time, or are derivatives of spectral and temporal domain features, such as rhythm, melody etc.
4. **Qualitative features**: Largely human-judged properties that can be assigned to a voice signal, such as voice quality and its related components—nasality, raspiness, breathiness, roughness etc.

1.2 A Look at the Landscape of Voice Studies

Human voice has been widely studied in the context of, and within the purview of many scientific fields. In fact, the range of studies that have been performed in this context over the past century is staggering. Within these fields, voice has been correlated to multiple parameters.

Table 1.1 shows some examples of the fields in the context of which voice studies have been performed, with one or more sample publications listed against all except those that are likely to include voice, or are obviously sound or voice related fields. This list is by no means exhaustive. There are many more fields that have performed investigations in the context of voice and speech. Collectively, these studies reveal a wealth of information that can be potentially brought to bear on the science of profiling.

1.2.1 Parameters That Have Been Correlated to Voice

Studies mentioned in Table 1.1, and others that exist abundantly in the literature, fall under two broad categories. In one, *perceptions* of various entities, such as facial appearance, body-size, benevolence, competence, dominance, dynamism, personality etc. have been correlated with voice. In the other, *measurements*, or expert ratings of multiple parameters that relate to the speaker's persona, and their speaking environment have been correlated with voice. This list overlaps with the one above, and includes a multitude of diseases such as psychiatric illnesses, physical illnesses, various genetic syndromes etc. and a very wide range of other parameters such as age, attractiveness, body size and shape, dominance, emotion, behavior, personality, intellectual capacity, deception, intoxication, drug use and abuse, exposure to chemicals and toxins, injuries, trauma, sexual bias and orientation, sexual behavior, sleep deprivation etc. In addition, the list includes myriad environmental factors such as atmospheric moisture, relative humidity, odors, pollution etc.

The parameters that have been found to influence voice, and the specific characteristics of voice that reflect them, are discussed in detail in Chap. 3.

Table 1.1 Examples of fields of science and voice related studies

Field [studies]	
1 Acoustics	2 Adenology [1]
3 Algedonics [2]	4 Anaesthesiology [3]
5 Anatomy [4]	6 Andragogy [5]
7 Angiology [6, 7]	8 Anthropobiology [8–10]
9 Anthropology [11, 12]	10 Aromachology [13]
11 Archaeology [14]	12 Astheniology [15, 16]
13 Audiology	14 Auxology [17]
15 Bioecology [18]	16 Biology [19]
17 Psychiatry [20, 21]	18 Biometrics
19 Bionomics [22]	20 Cardiology
21 Catacoustics	22 Catechectics
23 Cell biology	24 Characterology [11, 23]
25 Chronobiology [24, 25]	26 Criminology [26]
27 Cinematology [27]	28 Demography [28]
29 Demology [29]	30 Dialectology
31 Dramaturgy [30, 31]	32 Ecology [32, 33]
33 Endocrinology [34, 35]	34 Enzymology [36]
35 Ephebiatrics	36 Epidemiology [37]
37 Epileptology [38]	38 Ergonomics [39]
39 Ethnology [40]	40 Ethology [41]
41 Forensics [42, 43]	42 Gastroenterology [44]
43 Genealogy	44 Gerontology [45]
45 Glossology	46 Harmonics
47 Heredity [46, 47]	48 Histopathology [48, 49]
49 Hydrodynamics [50, 51]	50 Hydrokinetics [52, 53]
51 Hydrometeorology [54]	52 Hymnology [55]
53 Hypnology [56]	54 Immunology [57, 58]
55 Immunopathology [59, 60]	56 Kalology [61]
57 Kinematics [62]	58 Kinesics [63]
59 Kinesiology [64]	60 Koniology [65]
61 Laryngology	62 Linguistics
63 Magnetics [66]	64 Mammalogy [67]
65 Mechanics [68]	66 Metaphysics [69, 70]
67 Microanatomy [71]	68 Musicology
69 Myology [72]	70 Naology [73, 74]
71 Nasology [75]	72 Neonatology [76, 77]
73 Nephrology [78]	74 Neurobiology [79]

(continued)

Table 1.1 (continued)

Field [studies]	
75 Neurology [80, 81]	76 Neuropsychology [82]
77 Odontology [83]	78 Olfactology [13]
79 Osteology [84]	80 Otology [85]
81 Otorhinolaryngology [86, 87]	82 Paedology
83 Peridontology	84 Parapsychology [88–91]
85 Pathology	86 Pedagogy [92, 93]
87 Pharmacology [94]	88 Pharyngology
89 Phenomenology [95]	90 Philology [96]
91 Philosophy [97, 98]	92 Phoniatrics
93 Phonology	94 Phraseology
95 Physics	96 Physiology
97 Primatology [99]	98 Prosody
99 Pseudology [100]	100 Psychobiology [101]
101 Psychogenetics [102]	102 Psychognosy [103, 104]
103 Psychology	104 Psychopathology [105]
105 Psychophysics [106]	106 Quinology [107]
107 Reflexology [108]	108 Rheumatology [109, 110]
109 Rhinology [85]	110 Semantics [111]
111 Sexology [112, 113]	112 Sociobiology [114]
113 Sociology [103, 115]	114 Somatology [116]
115 Sophiology [117]	116 Stomatology [118, 119]
117 Teratology [120]	118 Thermokinetics [121]
119 Thermology [122]	120 Threpsology [123, 124]
121 Tonetics [125]	122 Toxicology [126]
123 Traumatology [127, 128]	124 Typhlology [129–131]
125 Victimology [132, 133]	126 Virology [134, 135]

1.3 Profiling Humans, by Humans

Section 1.2.1 above lists many parameters that have been correlated to voice. This is not necessarily a list of *judgments* we make from voice—a subject that has been investigated in many studies. These in turn must be differentiated from studies on what (primarily emotions and thoughts) can be *evoked* in humans by voice, although both are related in obvious ways.

1.3.1 Judgments Made from Voice

Even elephants can judge age, ethnicity and gender of humans from their voices [136].

It is well known that we make myriad judgments about people based on voice. In the field of speech perception, this aspect has been especially well studied. Interestingly, the ability to make judgments about people from their voice differs between the genders [137].

Of the judgments made, many are made independently of the co-occurrence of other judgments (such as height and education, which have no correlation with one another), while some are conditioned on other parameters. For example, the judgment of gender is conditioned on age (it is difficult to tell the gender of very young children from their voice). The properties of the voice signal that play the greatest role in allowing us to make these judgments are those related to *voice quality* [138]. Voice quality is not a single entity—rather it is a loosely defined term given to a collection of entities that comprise it. We discuss voice quality and its measurement in detail in later chapters of this book. The list below shows a number of judgments that are made based on the perception of voice *quality* alone, as in [139]. Note that the categorizations here differ from what we will use in later chapters.

- **Physical characteristics**: age, appearance (height, weight, attractiveness), dental and oral/nasal status, health status, fatigue, identity, intoxication, race, ethnicity, gender, sexual orientation, smoking habits.
- **Psychological characteristics**: arousal (relaxed/hurried), competence, emotional status/mood, intelligence, personality, psychiatric status, stress, truthfulness.
- **Social characteristics**: education, occupation, regional origin, role in conversational setting, social status.

Because humans ubiquitously and subconsciously judge personality from voices, people with voice disorders often fear that their personalities may be incorrectly portrayed by their voice, and refrain from speaking in many instances. This is a well-known problem encountered in people with voice disorders. Of course, human hearing is not perfect, and many of the impressions we form of other people's characteristics based on even normal voices are inaccurate. This is also something that we experience quite often in today's world—when we talk to strangers over phone and meet them later only to discover that they don't fit all of the impressions we formed of them from their voice.

Humans also make significant judgments about the *environment* of the speaker from their voice. For example, we can easily tell whether a recording has been transmitted over the phone, or recorded in close (physical) vicinity of the speaker. We "hear" echoes and reverberations in the acoustic signal, and its effect on voice, and are able to gauge the size of the enclosure within which it was produced to coarse degrees from voice (small space, large hall etc.). Often when we hear the voice of a person calling from a moving vehicle (e.g. car, train, boat), from the manner in which the voice "shakes" (in addition to other acoustic cues) we are able to tell that the person is on a moving vehicle, and are often able to identify the broad category of the vehicle. Many of the judgments that people make about their environment are not well documented or studied. Nevertheless, it is clear that the voice signal carries cues about the environment as well.

Human abilities to judge from voice are thought to arise from a long evolutionary process, wherein they were germane to survival. In animals, such judgment co-evolved with the successful use of vocal cues to convey different *impressions*, such as those of danger, interest, dominance. Studies have found that many animal species use vocal cues to indicate situations that are threatening, or relevant to mating and social relationships—all of which are important for survival [140, 141]. The evolutionary hypothesis is further augmented by the observed co-development of abilities to detect and ignore irrelevant or unreliable vocal signals [142, 143]. A case in point is the common observation that urban humans today seem to have lost the acuity with which they note and differentiate natural sounds—such as those that may indicate wind speed, approaching weather conditions etc. Presumably this is because the modern urban human is well protected against the vagaries of nature, and these sounds are no longer as relevant to their survival as they were in the past centuries. On the other hand, an urban driver may be acutely aware of even faint sounds made by their vehicle on the road as it runs over obstacles, and may be able to judge the nature of the obstacle by the sounds. This shift in auditory awareness also affects the judgments made of other people within a crowded urban environment. Often, people lose the ability to "tell" much about the new people they talk to each day. We must however note that while these observations are commonly made, and support the evolutionary hypothesis, formal studies have not been performed to establish the connections between living styles and loss of voice-based judgment in urban humans.

Of the voice signal characteristics that have been found to be relevant to human judgment, some examples are tone and the sub-components of voice quality (explained in Chap. 6). Alterations in these have been observed to convey sarcasm, irony, disapproval, contempt and other elements of interaction [144–146]. Alterations in *prosody* have been related to various affective cues in vocal interactions [147]. *Speaking rate* has been related to the judgment of competence [148, 149] and other personality characteristics [150, 151]. *Loudness* has been reported to convey perceptions of dominance and dynamism, among other things [152]. *Pitch variations* are thought to convey impressions of benevolence [153]. These are just a few examples. There is a wealth of information in the literature about signal-level characteristics in relation to the judgments people make about other people. Later chapters in this book give more details.

1.3.2 Reactions Evoked by Voice

Voice evokes reactions in the speaker as well as in the listeners. Hearing oneself speak activates a feedback mechanism that is important for proper production and delivery of speech. It modifies the activity in the brain's auditory complex. Hampered audiovisual feedback mechanisms have been implicated in voice disorders such as aphasia, stuttering and schizophrenic voice hallucinations [154]. Often actors report that even the emotions that they merely portray through voice modify their mood, breathing, heart-rate etc. A very interesting perspective on the human voice is given in [155], in which audible expression or vocalization of expression is analyzed in this context.

In listeners, voice evokes many kinds of reactions such as changes in feelings, mood and emotions. These in turn trigger physiological (such as hormone surges), psychological and behavioral changes. The first recorded study that attempted to categorize such effects was carried out in the 2nd century AD by Julius Pollux, a Greek grammarian, scholar and rhetorician from Naukratis, Egypt. According to historical accounts, Emperor Commodus appointed him a professor-chair of rhetoric in Athens at the Academy—*on account of his melodious voice*! Julius Pollux was the author of "Onomasticon," which still survives, and is, according to the Encyclopedia Britannica, "a Greek dictionary in ten books, each dedicated to Commodus, and arranged not alphabetically but according to subject-matter." His study of voice (vocal quality) from the listeners' perspective is reported to be part of these books.

The reactions evoked by voice in humans (and animals) continue to be studied and recorded. It is well known today, for instance, that the sound of music evokes activity in the brain, and consequently evokes emotions in listeners [156]. The same is true for the sound of human voice, and is supported by a multitude of studies e.g. [157–161]. Characteristics such as speech rate and loudness have been found to affect psychological states such as fear, sadness, anxiety, depression etc. [162]. Amplitude, frequency, pitch range, speed, inharmonicity etc. have been found to be used in conveying a sense of urgency [163]. A large number of reactions have been shown to be evoked by alterations in voice quality, prosody and energy. The *amygdala*, the brain's integrative center for emotions situated to the anterior of the hippocampus, is affected significantly by voice [164, 165], as is the pre-frontal cortex [166] and other areas of the brain, as these studies indicate. The amygdala also controls emotional behaviors, such as motivation. In fact, the net effect of voice on the brain is profound in many ways, and the pathways to evoking reactions in the limbic and nervous systems are well established today. In the light of these, it is only reasonable to believe that the features in voice that carry the necessary cues, if found, could indicate the presence of changes (such as hormone levels) that are brought about by the known reactions evoked by them. This chain of causes and effects is very relevant to profiling.

1.4 Computational Profiling

Computational profiling—what this book is largely about—is essentially an informed transfer of the process of making judgments about humans into the computing domain. From the studies mentioned above and in the literature, it is clear that human judgments are constrained by the limitations of human hearing, the brain's interpretative abilities, and the physical and mental states of the listener. As a result, they are often inaccurate. They are also inconsistent—they may be different at different times, although the voice signal continues to have the same information content.

Machines, of course, do not have these drawbacks. Once the process is computationalized, most ambiguities of repeatability and accuracy that play up in humans, are expected to vanish. In addition, it may be possible to extract, process and analyze much *finer* cues in spectro-temporal and other transform domains, more accurately, faster and at a much greater scale (both in terms of number of cues processed, and the number of signals processed for comparison, learning or other purposes). However, for computational profiling, there are many associated challenges that need to be addressed. Some of these confound even humans today, and are not well researched. Before we discuss these, let us review the basic process of computational profiling, as envisioned today.

1.4.1 The Basic Process of Computational Profiling

The basic process of computational profiling could be knowledge driven, data driven, or a combination of both.

In a purely knowledge-driven approach, voice characteristics that are known to correlate well to the parameters to be deduced are quantified, are extracted from the voice signal. They are then mapped to the parameters within ranges that are statistically significant. The correlations are established first because the accuracy of the profile is dependent on the degree of correlation. It also depends on the accuracy with which the relevant characteristic(s) can be measured. In addition to *how* to measure them, it is equally important to know *where* (in the signal) to find them, to measure them from—especially if the signal is degraded or noisy.

In a data-driven process, one could take a bottom-up or a top-down approach. The conceptual difference between the two is depicted in Fig. 1.1. In the bottom-up approach, The signal or its transform is processed to identify the patterns or signatures relevant to the parameter to be deduced. These are then either explicitly measured or extracted, or secondary features that capture them (e.g. prosody) are extracted. Subsequently, the relationship between these and the parameter to be deduced is learned. The identity of the parameter is needed during the learning process only. When a new voiceprint is presented for profiling, the same process is followed, except now the learned relationship is used to predict the presence/absence or *degree* of manifestation of the parameter. A degree, also called a *rating*, is computed based on some numeric values computed by the regression mechanism employed during learning and prediction.

In the top-down process, on the other hand, the signal or its transform is directly processed through a mechanism that learns to relate the signal or its transform to the parameter, while also learning the most appropriate features for this, and seamlessly isolating, representing and using the patterns that are most relevant to the deduction process. The identity of the parameter must be known for this, but multiple parameters can be processed simultaneously. The artificial intelligence (AI) based profiling processes described in Part II of this book follow the top-down approach. In such an approach, the discovery of features and patterns is dependent on the design or engineering of the profiling system, and can either be entirely data-driven, or be constrained by known factors in a knowledge-based fashion. This would then represent

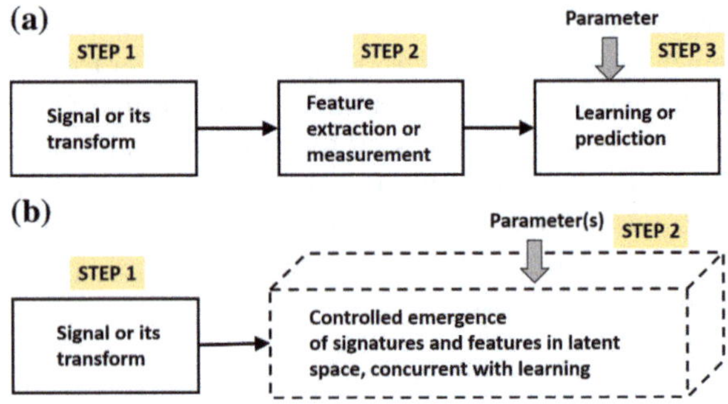

Fig. 1.1 The (**a**) top-down and (**b**) bottom-up approaches for computational profiling

a "combination" system, that is both data and knowledge driven. We will discuss both these approaches for profiling in detail in later chapters.

Note that as an alternative, an approach that may succeed in approximating some profile parameters *by proxy* is possible. This can be done by using existing voice matching techniques in an indirect way, wherein a voiceprint may be compared with those in an existing database, and known profile parameters of the speaker of the matching voice may be assigned to the speaker of the voiceprint. We will refer to this as "*phantom profiling*," but will not discuss this in detail in this book. Phantom profiling may be done for first-level approximations. It is not true profiling in that the parameters are not directly deduced from voice, and a given speaker's deduced parameters are likely to be shared with those of the matched speaker.

1.4.2 The Basis for Computational Profiling

At the core of computational profiling is the hypothesis that if *any* parameter influences the speaker's voice or its production mechanism, regardless of the degree of its influence, its effects on the voice signal must be discoverable, and the parameter must be deducible from the effects. The *basis* of profiling is the fact that the voice signal is modulated and modified in different ways and at different scales by different factors. During its production, it is modulated by the factors that influence the speaker's voice production mechanism. Once produced, and before it is captured for communication or storage, it is modulated and modified by the environment, and subsequently by the devices that capture it, media that store it, and channels that communicate it. All of these factors alter the signal in perceptible and imperceptible ways. We will refer to these alterations as "signatures" of the influencing factors (or parameters). Signatures of profile parameters can appear at multiple scales of resolution in the voice signal.

An example is shown in Fig. 1.2, which shows the influence of a common caffeine beverage on an individual, under highly controlled conditions. The recording was done on the same day, and the recording device and conditions across the two are

The spectrogram of the sound /EY/ uttered by 32 year old, 5'8" tall white Caucasian male in two different physical states.The utterances are normalized for energy, and spliced into the same figure to maintain scale and for visual clarity. The left instance is captured when the person is in the normal state. The right one is captured after the person had a significant extra dose of a caffeine beverage.

Fig. 1.2 An example of the influence of caffeine on voice

identical. The influence of this beverage (also noted in other studies, e.g. [167]) is clearly visible in Fig. 1.2. Note that not all influences may show up so clearly in standard visual representations of speech. In fact, many signatures may not manifest in visual representations at all. For profiling, the presence of these must be determined, and their degree measured to deduce (in this specific example case) the time and/or amount of beverage consumed. In general, for profiling for any parameter, its signatures must be isolated from a voice sample and used to infer its type and/or degree of manifestation on the human. In addition, these signatures must be *disambiguated* from the signatures of other concurrent factors that may have similar influence on voice. Moreover, recording conditions may not always be so ideal and controlled, and multiple *types of factors* may influence speech at the same time. These complications make profiling an extremely difficult task.

1.4.3 Key Challenges in Computational Profiling

Of the key challenges in profiling, the most important ones (outside of computational complexity and algorithmic issues) are: (a) the necessity to deal with the propensity of the voice signal to vary, coupled with the ability of humans to mask or disguise their voice, and (b) the propensity of the parameters to be deduced to vary in their degree or intensity. For simplicity, we refer to these as *human variability* and *parameter variability* respectively.

1.4.3.1 Inherent Human Variability

The human voice is the outcome of an extremely complex process that is not deterministic. At any given time, under any given parameter conditions, the same sequence

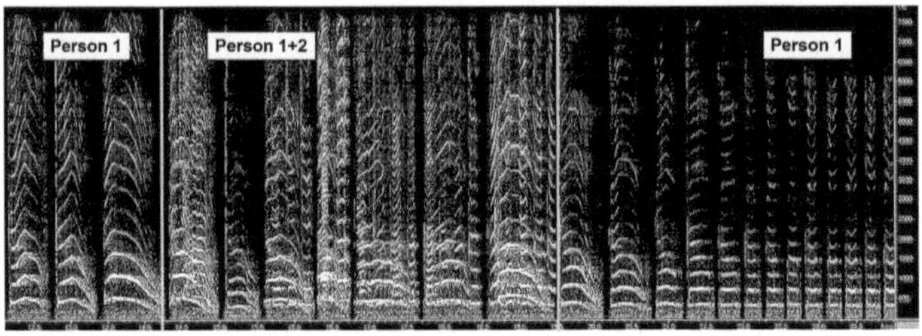

(a) A narrow-band spectrogram of two children of the same age, height, schooling, language and ethnicity saying the word "go" repeatedly while playing a game. The recording is in its original continuous form, and thus all physical and environmental factors can be considered to be equal and constant across the recording for the two speakers. The instantaneous differences in each rendering are entirely attributable to inherent variability in response to emotion (levels of excitement in this case). These manifest very differently for the two speakers, as seen in the central portion of the figure that allows for close comparison.

(b) Narrow-band spectrograms of the voice of the same person while a) talking normally, b) loudly, c) shouting, d) screaming, e) whispering, f) crying and g) laughing.

Fig. 1.3 An illustration of multi-resolution, instantaneous and cross-human variability in voice

of sounds uttered by the same speaker may be different in spectral and temporal aspects. An example is shown in Fig. 1.3.

This inherent variability of the human voice is amplified by changes in human parameters, such as emotions, physical and mental state, social circumstances, environmental circumstances etc. While it is this variability that allows for profiling for instantaneous states of the human, not all of the variations are entirely attributable to parametric conditions. For example, if a person in terror is screaming, some aspects of the signal produced are attributable to the nature of the vocal mechanism of the person, and not to fear alone. This allows for the deduction of relatively more persistent characteristics in the person, such as age, ethnicity etc. The term *inherent variability* refers to range of variations that any person's vocal mechanism can sustain, voluntarily or involuntarily. On the other hand, voluntary changes in voice done to convey impressions of parameters that are not in evidence at the time, such as a different personality, are referred to as *voice disguise*.

Comparative studies to gauge the **range of variability** of human voice across the human population have not been done in the literature. Some variations are common across humanity, for example, extreme factors, imminent dangers, the need to be secretive etc. result in voluntary but natural variations in voice such as whispering, shouting, screaming etc.

Inherent variations in voice happen in *response* to multiple types of stimuli. Often these responses are instantaneous, as illustrated in Fig. 1.3a. Many stimuli have longer-term effects as well, which can manifest at different resolutions in the voice signal. For example, Fig. 1.3b shows the spectrograms of the voice of the same person while talking, shouting, whispering, screaming, laughing and crying. These are enacted, but show the key features of voice in these different states.

While the range of inherent variability of voice has not been comparatively studied, the *mechanisms* underlying many of these have been studied and are known. For example, whispering happens via suppression of the process of phonation (explained in Chap. 2), by keeping the vocal cords separated while speaking all sounds. Tonal variations and change of prosody while speaking involve the selective emphasis and de-emphasis of several voice components, including energy distribution across frequencies, loudness, pitch etc. Shouting and screaming involve extreme alterations of pitch.

The challenges in profiling that are related to the natural variability of the human voice can be expressed in the form of questions that remain unanswered to-date, of which, of course, the most sweeping one is—what *is* the range, nature and extent of possible variations in the human voice? Related to this is a slew of other questions, of which some important ones are:

1. In any given state where only one stimulus is allowed to vary, can a speaker produce a continuum of voice variations as characterized by the values of one or more measurable signal characteristic(s), or only a subset of quantized values?
2. If the latter, how are they related to the values in the speaker's "normal" voice? Is there a clear linear or nonlinear relationship that can be expressed mathematically?
3. Is there a relationship between the range of possible variations in response to one stimulus, and co-variation of other stimuli or speaker parameters?
4. Does the variability depend on a person's native language, or overall linguistic training and abilities?
5. Does the variability depend on a person's cognitive abilities?
6. Is the variability affected by environmental factors? If so, which ones?
7. Is there a way to normalize the variability for analysis? What would be the appropriate operations for this?

Later in this book we will discuss these in different contexts.

1.4.3.2 Voice Disguise: Impersonation, Masking, and Illusions

Voice disguise represents a different facet to voice variability. Voluntary alterations to voice may not necessarily exhibit the same range of variations as those encountered in natural, involuntary responses to stimuli. The term *voice disguise* applies to any situation where the voice signal is altered to hide the true identity of the originator of the voice. In its normal usage, the term "voice disguise" carries an undertone of deliberation in the act of voice alteration. Thus, inadvertent distortion or attenuation of voice over noisy channels, for example, is generally not considered to be voice disguise.

Voice disguise and impersonation: Humans are able to modify their voice to sound like persons other than themselves (fictional or real persons). This is generally referred to as *impersonation*, but when the voice is disguised as that of a real person, the act of disguise is also termed as *mimicry*. The impression of a different originator for the voice can be created in different (complex) ways—for example, by altering voice quality, vocal production, speaking style or demeanor—each in isolation, or jointly.

When voice is disguised, the impressions we form from it, i.e. our judgments about the speaker, are impaired and more likely to be incorrect—in fact, creating incorrect impressions is the goal of successful voice disguise. For profiling, the elements of voice that change under disguise are likely to be unreliable. The challenge for accurate profiling is then to find the aspects of voice that are not under voluntary control of the speaker, and do not change with disguise, and derive conclusions based on those aspects. This problem is compounded by that of *in-vacuo* detection of voice disguise—defined as the problem of determining whether the voice is disguised *in the first place*. If no reference is available, an isolated voiceprint that represents disguised voice may not appear to be disguised. On the other hand, if every voice is treated as disguised by default, then valuable evidence for profiling (in the case of voiceprints that are not disguised) may be missed.

When comparators, or reference voice samples are present for analysis, voice disguise can be detected (or claimed disguise can be disproved) by finding the differentiators between reference and disguised voices using appropriate techniques. However, the design of such techniques remains an open problem.

Voice masking: Voice masking is an aspect of voice disguise wherein "natural modes," or the true nature, of voice is suppressed, e.g. a man imitating a bird-call, or the sound of a musical instrument, a train etc. Voice can also be masked using external aids, such as physical masks that muffle voice, or electronic voice transformation devices. As long as natural human *speech* is used at the core, the possibility of successful profiling exists (albeit to different degrees for different kinds of masking). However, if voice or speech is synthesized completely, profiling is likely to (and should, ideally) fail. Currently, it is not possible, without borrowing elements from real voices, to synthesize human voice or speech in all its micro-nuances accurately enough to elude detection (as having been synthetically produced). However, in-vacuo detection of synthetic voices, just as in the case of voice disguise, is a problem that must be addressed in the context of profiling.

Illusions: Another form of voice masking is to create voice illusions. While they may not confound computational procedures, auditory illusions hamper the judgments we make from voice. Some of these are created as a result of the masking effect of concurrent modalities of communication, such as facial expressions along with voice. A common example of an illusion created by visual detractors is the *McGurk effect* [168], wherein the same sound may be perceived differently by a listener when the lips of the speaker are concurrently observed to mime different sounds. For example, the sound /B AA/ could be heard as "baa" if the listener is shown a video of lips forming a closure at the beginning of the sound sequence, and as "faa" if the listener sees a video wherein the lower lip and the upper teeth form a closure instead. Another form of auditory illusion is often encountered when we seem to hear voices in random sounds, or noise. This illusion is called *Auditory pareidolia*. Even lossy compression of audio signals has been known to cause spurious psychoacoustic effects. Examples of other auditory illusions are *Risset's effect*, *Shepard's effect*, *Glissando effect*, *Binaural beats effect*, *Franssen effect*, *Constant spectrum melody illusion*, *Octave illusion* or *Deutsch's High-Low Illusion* [169], *Speech-to-song illusion* etc.

Illusions occur because our brain processes auditory signals in a selective manner. The reasons for this are multiple and have to do with the brain's inherent ability to process stimuli, which is dependent on the neuronal rates of firing, among other factors. Many studies have been performed to explore the brain's selective processing of auditory stimuli. A well-known example is that of Deutsch's experiment [170]. In this simple experiment, when subjects were presented with different sequences of two tones concurrently in both ears, they selectively reported hearing only the ones in one ear (the choice of the ear was also consistent with their "handedness"). There have been hundreds of other studies in psychoacoustics that establish the bounds on what we hear, and choose *not* to hear.

Illusions can of course be utilized to effectively mislead human judgments made from voice.

Ventriloquism: Ventriloquism combines two different aspects of voice disguise, masking *and* illusions. Often also called *voice projection*, it involves changing the characteristics of the voice signal in a manner that the perception of geometrical location of the origin of the voice (where the voice is coming from), and the identity of the originator (who is speaking), are both deliberately misled or misrepresented. The result is that the person creating this impression may successfully cause the listener to think that another person or entity in a different location (within hearing range or proximity of the listener) is speaking in *their* natural voice, to the listener. Ventriloquism has been relatively less studied so far, presumably for lack of representative data, since this is not an easy skill to perform. Some studies however exist that are relevant to the context of profiling, e.g. [171].

The ability of humans to create impressions does pose a challenge to profiling—how can we profile accurately in the presence of such alterations of perceptual qualities of voice, and in the presence of disguise, masking and illusions as mentioned above? Of course, a significant advantage with computational profiling is that while human perception may be confounded by our own physical and mental conditions, this is not likely to happen with computing machines that take such phenomena into account.

1.4.3.3 Parameter Variability

In the section above, we discussed briefly how the human voice can have inherent variabilities, and how these can be used to produce different desired effects. The parameters that affect a speaker's voice may also have inherent variations over time. Parameter variability refers to the ranges within which each profile parameter can vary. This is different from the variability in voice as a result of the *effects* of these parameters, which we have discussed above. For example, humans can judge race from voice, *on an average*. However, the extent to which the average race prototype (in the anthropological sense) is expressed in each individual belonging to that race may vary. In other words, the "race" parameter is a variable one in itself. Continuing with this example, let us say that we know what developmental factors and genetic caprices can cause the manifestation of average racial characteristics in different individuals to differ. The question then is—to what extent *can* they differ? If this is reasonably well known, profiling algorithms can be better constrained and guided. Conversely, not accounting for such inherent variability in parameter ranges could increase the likelihood of any data-driven mechanism (for profiling) to be erroneous. It is important to keep in mind that the error rates associated with such processes are highly likely to be correlated with the frequency with which norms are superseded for any parameter.

Different "categories" of parameters are variable to different general extents. Some parameters are especially more susceptible to variation. For example a person's dialect and accent and both parameters that help place the person's geographical origin (also a profile parameter). Regional dialect and accent effects can however dissolve with shifts in time and location of speakers. The style of speaking could be changed, affected or altered by prolonged exposure to new cultures, new languages and sounds therein, by gaining maturity in vocal expression of thoughts and many other factors. Other parameters like emotion, behavior, personality etc., which are highly subjective in definition and only roughly follow any specific set of norms, have a wide variation even with the same "category."

In addition, since many parameters influence an individual at the same time, and since the subset of parameters that affect an individual may vary at different times, profiling for any given parameter can only be accurate if these concurrent parameters are identified, and their effects are isolated or accounted for in the prediction of the parameter of interest. *Parameter disambiguation* is thus a crucial aspect of and a key challenge in successful profiling, and is made more complex by parameter variability.

1.4.3.4 Tiers of Information in Voice

The fact that voice varies in response to changes in parameters (or influencing factors) is what enables profiling. As the influences disappear, so do their signatures in voice. However, there are some signatures that pervade voice at all times, for the lifetime of the speaker. These are parameters like race, ethnicity, (adult) height etc., which are invariable for the lifetime of the speaker. It is therefore useful to think of the existence of at least two tiers of information in the voice signal: short-term influences and long-term influences. Eventually, in a computational setting, the type, reliability and accuracy of profiling procedures would depend on which tier of information they are based on.

1.4.3.5 The Search for Signatures of Profile Parameters

From existing literature, of which we have a small sampling in this chapter, it is clear that any search for the signatures of a profile parameter of interest, be it a broad one (such as personality comprising many traits) or a narrow one (comprising one or a few traits such as age), would benefit greatly by drawing from prior studies. The search for signatures of profile parameters begins by identifying the characteristics of voice that have been found to be relevant to the parameter in question. Once identified, these entities must be measured, or quantified. This immediately runs into problems in the case of many characteristics, such as voice quality sub-features (e.g. nasality, roughness etc.), for many of which no objective measurements are available or even obvious. Once measured, the next step entails finding representations of these within which the signatures appear as patterns that are clearly evident, identifiable or discoverable. The patterns may be extremely fine (micro-patterns) in many respects, and may require the design of techniques that can measure them with high precision. Locating them may also be non-trivial, since these may appear to different degrees in different sounds. In subsequent chapters, all of these issues are addressed in greater detail.

References

1. Faulkner, R. B. (1913). *The tonsils and the voice in science, surgery, speech and song*. Pittsburgh, PA: Presbyterian Book Store.
2. Beer, S. (1969). The prerogatives of systems. *Management Decision*, *3*(2), 4–12.
3. Bullough, A., & Craig, R. (2001). Anaesthesia for the professional singer. *European Journal of Anaesthesiology*, *18*(6), 414–416.
4. Zemlin, W. R. (1982). Speech and hearing science: Anatomy and physiology. *Otology & Neurotology*, *4*(2), 186–187.
5. Dalton, J. C., & Keegan, L. C. (2017). Using speech analysis to unmask perceptual bias: Dialect, difference, and tolerance. *Perspectives of the ASHA Special Interest Groups*, *2*(19), 9–18.
6. Malinow, K. L., Lynch, J. J., Thomas, S. A., Friedmann, E., & Long, J. M. (1982). Automated blood pressure recording: The phenomenon of blood pressure elevations during speech. *Angiology*, *33*(7), 474–479.
7. Orlikoff, R. F., & Baken, R. J. (1989). Fundamental frequency modulation of the human voice by the heartbeat: Preliminary results and possible mechanisms. *The Journal of the Acoustical Society of America*, *85*(2), 888–893.
8. Weidman, A. (2014). Anthropology and voice. *Annual Review of Anthropology*, *43*, 37–51.
9. Gualandi, A. (2013). "Sensing" voices: A theoretical comparison. *Phenomenology and Mind*, *4*, 94–102.

10. Pieper, J. (1969). The meaning of "God speaks". *The New Scholasticism, 43*(2), 205–228.
11. Agha, A. (2005). Voice, footing, enregisterment. *Journal of Linguistic Anthropology, 15*(1), 38–59.
12. Keane, W. (1999). Voice. *Journal of Linguistic Anthropology, 9*(1/2), 271–273.
13. Millot, J. L., & Brand, G. (2001). Effects of pleasant and unpleasant ambient odors on human voice pitch. *Neuroscience Letters, 297*(1), 61–63.
14. Levin, T. Y. (2003). "Tones from out of nowhere": Rudolph Pfenninger and the archaeology of synthetic sound. *Grey Room, 12*, 32–79.
15. Linville, S. E. (1996). The sound of senescence. *Journal of Voice, 10*(2), 190–200.
16. Kendall, K. (2007). Presbyphonia: A review. *Current Opinion in Otolaryngology & Head and Neck Surgery, 15*(3), 137–140.
17. Hodges-Simeon, C. R., Gurven, M., Cárdenas, R. A., & Gaulin, S. J. (2013). Voice change as a new measure of male pubertal timing: A study among Bolivian adolescents. *Annals of Human Biology, 40*(3), 209–219.
18. Browne, J. V. (2017). Recorded maternal voice, recorded music, or live intervention: A bioecological perspective. *Early vocal contact and preterm infant brain development* (pp. 183–201). Cham, Switzerland: Springer International Publishing.
19. Jones, J. A., & Munhall, K. G. (2005). Remapping auditory-motor representations in voice production. *Current Biology, 15*(19), 1768–1772.
20. Darby, J. K. (1981). Speech and voice studies in psychiatric populations. *Speech evaluation in psychiatry* (pp. 39–51). New York: Grune and Stratton.
21. Cannizzaro, M., Harel, B., Reilly, N., Chappell, P., & Snyder, P. J. (2004). Voice acoustical measurement of the severity of major depression. *Brain and Cognition, 56*(1), 30–35.
22. Johnson, H. M. (1912). The talking dog. *Science, 35*(906), 749–751.
23. Moses, P. J. (1941). Social adjustment and the voice. *Quarterly Journal of Speech, 27*(4), 532–537.
24. Warner, R. M. (1979). Periodic rhythms in conversational speech. *Language and Speech, 22*(4), 381–396.
25. Artkoski, M., Tommila, J., & Laukkanen, A. M. (2002). Changes in voice during a day in normal voices without vocal loading. *Logopedics Phoniatrics Vocology, 27*(3), 118–123.
26. Vrij, A., & Mann, S. (2004). Detecting deception: The benefit of looking at a combination of behavioral, auditory and speech content related cues in a systematic manner. *Group Decision and Negotiation, 13*(1), 61–79.
27. Satz, A., & Wood, J. (Eds.) (2009). *Articulate objects: voice, sculpture and performance.* Bern, Switzerland: Peter Lang.
28. Van Borsel, J., De Cuypere, G., & Van den Berghe, H. (2001). Physical appearance and voice in male-to-female transsexuals. *Journal of Voice, 15*(4), 570–575.
29. Puts, D. A., Gaulin, S. J., & Verdolini, K. (2006). Dominance and the evolution of sexual dimorphism in human voice pitch. *Evolution and Human Behavior, 27*(4), 283–296.
30. Kingson, W. K., & Cowgill, R. (1950). *Radio drama acting and production: a handbook.* New York: Rinehart & Company.
31. Martin, J. (1991). *Voice in modern theatre*. UK: Taylor & Francis.
32. Ohlsson, A. C. (1988). Voice and work environment: Towards an ecology of vocal behaviour. Doctoral dissertation. Department of Logopedics and Phoniatrics, Gothenburg University, Sweden.
33. Vilkman, E. (1996). Occupational risk factors and voice disorders. *Logopedics Phoniatrics Vocology, 21*(3–4), 137–141.
34. Pedersen, M. F., Møller, S., Krabbe, S., & Bennett, P. (1986). Fundamental voice frequency measured by electroglottography during continuous speech. A new exact secondary sex characteristic in boys in puberty. *International Journal of Pediatric Otorhinolaryngology, 11*(1), 21–27.
35. Hollien, H., Green, R., & Massey, K. (1994). Longitudinal research on adolescent voice change in males. *The Journal of the Acoustical Society of America, 96*(5), 2646–2654.

36. Nakarat, T., Läßig, A. K., Lampe, C., & Keilmann, A. (2014). Alterations in speech and voice in patients with mucopolysaccharidoses. *Logopedics Phoniatrics Vocology*, *39*(1), 30–37.
37. Zausmer, E. (1951). Speech defects resulting from bulbar poliomyelitis. *Physical Therapy*, *31*(7), 262–265.
38. Hara, K., Ohta, K., Miyajima, M., Hara, M., Iino, H., Matsuda, A., et al. (2012). Mismatch negativity for speech sounds in temporal lobe epilepsy. *Epilepsy and Behavior*, *23*(3), 335–341.
39. Vilkman, E. (2004). Occupational safety and health aspects of voice and speech professions. *Folia Phoniatrica et Logopaedica*, *56*(4), 220–253.
40. Devereux, G. (1949). Mohave voice and speech mannerisms. *Word*, *5*(3), 268–272.
41. Young, G., & Décarie, T. G. (1977). An ethology-based catalogue of facial/vocal behaviour in infancy. *Animal Behaviour*, *25*, 95–107.
42. Hollien, H. (2013). *The acoustics of crime: the new science of forensic phonetics*. Berlin: Springer Science & Business Media.
43. Harnsberger, J. D., Hollien, H., Martin, C. A., & Hollien, K. A. (2009). Stress and deception in speech: Evaluating layered voice analysis. *Journal of Forensic Sciences*, *54*(3), 642–650.
44. Hamdan, L., Sharara, A. I., Younes, A., & Fuleihan, A. N. (2001). Effect of aggressive therapy on laryngeal symptoms and voice characteristics in patients with gastroesophageal reflux. *Acta Oto-Laryngologica*, *121*(7), 868–872.
45. Ramig, L. A., & Ringel, R. L. (1983). Effects of physiological aging on selected acoustic characteristics of voice. *Journal of Speech, Language, and Hearing Research*, *26*(1), 22–30.
46. Sataloff, R. T. (1995). Genetics of the voice. *Journal of Voice*, *9*(1), 16–19.
47. Matheny, A. P., Jr., & Bruggemann, C. E. (1973). Children's speech: Hereditary components and sex differences. *Folia Phoniatrica et Logopaedica*, *25*(6), 442–449.
48. Hirano, M., & Kurita, S. (1986). Histological structure of the vocal fold and its normal and pathological variations. *Vocal fold histopathology: a symposium*. San Diego: College Hill Press.
49. Stemple, J. C. (1993). Voice research: So what? A clearer view of voice production, 25 years of progress; the speaking voice. *Journal of Voice*, *7*(4), 293–300.
50. Zhao, S., Bu, F., Sun, Y., & Han, C. (2003). Study on HNR in transmitted sound signals. In *Proceedings of the International Conference on Natural Language Processing and Knowledge Engineering (NLP-KE)* (pp. 580–584). Beijing, China: IEEE.
51. Vencovský, V. (2014). Modeling roughness perception for complex stimuli using a model of cochlear hydrodynamics. In *Proceedings of International Symposium of Musical Acoustics* (pp. 483–488). European Acoustics Association.
52. Andreassen, M. L., Smith, B. E., & Guyette, T. W. (1991). Pressure-flow measurements for selected oral and nasal sound segments produced by normal adults. *The Cleft Palate-Craniofacial Journal*, *28*(4), 398–407.
53. Olson, H. F. (1957). *Acoustical engineering*. New York: Van Nostrand Reinhold Company.
54. Feder, R. J. (1984). *The professional voice and airline flight*. Los Angeles, California: Sage Publications.
55. Bing, P. (1993). Impersonation of voice in Callimachus' hymn to apollo. *Transactions of the American Philological Association*, *123*, 181–198.
56. Levbarg, J. J. (1939). Hypnosis: A potent therapy for certain disorders of voice and speech. *Archives of Otolaryngology*, *30*(2), 206–211.
57. Blager, F. B., Gay, M. L., & Wood, R. P. (1988). Voice therapy techniques adapted to treatment of habit cough: A pilot study. *Journal of Communication Disorders*, *21*(5), 393–400.
58. Zelcer, S., Henri, C., Tewfik, T. L., & Mazer, B. (2002). Multidimensional voice program analysis (MDVP) and the diagnosis of pediatric vocal cord dysfunction. *Annals of Allergy, Asthma & Immunology*, *88*(6), 601–608.
59. Kent, R. D. (2004). Development, pathology and remediation of speech. In *Proceedings of the MIT International Conference: From Sound to Sense* (pp. B148–B163). MA, USA: Cambridge.

60. Brewer, D. W. (1989). G. Paul Moore lecture voice research: the next ten years. *Journal of Voice, 3*(1), 7–17.
61. Feinberg, D. R., Jones, B. C., DeBruine, L. M., Moore, F. R., Smith, M. J. L., Cornwell, R. E., et al. (2005). The voice and face of woman: One ornament that signals quality? *Evolution and Human Behavior, 26*(5), 398–408.
62. Luck, G., & Toiviainen, P. (2008). Exploring relationships between the kinematics of a singer's body movement and the quality of their voice. *Journal of Interdisciplinary Music Studies, 2*(1), 173–186.
63. Gentilucci, M., & Dalla Volta, R. (2007). The motor system and the relationships between speech and gesture. *Gesture, 7*(2), 159–177.
64. Hoit, J. D. (1995). Influence of body position on breathing and its implications for the evaluation and treatment of speech and voice disorders. *Journal of Voice, 9*(4), 341–347.
65. Sataloff, R. T. (2001). Essay the impact of pollution on the voice. *Voice and Speech Review, 2*(1), 247–253.
66. Pascual-Leone, A., Gates, J. R., & Dhuna, A. (1991). Induction of speech arrest and counting errors with rapid-rate transcranial magnetic stimulation. *Neurology, 41*(5), 697–702.
67. Zimmermann, E., Leliveld, L. M. C., & Schehka, S. (2013). Toward the evolutionary roots of affective prosody in human acoustic communication: A comparative approach to mammalian voices. *Evolution of emotional communication: from sounds in nonhuman mammals to speech and music in man* (pp. 116–132). Oxford: OUP.
68. Eidsheim, N. (2012). Voice as action: Toward a model for analyzing the dynamic construction of racialized voice. *Current Musicology, 93*, 9.
69. Gunn, J. (2008). Speech is dead; long live speech. *Quarterly Journal of Speech, 94*(3), 343–364.
70. Kivy, P. (1994). Speech, song, and the transparency of medium: A note on operatic metaphysics. *The Journal of Aesthetics and Art Criticism, 52*(1), 63–68.
71. Sanders, I. (2003). The microanatomy of vocal fold musculature. In J. S. Rubin, R. T. Sataloff, & G. S. Korovin (Eds.), *Diagnosis and treatment of voice disorders* (pp. 49–68). Clifton Park, New York: Delmar Learning.
72. Junqueira, P., Marchesan, I. Q., de Oliveira, L. R., Ciccone, E., Haddad, L., & Rizzo, M. C. (2010). Speech-language pathology findings in patients with mouth breathing: Multidisciplinary diagnosis according to etiology. *International Journal of Orofacial Myology, 36*, 27–32.
73. de Carvalho, A. P. O. (1994). *Influence of architectural features and styles on various acoustical measures in churches*. Ann Arbor, Michigan: UMI Company.
74. Galindo, M., Zamarreño, T., & Girón, S. (2005). Acoustic analysis in Mudejar-Gothic churches: Experimental results. *The Journal of the Acoustical Society of America, 117*(5), 2873–2888.
75. Niedzielska, G. (2005). Acoustic estimation of voice when incorrect resonance function of the nose takes place. *International Journal of Pediatric Otorhinolaryngology, 69*(8), 1065–1069.
76. Mastropieri, D., & Turkewitz, G. (1999). Prenatal experience and neonatal responsiveness to vocal expressions of emotion. *Developmental Psychobiology: The Journal of the International Society for Developmental Psychobiology, 35*(3), 204–214.
77. Ecklund-Flores, L., & Turkewitz, G. (1996). Asymmetric headturning to speech and nonspeech in human newborns. *Developmental Psychobiology, 29*(3), 205–217.
78. Smith, A., McGavran, L., Robinson, J., Waldstein, G., Macfarlane, J., Zonona, J., et al. (1986). Interstitial deletion of (17)(p11.2p11.2) in nine patients. *American Journal of Medical Genetics Part A, 24*(3), 393–414.
79. Fitch, R. H., Miller, S., & Tallal, P. (1997). Neurobiology of speech perception. *Annual Review of Neuroscience, 20*(1), 331–353.
80. Tobey, E. A., & Rampp, D. L. (1987). Neurological correlates of speech. *Textbook of developmental pediatrics* (pp. 153–166). Boston: Springer.
81. Myers, R. E. (1976). Comparative neurology of vocalization and speech: Proof of a dichotomy. *Annals of the New York Academy of Sciences, 280*(1), 745–757.

82. David, A. S. (1999). Auditory hallucinations: Phenomenology, neuropsychology and neuroimaging update. *Acta Psychiatrica Scandinavica, 99*(s395), 95–104.
83. Fawcus, R. (1968). *Dental problems in speech pathology*. Thousand Oaks, California: SAGE Publications.
84. Békésy, G. V. (1949). The structure of the middle ear and the hearing of one's own voice by bone conduction. *The Journal of the Acoustical Society of America, 21*(3), 217–232.
85. Hillman, R. E., Heaton, J. T., Masaki, A., Zeitels, S. M., & Cheyne, H. A. (2006). Ambulatory monitoring of disordered voices. *Annals of Otology, Rhinology & Laryngology, 115*(11), 795–801.
86. Paulauskiene, I., Lesinskas, E., & Petrulionis, M. (2013). The temporary effect of short-term endotracheal intubation on vocal function. *European Archives of Oto-Rhino-Laryngology, 270*(1), 205–210.
87. Back, G. W., Nadig, S., Uppal, S., & Coatesworth, A. P. (2004). Why do we have a uvula?: literature review and a new theory. *Clinical Otolaryngology, 29*(6), 689–693.
88. McAdams, E. E. (2011). Electronic voices: contact with another dimension? *The Journal of Parapsychology, 75*(1), 157.
89. Barušs, I. (2001). Failure to replicate electronic voice phenomenon. *Journal of Scientific Exploration, 15*(3), 355.
90. Dunning, A., Woodrow, P., & Hollenberg, M. (2008). Ghosts in the machine. In *Proceedings of the Sixteenth International Conference on Multimedia (ICME)* (pp. 1125–1126). Hannover, Germany: ACM.
91. Evrard, R. (2014). From symptom to difference: 'Hearing voices' and exceptional experiences. *Journal of the Society for Psychical Research, 78*(916), 129–148.
92. Ware, C. (1998). *Basics of vocal pedagogy: The foundations and process of singing*. New York: McGraw-Hill.
93. Sataloff, R. T. (2006). *Vocal health and pedagogy, volume II: advanced assessment and practice*. San Diego: Plural Publishing.
94. Thompson, A. R. (1995). Pharmacological agents with effects on voice. *American Journal of Otolaryngology, 16*(1), 12–18.
95. Thomas, P., Bracken, P., & Leudar, I. (2004). Hearing voices: A phenomenological-hermeneutic approach. *Cognitive Neuropsychiatry, 9*(1–2), 13–23.
96. Collins, D. E. (2001). *Reanimated voices: Speech reporting in a historical-pragmatic perspective* (Vol. 85). Amsterdam, The Netherlands: John Benjamins Publishing.
97. Rush, J. (1900). *The philosophy of the human voice: embracing its physiological history*. Whitefish, Montana: Kessigner Publishing.
98. Cavarero, A. (2005). *For more than one voice: toward a philosophy of vocal expression*. Redwood City, California: Stanford University Press.
99. Fitch, W., & Hauser, M. D. (1995). Vocal production in nonhuman primates: Acoustics, physiology, and functional constraints on "honest" advertisement. *American Journal of Primatology, 37*(3), 191–219.
100. Hollien, H., Geison, L., & Hicks, J. W. (1987). Voice stress evaluators and lie detection. *Journal of Forensic Science, 32*(2), 405–418.
101. Kent, R. D. (1984). Psychobiology of speech development: Coemergence of language and a movement system. *American Journal of Physiology-Regulatory, Integrative and Comparative Physiology, 246*(6), R888–R894.
102. Tuomi, S. K., & Fisher, J. E. (1979). Characteristics of simulated sexy voice. *Folia Phoniatrica et Logopaedica, 31*(4), 242–249.
103. Sapir, E. (1927). Speech as a personality trait. *American Journal of Sociology, 32*(6), 892–905.
104. Addington, D. W. (1968). The relationship of selected vocal characteristics to personality perception. *Speech Monographs, 35*(4), 492–503. Taylor & Francis Online.
105. Sapir, S., & Aronson, A. E. (1990). The relationship between psychopathology and speech and language disorders in neurologic patients. *Journal of Speech and Hearing Disorders, 55*(3), 503–509.

106. Green, K. P., Kuhl, P. K., Meltzoff, A. N., & Stevens, E. B. (1991). Integrating speech information across talkers, gender, and sensory modality: Female faces and male voices in the McGurk effect. *Perception & Psychophysics*, *50*(6), 524–536.
107. Williams, E. J. (2010). The effects of quinine-induced hearing loss on speech and other psychophysical tasks. Arizona State University, ProQuest Dissertations Publishing.
108. Love, R. J., Hagerman, E. L., & Taimi, E. G. (1980). Speech performance, dysphagia and oral reflexes in cerebral palsy. *Journal of Speech and Hearing Disorders*, *45*(1), 59–75.
109. Hunter, A., Morgan, A. W., & Bird, H. A. (1998). A survey of Ehlers-Danlos syndrome: Hearing, voice, speech and swallowing difficulties: Is there an underlying relationship? *British Journal of Rheumatology*, *37*(7), 803–804.
110. Amernik, K. (2007). Glottis morphology and perceptive-acoustic characteristics of voice and speech in patients with rheumatoid arthritis. *Annales Academiae Medicae Stetinensis*, *53*(3), 55–65.
111. Lieberman, P. (1963). Some effects of semantic and grammatical context on the production and perception of speech. *Language and Speech*, *6*(3), 172–187.
112. Spencer, L. E. (1988). Speech characteristics of male-to-female transsexuals: A perceptual and acoustic study. *Folia Phoniatrica et Logopaedica*, *40*(1), 31–42.
113. Zacher, V., & Niemitz, C. (2003). Why can a smile be heard? A new hypothesis on the evolution of sexual behaviour and voice. *Anthropologie*, *41*(1–2), 93.
114. Puts, D. A., Doll, L. M., & Hill, A. K. (2014). Sexual selection on human voices. In V. A. Weekes-Shackelford & T. K. Shackelford (Eds.), *Evolutionary perspectives on human sexual psychology and behavior* (pp. 69–86). New York: Springer.
115. Luke, C. (1994). Women in the academy: The politics of speech and silence. *British Journal of Sociology of Education*, *15*(2), 211–230.
116. Mendieta, E. (2015). Somatology of aurality. *Technoscience and postphenomenology: The Manhattan papers* (p. 239). Lanham: Lexington Books.
117. Jones, C. T. (2011). Prelude to the new world: The role of voice in early Pennsylvanian mysticism. *Eighteenth-Century Studies*, *44*(3), 331–343.
118. Jindra, P., Eber, M., & Pesak, J. (2002). Spectral analysis of syllables in patients using dentures. *Biomedical Papaers - Palacky University in Olomouc*, *146*(2), 91–94.
119. Petrović, A. (1985). Speech sound distortions caused by changes in complete denture morphology. *Journal of Oral Rehabilitation*, *12*(1), 69–79.
120. Shprintzen, R. J., & Siegel-Sadewitz, V. (1982). The relationship of communication disorders to syndrome identification. *Journal of Speech and Hearing Disorders*, *47*(4), 338–354.
121. Duffey, M. R. (2007). The vocal Memnon and solar thermal automata. *Leonardo Music Journal*, *17*, 51–54.
122. Sandage, M. J., Connor, N. P., & Pascoe, D. D. (2014). Vocal function and upper airway thermoregulation in five different environmental conditions. *Journal of Speech, Language, and Hearing Research*, *57*(1), 16–25.
123. Fung, K., Lyden, T. H., Lee, J., Urba, S. G., Worden, F., Eisbruch, A., et al. (2005). Voice and swallowing outcomes of an organ-preservation trial for advanced laryngeal cancer. *International Journal of Radiation Oncology, Biology, Physics*, *63*(5), 1395–1399.
124. Sataloff, R. T. (2017). *Professional voice: The science and art of clinical care (3-volume set)*. San Diego: Plural Publishing.
125. Beach, D. M. (1923). The science of tonetics and its application to Bantu languages. *Bantu Studies*, *2*(1), 75–106.
126. Comella, C. L., & Pullman, S. L. (2004). Botulinum toxins in neurological disease. *Muscle & Nerve: Official Journal of the American Association of Electrodiagnostic Medicine*, *29*(5), 628–644.
127. Walker, J. S., Eakes, G. G., & Siebelink, E. (1998). The effects of familial voice interventions on comatose head-injured patients. *Journal of Trauma Nursing*, *5*(2), 41.
128. Andrew, E. M., Gray, N. S., & Snowden, R. J. (2008). The relationship between trauma and beliefs about hearing voices: A study of psychiatric and non-psychiatric voice hearers. *Psychological Medicine*, *38*(10), 1409–1417.

129. Stent, A., Syrdal, A., & Mishra, T. (2011). On the intelligibility of fast synthesized speech for individuals with early-onset blindness. In *The Proceedings of the Thirteenth International Conference on Computers and Accessibility (SIGACCESS)* (pp. 211–218). Dundee, Scotland: ACM.
130. Bull, R., Rathborn, H., & Clifford, B. R. (1983). The voice-recognition accuracy of blind listeners. *Perception, 12*(2), 223–226.
131. Brieland, D. M. (1950). A comparative study of the speech of blind and sighted children. *Communications Monographs, 17*(1), 99–103.
132. Vrij, A., & Fischer, A. (1997). The role of displays of emotions and ethnicity in judgments of rape victims. *International Review of Victimology, 4*(4), 255–265.
133. Israel, L., Bello, C., Cottes, J., Age, B., Zobl, E., Cantor, L. L., et al. (2003). The voice behind the speech: Young women in UN conventions. Women in action - Rome then Manila (pp. 9–13).
134. Boyd, T. F., Watson, D. C., & Matthews, H. R. (1982). Oesophageal resection in patients with oesophageal speech. *British Medical Journal (Clinical research ed.), 285*(6348), 1081.
135. Ilmarinen, T., Nissilä, H., Rihkanen, H., Roine, R. P., Pietarinen-Runtti, P., Pitkäranta, A., et al. (2011). Clinical features, health-related quality of life, and adult voice in juvenile-onset recurrent respiratory papillomatosis. *The Laryngoscope, 121*(4), 846–851.
136. McComb, K., Shannon, G., Sayialel, K. N., & Moss, C. (2014). Elephants can determine ethnicity, gender, and age from acoustic cues in human voices. *Proceedings of the National Academy of Sciences, 111*(14), 5433–5438.
137. Hall, J. A. (1978). Gender effects in decoding nonverbal cues. *Psychological Bulletin, 85*(4), 845.
138. Laver, J. (1980). *The phonetic description of voice quality* (Vol. 31, pp. 1–186). Cambridge studies in linguistics London London: Cambridge.
139. Kreiman, J., Vanlancker-Sidtis, D., & Gerratt, B. R. (2008). Perception of voice quality. *The handbook of speech perception* (p. 338). New York: Wiley.
140. Seyfarth, R. M., & Cheney, D. L. (1986). Vocal development in vervet monkeys. *Animal Behaviour, 34*(6), 1640–1658.
141. Cheney, D. L., & Seyfarth, R. M. (1980). Vocal recognition in free-ranging vervet monkeys. *Animal Behaviour, 28*(2), 362–367.
142. Cheney, D. L., & Seyfarth, R. M. (1988). Assessment of meaning and the detection of unreliable signals by vervet monkeys. *Animal Behaviour, 36*(2), 477–486.
143. Seyfarth, R. M., & Cheney, D. L. (1980). The ontogeny of vervet monkey alarm calling behavior: A preliminary report. *Zeitschrift für Tierpsychologie, 54*(1), 37–56.
144. Ross, E. D. (1993). Nonverbal aspects of language. *Neurologic Clinics, 11*(1), 9–23.
145. González-Fuente, S. (2017). Audiovisual Prosody and Verbal Irony (Doctoral dissertation, Ph. D. dissertation, Universitat Pompeu Fabra, Barcelona.
146. Van Lancker, D., Canter, G. J., & Terbeek, D. (1981). Disambiguation of ditropic sentences: Acoustic and phonetic cues. *Journal of Speech, Language, and Hearing Research, 24*(3), 330–335.
147. Ross, E. D. (2013). Affective prosody. *Behavioral neurology & neuropsychiatry* (pp. 184–198). Cambridge: Cambridge University Press.
148. Brown, B. L. (1980). Effects of speech rate on personality attributions and competency evaluations. In H. Giles, W. P. Robinson, & P. Smith (Eds.), *Language: Social psychological perspectives* (pp. 293–300). Elmsford, New York: Pergamon.
149. Feldstein, S., Dohm, F. A., & Crown, C. L. (2001). Gender and speech rate in the perception of competence and social attractiveness. *The Journal of Social Psychology, 141*(6), 785–806.
150. Smith, B. L., Brown, B. L., Strong, W. J., & Rencher, A. C. (1975). Effects of speech rate on personality perception. *Language and Speech, 18*(2), 145–152.
151. Brown, B. L., Strong, W. J., & Rencher, A. C. (1973). Perceptions of personality from speech: Effects of manipulations of acoustical parameters. *The Journal of the Acoustical Society of America, 54*(1), 29–35.

152. Burgoon, J. K., Birk, T., & Pfau, M. (1990). Nonverbal behaviors, persuasion, and credibility. *Human Communication Research, 17*(1), 140–169.
153. Ray, G. B. (1986). Vocally cued personality prototypes: An implicit personality theory approach. *Communications Monographs, 53*(3), 266–276.
154. Curio, G., Neuloh, G., Numminen, J., Jousmäki, V., & Hari, R. (2000). Speaking modifies voice-evoked activity in the human auditory cortex. *Human Brain Mapping, 9*(4), 183–191.
155. Appelbaum, D. (1990). *Voice*. Albany, New York: State University of New York Press.
156. Koelsch, S. (2014). Brain correlates of music-evoked emotions. *Nature Reviews Neuroscience, 15*(3), 170.
157. Belin, P., Zatorre, R. J., Lafaille, P., Ahad, P., & Pike, B. (2000). Voice-selective areas in human auditory cortex. *Nature, 403*(6767), 309.
158. Belin, P., Zatorre, R. J., & Ahad, P. (2002). Human temporal-lobe response to vocal sounds. *Cognitive Brain Research, 13*(1), 17–26.
159. Grandjean, D., Sander, D., Pourtois, G., Schwartz, S., Seghier, M. L., Scherer, K. R., et al. (2005). The voices of wrath: Brain responses to angry prosody in meaningless speech. *Nature Neuroscience, 8*(2), 145.
160. Warren, J. E., Sauter, D. A., Eisner, F., Wiland, J., Dresner, M. A., Wise, R. J., et al. (2006). Positive emotions preferentially engage an auditory-motor "mirror" system. *Journal of Neuroscience, 26*(50), 13067–13075.
161. Morris, J. S., Scott, S. K., & Dolan, R. J. (1999). Saying it with feeling: Neural responses to emotional vocalizations. *Neuropsychologia, 37*(10), 1155–1163.
162. Siegman, A. W., & Boyle, S. (1993). Voices of fear and anxiety and sadness and depression: The effects of speech rate and loudness on fear and anxiety and sadness and depression. *Journal of Abnormal Psychology, 102*(3), 430.
163. Hellier, E., Edworthy, J., Weedon, B., Walters, K., & Adams, A. (2002). The perceived urgency of speech warnings: Semantics versus acoustics. *Human Factors, 44*(1), 1–17.
164. Sander, K., & Scheich, H. (2001). Auditory perception of laughing and crying activates human amygdala regardless of attentional state. *Cognitive Brain Research, 12*(2), 181–198.
165. Fecteau, S., Belin, P., Joanette, Y., & Armony, J. L. (2007). Amygdala responses to nonlinguistic emotional vocalizations. *Neuroimage, 36*(2), 480–487.
166. Fecteau, S., Armony, J. L., Joanette, Y., & Belin, P. (2005). Sensitivity to voice in human prefrontal cortex. *Journal of Neurophysiology, 94*(3), 2251–2254.
167. Akhtar, S., Wood, G., Rubin, J. S., O'Flynn, P. E., & Ratcliffe, P. (1999). Effect of caffeine on the vocal folds: A pilot study. *The Journal of Laryngology & Otology, 113*(4), 341–345.
168. McGurk, H., & MacDonald, J. (1976). Hearing lips and seeing voices. *Nature, 264*(5588), 746.
169. Deutsch, D. (1974). An auditory illusion. *Nature, 251*(5473), 307–308.
170. Deutsch, D., & Roll, P. L. (1976). Separate "what" and "where" decision mechanisms in processing a dichotic tonal sequence. *Journal of Experimental Psychology: Human Perception and Performance, 2*(1), 23.
171. Connor, S. (2000). Dumbstruck: A cultural history of ventriloquism. Oxford: Oxford University Press (on demand).
172. Allport, G. W., & Cantril, H. (1934). Judging personality from voice. *The Journal of Social Psychology, 5*(1), 37–55.

Chapter 2
Production and Perception of Voice

The goal of this chapter is to present the human speech production process in sufficient detail for the reader to understand why profiling should be possible, and to provide sufficient information to reason about the effects of different parameters on voice, so that profiling efforts may be better guided. The details are *sufficient*, but not complete since the area is too vast to be covered within one chapter of this book.

The human speech production process has been extensively researched and the current scientific understanding of it is very deep, though many mysteries in this context still remain to be solved. In a nutshell, during speech production, the articulators in the vocal tract, namely the velum, tongue, lips, jaw etc. move to change the shape of the vocal tract, creating resonant chambers of different dimensions separated by constrictions of different shapes. Air from the lungs, which may pass through directly as a turbulent flow or be modulated into a sequence of periodic pulses by the vocal folds, creates resonances (and anti-resonances) within these chambers, resulting in an aural sound with spectral peaks and valleys. Each sound takes its perceptual identity from the canonical spectral pattern associated with it that results from this process. While in the overall this sounds simple, the voice production process is extremely complex at the bio-mechanical and physiological levels. This chapter will provide a glimpse of the level of that complexity.

In addition to this, in this chapter we will also briefly touch upon the nature of sounds produced, and their general categorization. Based on such information, we will then reason *why* the human voice is unique (and *how* unique it is, in a rough quantitative sense). Finally, we will discuss some key principles of speech perception by humans. This is important in understanding how humans make qualitative judgments from voices, but does not play a great role in computational profiling, since machines are not bound to the rules of human perception (though they can benefit from some aspects of it). In the section on perception, therefore, we will focus only on a few concepts that may be useful in navigating through later chapters on computational profiling.

R. Singh, *Profiling Humans from their Voice*,
https://doi.org/10.1007/978-981-13-8403-5_2

2.1 The Vocal Tract

Voice is produced in the *vocal tract* of humans, which comprises the passageways and chambers connecting the lungs to the external environment. The key anatomical details of the vocal tract are shown in Fig. 2.1a. The figure is a schematic diagram of the cross-section of the vocal tract, which is also considered to be the upper portion of the respiratory tract.

The *oral cavity* is the chamber defined by our mouth. It is the region between the velum and the lips, delimited by the tissue of the inner cheeks at the sides, and the palate on top. It includes the teeth and tongue. The *nasal cavity* is the chamber that lies behind the nostrils, with the palate at the bottom and the pharynx at the back. The *pharynx* is the back region of the nasal and oral cavities (called the *nasopharynx* and *oropharynx* respectively) until the region of the *epiglottis*, which is a flap of cartilage situated a little below the position of the root of the tongue. The epiglottis depresses to cover the opening of the windpipe during swallowing. The pharynx in the region below the epiglottis is the *hypopharynx*, which gives way into the esophagus. Just below the epiglottis is the *larynx*, which houses the vocal cords. Positioned below the oropharynx, parallel to the hypopharynx and just above the trachea, as shown in Fig. 2.2a, the larynx gates the passage of airflow between the trachea and the mouth or nose during respiration. Within the larynx are two vocal cords, also called *vocal folds*, since they are membranes rather than cord-like structures in reality. The area between the vocal folds is called the *glottis*. The pharynx also extends to below the larynx, into what is called the sub-glottal region, where it is called the *laryngeal pharynx*. The part of the pharynx immediately above the glottis (i.e. in the supra-glottal region) forms part of the hypopharynx.

Fig. 2.1 Anatomy of the vocal tract

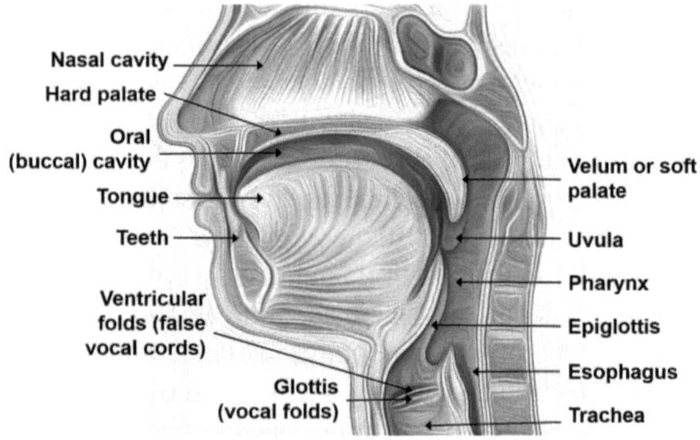

Key components of the vocal tract

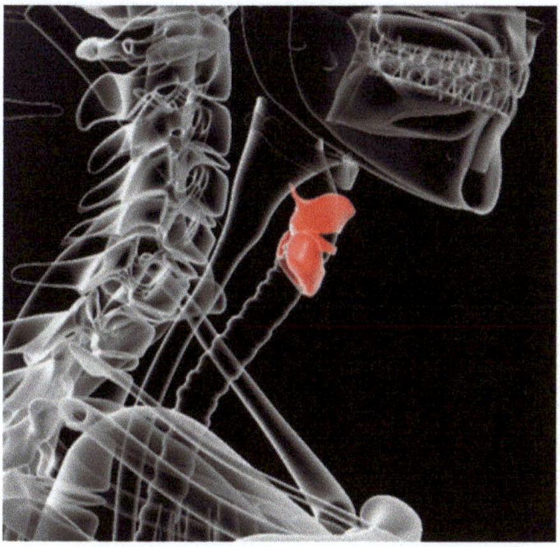

(a) Position of larynx in the vocal tract

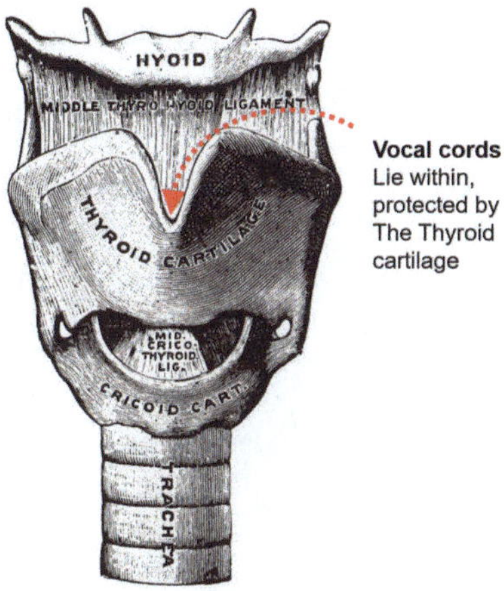

(b) Cartilages of the larynx and the location of the vocal cords

Fig. 2.2 Position of the larynx and its cartilages

2.1.1 Anatomy of the Larynx and Vocal Folds

Voice originates in the larynx, or the "voice box," as it is commonly known. The position of the larynx in the vocal tract is shown in Fig. 2.2a.

Structurally, the larynx comprises the vocal folds, and three cartilages—the top of the cricoid cartilage, the arytenoid cartilages (AC), and the thyroid cartilage (TC) (also known as "Adam's apple"). AC and TC lie on top of the cricoid cartilage. These are shown in Fig. 2.3a. The vocal folds lie in a front-back position (anterior-posterior position) in the larynx, they are attached to (and stretched between) TC in front, and AC at the back. AC and TC are also shown labeled with ovals in Fig. 2.3b, which shows the vocal folds and the muscles that control it's movements. Together, the three cartilages, and the muscles that are attached to them, can adjust the shape, position and mechanical motion of the vocal folds. The area between the vocal folds is called the *glottis*. The area of the vocal tract below the vocal folds is *sub-glottis* (or the sub-glottal region), and the area above is the *supra-glottal* (also called the *epiglottal*) region.

The vocal folds move in multiple ways. They can *abduct*, or move apart from each other, they can stretch, becoming longer and shorter along their major linear axis, and they can *adduct*, or come closer together. The term *vocal fold approximation* is often used to refer to the coming together of the vocal cords, causing glottal closure. Abduction allows air to move in and out of the lungs during respiration. Adduction is required for producing sound. The process of cyclic abduction and adduction comprises the "vibration" of the vocal folds. This process, called *phonation*, produces a sound wave in the vocal tract.

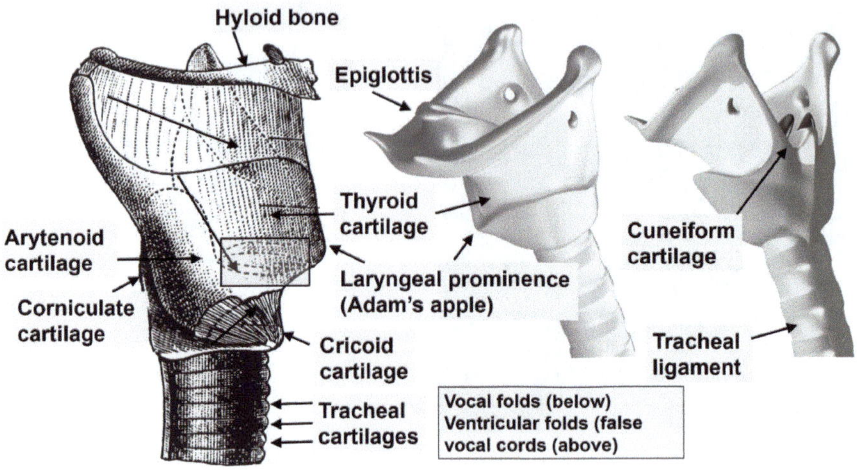

(a) Structure of the larynx

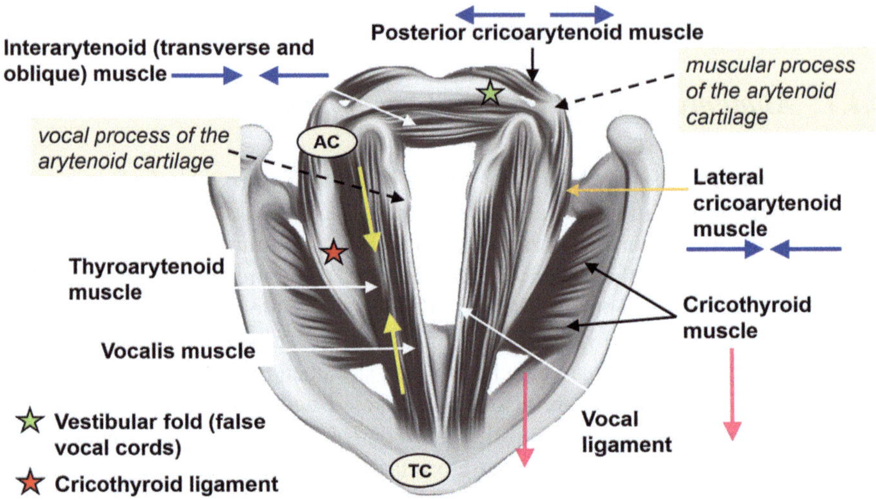

(b) Muscles in the larynx that control the vocal folds

Fig. 2.3 Anatomy of the vocal folds

The larynx has several muscles. Some facilitate the motion of the vocal folds. Others control the larynx outside the context of voice production, such as the movements and responses needed for respiration, swallowing etc. Five muscles in the larynx control the motion of the vocal folds. These are shown in Fig. 2.3. Their functions are briefly described below [1]:

1. **Cricothyroid (CT) muscles**: These control the front-back expansion of the glottis by facilitating longitudinal motion. The net result of their activity is the lengthening and shortening of vocal cords.
2. **Thyroarytenoid (TA) muscles**: These cause medial compression of the vocal fold, they thicken and thin out the vocal folds as they contract and stretch along the direction of their longitudinal axis (as shown by the overlaid arrows in Fig. 2.3b). Through this action, they can control, to some extent, the opening and closure of the glottis. The inner longitudinal part of this muscle (also called *internal thyroarytenoid*, or *inferior thyroarytenoid* and labeled as *vocalis* in Fig. 2.3b, lies parallel to, and attached to the vocal cords, and directly controls their vibration.
3. **Interarytenoid (IA) muscles**: These comprise a muscle bundle in two geometric configurations—a transverse and two oblique (forming a cross band) on the transverse bundle. The transverse and oblique bands together connect the posterior portions of the two arytenoid cartilages (see Fig. 2.3b) and can completely close the back portion of the glottis. This is done by reducing the distance between the arytenoid cartilages. However, they may not be able to close the mid and front part of the glottis. To close those regions, the TA and the LCA (described below) muscles must act in concert. To cause *complete* glottal closure, the action of all three muscles—IA, TA and LCA—is needed. Closure of only the back portion of the glottis can also effectively reduce the "active" length of the vocal folds, and thus the transverse arytenoid muscles also regulate vocal fold *length* along with the CT muscles.
4. **Lateral cricoarytenoid (LCA) muscles**: These muscles close (adduct) the vocal folds by internally rotating the arytenoid cartilages.
5. **Posterior cricoarytenoid (PCA) muscles**: These muscles move apart (abduct) the vocal folds by *externally* rotating the arytenoid cartilages. The muscles are critically responsible for abduction of vocal folds for respiration.

These five muscles (collectively called *intrinsic laryngeal muscles* in the medical literature) are also relatively better equipped to handle calcium fluctuations in the body (i.e. are better calcium buffers) [2, 3]—a capacity that allows them to work longer and move fast (as required in phonation). The other group of muscles in the larynx are collectively called the *extrinsic laryngeal muscles*. These keep the larynx in place as a whole, and are responsible for its positioning in the vocal tract. The extrinsic laryngeal muscles (not described here) that depress the larynx are the Sternothyroid, Omohyoid and Sternohyoid muscles. Those that elevate the larynx are: Thyrohyoid, Digastric, Stylohyoid, Mylohyoid, Geniohyoid, Hyoglossus and Genioglossus. In addition, there are muscles called the *inferior constrictor muscles* in the pharynx that control swallowing, but can affect the positioning of the larynx as well.

If the overall relative positioning of the larynx changes, the positioning of the vocal cords with reference to the overall vocal tract also changes, affecting the dimensions of the sub– and epiglottal regions. Even minor changes in the positioning can cause changes in the spectral characteristics of the voice signal. The vocal cords themselves have a layered structure, comprising the vocal ligaments and tissues around them.

The vocal ligaments are two strong bands of elastic tissue within the vocal folds. They are attached in front to the angle of the thyroid cartilage, and on the back to the vocal process of the arytenoid. The vocal process (shown in Fig. 2.3b) is the front angle of the base of the arytenoid cartilage. Like most muscles and ligaments in the larynx, the arytenoids are also paired cartilages. Each has a *medial* and a *lateral process* (shown by the dashed arrows in Fig. 2.3b). In Anatomy, the term "process" is used to refer to a projection of tissue from a larger entity.

2.2 The Bio-mechanical Process of Voice Production

Voice, and speech sounds, may be produced while we exhale or inhale. Most speech sounds are however produced during exhalation. Speech sounds that are produced as we inhale are called *ingressive* sounds. Within the category of ingressive sounds, there are three kinds: (a) *lingual ingressive or velaric ingressive*, wherein the airflow is through the mouth, and the tongue and the velum play a key role in shaping the sound, (b) *glottalic ingressive*, where the sound source is in the glottis, and (c) and *pulmonic ingressive*, where the sound is produced by turbulence below the glottis (in the sub-glottal region). Since ingressive sounds are rare, we will not focus on these any further.

The rest of the explanation in this chapter is in the context of speech produced as we exhale. Such speech is called *egressive* speech. For most of humankind, egressive sounds dominate speech production. As we speak, air is pumped out from the lungs through the trachea, and passes through the vocal folds in the larynx. Pressure differentials in the larynx cause the vocal folds to rapidly and cyclically open and close, causing air to be released through the vocal folds in a pulsing manner. The pulses are rapid—at the rate of several hundred times per second. This vibration creates sound waves. This sound is called the *excitation signal*—it excites the vocal tract resonances in the chambers of the vocal tract.

As we speak, we produce different kinds of sounds, which we string together in sequence to form words and sentences. Each distinct sound is characterized by its own specific spectral patterns, which are created by shaping the vocal chambers differently using our articulators (the tongue, lip, jaw, teeth, velum etc., as shown in Fig. 2.1). The process of production of sounds by the vibration of the vocal folds is called *phonation*. Of course, not all sounds are produced in this manner. Some sounds are produced through airflow turbulence, friction and other mechanisms in various parts of the vocal tract, and may not involve the vibration of vocal folds. We will discuss the different categories of sounds based on their production mechanism later in the chapter.

In this section we bifurcate the concept of sound production in the vocal tract into *voice production* and *speech production*. We will focus on voice production in this section, and on speech production in the next. The distinction between the two is that voice production (for our purposes) is the basic bio-mechanical process that produces sound waves. This process is however extremely complex and involves the coordinated action of the brain (nervous system) and the musculoskeletal system [4]. Speech production is even more complex. It involves mechanisms that are involved in modulating the sound produced into distinct *meaningful* forms at a cognitive level. To clarify this a little further here, we note that speech production can be viewed as a three-level process: (a) cognitive (b) social (c) bio-mechanical and physiological. At the cognitive level, a person's thought processes are involved in forming words and sentences, concurrently with changes in the quality of voice, to convey meaning. This aspect of speech production is driven by a conscious desire to communicate information. At a social level, what is learned by the speaker over their lifetime comes into play. This set includes more than just proficiency in a language. It includes speaking styles and mannerisms that are used in different social contexts, and other modes of vocal expression. There is abundant literature on research focused on these influences on speech.

The bio-mechanical aspects of speech production are the same as those of voice production, but the brain plays a much greater role in it, also focusing on the rapid movements and placements of the articulators, and fine-tuning the characteristics of the rest of the vocal tract as a person speaks. At a slightly deeper level of explanation (following [4]), the formation of voluntary voice and speech begins in the brain's cerebral cortex. Input from several other areas in the brain, including those of artistic expression, cognitive thought, planning and hearing, contribute to planned vocalization. This "plan" is then conveyed to the brain's precentral gyrus in the motor cortex, which transmits instructions to the motor nuclei in the brain stem and spinal cord, through which the neurotransmitters necessary for coordinated activity of the muscles in the larynx, chest and abdominal regions, the vocal cords and all the articulators are transmitted. Simultaneously a feedback mechanism is activated from the ears through the auditory nerve and from visual input (if present), to allow the brain to adjust the vocalizations and fine tune them. In the adjustment process, several areas of the brain are again involved. Since voice is a central part of speech, the very same areas are also activated during the process of voice production, some to a lesser extent if there is no meaningful content in speech in the form of language. Both voice and speech production are affected if any of the areas of the brain that play a role in vocalization are impaired. It is important to keep in mind that even our wordless vocalizations are mostly done with a purpose, and that involves cognitive thought.

Let us return to voice production. Let us first understand the phenomenon of phonation, or the vibration of the vocal folds.

At the bio-mechanical level, phonation happens as a result of a specific pattern of events in the glottal region. The vocal folds are three-dimensional membranes that are set into vibratory motion as a result of a complex interplay of forces in the vocal tract. These relate to (a) pressure balances and airflow dynamics within the regions above and below the glottis (the supra-glottal and sub-glottal regions respectively), and (b) muscular control within the glottis and the larynx. These two factors are in turn dependent on and controlled by myriad factors within the body. Although (a) and (b) are deeply interrelated, for simplicity we discuss each of these separately.

2.2.1 *Macro-mechanics of Vocal Fold Vibration*

Let us now take a closer look at the vibrations of the vocal folds. During phonation, they vibrate hundreds of times per second. The vocal fold vibrations are not individually controlled, nor do they happen due to repetitive nerve simulation [5], as was previously thought. Rather, they are *self-sustained*, as in a physical system that exhibits self-sustained oscillations under the right balance of different opposing forces acting on it. This happens as a consequence of a chain of events in the laryngeal region [6].

First, the vocal folds close. Then the lungs contract, pushing out air and causing an increase in pressure just below the glottis. When this subglottal pressure crosses a threshold, the vocal folds are pushed apart, and air flows out creating a negative intraglottal pressure (with reference to atmospheric air pressure). As a result, the negative Bernoulli pressure causes elastic recoil that causes the glottis to close again. This is one of the several factors that contribute to the self-sustained nature of the vocal fold vibrations.

When the vocal folds vibrate, they release air into the vocal tract in a pulsed fashion. These are called *glottal flow pulses*. These are further discussed in Sect. 2.3.3.1.

The exact physics of the airflow through the glottis during phonation has in fact been very finely studied, e.g [7–12]. It is broadly described below.

2.2.1.1 Pressure Balances and Airflow Dynamics Across the Glottis

The balance of forces necessary to cause self-sustained vibrations is created by two physical phenomena that play a significant role in it: the Bernoulli effect and the Coandă effect. Both of these are illustrated schematically in the context of the vocal tract geometry in Fig. 2.4a. The Bernoulli effect (or Bernoulli's principle) applies to fluid flow or airflow, and states that an increase in the velocity of a fluid occurs simultaneously with a decrease in its pressure or its potential energy. This follows the law of conservation of energy, and can be explained easily by considering the flow of a fluid (air, in our case) moving through any channel of cross sectional area A_1 into a connected channel of cross-sectional area A_2. From the principle of conservation of energy we have:

$$p_1 + \frac{1}{2}\rho v_1^2 + \rho g h_1 = p_2 + \frac{1}{2}\rho v_2^2 + \rho g h_2 \tag{2.1}$$

where p_1 and p_2 are the pressure, v_1 and v_2 are the flow velocities in the two channels, and h_1 and h_2 are their height from sea level, respectively. ρ is the density of the fluid, and g is the gravitational constant. The three terms on each side of the equation above are in fact the pressure (force per unit volume), the kinetic energy (per unit volume) and the potential energy (per unit volume) of the fluid flowing through each channel. It is clear from Eq. 2.1 that for the same volume, as the kinetic energy through any channel increases, the pressure in the channel must decrease. The same holds for potential energy.

The Coandă effect is a consequence of Bernoulli's principle. It can be easily understood in the context of an air jet blown into the atmosphere. The reason why the shape of the jet remains tubular for some distance is that the atmospheric air pressure is higher than that of the jet stream, and compresses it into a tubular form from all sides. If a surface intervenes on any side the compressive force exerted by the higher-pressure air on one side is greater than that from the other, and the jet stream bends and adheres to the intervening surface. This effect is what causes paper held in a horizontal planar orientation to flap periodically as air is blown parallel to its upper surface. Figure 2.4a illustrates this effect in the right panel.

The process of phonation begins with the closing of the glottis. This closure is voluntary, and facilitated by the laryngeal muscles. Once closed, the muscles do not actively play a role in sustaining the vibrations. Air pressure builds up behind the closed glottis, and when it reaches a threshold, it pushes the glottis apart, causing it to open. As this process begins, air rushes out of the narrow glottal opening into the much wider supra-glottal region. From the perspective of airflow, the glottis thus forms a *flow separation plane*.

The expansion of air in this region, and the low-pressure created in the vicinity of the glottis through the Coandă effect induced entrainment (see Fig. 2.4a), cause a lowering of pressure close to the glottis, and cause a net downward force on the glottis. At the same time, lowered pressure in the glottal region due to the Bernoulli effect that ensues from the high-velocity air volume flow through the glottis exerts a negative force on the glottis. As mentioned above, the negative Bernoulli pressure also causes elastic recoil, causing it to begin to close again. The closing reduces the volume flow through the glottis, diminishing the downward forces acting on it. Together with increased pressure buildup in sub-glottal region, this causes the glottis to open again. This chain of oscillations continues in a self-sustained fashion throughout phonation, until voluntary muscle control intervenes to alter or stop it. The intervention occurs in accordance with the needs of the subsequent sound (e.g. it may require a different pattern of vibration, or no vibration at all), or as the respiratory volume of air in the lungs is exhausted. Figure 2.4b illustrates the balance of forces through each cycle of phonation.

A couple of points about the elastic recoil of the vocal folds are worth mentioning here. The material of the vocal folds is anisotropic and viscoelastic, and these properties cause the vocal folds to resist deformation in response to an applied force. This resistance is nonlinear and its role in this process is supported by a multitude of studies [13–15] that have charted the mechanical properties of the vocal folds through techniques such as shear rheometry [13–15], surface wave method [16], indentation [17–19] etc.

The turbulence that occurs in the supra-glottal region (shown in Fig. 2.4) also warrants special mention. Its features depend on the geometry of the region. The pressure differences across the glottis play a critical role in its formation [20, 21]. During phonation, as the glottis opens and air escapes from the glottis, it forms a jet above it. Air jets have been studied in many contexts, and some of the character-

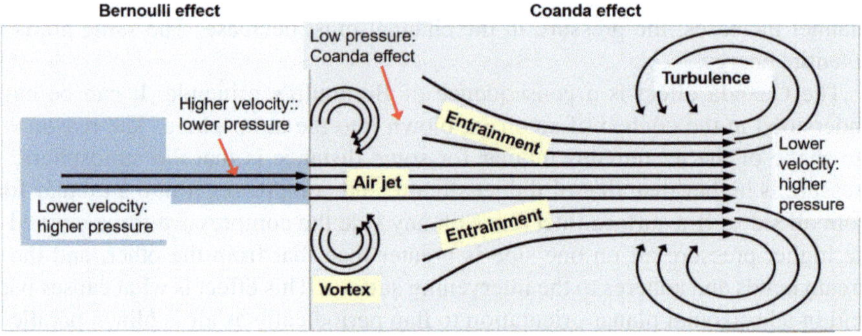

(a) Bernoulli and Coandă effects in the context of the vocal tract. The rectangle on the left represents the sub-glottal region. It leads into the narrower glottal region, and again into a wider supra-glottal region. The jet of air that emerges from the glottal opening it depicted by the longer lines going from the left to right. As the jet-stream emerges, lowered pressure in it (Bernoulli effect) causes surrounding air at higher pressure to move towards it and flow alongside (entrainment with Coănda effect). This movement causes vortices of low pressure in the regions shown. Turbulence is caused when the jet stream breaks up and rolls back.

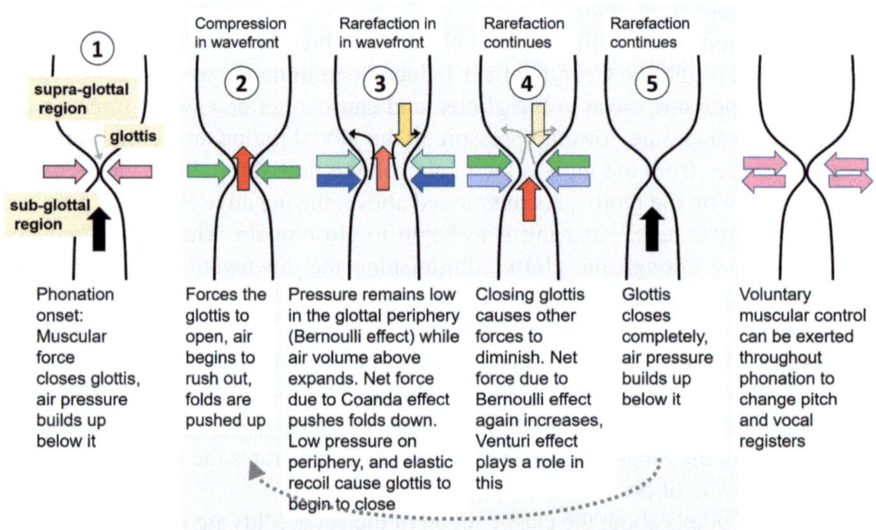

(b) The balance of forces through one cycle of the self-sustained vibrations of the vocal cords. The color codes for the arrows depict net forces due to the following: Pink – muscular; Green – Bernoulli effect; yellow – Coandă effect; blue – vocal fold elasticity and other factors; Black and Red – air pressure. Lighter shades of each color depict lower forces.

Fig. 2.4 Schematic illustration of the balance of forces in vocal fold vibration

istic features of these are the formation of a vortex, which rolls up, then sheds and turns into turbulent and dispersed air. This turbulence causes disturbance to the jet emerging from the glottis in the next opening cycles, often diverting the jet coming through towards the walls of the supra-glottal region. This diversion may alternate, or oscillate, causing *jet flapping* [22–25]. The turbulence also affects the regions further up in the vocal tract. Under extreme conditions, the airflow separates inside the glottis, creating additional convection air-streams and vortices in the intraglottal region [26]. All of these make turbulence an additional source of sound in the vocal tract, which influences its resonances and anti-resonances. These effects are concurrent with those of phonation.

2.2.1.2 Muscular Control and Other Factors

The length of the vocal folds at rest varies across people. Vocal folds typically vary between 11 and 15 mm in length for adult women and 17–21 mm for adult men. The vocal folds are about 10–80 times more elastic along their length as compared to transversely, and this ratio varies between people [15, 27–30].

Each vocal fold has multiple layers. In the medical literature there appears to be some debate about the division of these and even the number of these. For our purposes, however, it is sufficient to adopt the view that the tissue of the vocal fold has three layers—an inner muscular layer that comprises the thyroarytenoid (TA) muscle, a middle soft tissue layer called the *lamina propria*, and an outermost epithelial layer. The lamina propria is about 1 mm thick, and is made up of an extracellular matrix comprised largely of the proteins collagen and elastin, which give it its elastic properties. This layered structure plays a key role in phonation. The spectral characteristics of the voice produced during phonation depend greatly on the combination of the material properties of these different layers [31]. The exact nature of these dependencies is not however fully understood at this time.

Other finer biological factors that play significant roles in vocal fold vibration include laryngeal muscle activation and fluid structure interactions within the glottis. These are not discussed here.

One of the most significant biological factors that affects the dynamics of phonation is *glottal closure*. As mentioned above, phonation requires rapid opening and closing of the glottal opening. Glottal closure refers specifically to the closing of the exact area between the vocal folds. Depending on the muscular and tissue structure and movements in the larynx, glottal closure can be partial or complete. In a healthy larynx, during normal phonation, it is complete. When this happens, for reasons that relate to the physics of wave propagation (explained later in this chapter), the voice signal produced has greater intensity, due to clear resonances in the vocal tract. This appears as increased energy in the spectrum of the voice signal, especially within the harmonics. As an aside—under pathological conditions that affect glottal closure, the energy in the higher harmonics is reduced, and the voice is perceived to be "weak."

Fig. 2.5 Schematic diagram showing the histological layers of the vocal folds

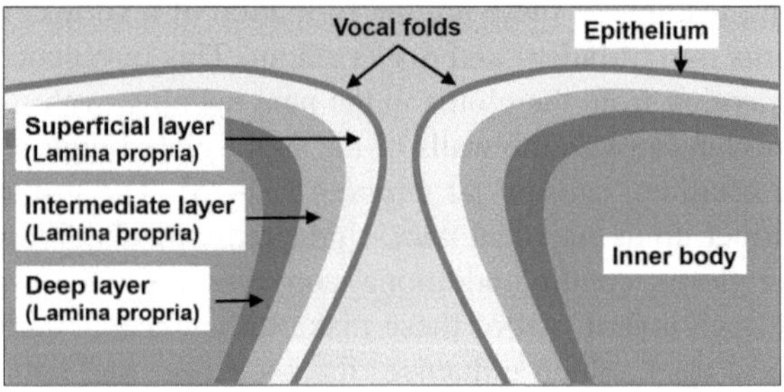

Layering of the Lamina propria. The relative thickness of the layers is not true to scale.

Many factors play a role in achieving complete closure of the glottis. Of these, two extremely important factors are the muscular control exerted by the LCA, IA and TA muscles, and the stiffness and geometry of the vocal structures. The bio-molecular composition of the vocal fold membranes also plays a critical role in this.

To understand this better, we focus especially on the *lamina propria*, which forms part of the layered structure of the vocal fold tissue. For reference, the histological layers of the vocal folds are as shown in Fig. 2.5. The lamina propria is a thin layer of connective tissue found in mucous membranes. It is part of the moist linings of the mucosa of the respiratory, gastrointestinal and other tracts in the body. It comprises a layer of loose areolar connective tissue below the epithelium, that is rich in different kinds of cells. Together with the basement membrane, it constitutes the mucosa. Depending on its location in the body, the different cells found in the lamina propria can include fibrous cells, plasma cells, mast cells, lymphocytes, leukocytes, macrophages etc. The lamina propria helps bind the epithelium to the underlying tissue and provides nutrition to the epithelium. In the vocal folds, the intermediate and deep layers of the lamina propria comprise collagen and elastin fibers. These fibers are thought to play a role in facilitating glottal closure during phonation. As a person ages, there is a loss of collagen and elastin in the body—and so also in the lamina propria of the vocal folds [32]. It is this loss that causes changes in glottal closure properties, and is one of the causes for changes in voice with age.

As these fibers are lost, the vocal folds become thinner, and glottal closure is adversely affected. Thin folds predominantly vibrate in a lateral and medial motion. As a result, they are easily moved apart by sub-glottal pressure and are unable to close completely during phonation. The glottal airflow is much "smoother"—the derivative of the glottal flow (discussed in Sect. 2.3.3.1) is lower than normal. This again causes the voice to be perceived as "weak." Abundance of collagen and elastin in the vocal folds has the opposite effect, and it for this reason that the voices of young people are rich in harmonics.

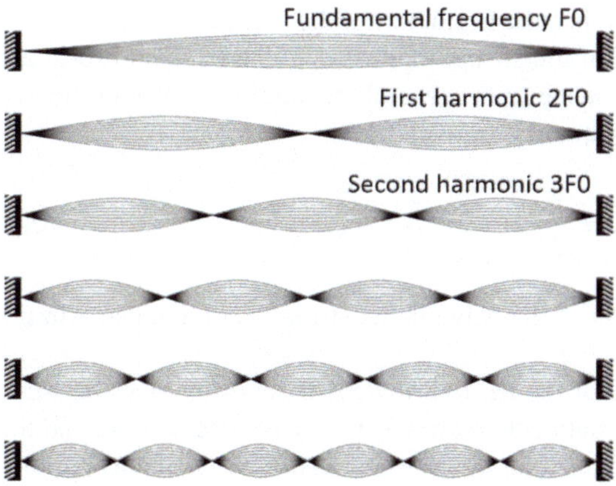

(a) Six natural modes of vibration of a string fixed at both ends

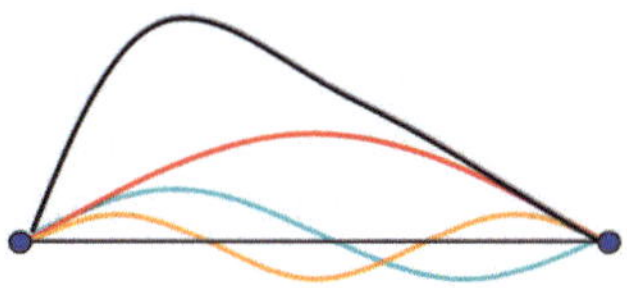

(b) An unnatural mode of vibration, where the string is plucked at $1/3^{rd}$ the distance from a fixed end. When a string fixed at both ends is not plucked in the center, waves of irregular/asymmetrical shape are produced (shown in black), which can be resolved into a superposition of several natural modes (shown in other colors) through Fourier analysis, explained in Chapter 4.

Fig. 2.6 Modes of vibration of a string

2.2.2 Micro-mechanics of Vocal Fold Vibration

Let us now return to the phenomenon of phonation, which gives rise to the glottal pulses discussed above. The best way to think about vocal folds, in order to understand the phenomenon of phonation, is to think of them as simple strings tethered at both ends, and parallel to each other.

Let us begin with a brief recap of the physics involved in their vibration. In what ways can such strings vibrate? From basic physics, we know that such strings have characteristic modes of vibration, each of whose pattern and frequency depends on the length of the string. The possible ways in which such a string can vibrate are called its *natural modes of vibration*. Since it is fixed at both ends, transverse waves produced along it must have zero displacement at the fixed end-points. With a little thought it is easy to see that this can happen if the wave on it has a wavelength that is twice the string's length, or of its length, or two-thirds its length, or one half and so on. The waveforms for natural modes are regular or symmetrical, as illustrated in Fig. 2.6a. The shape of the waveform may differ depending on where it is plucked, as in Fig. 2.6b.

On such a string, a transverse wave of amplitude A will bounce back and forth at the fixed ends, creating standing waves. The displacement y at any time t at position x is given by

$$y(x, t) = Asin(kx + 2\pi f) + Asin(kx - 2\pi ft) \tag{2.2}$$

where f is the frequency. At the nth natural mode the shape of the waveform is given by

$$y(x, t) = Asin\left(\frac{n\pi}{L}x\right) cos(2\pi ft) \tag{2.3}$$

where L is the length of the string. When the frequency of the standing wave is exactly equal to one of the natural modes of the string, the waves going back and forth are exactly in phase. This causes resonance to occur, and the amplitude of vibration of the string is highest. A resonant frequency is given by:

$$f_n = \frac{n\nu}{2L}, \nu = \sqrt{\frac{\tau}{\rho}} \tag{2.4}$$

where ν is the speed of propagation of the wave in the string, τ is the tension in the string, and ρ is its linear density.

At frequencies other than those of the natural modes, phase differences between the waves traveling back and forth between the fixed ends cause attenuation, and the displacements seen are much smaller than those associated with resonant frequencies. The lowest frequency that is a natural mode of vibration for the string is its fundamental frequency. Other natural-mode frequencies are its integral multiples, or its *harmonics*.

Let us now extend this simple string analogy to vocal cords. In the case of the vocal cords, the fundamental frequency of vibration is called the **pitch**, or **F0** by convention. If the vocal cords were simply two strings placed parallel to each other, and if they were independent of each other and each were caused to vibrate symmetrically by a stimulus, they could vibrate independently in any of their natural modes. If, coincidentally, they vibrated in the same natural mode, then their frequencies would be *synchronized*. This is schematically illustrated in Fig. 2.7a.

Synchronization of frequencies is extremely important in maintaining the self-sustained vibrations of the vocal folds during phonation. This is exactly what the muscles that connect the two vocal cords do in a normal person as they speak—they facilitate synchronized modes to induce phonation. This happens under the joint control of the IA/TA/LCA muscles in the larynx. The synchronization is done in the easiest possible manner—normally using the frequencies that require the least effort to synchronize. Note that this is a very simplified explanation. In reality the physics is much more complex: the vocal cords are three dimensional structures. Their natural modes are their *eigenmodes*, and the corresponding frequencies are the *eigenfrequencies*.

Figure 2.7 shows the modes of vibration of planar two-dimensional structures—a rectangular membrane and a circular one, that are fixed at their perimeters. Two dimensional structures are better approximations to the vocal folds, but even at that we must consider that in the case of the vocal folds, one edge is not tethered and their geometry is not as regular as the structures shown in Fig. 2.7. In three dimensions these factors make their modes much more complex. In addition, although they lie parallel to each other, the two sections are free to vibrate in different modes at any given instant of time. Under such conditions, their vibrations may become synchronized only if their individual frequencies of vibration are within a close range of each other.

When the vocal cords are simultaneously excited and synchronized for any mode of vibration (natural or unnatural), as shown in Fig. 2.7a, their vibration causes a

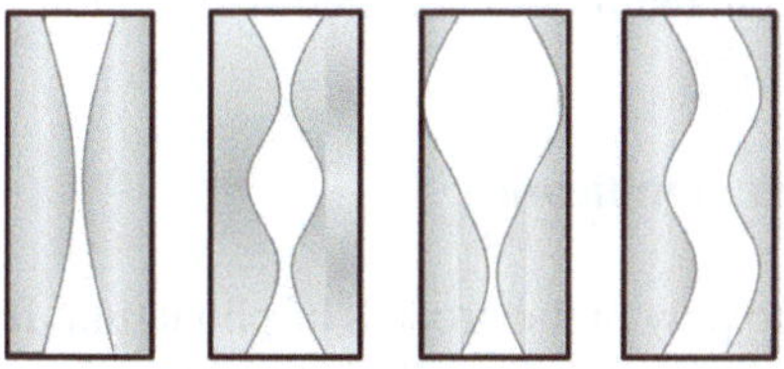

(a) Some synchronized modes of two 1-D parallel strings. The gray shading on the inside exemplifies how the analogy may extend to vocal cords when viewed as a pair of 2-D membranes, each tethered on three sides. The modes are symmetric in the first three panels and asymmetric in the last.

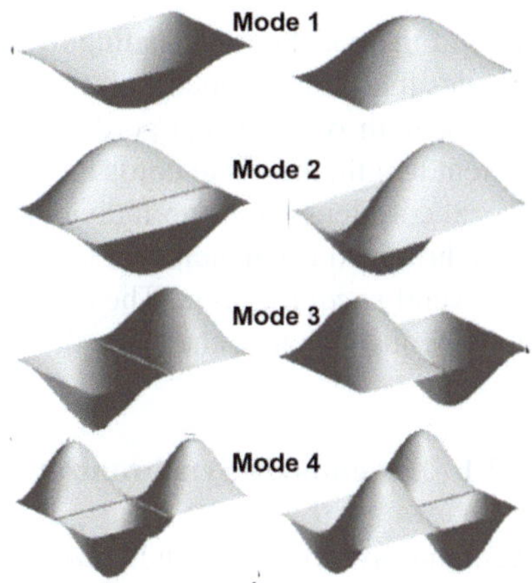

(b) Four modes of vibration of a 2-D rectangular planar membrane. Each pair shows a snapshot of the vibration. The standing waves along the two orthogonal axes are independent of each other and the pattern of vibration of the mode can be viewed as their combined effect.

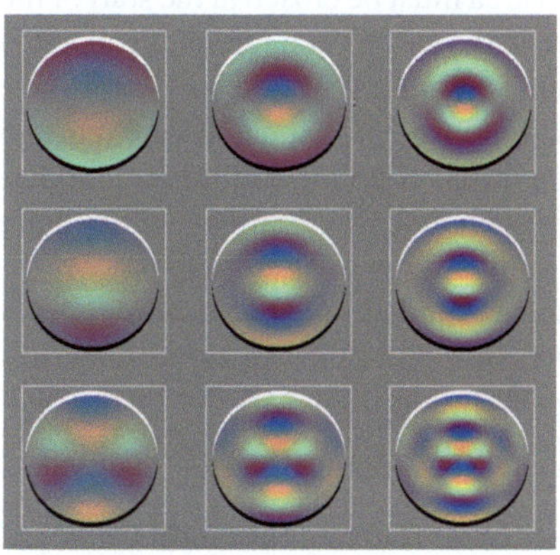

(c) Eight modes of vibration of a 2-D circular membrane.

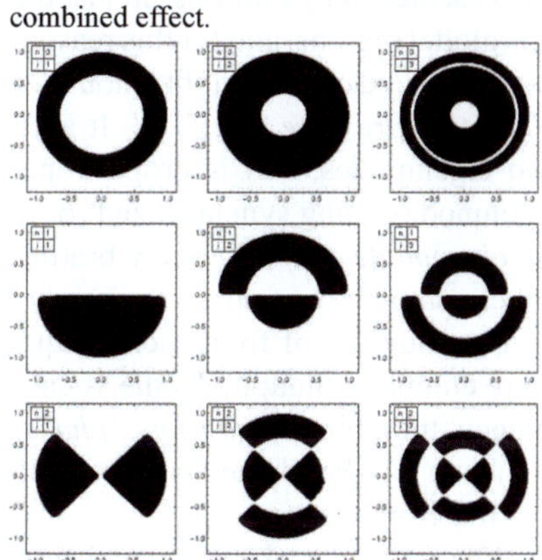

(d) Codified vibrational modes: the eight modes of vibration of a circular membrane shown schematically. Black denotes elevation, white denotes simultaneous depression.

Fig. 2.7 Modes of vibration of one- and two-dimensional structures tethered at their perimeters

rapid and periodic release of air through the glottal opening. The temporal patterns of glottal opening and closure, and the vibratory patterns are *very specific* to the geometry of the glottal region. This has spectral consequences (discussed in later chapters of this book). There are many modes that can be synchronized for a given length of the vocal cords, leading to the possibility of production of an entire range of modes that can be synchronized dynamically during phonation.

The length of the vocal cords can be changed through muscular control. For a different length, the vocal cords have a different set of vibrational modes. The range of vocal cord lengths for an individual is bounded but infinite. These, and the range of synchronized frequencies for each, are coarsely clustered together (by definition) into several *vocal registers*. The process of voluntarily controlling the vocal cords to simultaneously change their length and synchronized modes is called *registration*.

2.2.2.1 Eigenmode Synchronization and Its Significance

Eigenmode synchronization has a critical role to play in the process of phonation. In the balance of forces required for vocal folds to vibrate in a self-sustained manner, for reasons that have to do with energy transfer from airflow to the vocal folds (not explained here), a *vertical phase difference* must be maintained across the vocal folds. This is achieved through eigenmode synchronization.

The synchronization of eigenfrequencies, however, requires energy. In the larynx, this is achieved by building up just the right amount of instantaneous pressure in the sub-glottal regions, and for this reason the glottal area must be closed at the start of the phonation cycle. This "right amount" of pressure in the sub-glottal region is called the *phonation pressure threshold*. It is the minimum pressure required to synchronize two eigenmodes, and is proportional to the difference in frequencies of the two eigenmodes being synchronized [6, 33], and the initial coupling of the eigenmodes. An *eigenmode coupling* of a vibrating system is such that all vibrations within it are either completely in phase or exactly 180° out of phase (as in a standing wave).

The coupling of frequencies required for self-sustained vocal fold vibrations is more complex, though. As the vocal folds synchronize and vibrate at the same frequency, they must *maintain a phase difference* between them [10, 34]. This phenomenon can be likened to how a standing wave results from the superposition of two traveling waves.

Why is synchronization necessary? If the two vocal folds vibrate at different frequencies, it would be sufficient to produce a voice signal. However, synchronization of eigenmodes that are close in frequency provides a great bio-mechanical advantage. When the frequencies are close, very low subglottal pressures are needed induce and sustain the vibrations [35, 36]. This allows different people to control how they induce the phonation vibrations, and which frequencies they choose to use.

Some facts to note about this mechanism are as follows:

1. The vocal folds have an infinite number of eigenmodes. To initiate phonation, any of these can be used. Which ones are exactly used and synchronized by the speaker depends on their frequency difference and coupling strength. This in turn is dependent on the physical (especially respiratory) state of the speaker.
2. For the same speaker, different eigenmode synchronizations result in pitch and voice quality changes.

3. The material properties of the vocal folds play a significant role in eigenmode synchronization—a slight change in their characteristics, such as elasticity, or thickness, or shape of the surface, can lead to entirely different eigenmode couplings and synchronization, resulting in voice quality changes, and changes in other spectral properties of voice.
4. Synchronization can also occur between frequencies that are integral multiples (ratios) of each other, e.g. 100 and 200 Hz, 50 and 150 Hz. When this happens, it causes subharmonic structures to appear in the spectrum, and causes *biphonation* [37, 38].
5. Eigenfrequencies of the vocal folds may become synchronized with the resonances (resonant frequencies) of the sub-glottal and supra-glottal regions.
6. If synchronization between coupled eigenmodes breaks down *during* phonation, chaotic vocal fold vibrations occur.
7. One synchronization pattern can give way to another during the process of phonation. This is called bifurcation [39]. The sudden change in vibration patterns is again perceptible as a change in voice quality (see Chap. 6).

In this microcosm of factors affecting and controlling vocal fold vibrations, there is infinite possibility and range of change and variation even within a single individual. This is the underlying reason for not only the uniqueness of human voice, but also for the readiness of voice to capture the most subtle changes in the body and external factors. It takes very little to cause these subtle but explainable changes in the human voice. We can subconsciously choose which eigenmodes our vocal folds must vibrate in and change them as we speak. They involuntarily and instantaneously respond to simply *any* factor that affects the body or mind, and thereby the muscular, nervous, endocrine and other systems in the body. It is even more important to note that phonation is only one way in which we produce sound. There are other ways as mentioned before, that cause modification and selective amplification and suppression of different characteristics of voice, mediated or caused by the other parts of the vocal tract. The playing ground for profiling thus gets proportionately diversified and widened.

An excellent account of glottal opening and closing events, along with high-speed videography of these events, is found in [40].

2.2.2.2 Vocal Registers

As we have discussed above, the vocal folds can vibrate in different geometrical configurations. Within any one configuration, a range of frequencies or tones may be produced. This range comprises a vocal register [41]. Based on each one's frequency range, and going from the lowest to the highest clusters of tones, the registers are vocal fry, modal (normal speaking frequencies), falsetto and whistle. These are discussed in Chap. 6 in the context of voice disguise.

The tones that comprise one register can be thought of as those that can be maintained in pure form (as pitch) and loudness equally well when they are rendered by the speaker [42]. In the process of registration, a speaker moves smoothly from one cluster of tones (by definition) to another. Singers perform this smoothly enough that we do not notice this, but the net effect is pleasant (usually). In the case of normal speakers, moving through registers may not be very smooth, but may convey emotion and other information.

2.3 The Sounds We Produce—An Articulatory-Phonetic Perspective

2.3.1 Sound Sources in the Vocal Tract

Phonation is an important source of sound in the vocal tract. However, it is not the only source. Another important source is the *turbulence sound* (or compression sound) produced at any constriction of the vocal tract. Turbulence sounds are of several types, of which two key types are frication sounds and aspiration sounds. These sounds do not have the periodic structure of the sound of phonation. Nevertheless, they comprise a wide range of frequencies. When we speak, we may use a combination of these sound production mechanisms to produce the voice that is heard.

Aspiration (whisper or breath-noise) induced sound is the result of modulated airflow through vocal folds that are closed for the most part, while allowing air to rush through a small opening near the transverse arytenoid muscles, between the vocal processes of the larynx.

Frication, or friction-induced sound, occurs by airflow channeled through constrictions in the upper vocal tract. The geometrical placement of these constrictions within the vocal tract can be controlled by our articulators. As the placement moves towards the mouth, the energy in higher frequencies increases (as expected from one-end open tube models of wave formation in physics).

Compression induced sounds also occur when there is a complete closure at some location in the vocal tract, causing air-pressure build-up behind it. The sudden release of this air causes a characteristic sound depending on the placement of the closure in the vocal tract.

All of this brings us to the point where we must now take a closer look at the different sound sources, and the different sounds produced as a result of their interactions, within the vocal tract. This discussion now falls within the purview of articulatory phonetics—the science and study of articulations and sound modulations within the vocal tract. We discuss this next.

2.3.2 Vocal Resonances

As discussed above, the sounds that we hear are produced in different ways in the vocal tract. The source for some is the vibration of the vocal folds (phonation), others may be produced by our articulations, such as lips smacking, tongue clicking on palate, constricted air-movement sounds made through the teeth or tongue (frication), air release sounds made when obstructions created in the vocal tract are suddenly released after building up air pressure behind them (compression), air turbulence caused in various other ways (including breathing) etc. All of these sounds resonate in the vocal tract. As sounds are produced, the movements of articulators change the shape and size of the resonant chambers in the vocal tract. This affects the resonant frequencies of the chambers, and the same excitation signal can be modulated to produce different sounds through this mechanism. As a result of the movement of articulators, both *resonances* and *anti-resonances* are created. These play a vital role in defining the identities of the sounds as we hear them, in an *articulatory-phonetic* sense.

For any sound, depending on the shape of the vocal tract at the time, resonances occur at specific frequencies (more generally a narrow band of frequencies around a central frequency, due to the soft tissue structure of the vocal tract). Since resonances maximize amplitude, frequencies around the resonance bands are heard the loudest.

When frequencies are dispersed, as in sounds of turbulence, and if the vocal tract is constricted at the time, *anti-resonances* are caused. These are the opposite of resonances, in that the energy in the frequencies within which they occur is minimized by destructive interference (as opposed to constructive interference in the case of resonances). While resonances appear as peaks in spectral displays of voice, anti-resonances appear as valleys. Anti-resonances are especially prominent in nasal sounds. In the case of nasal sounds, additional resonances are produced in the nasal chambers. Since both the nasal and oral branches of the vocal tract are involved, some of these cancel out the patterns that may have otherwise been seen as oral resonances, producing prominent anti-resonances with bands of almost zero energy (or *spectral holes*).

The resonant frequency bands in a sound produced in the vocal tract are called *formants* [43]. Within a band, only the central frequency is quantitatively designated as the formant value. By convention, formants are numbered sequentially, starting from the lowest frequency, as F1, F2, F3, ... An example is shown in Fig. 2.8c.

Formants contribute to the intelligible content of speech—they help discriminate between different sounds in speech. The distinct identity of articulatory-phonetic *units* is derived largely (though not entirely) from the arrangement of these spectral peaks relative to one another. In fact, simply generating sounds with the appropriate arrangement of spectral peaks and excitation can result in renditions that are perceived as articulatory phonetic units [44].

The fact that the identity of a sound unit is largely coded in the *positions* of the spectral peaks has the consequence that its identity survives a variety of distortions, such as additive noise, distortions introduced by recording channels etc., which generally affect the heights and shapes of the spectral resonances, but not their positions. This characteristic is also exploited by many speech coding schemes [45] which attempt to ensure that spectral peak positions are retained.

2.3.3 *The Acoustic Theory of Voice Production*

The sounds that we hear as someone speaks—or the voice that we produce—is not just the result of vocal fold vibration. As mentioned earlier, it is in fact a result of multiple phenomena that co-occur in the vocal tract as we speak. These are best explained through the simplified *source-filter* model of voice production, which was devised to help explain and quantify the process of voice production.

The vibration of the vocal folds creates an acoustic pressure wave at the glottis. This sound, also called a **monopole source**, does not have much energy in itself, since the vocal folds are very small in size and so is the volume of air displaced by their vibration. Note that the sounds considered as the monopole source also include a few other weak sources of sound: in humans the vocal folds vibrate at a frequency close to an in-vacuo vocal fold resonance [46, 47]. This resonance phenomenon of the vocal folds amplifies the sounds somewhat, adding to the acoustic energy in the monopole source.

The sound is further amplified by a **dipole source**, which is ascribed to the force applied to the airflow due to the opening and closing of the glottis. This source is responsible for the energy in the harmonics of the voice signal. If the false vocal folds are closed, an additional dipole source arises as airflow from the glottis impinges on those [12, 48].

The third component, called the **quadrupole source**, is what causes sounds to be loud enough to be heard. The air movement caused by the monopole and dipole sources causes turbulence, and together all sources cause resonances in the pharyngeal regions and in the upper vocal tract, including the oral and nasal chambers. Resonances amplify the energy in some frequency bands of the spectrum of the voice signal. These resonant frequencies (formants) are largely what we hear. The exact frequencies at which these resonances occur are characteristic of the dimensions, reflectivity, absorptivity etc. of the tissues and walls of the vocal tract.

2.3.3.1 Glottal Flow Waveform and the Pitch

The dipole source, i.e. the opening and closing of the glottis is in fact what causes pulses of airflow that comprise the *glottal pulses*, also called *pitch pulses* or *vocal tract excitation pulses*. As the glottis opens, the volume of airflow increases and when it is fully open, it peaks. Thereafter, as the glottis closes, it decreases and stops when the glottis is fully closed.

The middle panel in Fig. 2.8a is an *electro-glotto-graph* (EGG) plot showing the dynamics of vocal fold vibrations for a vowel. An EGG is a throat mounted device that measures the movements of the vocal folds. Below this is an estimate of the actual glottal airflow obtained through *inverse filtering* [49]. We see from this figure that when the contact between the vocal folds is high, i.e., the vocals folds are close together and the area of the open glottis is small or absent, the amplitude (or energy) of the air pulse in the vocal tract is low. As the glottis opens and closes during phonation, the airflow volume in the region rises and falls. At the epiglottis (which is above the glottis) pressure is lowest as the air is rushing through, and highest when the glottis is closed. Due to the manner in which the glottis opens and closes, the shape of the pitch pulse waveform (see Fig. 2.8b) is usually skewed to the right. The time derivative of airflow at the time of opening of the glottis is thus lower in value than that at the closing.

In the time derivative of the airflow, the instant of the most negative peak is considered the instant of main excitation of the vocal tract. This peak value is called the maximum flow declination rate (MFDR). The typical structure of a glottal flow waveform is shown in Fig. 2.8b.

On close examination of Fig. 2.8c, which shows some pitch pulses and the distribution of acoustic energy in them, we can see that the excitation signal produced during phonation has a *quasi-periodic* structure, i.e. it is not *exactly* the same from cycle to cycle.

2.3.3.2 The Source-Filter Model of Voice Production

The source-filter model of voice production is a highly simplified model [51] that assumes that the excitation signal produced at the glottis is a *source* signal that is modulated by a *filter*, represented by the vocal tract above the glottis. The nature of this modulation is dependent on the *response* of the filter (the manner in which it responds to an impulse or a spike stimulus). The notion of a filter and its response are both explained in detail in Chap. 4. The response of this filter (or the system that is assumed to be the filter) is called the **vocal tract transfer function**.

The source-filter view of voice production is a widely used model that is used to characterize many of its observed properties. Figure 2.9 illustrates this model.

The nature of the quadrupole source, and the manner in which it defines the voice can be understood (and also quantified) by an analysis of the actual nature of the transfer function of the model. For the derivation of the transfer function of the source-filter model, the vocal tract (filter) is approximated by physical models, called the *acoustic models of the vocal tract*. The resonance characteristics of these have exact analogies to the transfer function of the source-filter model—so much so that very often, the two are used interchangeably, or even in parallel, within the same analysis of vocal acoustics. The transfer function of the source-filter model may be derived through exact analogies drawn between an electrical circuit representing it, and a corresponding acoustic model of the vocal tract (and vice-versa).

The details of these parallel analyses are not discussed in this chapter. However, the sections below discuss the acoustic models of the vocal tract in some detail. Analogies to the source-filter model of the vocal tract are pointed out for future reference where appropriate, and the subject is revisited from a signal-processing perspective in Chap. 5.

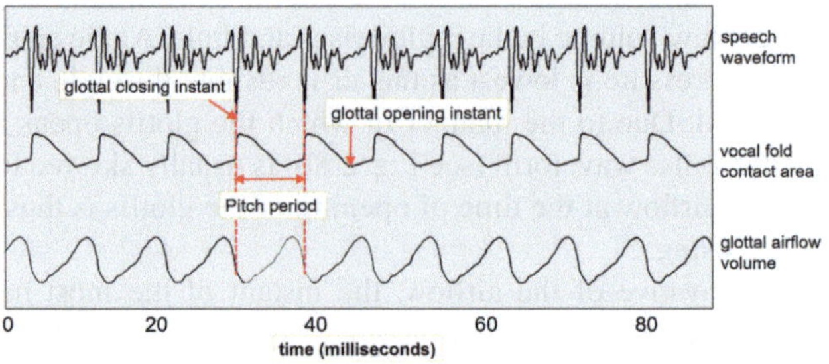

(a) **Top:** The speech waveform (air pressure vs. time) **Middle:** A laryngograph or electroglottograph signal showing contact between vocal folds vs. time. **Bottom:** Amplitude of air pulse exciting vocal tract. The amplitude scales are arbitrary for all waveforms shown. The glottal opening instant is defined as the instant when the glottis is *fully* open. However, the true opening instant is shortly after the glottis closes when the vocal folds begin to move apart. This is shown in the figure below.

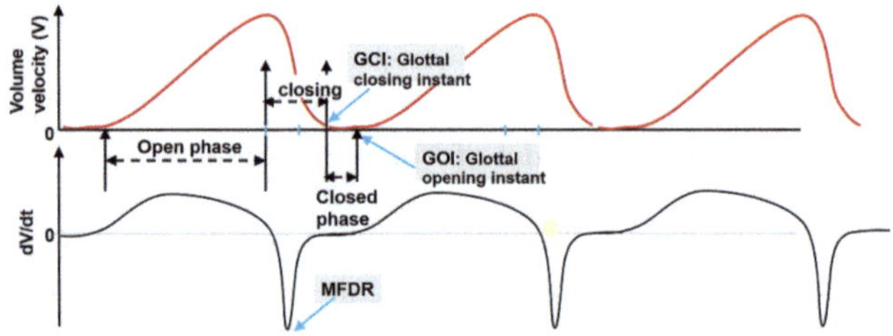

(b) **Top:** The volume velocity (the volume of air pushed out through the glottis per unit time) waveform, also called the pitch pulse waveform. **Bottom:** its time derivative. As glottis opens, pressured air from below the glottis rushes out. The rate of increase of the volume of air pushed out is positive some time. As the pressure in the subglottal region falls, this decreases. Once the glottis begins to close, there is a sharp change from positive to negative in the time derivative of the volume velocity. The derivative follows the pattern of decline of volume velocity, becoming more gradual and almost zero in the closed phase of the glottis [52]. The instant of the most negative peak is considered the instant of main excitation of the vocal tract. This peak value is called the maximum flow declination rate (MFDR).

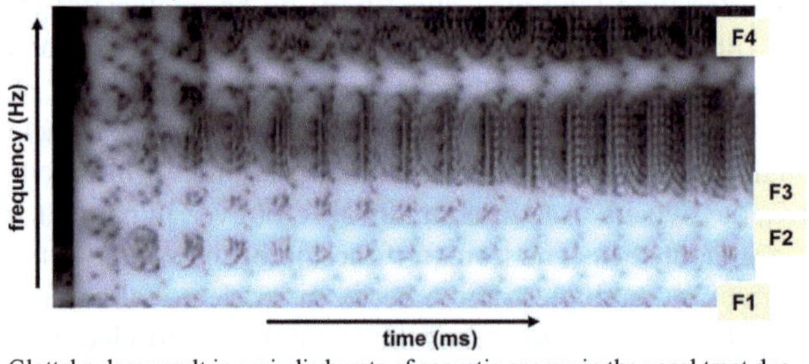

(c) Glottal pulses result in periodic bursts of acoustic energy in the vocal tract due to its resonances. Each burst occurs during a glottal *closure* period, as the vocal tract. These are the pitch pulses seen on a wideband frequency vs. time plot (spectrogram) of the speech signal. This figure shows a closeup of pitch pulses corresponding to the vowel /aa/, as the vocal fold vibrations begin and some time into the vowel. The total duration of the horizontal axis shown is about 200 ms. The high-energy (brighter white) bands are the formants, marked F1-F4.

Fig. 2.8 Illustration of the dynamics of vocal fold vibration

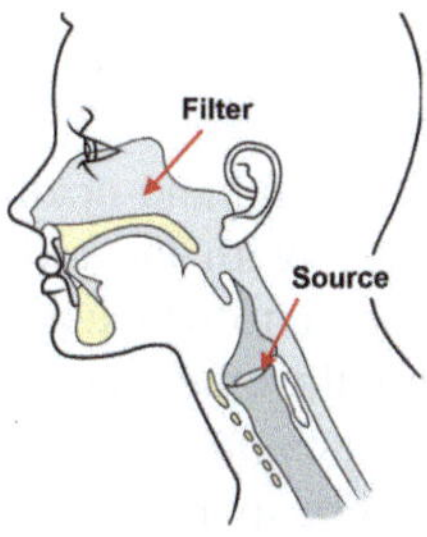

(a) The entire light gray area comprising the nasal and oral cavities, and the pharynx are the filter. The glottis or larynx is the source.

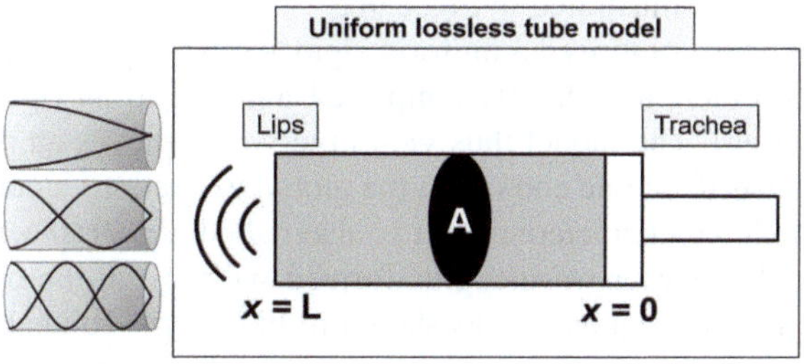

(b) The uniform lossless tube model. When the tube is closed at one end, only odd harmonics are seen.

Fig. 2.9 The essential elements of the source-filter model of the vocal tract

2.3.3.3 The Acoustic Models of Voice Production

Of the possible acoustic models of voice production, *acoustic tube models* are the most widely used ones in the scientific community. These attempt to explain vocal resonances through the physics of sound propagation through *tubes* that approximate the shape of the vocal tract. Specifically, these models approximate the vocal tract with one or a series of cylindrical channels to analyze the dynamics of sound propagation through it, and its other characteristics. They are of several types, depending on how they approximate the geometrical shape of the vocal tract. The key types are:

- **Uniform cylinder model**: uses a single cylinder
- **Multistage model** or **Concatenated tube model**: uses multiple concatenated cylinders that best fit the vocal tract shape
- **Multistage integrable model**: uses multiple short concatenated cylinders of equal length

These are further divided into **lossless** and **lossy** versions, depending on whether or not they consider the losses (to sound wave propagation) induced by various factors within the vocal tract. Lossless versions of these assume that there is no loss of acoustic energy within the vocal tract, while lossy versions assume different sources of loss within the vocal tract, such as lip radiation, absorption of sound by the walls of the vocal chambers etc. In the discussion below we focus largely on the lossless versions of these models.

2.3.3.4 The Uniform Lossless Cylinder Model of the Vocal Tract

One of the simplest and most widely used acoustic models of the vocal tract is the *uniform lossless cylinder model*, also called the *uniform lossless tube model*.

The uniform lossless cylinder model, and its correspondence to the source-filter model is illustrated in Fig. 2.9b.

This comprises a uniform rigid lossless cylinder closed at one end and open at the other (as a highly simplified analogy to an open vocal tract with the glottis closed). The model thus views the vocal tract as an acoustic tube into which air is "pushed" at one end (from the glottal excitation), emerging from the other end with modified characteristics. It is also called a *quarter wavelength resonator* (alluding to the quarter wavelengths formed within it at the open end, corresponding to its resonant frequencies, as shown in Fig. 2.9b). To analyze the effect of transmission of sound through this, let us consider what happens in the tube.

Let L be the total length of the tube. In practice, L is expected to be about 17 cm, which is approximately the length of the average adult human vocal tract.

In order to understand the acoustic behavior of this tube, let us consider a sound wave of a single frequency f (or a single wavelength), produced at the glottal source noting that the voice signal can be expressed as superposition of many such frequency components (as discussed in Chap. 4).

Let $U(x,f)$ be the amplitude of any frequency component f at position x in the tube, measured from the closed (source) end, where $x = 0$. $U(0,f)$ is the amplitude of frequency component f at the excitation source end of the tube, while $U(L,f)$ is the amplitude of the corresponding component in the final voice signal. The *transfer function* of the vocal tract for frequency f is given by the ratio

$$T(f) = \frac{U(L,f)}{U(0,f)} \tag{2.5}$$

$T(f)$ informs us about how the vocal tract modifies individual frequencies in the signal. To understand the nature of the output signal, we must understand $T(f)$. The expression in Eq. 2.5 is only a generic one for $T(f)$. Let us now consider more explicit expressions for it.

Let $u(x, t)$ be the *volume velocity* of air at position x along the tube, at time t. In other words, at each location x in the tube the air travels as a longitudinal wave with a time-varying volume velocity $u(x, t)$ which creates a pressure $p(x, t)$. Since $x = 0$ represents the glottal end of the tube, and $x = L$ represents the lips end of the tube, $u(0, t)$ is the glottal excitation at the entry to the tube. $u(L, t)$ is the volume velocity of the air—i.e. the acoustic "signal"—coming out of the mouth, which is the voice sound we hear.

As the wave travels down the tube, it eventually encounters the end where a sudden change occurs in the transmission medium. At this point, while some of the wave moves on into the new environment, some of it is *reflected* back into the tube, now traveling backwards into the tube. The fraction of the wave amplitude that is reflected into the tube is called the *reflection coefficient* at the end of the tube, and is generally denoted by the symbol r.

When the reflected wave returns to the entry point of the tube it encounters a similar change of medium and is reflected again. In fact it is this repeated reflection that creates resonances; if the end of the tube were perfectly "matched" to the subsequent medium and there were no reflections, there would be no resonance.

Representing the cross-sectional area of the tube as A, the density of air as ρ and the velocity of sound as c, it can be shown that for frequency components whose wavelengths are long compared to the length of the vocal tract (i.e. for wavelengths greater than ~8.5 cm, corresponding to frequencies less than 4000 Hz) the mathematical relation between the pressure and the air volume velocity in a lossless uniform tube is given [52] by

$$-\frac{\partial u(x,t)}{\partial x} = \frac{A}{\rho c^2}\frac{\partial p(x,t)}{\partial t}$$
$$-\frac{\partial p(x,t)}{\partial x} = \frac{\rho}{A}\frac{\partial u(x,t)}{\partial t} \tag{2.6}$$

The solution has two components. The velocity $u(x, t)$ is the combination of a *forward* component $u^+(t)$, and a reflected component (from the open end) $u^-(t)$. The net velocity at any position x is given by

$$u(x,t) = u^+(t - x/c) - u^-(t + x/c) \tag{2.7}$$

where x/c is the time for the signal to propagate to x, and the signs of x/c in the two components account for the fact that $u^+(t)$ and $u^-(t)$ travel in opposite directions. The pressure of the acoustic wave is related to the volume velocity, but differs from volume velocity in that when air propagating in opposite directions meets at any location, the pressure contributions of the two add, giving

$$p(x,t) = \frac{\rho c}{A}\left(u^+(t - x/c) + u^-(t + x/c)\right) \tag{2.8}$$

We now take the following boundary conditions into consideration: the velocity at the closed end is 0, since no air is crossing the boundary, i.e. $u(0, t) = 0$. More generally, $u(0, t) = u_g(t)$, the volume velocity at the glottal source (which is also 0 for a closed glottis). Similarly, at the open end, the contribution of the air from the tube to the external pressure is zero, i.e. $p(L, t) = 0$.

Combined with these boundary conditions, Eq. 2.6 leads to the following relation [52]:

$$T(f) = \frac{U(L,f)}{U(0,f)} = \frac{1}{\cos\left(\frac{2\pi Lf}{c}\right)} \tag{2.9}$$

This also corresponds to having a reflection coefficient of $r = 1$ at the open end and $r = -1$ at the closed end of the tube.

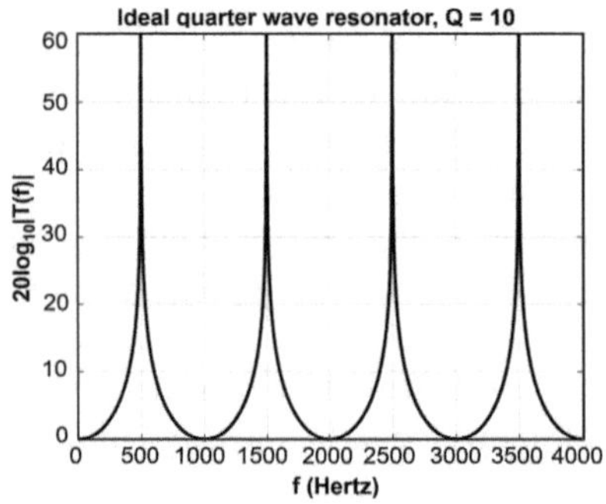

(a) The transfer function of an ideal tube – a quarter wave resonator – that emulates the vocal tract. The peaks are shown clipped, and are infinite in height in reality. Assuming the speed of sound to be 340 m/s and length of the average vocal tract to be 17 cm, the resonant frequencies are seen to be 500 Hz, 1500 Hz, 2500 Hz .... These are also called *formants*.

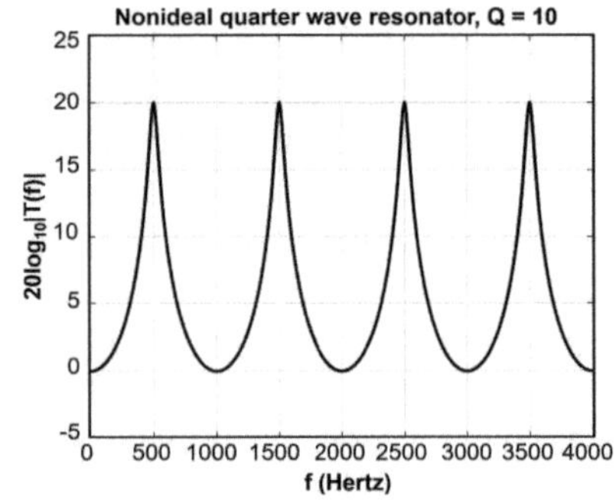

(b) The transfer function of a non-ideal tube – a quarter wave resonator – that emulates the vocal tract more closely. The peaks are finite in height. The vocal tract is not lossless inherently and in addition, sound itself causes changes in it – energy is lost through radiation, absorption, changes in viscosity, temperature etc.

Fig. 2.10 Transfer function of the vocal tract

Figure 2.10 shows the magnitude of the transfer function $|T(f)|$ as a function of frequency. As we can see, the response shows a sharp (infinite in reality, truncated for visualization) peak at all frequencies $f = \frac{(2n+1)c}{4L}$, where n is an integer. These are the resonances of the tube.

When the tube is not lossless, if the glottis end is not perfectly closed, and the pressure at the open end is almost 0, i.e. the reflection coefficients at the two ends are $r \approx -1$ and $r \approx 1$ respectively, the magnitude of the transfer function changes to

$$|T(f)| = \frac{1}{\sqrt{K^2 + \cos^2\left(\frac{2\pi Lf}{c}\right)}} \tag{2.10}$$

The constant K in this equation accounts for the losses incurred in the vocal tract. These losses are due to various factors such as viscous friction at the walls of the tube, thermal losses due to heat conduction through the walls, vibration of the walls as a result of pressure changes and wall elasticity, radiation from the lips etc. We do not derive the exact formulation for K here.

As we can see in Fig. 2.10, this results in increased bandwidths at the peaks, while reducing their height. Additional effects such as thermal and friction losses on the tongue will further reduce the heights of the peaks and increase their bandwidth, but not change the positions of the peaks. Note that the equations above view the signal in the tube as a continuous-time signal (discussed further in Chap. 4).

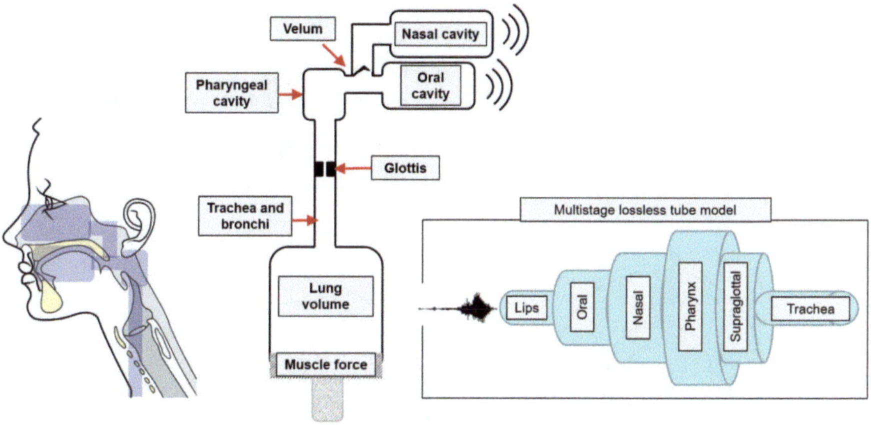

(a) The motivation for the structure of the multistage acoustic-tube model: the model approximates the dimensions of the various parts of the vocal tract that comprise the quadrupole sound source. This is further simplified as the model shown on the right, which comprises a (variable) set of concentric lossless cylinders.

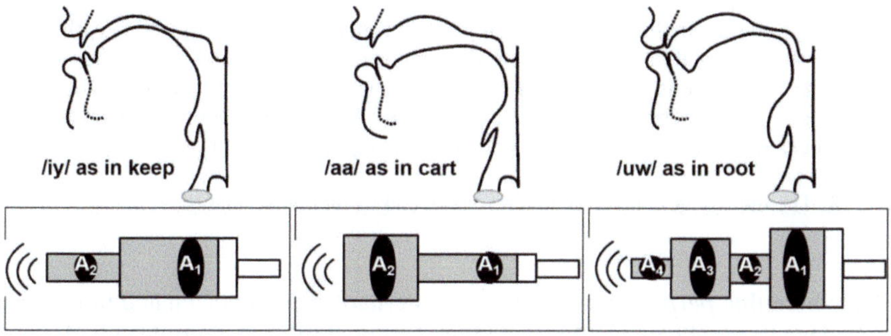

(b) The multistage model is adaptive to vocal tract configurations. It explains the resonance patterns of each sound as the resonance patterns of the corresponding lossless tube configuration that best approximates the vocal tract when the sound is produced.

Fig. 2.11 Alternative representations of the multistage model

2.3.3.5 The Multistage Lossless Tube Model of the Vocal Tract

The *multistage lossless tube model* of the vocal tract extends the uniform cylinder acoustic tube model. It is also called the *concatenated tube model* of the vocal tract. This model, and its equivalence to the source-filter model is illustrated in Fig. 2.11a. The model approximates different parts of the vocal tract with multiple concentric lossless tubes whose relative proportions (lengths and diameters) follow those of their corresponding counterparts in the vocal tract. In its most simplified version, only the relative lengths and diameters of the tubes change.

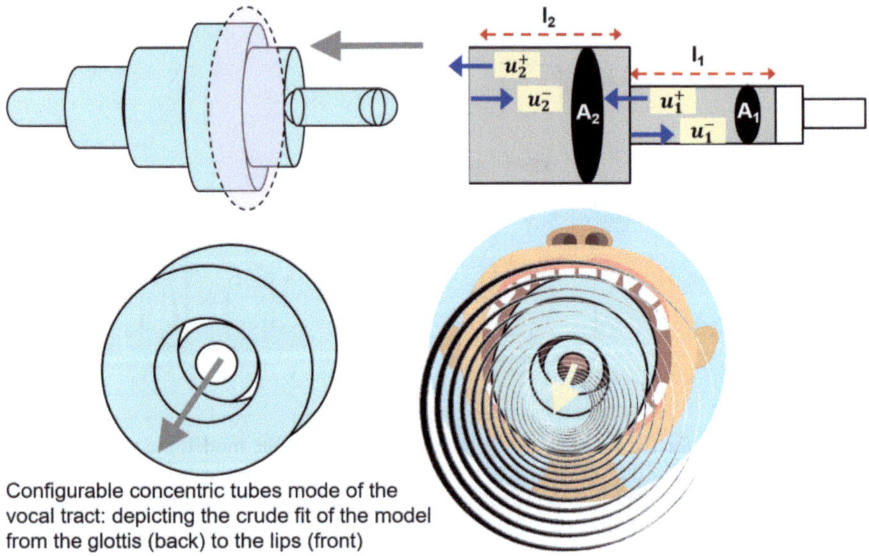

The highlighted area in the model on the top left is considered on the top right. Within it are shown the forward and backward (reflected) components of the sound propagating through each of the two sections of the model. The sudden changes in physical dimensions at the boundaries cause some part of the sound to be reflected back. Solving for the transfer functions of the concatenated tube model is done piecewise, using such breakups.

Fig. 2.12 The multistage concatenated tube model and its piecewise analysis

Note here that *perturbation models* are alternatives to concatenated tube models that achieve spectral patterns that are equivalent or similar to multiple cylinder models, by introducing internal notches or perturbations in a uniform cylinder [53]. This creates areas of varying cross-sectional area within the uniform cylinder, simulating the concatenation of acoustic tubes. We do not discuss perturbation models in this chapter.

2.3.3.6 The Multisage Integrable Model of the Vocal Tract

A more sophisticated model is represented by the *multistage integrable model* which comprises a large number of tubes of equal lengths but varying cross-sectional area, which integrate into the shape of the vocal tract. This model can potentially approximate the contours of the vocal tract much more closely while being mathematically tractable. In this section we focus on the lossless version of this model. Its key elements are depicted in Figs. 2.12 and 2.13.

The model comprises N concentric tubes. Let x be a length variable, representing the distance measured from the beginning of any tube along the direction of sound

The multistage integrable model depicted in (a) and (b) approximates the vocal tract using a large number of connected cylinders of the same length Δx. The panels (a) and (b) show an inner section of a longer flexible pipe with multiple segments. Their cross sectional areas may however vary. Sound propagates through each tube in time $\tau = \Delta x/c = L/Nc$, where L is the total length of the assembly of tubes, N is the number of tubes, and c is the velocity of sound. This model is a much closer approximation to the (moving) vocal tract The surface within this model is assumed to be smooth, while in reality the inside of the vocal tract is not so smooth, as depicted in (c). Modeling such fine variations may require a handover to artificial intelligence techniques in the future.

Fig. 2.13 Illustration of the multistage integrable model

propagation in the vocal tract. In this model, each tube is of the same length $\varDelta x$. Let the cross-sectional area of the kth tube be A_k.

As before, ρ and c are the density of air in the tube, and the velocity of sound respectively. Let $p_k(x, t)$ and $u_k(x, t)$ denote the sound pressure and volume velocity respectively, in the kth tube at position x and time t. From the basic equations of wave propagation (not given here), the derived wave propagation equations for the kth tube in this model are (see Fig. 2.12a for reference):

$$p_k(x,t) = \frac{\rho c}{A_k}\left[u_k^+\left(t - \frac{x}{c}\right) + u_k^-\left(t + \frac{x}{c}\right)\right] \quad 0 \le x \le \varDelta x \tag{2.11}$$

$$u_k(x,t) = u_k^+\left(t - \frac{x}{c}\right) - u_k^-\left(t + \frac{x}{c}\right) \quad 0 \le x \le \varDelta x \tag{2.12}$$

where $u_k^+(t - \frac{x}{c})$ is the volume velocity of the wave traveling forward in the kth tube and $u_k^-(t + \frac{x}{c})$ is the volume velocity of the wave traveling backward in the kth tube.

Although not explicitly shown, the multistage integrable model of Fig. 2.13 has a configurable overall external shape, which can closely approximate the overall shape assumed by the vocal tract in producing different sounds. This model best explains the structure of the pressure wave that comprises the voice signal. To understand how, we must understand the volume flow dynamics within it more closely.

Figure 2.14a shows the volume flow dynamics at the junction of any two cylinders in the multistage integrable model. The overall sound source, and volume flow direction from it, is indicated in the figure by the thick arrow on the top left. Volume flows for the two cylinders are denoted by color-coded thick arrows.

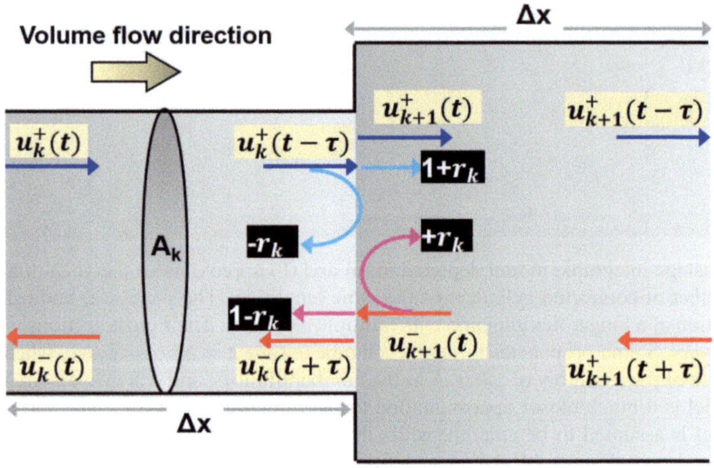

(a) The flow velocity components are shown for two adjacent tubes in the model.

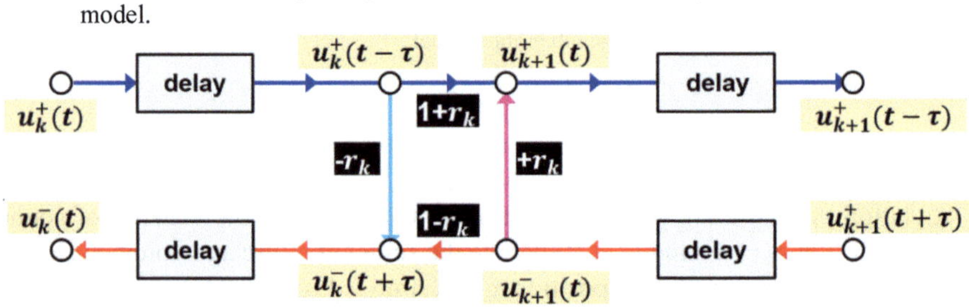

(b) A schematic version of the flow components across the junction of adjacent tubes in the multistage integrable model. Solutions for the model expressed in this form are obtained through signal processing methods by analyzing analogous digital circuits.

Fig. 2.14 Volume flow velocities for the multistage integrable model

At any given time t, let us consider the flow at the boundaries of the two cylinders. For the kth cylinder, the various components of the volume flow are listed below:

- At the entry point where the distance $x = 0$:
  - $u_k^+(t)$ is the component of volume velocity going forward into the cylinder.
  - $u_k^-(t)$ is the component of volume velocity from the reverse flow direction, which results from reflection at the junction, and the reverse flow from the $k+1$th cylinder.
- At the junction where the distance $x = \Delta x$, part of the forward propagating wave goes on ahead, while a part of it is reflected back. Similarly, part of the backward propagating wave (from reflections) goes on, while part of it is reflected back (going ahead in the direction of the forward propagating wave):

- $u_k^+(t-\tau)$ is the component of volume velocity going forward into the cylinder—this is the volume that entered the cylinder at time $t-\tau$.
- $u_k^-(t+\tau)$ is the component of volume velocity from the reverse flow direction—this is the reflected component of the volume that arrives at the entry to the cylinder at time $t+\tau$.

The same considerations and notations hold for the volume flows at the entry point and the forward junction of the $(k+1)$th cylinder.

At the junction, the continuity constraints stipulate that both the pressure and volume flow must be continuous. Accordingly, the boundary conditions at the junctions ($x = \Delta x$) are:

$$p_k(\Delta x, t) = p_{k+1}(0, t) \tag{2.13}$$

$$u_k(\Delta x, t) = u_{k+1}(0, t) \tag{2.14}$$

Substituting from Eqs. 2.11 (and recalling that $\tau = \frac{\Delta x}{c}$) we get:

$$\frac{A_{k+1}}{A_k}[u_k^+(t-\tau) + u_k^-(t+\tau)] = u_{k+1}^+(t) + u_{k+1}^-(t) \tag{2.15}$$

$$u_k^+(t-\tau) - u_k^-(t+\tau) = u_{k+1}^+(t) - u_{k+1}^-(t) \tag{2.16}$$

This results in the following relationships for the flow volumes at the junction:

$$u_{k+1}^+(t) = (1+r_k)u_k^+(t-\tau) + r_k u_{k+1}^-(t) \tag{2.17}$$

$$u_k^-(t+\tau) = -r_k u_k^+(t-\tau) + (1-r_k)u_{k+1}^-(t) \tag{2.18}$$

where r_k is the reflection coefficient for the k_{th} junction:

$$r_k = \frac{A_{k+1} - A_k}{A_{k+1} + A_k}, \quad |r_k| \le 1 \tag{2.19}$$

As an aside—if the model were modified such that each tube had a different length, the expressions in Eq. 2.17 would include τ_k instead of τ in all instances.

For small Δx, the model can be extended by adding the contributions from other tubes in a similar fashion to give the total volume flow at any instant t at any position x in the tube. In an extended model, the junction flows are updated every τ seconds, and therefore the model correctly approximates the characteristics of a signal sampled at a rate of $2/\tau$. The digital version of this model allows us to easily solve for its transfer function and resonance characteristics [54]. We do not discuss the digital version of the model here.

2.3.4 Ariculatory-Phonetic Units of Speech

Articulatory-phonetics is a branch of science that deals with the study of sounds produced by the vocal tract. **Phonemes** are articulatory-phonetic units of speech associated with "canonical" arrangements of the vocal tract that result in sounds with distinctive spectral (and spectral resonance) patterns. These canonical arrangements give each phoneme its specific identity. In addition, phoneme identity is also dependent on the source of sound in the vocal tract. In articulatory-phonetics, phonemes are in fact categorized based on three criteria: the locations of the constrictions, the arrangement of the chambers, and the manner of excitation (including vocal fold activity and other sound sources) associated with them. Phonemes that are categorically similar in all these respects and yet enunciated differently in different contexts are called *allophones*. For example in American English the phoneme /p/ as in "pin" and "spin" are two versions of the same sound that is categorized as a voiced stop consonant (explained later in this section). The first instance involves aspiration, while there is no breath release associated with the second instance.

In the discussion below, we give a specific example of this characterization for the sounds produced in American English using a specific symbol set of phonemes used by the CMU Sphinx automatic speech recognition (ASR) system [55]. Figure 2.16 shows the key membership of this specific phoneme set, along with their IPA (International Phonetic Alphabet) symbols. The choice of this set was made solely for ease of illustration and readability by those not familiar with the comprehensive standardized IPA system that is used for representation of articulatory sounds in scientific studies. With this in reference, the following is a brief description of the phoneme categories that are shown in Fig. 2.16.

The sounds we produce for speech are divided into two broad categories: *vowels* and *consonants*. Vowels are phonemes that involve only phonation as the sound source, and are produced in different open configurations of the vocal tract. Consonants involve different combinations of the other sounds sources mentioned earlier in this chapter. These are further discussed below.

2.3.4.1 Vowels

In the production of vowels, the vocal tract is open from the glottis to the lips. The relative areas of different parts of the vocal tract however differ between different vowels. This is illustrated in Fig. 2.11b which also shows the corresponding acoustic tube model approximations to the vocal tract for the specific configurations shown. The transfer function of these models allow us to estimate the specific resonance frequencies (formant frequencies) of the corresponding vowels. For example, for the vowel /iy/ as shown in Fig. 2.11b, the two tube approximation can be solved by assuming it to be the sum of two independent tubes—a quarter wave resonator closed at one end, adjacent to the source, and a half wave resonator comprising a tube that is open at both ends. These are shown in Fig. 2.15. The decoupling holds

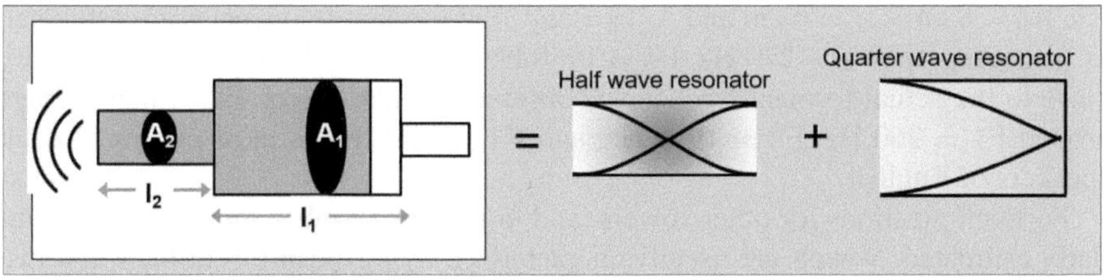

The vocal tract configuration for the vowel /iy/ (as in "sleep") can be approximated by the two-tube model shown on the left. It can be further approximated as the sum of a half-wave resonator and a quarter-wave resonator.

Fig. 2.15 A two-tube model for the phoneme /iy/

true especially for the case where the cross-sectional area of one tube is much larger (or much smaller) than of the other.

If the areas of the two cylinders in the model (counting from the one closest to the source) are A_1 and A_2, and their lengths l_1 and l_2 respectively, then for the quarter-wave resonator the **acoustic compliance** is given by $C_1 = A_1 l_1 / \rho c^2$ and for the half-wave resonator, the **acoustic mass** is given by $M_2 = \rho l_2 / A_2$ [56]. These are taken into consideration in deriving the expressions for the resonant frequencies of the two tubes respectively.

For the quarter-wave resonator, the resonant frequencies are given by

$$f_n = \frac{c}{4l_1}(2n-1) \quad n = 1, 2, \ldots \tag{2.20}$$

For the half-wave resonator, the resonant frequencies are given by

$$f_n = \frac{c}{2l_2} n \quad n = 0, 1, 2, \ldots \tag{2.21}$$

The formant frequencies of the two-tube model are given by

$$f_n = \frac{c}{2\pi}\left[\frac{A_2}{A_1 l_1 l_2}\right]^{1/2} \quad n = 1, 2, \ldots \tag{2.22}$$

At low frequencies (usually only the lowest), this can be shown to be independent of the speed of sound as

$$f = \frac{1}{2\pi}\left[\frac{1}{C_1 M_2}\right]^{1/2} \tag{2.23}$$

f is now called the *Helmholtz frequency* of the resonators, or the *Helmholtz resonance*.

In the current example, given specific lengths and areas for the two tubes, we can compute the expected formant frequencies for /iy/. For example, if we assume $l_1 = 9$

cm, $A_1 = 8$ cm^2, $l_2 = 6$ cm and $A_2 = 1$ cm^2, the formant frequencies are estimated as (rounding up) F1 = 270 Hz, F2 = 1950 Hz, and F3 = 2920 Hz. These are very close to the actual formant frequencies observed for the vowel /iy/, which average around F1 = 260 Hz, F2 = 1910 Hz, and F3 = 2920 Hz in most groups of adult speakers of English.

Formant positions for other vowels (and in fact for all other sounds) can be similarly estimated. Vowels are usually characterized by the positions of their first two formants, F1 and F2. However, vowels generally have prominent F1, F2 and F3. The distance between F1 and F2 is wider in the *front* vowels, and very small in *back* vowels. Vowels such as /iy/ and /y/ are differentiated via their F3. In singing voices, F3, F4, F5, ... are sometimes made prominent by the singers by vocal fold register manipulations that shift pitch and throw more energy into the higher frequencies, and/or by lowering the larynx while elevating the tongue blade, so that they can be heard over the orchestral accompaniment. This is necessary because musical instruments produce wider frequency ranges than humans do in normal speech, and the energy in higher frequencies produced by the instruments generally tends to mask the energy in the weaker higher formants of speech.

In fact, vowel *height* and vowel *backness* are two key *positional* variables that are involved in the categorization of vowels.

Height refers to the vertical position of the tongue relative to the palate or jaw. The frequency of the formant F1 is inversely proportional to the height of the tongue. For example, the sounds /iy/ and /uw/ are produced with the tongue position high in the mouth. Their formants F1 are lower in frequency and they are called *high vowels*. Conversely, vowels like /aa/ are produced with the tongue position lower in the mouth. Their F1 is higher, and they are called *low vowels*.

Backness refers to the horizontal position of the tongue relative to the back of the mouth. Backness is defined by the position of the formant F2. When the tongue is positioned forward in the mouth, as in /iy/, F2 is higher in frequency. Such vowels are called *forward vowels*. *Back vowels* such as /uw/ have lower F2 values because the tongue is positioned towards the back of the mouth.

In addition to height and backness, there are other finer characteristics used for the categorization of vowels, such as roundedness, nasalization, degree of phonation, tongue-root retraction etc. We will not discuss those distinctions here. The vowels shown in Fig. 2.16b are categorized based on their height (how high the tongue is), backness (which part of the vocal tract is pivotal in its production), laxness (rigidness or relaxedness of configuration) and roundedness (whether or not the lips are rounded). Figure 2.16b also shows four additional phonemes: three *Diphthongs*: /ay/ as in *side*, /aw/ as in *how* and /oy/ as in *toy*, and the semivowel /er/ as in *surround*. More detailed accounts of articulatory-phonetic categorizations of sounds produced in the world's languages are given in many articles in the literature, e.g. [57].

2.3.4.2 Consonants

Consonants are exemplified in Fig. 2.16a. By **manner of articulation**, Fig. 2.16a categorizes consonants into five groups, namely *Plosives, Fricatives, Affricates,*

MANNER	VOICING	Bilabial	Labio-dental	Inter-dental	Alveolar	Palatal	Velar	Glottal
		PLACE OF ARTICULATION						
Stop	Voiced	b			d		g	
	Unvoiced	p			t		k	
Fricative	Voiced		v	dh (ð)	z	zh (ʒ)		
	Unvoiced		f	th (θ)	s	sh (ʃ)		hh (h)
Affricate	Voiced					jh (dʒ)		
	Unvoiced					ch (tʃ)		
Nasal	Voiced	m			n		ng (ŋ)	
Liquid(L)	Voiced				l			
Liquid(R)	Voiced					r		
Glide	Voiced					y (j)	w	

(a) Classification of consonants based on the involvement of different parts of the vocal tract (place of articulation) and the manner of their movements (presence of absence of voicing, and the manner or articulation).

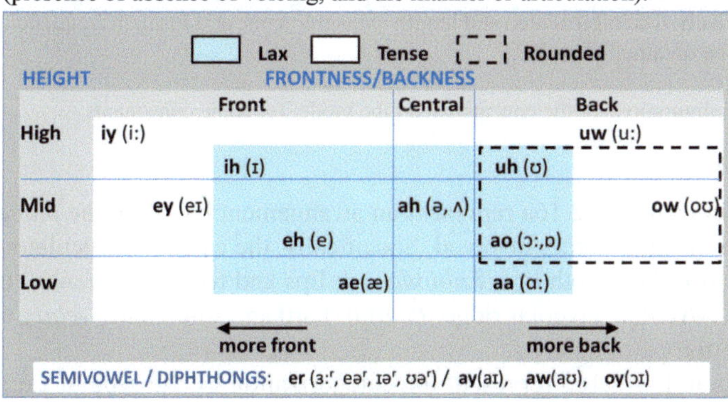

(b) Vowels in American English, referenced from [60] to represent the specific vowels modeled by the CMU Sphinx ASR system. The IPA symbols for these phonemes are also shown in these charts. Plain alphabetical symbols are used for easier reading.

Fig. 2.16 An example of categorization of vowels and consonants in American English

Nasals, Liquids and Glides. The key characteristics associated with each of these are—*Plosives*: the sound source is compression—or the complete stoppage of airflow, followed by sudden release of air; *Affricates*: The mechanism is the same as plosives, except that the release of air is gradual, and the sound that is produced is that of turbulent airflow through a constriction; *Fricatives*: creation of turbulent airstream through a construction at some position in the vocal tract; *Liquids*: The sound source is phonation accompanied by additional vibrations created with the forward parts of the tongue, with airflow along the sides or top of the tongue; *Nasals*: complete airflow obstruction and release through the nose; *Glides*: stricture between the roof of the mouth and the tongue.

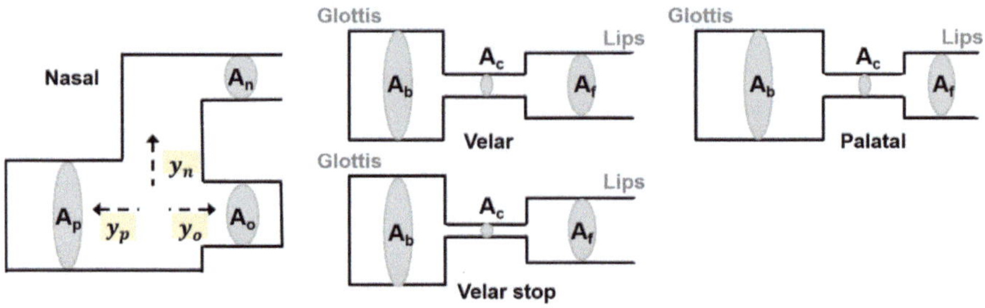

Left: Four-tube model of a nasal consonant. For nasal consonants the model is branched. The subscripts *o*, *p* and *n* stand for oral, pharyngeal and nasal respectively. *y* is acoustic impedance. **Right:** Three-tube models for a velar, velar stop and palatal consonants. The placement of the narrow tube (representing a constriction in the vocal tract) in the tube model is specific to the type of consonant under consideration. A_b,A_c,A_f,A_p,A_o,A_n stand for the cross sectional areas of the back, central, front, pharyngeal, oral and nasal regions of the vocal tract respectively. The typical area and length measurements, and formant frequencies of these models can be obtained from [62].

Fig. 2.17 Configurations of the concatenated tube model for some consonants

The columns in Fig. 2.16a represent an arrangement based on the *place* of articulation, and the articulators involved. Specifically, the category:articulator pairs are as follows—*Bilabial*: both lips, *Labiodental*: lips and teeth, *Interdental*: upper and lower teeth, *Alveolar*: alveolar ridge, *Palatal*: hard and soft palate, *Velar*: velum and *Glottal*: glottis.

The rows in Fig. 2.16a represent the categorization by the type of excitation, or the type of sound source. When the sound source is phonation, the consonants are categorized as *Voiced*, and as *Unvoiced* otherwise.

The formant frequencies of consonants can be computed (as explained above in the case of vowels) by computing the resonances of the corresponding acoustic tube models. Models for consonants include the specific constrictions in the vocal tract that are characteristic of each consonant. In the case of nasal consonants they include the alternate branchings of passages in the vocal tract, as shown in Fig. 2.17. Both constrictions and alternate branchings can result in *anti-resonances* (frequencies that are almost or completely canceled out), in addition to resonances (formants, or frequencies that are reinforced). Anti-resonances are discussed further in Chap. 5. Constrictions are also characterized by abrupt discontinuities in spectro-temporal patters that correspond to their formation and release [59].

In the case of vowels, the first two formants F1 and F2 are thought to be enough to disambiguate the vowel. Consonants too have characteristic formant patterns that help disambiguate them. For example, nasal consonants have an additional formant around 2500 Hz. Nasal consonants also often have their third formant canceled out by anti-resonances resulting from the opening of the nasal passageway. The liquid phoneme /l/ typically has a formant at around 1500 Hz. The phoneme /r/ is characterized by a very low third formant (below 2000 Hz). Plosives (also to some degree, fricatives) modify the placement of formants in the surrounding vowels, with the exception of the alveolar stops /t/ and /d/ that do not do so. Formants are often not clearly discernible in fricative sounds such as /ch/, /jh/, /hh/ etc. due to the noisy nature of these sounds; they are nevertheless present. This is also the case for stop sounds such as /t/, /p/, /k/ etc., which are associated with a complete closure of the vocal tract at specific geometric locations in it. Although there is a consequent cessation of energy in the output signal, the ensuing sound may nevertheless have observable resonant frequencies. For example, in Fig. 7.4b (see Chap. 7), we observe that resonances are discernible for almost all phonemes, regardless of whether they are voiced or not. In general, even the unvoiced phonemes such as /t/, /p/ and /s/ show traces of resonant structure, since the vocal tract continues to resonate through their production period. The bilabials /b/ and /p/ show a lowering of formants as compared to others. The velar phonemes /k/ and /g/ show F2 and F3 coming together in a "velar pinch". Many such generic formant characteristics can be found for consonants in all languages.

2.4 The Uniqueness of Human Voice

The subject of uniqueness of any biometric entity—including DNA and fingerprints—has been a controversial one [61, 62]. Uniqueness is difficult to prove, and is only accepted through *plausible* evidence and reasoning based on it. Nevertheless, the conjecture that no two individuals are alike is widely accepted in the context of DNA and fingerprint biometrics.

With the right approach, gauging the uniqueness of human voice is quite simple: we merely work with the assumption that for two voices to be exactly the same, all parameters that influence voice production must be the same.

Having discussed the bio-mechanics of voice production earlier in this chapter, we can see why it is extremely unlikely that the vocal tracts of two different individuals will undergo the exact same micro-movements—they depend on too many intricate factors, down to the bio-molecular levels of the layers of the vocal folds—to be exactly replicable (as a process) completely by chance. However, even while this is intuitively and conceptually clear, it is difficult to obtain a clear quantitative figure for the uniqueness of human voice based on a process model of the bio-mechanics of the vocal tract. Especially since the influence of the different factors involved on each other, and the net effect of those influences on voice is also not well understood.

It is much simpler to approach the problem from a structural perspective, and from coarse considerations related to the more fossilized aspects of voice production.

The geometry of the vocal tract affects voice. For the vocal tracts of two individuals to be geometrically exactly equal, they must be at least belong to the same skeletal type, since bone structure determines the shape and size of the vocal chambers, nose, jaws, chest cavities etc. The reasoning we will begin with is coarse yet statistically supportable: we find the probability of two speakers being of the same skeletal type, within the same age range (of 10 years), being of the same gender, and then factor in the probability of both having *exactly* the same physical and physiological parameters that are known to influence voice. This probability will then indicate the chance of two voices being exactly the same.

From the literature, the list of physical and physiological factors that influence voice is very long. Let us only consider a small set S of obvious factors: $S = \{$jaw shape, jaw size, size of tongue, position of tongue root relative to jaw, dental shape, dental closure, nasal size, musculature of larynx, muscle physiology, size of vocal cords, composition of vocal cord layers, cartilage placements and size, diameter of trachea, lung volume, lung capacity, dimensions of pharynx (3D, including tissues), dimensions of oral cavity (3D), dimensions of nasal cavity (3D)$\}$. Let us assume that the elements of S co-occur, but independently influence voice (e.g. nasal size and jaw shape exert independent influences on voice). Let us further generously assume that for any element of S, one-fourth of a given population has *exactly* the same value. Thus the probability that two people in a population have exactly the same S in its incomplete form as listed above, would be $0.25^{18} = 1.4551915e^{-11}$.

From anthropological literature, we follow the most widely accepted division of humanity in terms of skeletal structure based on skull shape and phenotypic traits (such as epicanthic folds, sinodonty, incisor shape, and a plethora of other factors discussed in Chap. 9). Scientifically, based on these, there are three major "races" of humankind: Mongoloid, Caucasoid and Negroid. Note that these divisions have nothing to do with skin color. For instance, Caucasoid structures are found in all ethnicities, as are Mongoloid and Negroid structures.

Let us assume that the occurrence of each type of skeletal structure is exactly one-third in this world. This is of course highly debatable—better estimates can be arrived at by considering the current ethnographic maps for the world. The probability of being within the same age group, and of the same gender in a given population is easier to estimate—we can simply follow the current world statistics (e.g. [63]: world statistics). Currently, the world's male:female ratio is 50.4:49.6. The distribution over age ranges (yrs) in ([0–14], [15–24], [25–64], [65+]) is (26.0%, 16.3%, 49.5%, 8.2%). Roughly estimated from these, the probabilities associated with matching skeletal type, gender and age (10 year range), independently of each other, are respectively 0.33, 0.5 and 0.2.

When we factor these into the figure above, the probability of two people having exactly the same voice works out to be $0.33 \times 0.5 \times 0.2 \times 1.4551915e^{-11} = 4.802132e^{-13}$ or 5 in 10 trillion, or **less than 1 in a trillion**. If we were willing and able to calculate this in consideration of all factors that influence voice (simply enumerating which is a gargantuan task), this probability would be tremendously lower.

2.4.1 The Problem of Identical Twins

In this context, the problem of identical twins must be addressed. Could the estimate above have been different if we simply considered the probability of two people being identical twins (The occurrence of identical twins in the world is approximately 1 in 350.)? Do identical twins in fact have *exactly the same* vocal parameters in all respects? The answer is "no." Identical twins are identical at birth, but epigenetic factors (diet, habits, disease, environmental and other influences) cause a rapid divergence in their physiological and mental processes with time. If we try to gauge uniqueness from the perspective of identical twins, the epigenetic variations between them in the context of all other factors mentioned above must also be considered, and this can be a complex task. For example, what is the probability that a pair of identical twins have exactly the same respiratory patterns? Clearly this must be conditioned on multiple epigenetic factors that have affected their development individually since their birth.

The path of identical twins has been taken up in the case of both DNA and fingerprints for exactly the same purpose as above—to examine their uniqueness across the human population [64]. In the case of voice, studies have shown that identical twins do *not* have similar voices in many respects, just as they do not have similar fingerprints or palmprints [65]. This dissimilarity in different biometric parameters develops even during the gestation period, influenced by many random factors. An example that demonstrates the fact that voices of identical twins are not the same is given by a study in which subjects were asked to perceptually identify the voices of identical twins from controlled trios of voices. The subjects made significant errors in identification. The study found that while there was "greater perceptual resemblance between the voices of identical twins than between voices without genetic relationship," the perceptual identification was largely based on fundamental frequency, and even that was not perfect [66]. Factor into this the limitations of human audition, which cause voices that are only slightly different from each other to be perceived to be the same. In fact, the voices of identical twins are *not* identical. There are no studies that have been able to assert that they are. In other words, the contention that each person in the world has a measurably unique voice has not been falsified by any study on the voices of identical twins so far. See for example the study of identical twins' voices in [67], which explicitly states this, and also shows that a simple analysis of the fourth formant in most vowels is able to discriminate between closely related similar-sounding speakers.

The analysis above is nevertheless clearly a gross approximation. There may be other pathways of reasoning that may better facilitate this, but they are likely to lead to similar conclusions. When all factors that influence voice are taken into account, and given the fact that voice varies *dynamically* (unlike facial features or fingerprints or DNA) on a daily basis in response to transient factors, it is reasonable to assume that it is practically almost impossible for two voices to be exactly the same in all respects.

2.5 Human Hearing and Perception of Voice

In Chap. 1, we discussed the judgments humans make about others from their voice, and also the reactions that voice is able to evoke in other people. All of these are made possible by specific sound patterns in voice that we are able to sense. But more than this, the fact that we are able to sense the voice signal itself is quite amazing. Human voice is produced by the vibration of tiny membranes within the larynx, and other mechanisms that are in themselves not very loud compared to other sounds in the environment. We not only hear it, we hear many subtle variations in it with clarity (when in good health). This ability of ours, to hear sound and its different gradations (or variations) is called sound *perception*. Speech and voice perception, come under the broader purview of **Psychoacoustics**—the science of human hearing.

Earlier in this chapter we have discussed voice production. Many aspects of voice production can be gauged in the time domain from the voice signal, such as glottal opening and closing, sound intensity or volume etc. To understand the mechanism of speech *perception*, however, we must look into the spectral (or frequency) domain. This is because the human auditory system resolves the voice signal into its frequency components, much like the Fourier analysis explained in Chap. 4.

We perceive voice by sensing its spectral content. We also sense many of its other properties, such as its loudness, intensity and pressure. It is important to understand the difference between these clearly—even for profiling purposes, at the outset.

2.5.1 *Loudness, Pressure, Intensity and Their Perception*

From basic acoustics, we know that sounds waves are longitudinal waves. Our vocal mechanism produces these *pressure* waves, which propagate outwards through the mouth as traveling waves of energy. To facilitate discussion and illustration of concepts, it is useful to think of the pattern of compression and rarefactions in the sound wave as a transverse wave, as shown in Fig. 2.18 with a spring as an example. The compressions and rarefactions in the spring are an analogy to those in a sound wave, while the transverse wave representation of it is only an abstraction and must *not* be thought of as a real transverse wave in the context of sound.

Loudness is a relative concept. It is the *human impression* or *human perception* of the strength of (or energy in) a sound wave. A sound, of which a voice signal is an example, may be said to be x times as loud as, or as faint as, another. This concept is shown in Fig. 2.19a, where the horizontal axis is time and the vertical axis is some linear scale of amplitude.

Loudness is measured in units called Phon or Sone (which is similar to Phon). As expected, Phon is measured on a relative scale. To understand this, we first note that the sensitivity of human hearing depends on the specific frequency we hear. A 1000 Hz tone (see Chap. 4 for explanation) at any selected amplitude will be heard at a specific loudness that our ear will perceive. For any other given tone, if we then adjust its amplitude to perceive the same loudness as the 1000 Hz tone, we will find its amplitude to be different from the amplitude of the 1000 Hz reference tone. When we repeat this for different frequencies, we obtain the "equal loudness" curve shown in Fig. 2.20. In Fig. 2.20a (top), the lower solid curve is the equal loudness curve for the specific amplitude selected for the 1000 Hz tone. Note that this entire curve

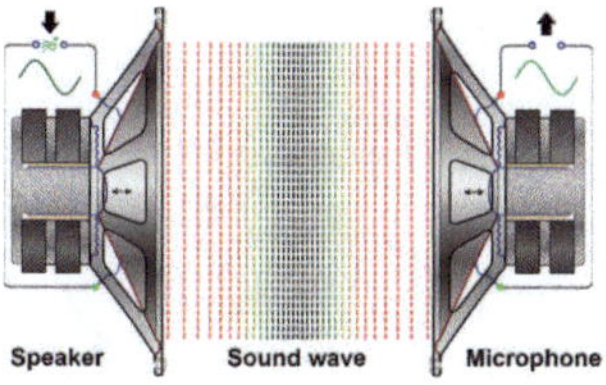

(a) Longitudinal sound wave: this illustration depicts a sound wave produced at a speaker and captured by a microphone. Only a single compression (and two rarefactions) is shown, in reality there are as many per second as the frequency of the sound wave, and their width is proportional to the wavelength of the sound wave, since its speed is constant in a medium under given conditions.

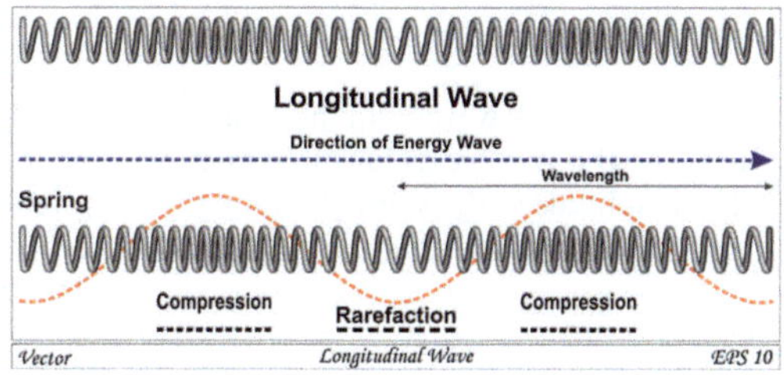

(b) Interpretation of sound as a transverse wave (for ease of explanation only).

Fig. 2.18 A longitudinal sound wave and its interpretation as a transverse wave

is relative to the 1000 Hz tone. Let us assume that its specific amplitude selected to generate this equal loudness curve was 10 dB (deciBel, explained later in this section). Figure 2.20a (middle) shows the amplitude of a 500 Hz tone on this equal loudness curve. At this point, the 500 Hz tone is said to have a loudness of *10* Phons, meaning "*it is as loud as a 10* dB *tone of 1000* Hz." Figure 2.20a (bottom) highlights a 200 Hz tone that has a loudness of 10 Phons. Figure 2.20b shows the equal loudness curves for many reference amplitudes at 1000 Hz, and at the threshold of human hearing, which is considered to be the beginning of the dB scale. The threshold of hearing corresponds to 0 dB. The point highlighted on the curve in Fig. 2.20b is in fact at 4 dB, which is close enough. To understand loudness and related concepts further, it is very important that we clearly understand the dB scale.

Like loudness, the dB scale is a relative one. It is defined with reference to what is perceived at 0 dB by the human ear. We cannot hear a 1000 Hz tone below the amplitude of 0 dB. That is the empirically determined threshold of human hearing.

This brings us to the concept of *intensity*. At the threshold of hearing, the surface power density of sound is measured (instrumentally) at the value of:

$$I_0 = 10^{-12}\,\mathrm{W/m^2}, or 10^{-16}\,\mathrm{W/cm^2} \tag{2.24}$$

A watt (unit of power) is equal to a rate of energy transfer of 1 joule per second. Intensity is expressed on the decibel scale. An intensity I is expressed in *decibels above* the standard threshold of hearing I_0 as:

$$I_{dB} :\leftrightarrow 10 log_{10} \frac{I_{measured}}{I_0} \tag{2.25}$$

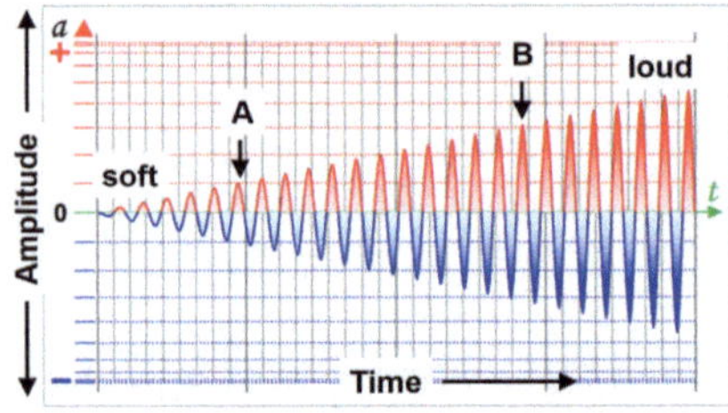

(a) Loudness: A wave with increasing loudness. At **B** it is twice as loud as it is at **A**. The amplitude scale is linear, although the grid lines shown are nonlinearly placed.

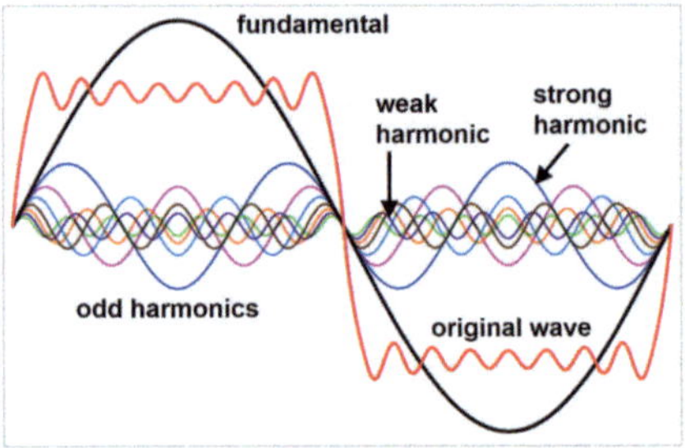

(b) Harmonics: A wave can be "composed" by a sum of pure sinusoids (or tones) of different amplitudes and phases. In this case only those with odd half-cycles are shown. The one with the lowest period is the fundamental, its frequency is the fundamental frequency F0 (or pitch). Depending on their amplitude, the harmonics may be weak or loud.

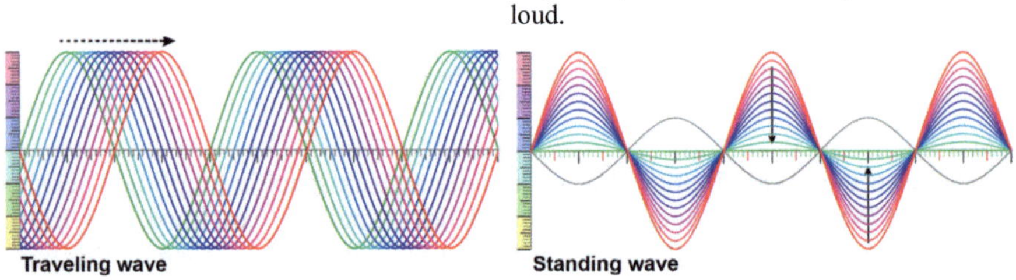

(c) Traveling and standing waves: Traveling waves are formed in the vocal tract since it is open at the oral and nasal openings. The vocal cords are however tethered at both ends, and their movements are those of standing waves.

Fig. 2.19 Loudness, pitch, harmonics and waves in the vocal tract

where I_{dB} is the intensity in decibels. Intensity I_{dB} is thus a measurement of sound intensity *over* the standard threshold of hearing. If $I_{measured}$ is 10,000 times I_0, then $I_{dB} = 40$ dB. A simple way to get an idea of the dB scale is to note that for a given value in dB, dividing it by 10 gives an idea of the order of magnitude of intensity difference between what is measured, and I_0. Thus, given a value of 20 dB for I_{dB}, we can gauge, without explicit calculation, that I, when measured explicitly, would be a hundred times greater than I_0. The unit Bel is named after Alexander Graham Bell. A *deci*Bel is one-tenth of a Bel.

In the equation above, the factor of 10 appears due to perceptual reasons. The human ear cannot differentiate between very fine levels in sound intensity. A difference of 1 dB is just about noticeable for the normal human ear. *It takes about 10 times the intensity of one sound, to be perceived as twice as loud as another*. In addition, a logarithmic unit can better span the range of human response, which is very wide, from the threshold of hearing at 1000 Hz to the threshold of pain at about ten trillion

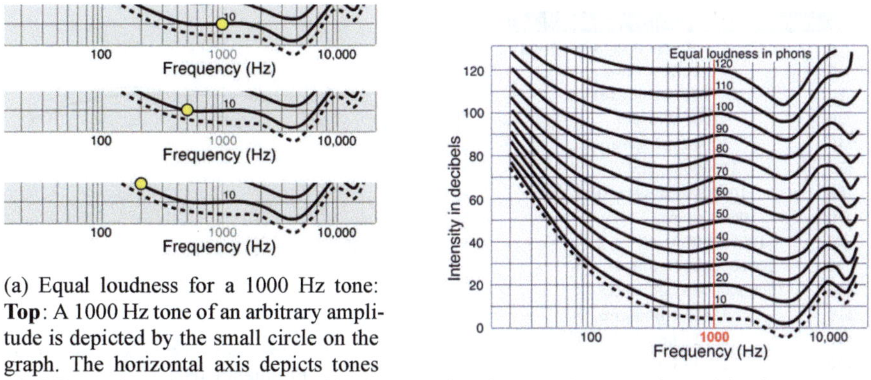

(a) Equal loudness for a 1000 Hz tone: **Top**: A 1000 Hz tone of an arbitrary amplitude is depicted by the small circle on the graph. The horizontal axis depicts tones of different frequencies on a logarithmic scale, and the vertical axis depicts amplitude. **Middle**: A 500 Hz tone of 10 Phons loudness **Bottom**: A 200 Hz tone with the same perceived loudness of 10 Phons.

(b) The complete set of equal loudness curves. The lower dashed curve is the equal loudness curve at the threshold of hearing. In (a), The circle at 1000 Hz is at 0 dB (in reality, approximately at 4 dB, which is close enough).

Fig. 2.20 Illustration of the concept of equal loudness

times its intensity. Figure 2.21a gives an idea of the span of the dB scale in terms of real-word sounds.

While the decibel scale as defined above is widely used in hearing and perception studies, it may also be further differentiated. The A-weighted dB scale, denoted dBA is sound intensity as specified after an "A" contour filter is applied to it. On such a filter, the regular dB measurements are adjusted for hearing *sensitivity*, which as we see from the equal loudness curves, is low for low and high frequencies as we move away from the frequency of 1000 Hz. On the dBA scale, low and very high frequencies are given less weight than on the standard decibel scale, as shown in Fig. 2.21b.

The dBA scale is often used in describing noise-induced hearing loss, in regulatory specification of environmental noise levels, and in other perceptual contexts where *hearing sensitivity* is especially important to consider, since it scales the intensities according to human hearing patterns (discussed below). The C-weighting of the dB scale gives the dBC scale, which has a slightly different purpose—it is used in contexts where peak levels are specially important, such as impact noises, explosions, gunfire etc. For the most part, this is the same as the standard dB scale.

In this context, it is important to know the difference between the often used terms *sound intensity level* (SIL), *sound power level* (SWL), and *sound pressure level* (SPL). A sound source produces sound power, or acoustic power. This is the sound energy constantly transferred per second from the sound source. As sound propagates through a medium, its intensity can be felt at any point in the space through which it propagates, but at any point, it would depend on the distance from the sound source. This *distance-dependent* measure of sound intensity away from the power source is called sound *pressure*, and is also measured in dB for obvious reasons. In practice,

dB SPL	relative loudness	Example
0	0	threshold of hearing
10	10	breathing
20	100	rustling leaves
30	1000	whisper
40	10000	refrigerator
50	100000	rainfall
60	1000000	conversation
70	10000000	car
80	100000000	truck
90	1000000000	hairdryer
100	10000000000	helicopter
110	1E+11	trombone
120	1E+12	police alert
130	1E+13	jet engine
140	1E+14	fireworks

(a) Some common sounds on the dB scale

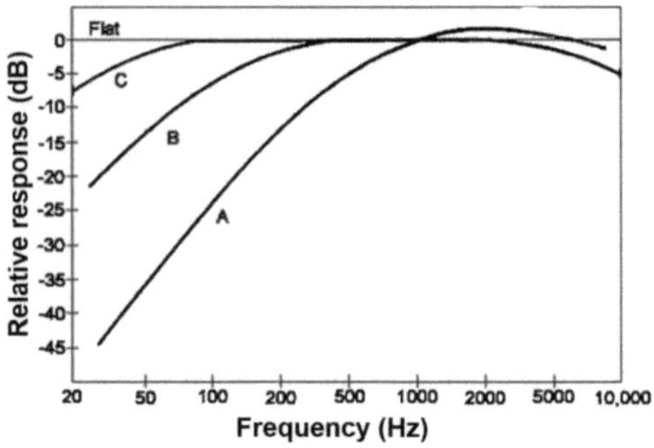

(b) dBA and dBC filters

Fig. 2.21 The decibel scale and the dBA and dBC filters

SIL, SWL and SPL are all sometimes loosely referred to as "sound level," which is an ambiguous term. We will refer to these as the sound intensity level L_I, sound pressure level L_P and sound power level L_W. At a distance of **one** meter from the sound source, L_I is assumed to be the same as L_P. The impedance of air at sea level is $Z = 400\,\text{N-s/m}^3(\text{Pa-s/m})$ Sound *power* can be expressed in watts, and as sound power level L_W in decibels.

Sound power level is related to sound pressure level through the following relation:

$$L_P = L_W - \left|10log_{10}\left(\frac{Q}{4\pi r^2}\right)\right| \tag{2.26}$$

Where Q is the *directivity index*. For a wavefront propagated as a perfect sphere from the source, its value is 1, for a half-sphere (a wavefront obstructed at the sound source from propagating in one direction, a good example is a sound source placed on a hard insulating floor, which cannot propagate through the floor), the value of Q is 2, when obstructed by two walls, it is 4, and obstructed by 3 walls, as in the corner of a cube, it is 8.

For unobstructed sound, since the sound pressure level falls off as r^2, at ten times the distance from the sound source, the sound pressure will drop to $1/100$th its value (which corresponds to a 20 dB fall-off with reference to the source), which is equivalent to being "spread" over 100 times the area of the source. Obstructed sounds are *focused* according to the relation above.

The sounds produced within the human vocal tract are obstructed by the walls of the vocal tract (which also absorb some energy), and the sound is thus focused. The intensity and pressure of voice radiated out from the mouth are affected by the nature of the obstructions, place of production and nature of the material of the walls of the vocal tract. At the hearing end, the pressure sensed by our ear falls off with

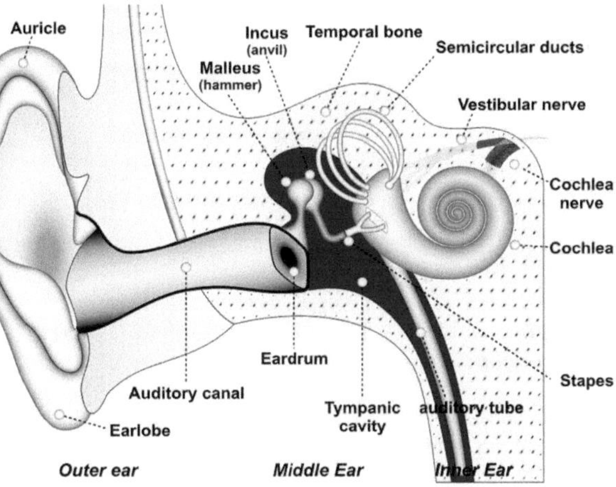

Fig. 2.22 The anatomy of the human ear

the square of the distance between the person speaking and the person listening (in case of voice), but the exact relationships in the case of voice are not as simple as those described above, outside of the medium between the speaker and the listener. The significance of these observations for profiling remains to be explored.

The threshold of hearing corresponds to the sound pressure level $L_P = 0$ dB at a frequency of 1 kHz. This is used as the reference *SPL*. The pressure associated with this, $P_0 = 20\,\mu\text{Pa} = 2 \times 10^{-5}$ Pa, where 1 Pascal = 1 Newton/sq.m (approx. 0.000145 pounds/sq. inch, or 9.9×10^{-6} atmospheres).

Figure 2.20b shows the complete set of equal loudness curves for the human ear. As an important takeaway from these curves, we must keep in mind that *the loudness of a sound does not linearly correlate with its sound (intensity) level.* The relation depends on the actual frequency of the sound wave. This is important in profiling, if it is based on measurements of energy or power in the voice signal as captured at a distant microphone. It is also useful to keep this in mind while perusing literature on voice production that is related to vocal intensity, e.g. [68].

2.5.2 Hearing the Spectral Content of Sound

Sensing the spectral content of sound is facilitated by the structure of the inner ear, or the Cochlea—shown in Fig. 2.22, which has a number of hair cells that vibrate in response to sound. Each hair cell vibrates in response to a band of frequencies around a central frequency in the speech spectrum. The pattern of responses of the hair cells to different frequencies is shown in Fig. 2.23.

Without going into details of the human ear, for which excellent and detailed accounts are available in the literature, we simply note that external sound causes

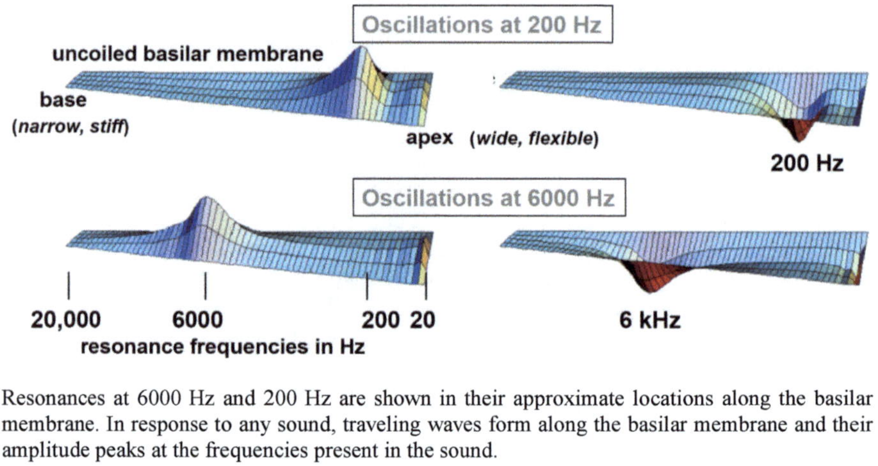

Fig. 2.23 Depiction of the uncoiled basilar membrane inside the cochlea

vibrations in the ear which are propagated to the Cochlea, and to the *basilar membrane* that lies inside it. This membrane is attached at one end to the base of the cochlea, while the other end is free to move within it. When it moves, like any object that is tethered at one end and free at another, traveling waves are created along its length. As shown in Fig. 2.23, the basilar membrane is narrower and less elastic at the base, and wider and more elastic at the apex of the cochlea. When a wave of some frequency impinges on it, it responds maximally (resonates) at a specific position along it [69]. Responses to lower frequencies occur near the apex, and to higher frequencies (increasingly) towards the base of the membrane. The resonance frequencies are spread out along the membrane in a nonlinear pattern.

The membrane is part of a larger structure in the cochlea called the *Organ of the Corti*. Within this, the basilar membrane lies at the base, the *tectorial membrane* [70] at the top, and between these are hair cells. As the basilar membrane undulates in response to a sound wave, the hair cells brush up against the tectorial membrane. This causes a series of chemical changes which eventually result in neurotransmitters triggering action potentials that are propagated down the auditory nerve to the brain. In all, there are four hair *bundles* that make contact with the tectorial membrane. Each bundle ends in cells called *stereocilia*. Three are closer to the apex (dedicated to lower frequencies), and one bundle is spaced out from this set towards the base of the basilar membrane (dedicated to higher frequencies). As a result, collectively the hair cells respond to differences in lower frequencies much more finely than to higher frequencies. This is shown in Fig. 2.24b. The responses, and their relay to the brain via the auditory nerve, is a process that is referred to as *auditory transduction*. The complex mechanical structure in our ear is able to transduce frequencies from about 20 to 20,000 Hz. This delimits and defines the human hearing range.

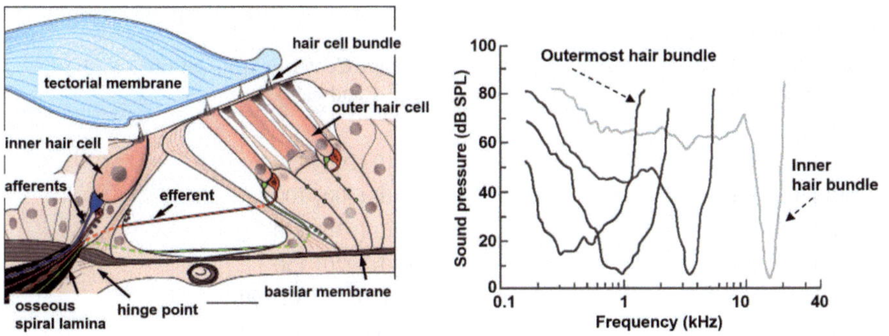

(a) Stereocilia in the Organ of the Corti

(b) Response of the four hair bundles (hair cell tuning) to different frequencies in sound.

Fig. 2.24 Organ of the Corti showing the hair bundles, and response of the hair bundles

Fig. 2.25 The Bark scale

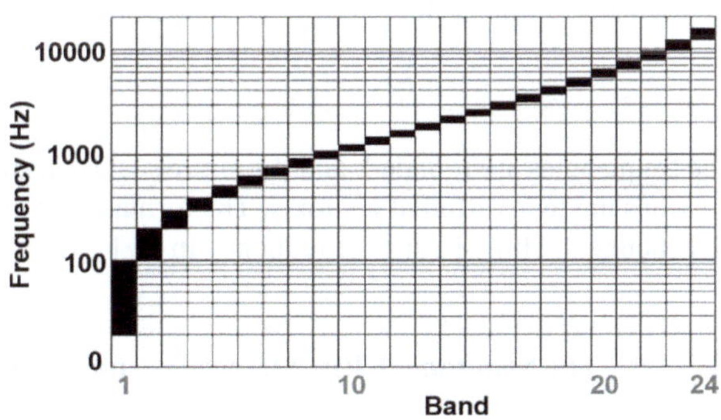

2.5.2.1 The Bark Scale

The Bark scale was proposed in 1961 by Heinrich Georg Barkhausen, a German Physicist [71]. Its purpose was to create a frequency scale on which equal distances were equally perceived by the human ear. The scale spans 24 *critical bands* of hearing. A critical band [72] is the band of frequencies within which a second tone will interfere with the perception of the first tone by *auditory masking* (not discussed in this chapter). The Bark scale is shown in Fig. 2.25.

There are several expressions for the Bark scale, of which we only give one below [73]:

$$f_B = \frac{26.81 \times f}{1960 + f} - 0.53 \begin{cases} \textit{if result} < 2;\ \textit{add}\ 0.15 \times (2 - \textit{result}) \\ \textit{if result} > 20.1;\ \textit{add}\ 0.22 \times (\textit{result} - 20.1) \end{cases} \tag{2.27}$$

Where f_B is the *critical band frequency*, or the Bark frequency. Table 2.1 lists the details of the Bark scale for quick reference. Note that this is only an approximation,

Table 2.1 The Bark scale

Number-(Center Freq.)-[Cut-off Freq.]-Bandwidth	Number-(Center Freq.)-[Cut-off Freq.]-Bandwidth
-()-[20]-	
1-(60)-[100]-80	2-(150)-[200]-100
3-(250)-[300]-100	4-(350)-[400]-100
5-(450)-[510]-110	6-(570)-[630]-120
7-(700)-[770]-140	8-(840)-[920]-150
9-(1000)-[1080]-160	10-(1170)-[1270]-190
11-(1370)-[1480]-210	12-(1600)-[1720]-240
13-(1850)-[2000]-280	14-(2150)-[2320]-320
15-(2500)-[2700]-380	16-(2900)-[3150]-450
17-(3400)-[3700]-550	18-(4000)-[4400]-700
19-(4800)-[5300]-900	20-(5800)-[6400]-1100
21-(7000)-[7700]-1300	22-(8500)-[9500]-1800
23-(10500)-[12000]-2500	24-(13500)-[15500]-3500

the frequencies are rounded, and the entire scale can shift depending on the approximations made in different contexts. The format used in this table is Number-(Center Frequency in Hz)—[Cut-off Frequency in Hz]—(Bandwidth in Hz)

2.5.2.2 The Semitone, ERB, Mel Scales, and the Gammatone Filter

Semitones are tied to the definition of octaves. An octave is double a given fundamental frequency. When octaves on a linear frequency scale are expressed on a logarithmic scale, the spacings between them emulate the pattern of human auditory responses to frequencies much better. A **semitone scale** is frequently used to gauge the perceptual consequences of differences in fundamental frequency (as of phonation). A frequency f (in Hz) on a linear scale can be expressed on a semitone scale as:

$$f_{Semitone} = 39.87\ log_{10}\left(\frac{f}{50}\right) \tag{2.28}$$

The **Equal Rectangular Bandwidth (ERB)** scale emulates the response at constant distances along the basilar membrane [74]. This scale is used where perceptual differences in fundamental frequencies need to be evaluated, e.g. in hearing aids [75]. A frequency on a linear scale f may be transformed to its equivalent f_{ERB} on the ERB scale as [76]:

$$f_{ERB} = 21.4\ log_{10}(1 + 0.00437 \times f) \tag{2.29}$$

The **Mel scale** [77] is a perceptual scale of fundamental frequencies (or pitch) that are perceived by human listeners to be equally separated from each other. The Mel scale is approximated by many functional forms, one of which is:

$$f_{Mel} = 2595 \, log_{10}\left(1 + \frac{f}{700}\right) \tag{2.30}$$

At 1000 Hz, most expressions for the Mel scale yield 1000 Hz (Mel frequency). This is called the *break frequency*. The ERB scale has a break frequency of 228.8 Hz [74].

All the expressions above can be thought of as *transformations* that warp the linear frequency axis into one that emulates human perception in some respect. This operation need not be done only in the frequency domain. Filters in the time domain can have a similar desired effect in the frequency domain. One such filter is the **Gammatone** filter [74, 78, 79]. This is a linear filter. Its impulse response is the product of a pure tone (sinusoid) and a Gamma distribution, and is given by:

$$f_{gammatone}(t) = at^{n-1}e^{-2\pi bt}\cos(2\pi ft + \phi) \tag{2.31}$$

where f is the center frequency, ϕ is the phase in radians, a is amplitude, n is the order of the filter, b is the filter's bandwidth in Hz, and t is time in seconds.

Other maps and scales also exit, such as the cochlear frequency-place map [80], which has a break frequency of 165.3 Hz. Many other concepts in perception that have not been discussed here are readily available in the literature, e.g. [81, 82]. In the next few chapters, the concepts discussed here will figure in understanding the different approaches that researchers have taken in the past in devising computational methods for processing voice.

2.6 The Human Breath

Breath plays an important role in controlling the dynamics of speech. Natural speech is produced as a person exhales. Most sounds in it are egressive in nature, and it is almost impossible to produce sustained speech during inhalation. As a person speaks, a specific volume of air is pushed out through the lungs and trachea into the vocal chambers, gated through the vocal folds in the glottis and modulated in different ways to produce speech. A speech *utterance* is produced within one exhalation burst, the prosody of speech, and its overall energy and entropy are distributed in characteristic ways until the next inhalation. This pattern is largely true for neutral speech, but varies with the expression of different emotions or when the speaker attempts to convey pragmatic content that is not encoded in the actual words spoken (e.g. a non-verbalized question).

During continuous speech, inhalation is generally sharp and rapid, and volumetrically anticipatory of the next burst of speech. The time within which this volume is inhaled is however much shorter than the duration over which speech is generated (and extremely short in excited speech). Consequently, *inhalation sounds* during continuous speech are accompanied by high turbulence. The intensity of such *intervocalic breath sounds* is high enough that the acoustic resonances within the vocal chambers of the speaker are prominent. A person's lung capacity is positively correlated with the intensity of the intervocalic breath sounds. It is also strongly correlated with the amplitude of the signal produced during standardized normal conditions, and to the length of the utterances produced by the person. Based on these, breath can be potentially used to gauge the dimensions of the vocal tract, the speaker's lung capacity, to identify speakers, differentiate between genders and other parameters (including lung volume and lung condition). The greatest advantage of using breath for profiling is that it is truly language agnostic.

The volume of air inhaled also depends on the air-intake capacity of the speaker's nasal and oral passageways, trachea and further inner structures leading to the lungs, and further varies with the speaker's physical needs for oxygen at the moment, and the oxygen carrying capacity of their blood. Since exhalation is volumetrically linked to inhalation, by association, the quality of the speech produced during exhalation also varies with all of these factors. As such, many parameters of the speaker's persona leave their signatures on the human breath sound as it occurs during continuous speech.

2.6.1 Physiological, Aerodynamic and Acoustic Perspectives of Breath

During respiration, the vocal folds function as a valve to regulate air pressure across the glottis, controlled by the muscles of the larynx. They open wide during exhalation (when atmospheric pressure is lower than the lung pressure, causing air to flow out) and come closer together during inhalation (when the atmospheric pressure is higher than the lung pressure, causing air to flow into the lungs).

The functional parameters of the lungs, their size and capacity, and the physiological demand for oxygen mainly define the breathing patterns of a person. Breathing is a multi-stage process.The volume of air mobilized during breathing differs over different breathing stages. Generally the maximum volumetric air storage capacity of human lungs is around 80 ml per kilogram of body weight for a healthy non-overweight individual. This is called the *Total Lung Volume*, or **TLC**. The two stages of normal breathing—expiration and inspiration of air, are not always volumetrically exactly equal. The volume of air exhaled or inhaled during normal breathing is called the *Tidal Volume*, or **TV**. Beyond this, the maximum air that can be exhaled by an individual without overtly stressing the respiratory system is called the *Expiratory Reserve Volume*, or **ERV**. The air that remains behind in the lungs, preventing them from collapsing completely (which may be fatal), is called the *Residual Volume*, or **RV**. The *Inspiratory Reserve volume*, or **IRV**, is the additional volume of air that

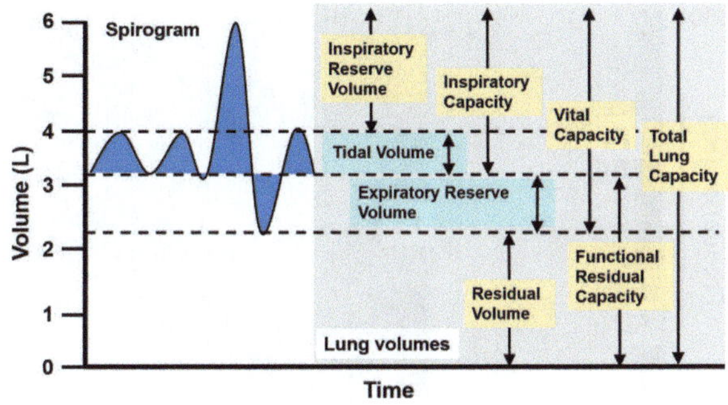

Fig. 2.26 Volumetric breakup of the human breath and its relation to various capacities of the lungs

can be inhaled beyond the tidal capacity to completely fill the lungs to capacity. The *Inspiratory Capacity*, or **IC**, is defined as the sum of TV and IRV, while the *Functional Residual Capacity*, or **FRC** is defined as the sum of ERV and RV. The*Vital Capacity* of the lungs, or **VC**, is defined as the sum of IC and ERV. It is also the maximum volume of air that can be exhaled after the deepest inhalation. Fig 2.26 shows the volumetric breakup of air intake during respiration (using a spirogram as reference), and its relation to the various lung capacities mentioned above. In practice, these capacities are measured using a spirometer.

In clinical settings, a spirometer is used to measure pulmonary function (lung function). This is measured as a ratio of *Forced Vital Capacity*, or **FVC**, defined as the maximum volume of air exhaled quickly (forcefully) after the deepest inhalation; and the*Forced Expiratory Volume* in *one second*, or **FEV1**, defined as the volume of air exhaled in one second within the FVC measurement process. Lung function is defined as the ratio FEV1/FVC. This is less than 80% for obstructive pulmonary diseases, but since FEV1 and FVC can both decrease proportionately for many conditions, including neurological disorders that affect muscle function, pneumonia, pulmonary fibrosis, pleural effusion, tuberculosis, sarcoidosis, silicosis, asbestosis etc., the absolute value of this ratio in any specific context requires informed clinical interpretation.

The relations between these volumetric breath parameters and the acoustic properties of breath sounds can be empirically established. The *ambient* (within the distance it is heard clearly) sound pressure level (SPL) generated by calm human breathing sound is approximately 10 dB. This increases significantly during normal speech. For whispered speech at close quarters, the SPL is about 20 dB, while normal conversation is associated with SPLs of approximately 60 dB with a distance of 1 m between the speakers. Breath sounds have proportionately higher SPLs in these contexts.

The SPL generated during calm breathing without speech or conversation is generally not audible to the human ear, unless it is at close quarters (less than 1m distant) from a person with normal hearing. Normal breathing sounds are therefore not audible within the silence regions of audio captured using far-field microphones placed at distances greater than 1m. However, when speech is captured over close-talking microphones, or recorded through devices that are held close to the mouth (or even

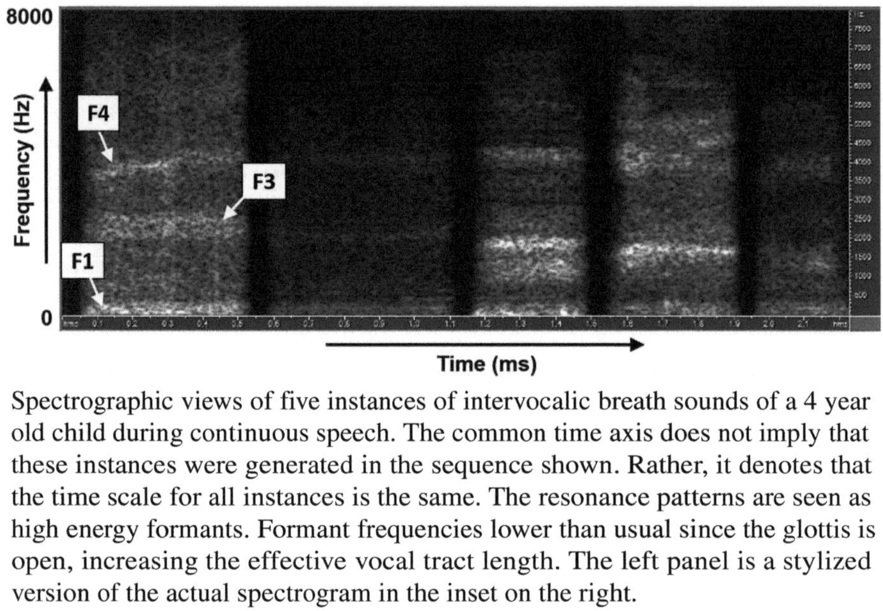

Fig. 2.27 An example of spectrographic patterns in breath sounds

worn within the mouth, such as the tooth microphone [83]), normal breathing sounds (outside of speech) may be captured and be audible to the human ear. These are directly associated with the speaker's tidal volume, and simply estimating the SPL from the recorded signal may give an idea of the function (and size, if healthy) of the speaker's lungs (and hence their body frame size, if healthy). Figure 2.27 shows the spectrogram of the normal breath sounds of a child. The resonances of this sound within the vocal tract are clearly seen.

For automated profiling, the most direct indication of lung function may be derived from breath sounds by specifically following the same exercise (deep breath intake followed by forced exhalation) as in clinical spirometry, with one modification to the procedure: to aid the computation of lung function through sound analysis alone, it might be best if breath were intentionally exhaled through rounded lips (in a whistling fashion, to create the maximum sound, or turbulence).

During animated speech, much of the available ERV is expected to be used up. During normal speech, usually TV is used up. The energy associated with the sounds produced using speech is spread over the signal until the end of the utterance, and followed by the intake of breath (episode of intervocalic breath). In excited speech, the volume of air used up during the production of a speech burst is close to TV+ERV, and this is the volume that is expected to be inhaled in intervocalic breath interspersion. However, when we compare the ratio of the duration between successive intakes of breath—i.e. the duration of speech that is resultant from the use of TV (or TV+ERV)—and the duration of such an inhalation event, we find that the latter is much shorter than the former. Since the entire volume TV or TV+ERV is inhaled in a much shorter time during an intervocalic breath episode, it associated with proportionately greater turbulence within the vocal tract. This creates prominent resonances (and anti-resonances if the nasal passageways are used). The spectral characteristics associated with each of these can be related to many profile parameters. Intervals between such sounds, and their temporal placement in continuous speech, also give away information. For example, when a person speaks normally, they have enough control over this process that they inhale at the end of a word, or at the beginning of a sequence of words [84]. When a person is physically exhausted, for example after or even during exercise, the urgent physical need of the body for oxygen may cause breath sounds to occur within a word, breaking up the word. This is seen often in children as they run and play vigorously or talk excitedly. It is also seen in people who have medical conditions that cause insufficient volumes of air or oxygen to enter the bloodstream [85–88].

There are many aspects of *speaking style* that can influence the volume of air inhaled, and the (usually short) duration over which it is inhaled. In steady, planned, read speech, such as in a newscast, the reader speaks at a steady rate, and the air intake during intervocalic periods is usually uniform throughout the speaking duration, with a slight escalation expected as the speaker gets physically exhausted with the effort of speaking. During spontaneous speech, however, the volumetric intake can be much more erratic [84], depending on the topic and tone of the conversation or monologue. If the person is incensed, the acoustic intensity of the breath is usually high, the breath durations are shortened, and the inter-breath time intervals may also be shorter than in planned speech. In a normal conversation, with more leisurely and amicable exchange of content, these parameters are more relaxed and approach that of planned speech. It is reasonable to expect, therefore, that the acoustic intensity, duration of the breath sounds and the inter-breath intervals may carry cues to differentiate between different speaking styles, and thereby emotions and other mental states of the speaker to some extent.

References

1. Titze, I. R., Luschei, E. S., & Hirano, M. (1989). Role of the thyroarytenoid muscle in regulation of fundamental frequency. *Journal of Voice*, *3*(3), 213–224.
2. Hermand, E., Lhuissier, F. J., Larribaut, J., Pichon, A., & Richalet, J. P. (2015). Ventilatory oscillations at exercise: Effects of hyperoxia, hypercapnia, and acetazolamide. *Physiological Reports*, *3*(6), e12446.
3. Yamagishi, M., Ishizuka, Y., Fujiwara, M., Nakamura, H., Igarashi, S., Nakano, Y., et al. (1993). Distribution of calcium binding proteins in sensory organs of the ear, nose and throat. *Acta Oto-Laryngologica*, *113*(sup506), 85–89.
4. Sataloff, R. T. (2017). Clinical anatomy and physiology of the voice. *Professional voice: The science and art of clinical care* (4th ed., pp. 157–196). California: Plural Publishing, San Diego.
5. Baer, T. (1981). Investigation of the phonatory mechanism. Status report on speech research SR-66 (pp. 35–54). New Haven: Haskins Laboratories.

6. Zhang, Z. (2009). Characteristics of phonation onset in a two-layer vocal fold model. *The Journal of the Acoustical Society of America, 125*(2), 1091–1102.
7. Flanagan, J., & Landgraf, L. (1968). Self-oscillating source for vocal-tract synthesizers. *IEEE Transactions on Audio and Electroacoustics, 16*(1), 57–64.
8. Ishizaka, K., & Flanagan, J. L. (1972). Synthesis of voiced sounds from a two-mass model of the vocal cords. *Bell System Technical Journal, 51*(6), 1233–1268.
9. Zhang, Z., Neubauer, J., & Berry, D. A. (2006). The influence of subglottal acoustics on laboratory models of phonation. *The Journal of the Acoustical Society of America, 120*(3), 1558–1569.
10. Zhang, Z., Neubauer, J., & Berry, D. A. (2007). Physical mechanisms of phonation onset: A linear stability analysis of an aeroelastic continuum model of phonation. *The Journal of the Acoustical Society of America, 122*(4), 2279–2295.
11. Zhao, W., Zhang, C., Frankel, S. H., & Mongeau, L. (2002). Computational aeroacoustics of phonation, part I: Computational methods and sound generation mechanisms. *The Journal of the Acoustical Society of America, 112*(5), 2134–2146.
12. Zhang, C., Zhao, W., Frankel, S. H., & Mongeau, L. (2002). Computational aeroacoustics of phonation, part II: Effects of flow parameters and ventricular folds. *The Journal of the Acoustical Society of America, 112*(5), 2147–2154.
13. Chan, R. W., & Titze, I. R. (1999). Viscoelastic shear properties of human vocal fold mucosa: Measurement methodology and empirical results. *The Journal of the Acoustical Society of America, 106*(4), 2008–2021.
14. Chan, R. W., & Rodriguez, M. L. (2008). A simple-shear rheometer for linear viscoelastic characterization of vocal fold tissues at phonatory frequencies. *The Journal of the Acoustical Society of America, 124*(2), 1207–1219.
15. Miri, A. K., Mongrain, R., Chen, L. X., & Mongeau, L. (2012). Quantitative assessment of the anisotropy of vocal fold tissue using shear rheometry and traction testing. *Journal of Biomechanics, 45*(16), 2943–2946.
16. Kazemirad, S., Bakhshaee, H., Mongeau, L., & Kost, K. (2014). Non-invasive in vivo measurement of the shear modulus of human vocal fold tissue. *Journal of Biomechanics, 47*(5), 1173–1179.
17. Haji, T., Mori, K., Omori, K., & Isshiki, N. (1992). Experimental studies on the viscoelasticity of the vocal fold. *Acta Oto-Laryngologica, 112*(1), 151–159.
18. Tran, Q. T., Gerratt, B. R., Berke, G. S., & Kreiman, J. (1993). Measurement of Young's modulus in the in vivo human vocal folds. *Annals of Otology, Rhinology and Laryngology, 102*(8), 584–591.
19. Chhetri, D. K., Zhang, Z., & Neubauer, J. (2011). Measurement of Young's modulus of vocal folds by indentation. *Journal of Voice, 25*(1), 1–7.
20. Scherer, R. C., Shinwari, D., De Witt, K. J., Zhang, C., Kucinschi, B. R., & Afjeh, A. A. (2001). Intraglottal pressure profiles for a symmetric and oblique glottis with a divergence angle of 10 degrees. *The Journal of the Acoustical Society of America, 109*(4), 1616–1630.
21. Li, S., Scherer, R. C., Wan, M., & Wang, S. (2012). The effect of entrance radii on intraglottal pressure distributions in the divergent glottis. *The Journal of the Acoustical Society of America, 131*(2), 1371–1377.
22. Kettlewell, B. Q. (2015). The influence of intraglottal vortices upon the dynamics of the vocal folds. Master's thesis, University of Waterloo, Canada.
23. Shinwari, D., Scherer, R. C., DeWitt, K. J., & Afjeh, A. A. (2003). Flow visualization and pressure distributions in a model of the glottis with a symmetric and oblique divergent angle of 10 degrees. *The Journal of the Acoustical Society of America, 113*(1), 487–497.
24. Kucinschi, B. R., Scherer, R. C., DeWitt, K. J., & Ng, T. T. (2006). Flow visualization and acoustic consequences of the air moving through a static model of the human larynx. *Journal of Biomechanical Engineering, 128*(3), 380–390.
25. Erath, B. D., & Plesniak, M. W. (2006). The occurrence of the Coanda effect in pulsatile flow through static models of the human vocal folds. *The Journal of the Acoustical Society of America, 120*(2), 1000–1011.

26. Mihaescu, M., Khosla, S. M., Murugappan, S., & Gutmark, E. J. (2010). Unsteady laryngeal airflow simulations of the intra-glottal vortical structures. *The Journal of the Acoustical Society of America, 127*(1), 435–444.
27. Hirano, M., Kakita, Y., & Daniloff, R. G. (1985). Cover-body theory of vocal fold vibration. In R. G. Daniloff (Ed.), *Speech science* (pp. 1–46). San Diego, California: College-Hill Press.
28. Alipour, F., & Vigmostad, S. (2012). Measurement of vocal folds elastic properties for continuum modeling. *Journal of Voice, 26*(6), 816-e21.
29. Kelleher, J. E., Siegmund, T., Du, M., Naseri, E., & Chan, R. W. (2013). Empirical measurements of biomechanical anisotropy of the human vocal fold lamina propria. *Biomechanics and Modeling in Mechanobiology, 12*(3), 555–567.
30. Xuan, Y., & Zhang, Z. (2014). Influence of embedded fibers and an epithelium layer on the glottal closure pattern in a physical vocal fold model. *Journal of Speech, Language, and Hearing Research, 57*(2), 416–425.
31. Hirano, M. (1974). Morphological structure of the vocal cord as a vibrator and its variations. *Folia Phoniatrica et Logopaedica, 26*(2), 89–94.
32. Hirano, M., Kurita, S., & Sakaguchi, S. (1989). Ageing of the vibratory tissue of human vocal folds. *Acta Oto-Laryngologica, 107*(5–6), 428–433.
33. Zhang, Z. (2010). Dependence of phonation threshold pressure and frequency on vocal fold geometry and biomechanics. *The Journal of the Acoustical Society of America, 127*(4), 2554–2562.
34. Horáček, J., & Švec, J. G. (2002). Aeroelastic model of vocal-fold-shaped vibrating element for studying the phonation threshold. *Journal of Fluids and Structures, 16*(7), 931–955.
35. Titze, I. R., & Strong, W. J. (1975). Normal modes in vocal cord tissues. *The Journal of the Acoustical Society of America, 57*(3), 736–744.
36. Berry, D. A. (2001). Mechanisms of modal and nonmodal phonation. *Journal of Phonetics, 29*(4), 431–450.
37. Mergell, P., & Herzel, G. H. (1997). Speech Communication, **22**(2–3), 141–154.
38. Berry, D. A., Zhang, Z., & Neubauer, J. (2006). Mechanisms of irregular vibration in a physical model of the vocal folds. *The Journal of the Acoustical Society of America, 120*(3), EL36–EL42.
39. Steinecke, I., & Herzel, H. (1995). Bifurcations in an asymmetric vocal-fold model. *The Journal of the Acoustical Society of America, 97*(3), 1874–1884.
40. Herbst, C. T., Lohscheller, J., Švec, J. G., Henrich, N., Weissengruber, G., & Fitch, W. T. (2014). Glottal opening and closing events investigated by electroglottography and super-high-speed video recordings. *Journal of Experimental Biology, 217*(6), 955–963.
41. Large, J. (1972). Towards an integrated physiologic-acoustic theory of vocal registers. *National Association of Teachers of Singing (NATS) Bulletin, 28*(3), 18–25.
42. Ware, C. (1998). *Basics of vocal pedagogy: The foundations and process of singing*. New York: McGraw-Hill.
43. Fant, G. (1967). Auditory Patterns of Speech. *Models for the perception of speech and visual form* (pp. 111–125). Cambridge, Massachusetts: MIT Press.
44. Pinto, N. B., & Childers, D. G. (1988). Formant speech synthesis. *IETE Journal of Research, 34*(1), 5–20.
45. Spanias, A. S. (1994). Speech coding: A tutorial review. *Proceedings of the IEEE, 82*(10), 1541–1582.
46. Švec, J. G., Horáček, J., Šram, F., & Veselý, J. (2000). Resonance properties of the vocal folds: In vivo laryngoscopic investigation of the externally excited laryngeal vibrations. *The Journal of the Acoustical Society of America, 108*(4), 1397–1407.
47. Ishizaka, K. (1988). Significance of Kaneko's measurement of natural frequencies of the vocal folds. In O. Fujimura (Ed.), *Vocal physiology: Voice production, mechanisms and functions* (pp. 181–190). New York: AT&T Bell Laboratories, Raven Press.
48. Zhang, Z. (2016). Mechanics of human voice production and control. *The Journal of the Acoustical Society of America, 140*(4), 2614–2635.
49. Rothenberg, M. (1973). A new inverse-filtering technique for deriving the glottal air flow waveform during voicing. *The Journal of the Acoustical Society of America, 53*(6), 1632–1645.

50. Alku, P. (2011). Glottal inverse filtering analysis of human voice production - a review of estimation and parameterization methods of the glottal excitation and their applications. *Sadhana, 36*(5), 623–650.
51. Fant, G. (2012). *Acoustic theory of speech production: With calculations based on X-ray studies of Russian articulations* (Vol. 2). Berlin: Walter de Gruyter.
52. Portnoff, M. R. (1973). A quasi-one-dimensional digital simulation for the time-varying vocal tract. Masters dissertation, Massachusetts Institute of Technology, Cambridge, USA.
53. Story, B. H. (2005). A parametric model of the vocal tract area function for vowel and consonant simulation. *The Journal of the Acoustical Society of America, 117*(5), 3231–3254.
54. Rabiner, L. R., & Schafer, R. W. (1978). *Digital processing of speech signals*. Englewood Cliffs, New Jersey: Prentice-Hall.
55. Lamere, P., Kwok, P., Gouvea, E., Raj, B., Singh, R., Walker, W., et al. (2003). The CMU SPHINX-4 speech recognition system. In *Proceedings of the International Conference on Acoustics, Speech and Signal Processing (ICASSP)* (Vol. 1, pp. 2–5). Hong Kong: IEEE.
56. Stevens, K. N. (2000). *Acoustic phonetics*. Cambridge, USA: MIT Press.
57. Ladefoged, P., & Maddieson, I. (1996). *The sounds of the world's languages* (Vol. 1012). Oxford, UK: Blackwell Publishers.
58. Labov, W., Ash, S., & Boberg, C. (2005). *The Atlas of North American English: Phonetics, phonology and sound change*. Berlin: Walter de Gruyter.
59. Stevens, K. N. (2000). Diverse acoustic cues at consonantal landmarks. *Phonetica, 57*(2–4), 139–151.
60. Fant, G. (1960). Acoustic theory of speech production: With calculations based on X-ray studies of Russian articulations (Vol. 2). Berlin, Germany: Walter de Gruyter.
61. Saks, M. J., & Koehler, J. J. (2008). The individualization fallacy in forensic science evidence. *Vanderbilt Law Review, 61*(1), 197.
62. Page, M., Taylor, J., & Blenkin, M. (2011). Uniqueness in the forensic identification sciences - fact or fiction? *Forensic Science International, 206*(1–3), 12–18.
63. http://www.worldometers.info/world-population/.
64. Jain, A. K., Prabhakar, S., & Pankanti, S. (2002). On the similarity of identical twin fingerprints. *Pattern Recognition, 35*(11), 2653–2663.
65. Sun, Z., Paulino, A. A., Feng, J., Chai, Z., Tan, T., & Jain, A. K. (2010). A study of multibiometric traits of identical twins. *Biometric technology for human identification VII* (Vol. 7667, p. 76670T). International Society for Optics and Photonics.
66. Van, W. G., Vercammen, J., & Debruyne, F. (2001). Voice similarity in identical twins. *Acta Oto-Rhino-Laryngologica Belgica, 55*(1), 49–55.
67. Loakes, D. (2006). A forensic phonetic investigation into the speech patterns of identical and non-identical twins. Doctoral dissertation, School of Languages, University of Melbourne, Australia.
68. Koyama, T., Kawasaki, M., & Ogura, J. H. (1969). Mechanics of voice production. I. Regulation of vocal intensity. *The Laryngoscope, 79*(3), 337–354.
69. Von Békésy, G., & Wever, E. G. (1960). *Experiments in hearing* (Vol. 8). New York: McGraw-Hill.
70. Reichenbach, T., & Hudspeth, A. J. (2014). The physics of hearing: Fluid mechanics and the active process of the inner ear. *Reports on Progress in Physics, 77*(7), 076601.
71. Zwicker, E. (1961). Subdivision of the audible frequency range into critical bands (Frequenzgruppen). *The Journal of the Acoustical Society of America, 33*(2), 248–248.
72. Fletcher, H., & Munson, W. A. (1933). Loudness, its definition, measurement and calculation. *Bell System Technical Journal, 12*(4), 377–430.
73. Traunmüller, H. (1990). Analytical expressions for the tonotopic sensory scale. *The Journal of the Acoustical Society of America, 88*(1), 97–100.
74. Moore, B. C., & Glasberg, B. R. (1983). Suggested formulae for calculating auditory-filter bandwidths and excitation patterns. *The Journal of the Acoustical Society of America, 74*(3), 750–753.

75. Fillon, T., & Prado, J. (2003). Evaluation of an ERB frequency scale noise reduction for hearing aids: A comparative study. *Speech Communication, 39*(1–2), 23–32.
76. Smith, J. O., & Abel, J. S. (1999). Bark and ERB bilinear transforms. *IEEE Transactions on Speech and Audio Processing, 7*(6), 697–708.
77. Stevens, S. S., Volkmann, J., & Newman, E. B. (1937). A scale for the measurement of the psychological magnitude pitch. *The Journal of the Acoustical Society of America, 8*(3), 185–190.
78. Holdsworth, J., Nimmo-Smith, I., Patterson, R., & Rice, P. (1988). Implementing a gammatone filter bank. *Annex C of the SVOS Final Report: Part A: The Auditory Filterbank, 1*, 1–5.
79. Lyon, R. F., Katsiamis, A. G., & Drakakis, E. M. (2010). History and future of auditory filter models. In *Proceedings the International Symposium on Circuits and Systems* (pp. 3809–3812). IEEE.
80. Greenwood, D. D. (1990). A cochlear frequency-position function for several species - 29 years later. *The Journal of the Acoustical Society of America, 87*(6), 2592–2605.
81. Zwicker, E., & Fastl, H. (2013). *Psychoacoustics: Facts and models* (Vol. 22). New York: Springer Science & Business Media.
82. Flanagan, J. L. (2013). *Speech analysis synthesis and perception* (Vol. 3). New York: Springer Science & Business Media.
83. Mersky, B. L. (1991). Method and apparatus for endodontically augmenting hearing. U.S. Patent 5,033,999.
84. Winkworth, A. L., Davis, P. J., Adams, R. D., & Ellis, E. (1995). Breathing patterns during spontaneous speech. *Journal of Speech, Language, and Hearing Research, 38*(1), 124–144.
85. Loudon, R. G., Lee, L., & Holcomb, B. J. (1988). Volumes and breathing patterns during speech in healthy and asthmatic subjects. *Journal of Speech, Language, and Hearing Research, 31*(2), 219–227.
86. Bellemare, F., & Grassino, A. (1982). Effect of pressure and timing of contraction on human diaphragm fatigue. *Journal of Applied Physiology, 53*(5), 1190–1195.
87. Pauluhn, J. (2006). Acute nose-only exposure of rats to phosgene. Part I: Concentration × time dependence of LC50s, nonlethal-threshold concentrations, and analysis of breathing patterns. *Inhalation Toxicology, 18*(6), 423–435.
88. Lucía, A., Carvajal, A., Calderón, F. J., Alfonso, A., & Chicharro, J. L. (1999). Breathing pattern in highly competitive cyclists during incremental exercise. *European Journal of Applied Physiology and Occupational Physiology, 79*(6), 512–521.

Chapter 3
Relations Between Voice and Profile Parameters

In Chap. 2 we saw *why* voice is susceptible to, and *how* it can respond to, intricate changes in the mechanisms of its production. In this chapter we will look at empirical evidence of this fact. We will take a closer look at *what* causes these changes—at the various bio-relevant and environmental parameters that have been observed to affect it. The purpose of this chapter is to give an overview of this topic, based on prior studies. However, the range of scientific studies in this context is vast. This chapter only represents a small sampling of the key findings and current understanding of the subject.

Traditionally, studies that have linked specific characteristics of voice and speech to the anatomy of the vocal tract, to the bio-mechanics of the voice production mechanism, to physiology, to neurological characteristics etc., have all been carefully done under controlled conditions under which the statistical validity of the findings could be supported. Their conclusions are therefore foundational, and the science of profiling can be confidently built upon them. Many of these have also been quantitative, and have used various physical measurement techniques that can gauge the differences in voice production (or articulation) as a function of changes in internal structures and other variables. The techniques include *electroglottography*, *electromyography*, *electropalatography*, *electromagnetic articulography*, *magnetic resonance imaging (MRI)*, *respiratory kinematic sensing*, *ultrasound*, *videostoboscopy* etc. Studies that link behavior, personality etc. to voice have also used various levels of quantitative diagnostic procedures, both invasive and non-invasive, augmented with careful expert observation of the subjects.

In this chapter, we will segregate these studies into physical, physiological, demographic, psychological, behavioral and environmental categories to minimize confusion. However, it is important to note at the outset that many of these studies are interlinked, and the categorizations do not imply causal separation or independence.

R. Singh, *Profiling Humans from their Voice*,
https://doi.org/10.1007/978-981-13-8403-5_3

3.1 Physical Parameters

Many studies have independently shown that the spectral characteristics of speech are affected by various physical attributes of the speakers, from their overall body structure to the structural details of the different parts of their vocal tract. It is no surprise then, that voice has also been correlated with the *perception* of physical parameters such as face [1], body size [2], height, weight, body shape etc. The two aspects are different in their significance to profiling, though. While human perceptions can guide the choices made in putting together the mechanisms for deducing physical parameters, the mechanisms themselves must be based on the direct causal relationships between these parameters and the observed voice signals.

The list below provides a small, selected set of physical parameters that have been reported to affect voice production. Some have been discussed in Chap. 2 in the context of voice production. A comprehensive list remains outside the logistic scope of this book.

3.1.1 External Physical Characteristics

1. **Gender**: Gender-based influences on voice are caused by factors that range from the skeletal to the cellular levels.

On a **macro**-level, morphological differences exist between the vocal tracts of males and females. The length of the vocal tract differs between the genders [3]. The relative proportions of the oral and pharyngeal cavities are different, since the bone structure differs between the genders. The angles of the thyroid cartilage (see Fig. 3.1) are significantly different, as is the placement of the larynx with respect to the overall vocal tract. These are some among many other factors that cause significant gender-related differences in voice. These gender-related differences however do not exist significantly in children, and begin to form at puberty as a result of the remodeling of the male vocal tract [4].

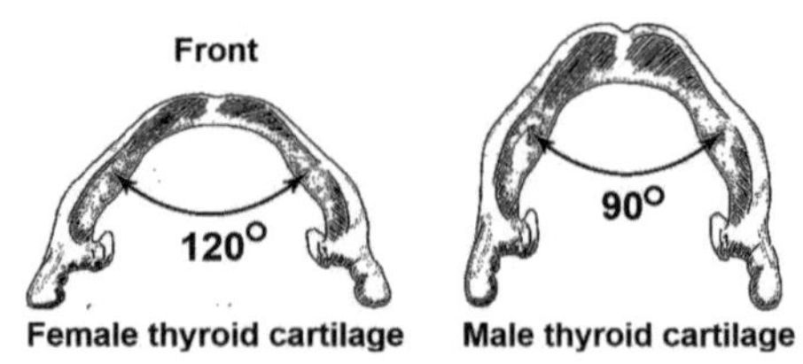

Fig. 3.1 Differences in the size and front angles of the thyroid cartilage in males and females

The smaller angle in men results in greater laryngeal prominence seen as the "Adam's apple," consequently longer vocal cords and lower fundamental frequencies.

Many characteristics of the voice signal are affected by gender, from long-term average spectra [5] to the voicing onset time [6, 7]. Gender-based variations of the proportions of the vocal tract also cause significant differences in formant frequencies. The average length of the vocal tract of an adult male varies from 17–20 cm. For a cylindrical vocal tract shape with a length of 17.5 cm, the formant frequencies occur at odd multiples of 500 Hz: $F1$:500, $F2$:1500, $F3$:2500 etc. Because adult females have shorter vocal tract lengths, their formant frequencies are 15% higher than adult males [8]. Children have short vocal tracts and as a consequence, the formant frequencies of children are about 40% higher than adult males. However, gender-based formant variations in children are not as marked as they are in adults. This is especially true of very young children.

On a **physiological** level, the degree of muscular control—or rather vocal tract inertia as evidenced by voicing onset times—is different on average for the two genders [9]. The hormonal differences between the genders also significantly affect their voice. During adolescence, the vocal characteristics of a woman's voice change [4, 10]. The mucosa in the sub-glottal and supra-glottal regions of the vocal tract changes with estrogen levels in the body. The fundamental frequency lowers slightly in comparison to pre-adolescent $F0$. Changes in women's voices happen throughout the menstrual cycles during their fertile life, and reflect the different stages of a woman's cycle [11]. Pre-menstrually, women's voices have increased vocal roughness [12], lose energy in higher frequencies and have increased hoarseness [10]. These changes are thought to be due to decreased estrogen levels that in turn cause changes in the lamina propria, and changes in the blood supply in the vocal fold area. They subside post-menstrually. In fact, it has been speculated in the medical literature that if the gender of the speaker is known, $F0$ variations could possibly indicate the hormonal state of the speaker, and their age [13]. Hormonal changes also affect men's voices—men with lower $F0$ have been found to have higher testosterone levels [14], though according to some studies, there is no correlation of this entity with body (or vocal tract, by association) size [15].

On a **micro**-level, as we discussed in Chap. 2, cellular and molecular compositions of the vocal folds affect voice in many ways. These compositions differ between the genders [16]. Although not measured in exact amounts, the amount of elastin in the vocal fold layers has been found to be different between men and women. Another molecule—Hyaluronic acid—is more abundant in the mid-portion of the lamina propria of the vocal folds of men then in women [17]. Hyaluronic acid is a carbohydrate polymer found in the connective, epithelial and neuronal tissues of the body. It is an important component of cartilages, and is responsible for their resilience against compression. Interestingly, its molecular weight decreases in the cartilage with age, but its *volume* increases with age. Thus it is also one of the factors that causes men and women's voices to *age* differently. The pitch of men's voices has been observed to increase with age, and in women to decrease with age.

Finally, even the *perception* of voice quality has been found to be different between genders [18]. Females "read" voice cues differently than males.

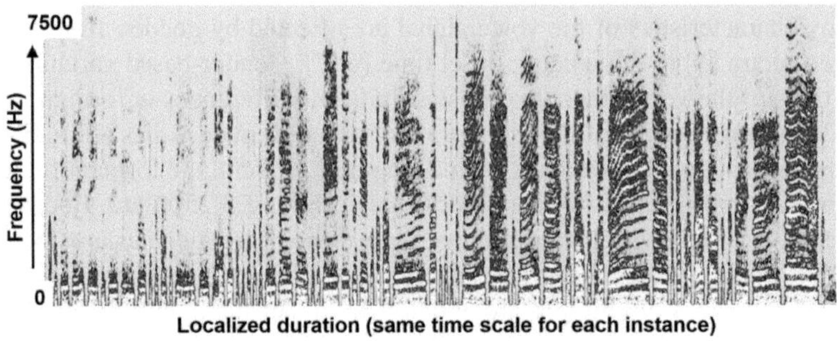

Fig. 3.2 The effect of body weight on the harmonic structure of speech

2. **Weight**: People who must be conscious of voice and body image, e.g. actors and professional singers, have often sought to understand the relationship between weight loss (or gain) and their voice. As a result, there have been many studies in the performing arts and medical domains, that have sought to map the effect of weight changes on voice. The literature on this suggests that weight gain or loss can, and do, affect the voice, but do so differently for men and women.

In women, increased weight is correlated to increased production of estrogen and to a lesser extent, to increased levels of testosterone in the body [19–21]. In men, increased weight is correlated to increased production of estrogen [22]. As a result, increased body weight causes a slight rise in $F0$ in men. The net effect of increased weight is opposite in women though, where increased testosterone levels result in lower $F0$ (the voice becomes slightly more masculine). These slight variations in $F0$ in response to hormone levels are not enough to fully characterize voice changes with weight, though.

The greater influence of weight on voice arises from a change in the dimensions of the vocal tract due to tissue thickening or enlargement. Studies carried out using magnetic resonance imaging of the vocal tract have found significant positive correlation between vocal tract morphology (length, shape and size of vocal chambers) and height and weight [4]. Figure 3.2 shows the spectrogram of 5 year old children saying the same word, of the same height, arranged in the order of increasing weights from left to right.

As discussed later in Sect. 6.4.1.3, the duration of glottal closure is largely determined by thickness of the vocal fold's medial surface. There is expected to be a correlation between this thickness and the weight of the person, although this fact has not been established. An increase in the thickness of the vocal folds results in voice that is richer in higher harmonics, a fact that is evident in Fig. 3.2. The effect of thickening of the vocal folds is visually clear in this—the vocalizations of heavier children are on average richer in harmonics. Thicker vocal folds also result in lower fundamental frequencies [23]. The figure is also suggestive of other relationships, such as a general correlation with the average *duration* of a word, but these have not yet been empirically examined.

Note that the term *vocal weight*, often used in the literature on music, is subjectively related to voice *quality* and is not to be confused with voice with reference to body weight, although there may be correlations between the two. "Vocal weight" in fact refers to the perceived "lightness" or "heaviness" of a singing voice. In classical western music such voices are described as "lyric" and "dramatic" respectively. Vocal weight is generally thought to be inversely related to vocal agility—the ability to traverse complex notes and registers, and directly related to their "carrying" capacity to audiences. Heavier voices are also described as "powerful" and "darker."

3. **Physical attractiveness**: In females, attractive faces have been correlated with attractive bodies [24]. This is extended to voice through other studies, that have found strong correlations between vocal attractiveness, and visual attractiveness in women (from the perspective of men). Men also seem to assess women's age, body size and (possibly) hormonal profile from women's voices [25], all of which have been correlated with voice on one hand, and attractiveness on the other. The links between facial attractiveness and voice that are suggested by such studies are also supported by more direct studies that have found that the *facial shape* (of women), and its attractiveness to men, are both highly correlated with the voices that men find attractive, suggesting that females who have an attractive face also have attractive voices (on the subjective attractiveness scales intuitively used by men, of course)—this is called the "redundant signal hypothesis" [26, 27]. These observations are also suggestive of the possibility that because voices are correlated with the geometry of the human face (see Chap. 9), the (albeit weak) ability to perceive attractive faces from voice has evolutionary groundings.

In most studies on attractiveness, the fundamental frequency $F0$ has been implicated as an important correlate [28]. Generally, voices with higher $F0$ are correlated to lower age in women as judged by men, and are strongly considered attractive by men. Larger women have lower $F0$ on average, and are found to be less attractive by men. This has been revealed in studies where a variety of voice types and facial profiles were presented to men [25]. The relation of $F0$ alone to men's attractiveness (from the perspective of women) is however more tenuous. In men, $F0$ is *not* such a reliable indicator of age, and age in turn is not a reliable indicator of their attractiveness to women (the opposite is not true). In men, manipulations of both $F0$ and formants *together* is judged to be more attractive [29] by women.

4. **Height**: Human judgments about height and weight from voice are not good. In one study that investigated this [30], less than 15% of the human judges could judge these correctly, and the judgment errors/accuracy were consistent across different acoustic conditions (and have even found to be robust to high-pass and low-pass filtering of speech [31]). The study concludes that humans follow wrong stereotypes about body size, and follow them consistently.

However, it is well known that height influences voice. Taller people have longer vocal tracts on average, and also have larger lungs on average. As a result, their voices have a greater degree of sub-glottal resonance. The positions of $F1$ and $F2$ are lower for taller people. Based on such observations, algorithmic (statistical) inference of height from voice has been attempted widely, and is accurate between ± 3 inches for height [32]. This presents empirical evidence of a positive correlation between height and voice but does not validate any specific causal hypotheses.

The causal hypotheses that have been formed are however quite simple: the body proportions of humans of different skeletal types are different. For each type, the average vocal tract dimensions are characteristic of each gender, within which the length of the vocal tract, especially, varies more or less linearly with height. The effects on the pitch, formants and other characteristics of voice within these groups, which vary roughly linearly with the factors mentioned, are also expected to be linear with height, as a consequence.

5. **Body size**: In the current context, body size refers to the largeness of the skeletal structure of people, and not to their weight. Larger people have greater lung volume and capacity within the same age group and within the same height range. Their voices, therefore, are expected to have greater vocal intensity (and therefore greater range, or *projection*). This is often true of opera singers with larger body proportions. There is evidence that the voices of larger people are richer in harmonic structure, and exhibit evidence of greater breath control, trans-glottal air flow and sub-glottal air pressure (during phonation).

The direct relationship between voice and body size, however, remains tenuous (although body size can be inferred from height and weight, both of which have stronger relationships to voice). One study [33] has found that within the same gender, there is only a weak correlation between body size and *formants*, and more within females than males. The study concludes that the pattern of development of the vocal tract (dimensions) is relatively free of skeletal size. In other studies on humans, body size has been negatively correlated with formant *dispersion* and pitch [34]. The conjecture is that this is attributable to the change in position of the human larynx relative to the vocal tract and its (spatial) placement in it, both of which correlate with body size.

There is an additional, rather indirect relationship between men's voices and their body size—adult males have higher levels of circulating testosterone than women and children, and the effect of the hormone is to lengthen the vocal folds and thicken them. This lowers $F0$. At puberty, the larynx also descends further down the vocal tract, lowers the formants and reduces formant dispersion. Thus higher $F0$, and more formant dispersion, are both more correlated with younger age in males, and *correspondingly smaller body sizes*. Voice was linked to (Kretschmerian) body *types* as early as 1940 [35].

6. **Eye color**: The first positive association of eye color to the voicing patterns of birds was made in 1949 [36]. In species of birds called Vireos, dark-eyed species ran their calls with almost no pauses, while light-eyed species included longer pauses in their calls. A possible correlation is mentioned in this context for the human case in [37], where the author cautiously states that (quote:) "*aspects of voice pattern other than frequency may also be worth studying as correlates of eye-color*." Of course, no human studies have been reported on this as yet and currently any relationship is merely speculative.

7. **Skin color**: We cannot discuss skin color without discussing race. Socially, skin color is associated with race, except that scientifically, *races do not exist*. Race is more of a social concept than a biological one. It is a fallacy that race can be maintained as a biological indicator of generic body-type and appearance, including skin color.

As explained later in Chap. 9, humankind can be divided into three groups based on their musculo-skeletal, and specifically skull structure [38, 39]. As a consequence, even some scientific studies that have implicated race as an indicator of voice type, have really only correlated specific skull types and/or skin color with voice. For example, a study conducted by researchers in linguistics [40] in 1993 found that race was identified with 60% accuracy by human listeners by listening to voice alone. On closer inspection, they found that voices of people with *darker skin* exhibited greater amplitude and frequency fluctuations and lower harmonics-to-noise ratios than the voices of *white skinned* people.

The correlation between skin color and voice is a matter of speculation at this time. It may simply be that since skin color and skull type are strongly correlated in most populations, the more direct relationships between skull structure and voice extend by proxy to skin color.

3.1.2 Morphology of the Vocal Tract

So far, we have discussed the relationship of voice to overall physical parameters of a speaker. The physical structure, or morphology, of the vocal tract also plays a highly significant role in defining the characteristics of voice produced in it. We discuss some aspects of this below.

1. **Nasal (and nasal chamber) shape and size**: The shape and size of the nasal chamber within the vocal tract plays a significant role in determining the characteristics of the voice produced. Even the *shape of the nose* matters. Asymmetry between the two nasal passages can cause extra pole-zero pairs and affect the spectral characteristics of sounds produced [41].

In this context (and to understand much of the explanations in later chapters) it is important to think of the vocal tract from the perspective of its response to a stimulus (such as an impulse of sound). The vocal tract can be viewed as a *system* (of differential equations) that responds to an excitation signal. Its response to a single impulse is the pattern of spectral characteristics generated as a result. This is called the *impulse response* or the *transfer function* of the system. The frequencies that are highlighted or appear with maximal energy are the resonances or the **poles**. The ones that vanish are the anti-resonances or **zeros**.

Conceptually, these are simple to understand and compute. Any transfer function can be written as a rational function $H(z)$ of a complex variable $z = re^{j\omega}$

$$H(z) = \frac{b_m z^m + b_{m-1} z^{m-1} + \ldots + b_1 z + b_0}{a_n z^n + a_{n-1} z^{n-1} + \cdots + a_1 z + a_0} \tag{3.1}$$

Which can be written in a factored form as:

$$H(z) = \frac{N(z)}{D(z)} = K\frac{(z-z_1)(z-z_2)\dots(z-z_{m-1})(z-z_m)}{(z-p_1)(z-p_2)\dots(z-p_{n-1})(z-p_n)} \tag{3.2}$$

The system does not respond when its transfer function is zero, and responds maximally when it is at its maximum (unbounded) value. These conditions can be stated as:

$$\lim_{z\to z_i} H(z) = 0 \tag{3.3}$$

and

$$\lim_{z\to p_i} H(z) = \infty \tag{3.4}$$

The polynomials $N(z)$ and $D(z)$ have real coefficients defined by the system's (differential or coupled differential) equations, and $K = b_m/a_n$. z_i and p_i are the roots of the equations $N(z) = 0$ and $D(z) = 0$. These are the zeros and poles of the system respectively.

Since the coefficients of $N(z)$ and $D(z)$ are real, the poles and zeros are also real or complex conjugate pairs. Generally, either $p_i = r_i$ or $p_i, p_{i+1} = r_i e^{\pm j\omega_i}$. Similarly for the zeros. The pole and zeros, together with K (called the gain constant) completely describe and define any system comprised of differential equations.

The nasal chambers often invoke the zeros of the transfer function of the vocal tract of an individual, and these depend on the size and shape of the nasal chambers. These zeros are anti-resonances, and appear as zero-energy regions, or *spectral holes* in the spectrogram (explained in Chap. 4) of the signal. The anti-resonances are also thought to be caused by sound absorption in the maxillary and sphenoidal cavities. Even a small opening of the nasal pharynx due to the retraction of the velum can cause higher formants ($F3 - F5$) to have higher energy [42], which give voice its nasal quality, in conjunction with the zeros in the spectrum that are *not* heard.

Nasal (and nasal chamber) shape and size thus determine the extent of *nasalence* of voice. Nasalence (or nasality of speech) is one of several facets of the characteristic of speech that is broadly referred to as *voice quality*. Back in 1968, a scientific publication declared a 93% accuracy in recognizing speakers based only on the nasal quality of their speech—The first sentence in this paper reads "*A method of automatic speaker identification based on the physiology of the vocal apparatus and essentially independent of the spoken message has been developed*" [43]. Nasalance can be quantitatively measured using *nasometers*, and studies have shown significant differences in nasalence between genders, people of different nativities etc. Nasal areas have been shown to be different for people of different "races" [44], and in fact are quite unique to individuals—we have all observed in our routine environments that different people have different nose shapes. Differences in nasal areas of humans have consequences on their voice, but these have not been systematically studied.

2. **Size and shape of dental arches**: Quoted directly from [45]: "*The most casual inspection of the general population will show how misleading any assumption of*

vocal tract equivalence must be. Even among the genetically related members of a family, there will almost always be differences in the size and shape of the dental arches, the palatal contour, the relationship of the upper and lower jaws, and the size of the larynx. All these factors have implications for an individual's speech production ..."

The most direct consequence of individual variations in dentition is evident in the placement of the front or anterior portion of the tongue, at rest or while speaking. In Chap. 2 we discussed how vowels are characterized and categorized by the placement of the tongue, and also how glides and fricatives are characterized by airflow around and over tongue. Variations in dentition affect the properties of the front vowels and front oral fricatives directly.

3. **Size and shape of palatal contour**: Following the same reasoning as above, the size and shape of the palatal contour significantly affects the production of palatal consonant sounds, and also affects other sounds by defining the overall shape of the upper oral chamber, and the lower nasal chamber.

4. **Physical setting of the upper and lower jaw in relation to each other**: The physical setting of the upper and lower jaws defines both the closure of the oral chamber, and how much (or how widely) it opens naturally. Sounds that are produced under both (partially open and closed, or constricted) configurations are affected by these settings. For example, the frequency of the first formant $F1$ is largely dependent on how wide the jaw is open, and there is a direct proportionality between the two [46, 47]. Figure 3.3 shows how $F1$ ad $F2$ vary with the size of the mouth (and jaw) opening. Other formants are also affected by these factors [48].

5. **Tongue size and tongue root position**: The size of the tongue and the position of its root varies across people of different ethnicities, and even between different individuals within small populations. Both these factors have a significant effect on voice quality [49]. The frequency of the second formant $F2$ is sensitive to the shape of the tongue. The frequency of the third formant $F3$ is sensitive to the cavity between the tongue tip and the lower incisors [8]. The volume of this cavity depends on the shape of the tongue, the dentition, shape and placement of jaws, and the position of the tongue root with respect to them. Together, the jaw and the tongue influence $F1$ and $F2$ in the manner shown in Fig. 3.3, wherein vowels that require different tongue heights and backness are enunciated with different jaw/mouth opening sizes.

6. **Pharynx size**: The pharynx is situated behind the nasal cavity, within the throat, behind the mouth and above the esophagus and larynx. It has three parts or sections: the nasopharynx (behind the nasal chamber), the oropharynx (behind the oral chamber) and the laryngopharynx (above the larynx and below the oropharynx). The relative proportions of these three, and the overall size of the pharynx play a significant role in many characteristics of the voice signal produced [50]. Specifically, reducing the volume of the pharyngeal cavity results in a rise in $F1$. If the mouth is open simultaneously, a sharp fall is observed in $F2$, and this brings $F1$ and $F2$ close together, increasing the intensity of the sound radiated outwards. This effect is used routinely by opera singers to increase their vocal output. In general, increased

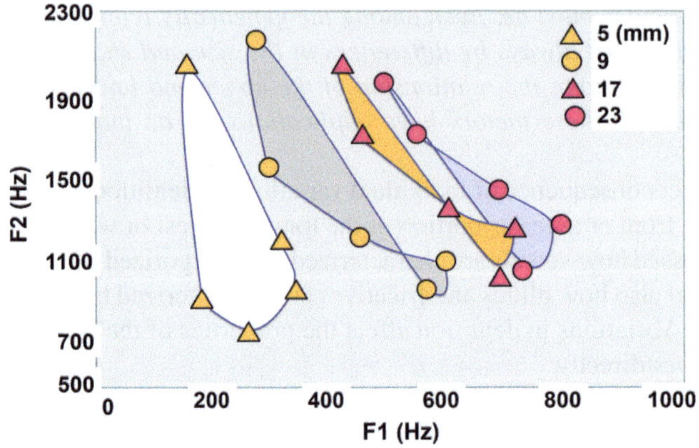

Effect of changing the tongue position from front to back and front again while cycling between the phonemes /iy/, /uw/, /aa/ for different jaw openings (after [8]).

Fig. 3.3 Effect of changes in jaw opening and tongue position on sounds

widening of the pharynx leads to decreased vocal intensity. Narrowing of the pharynx gives a twangy quality to the voice [51].

7. **Lip shape and size**: Lips play a significant role in the production of different speech sounds, and their characteristics. Lip kinematics are well studied and have a strong influence on phoneme identity [52]—so much so that in articulatory phonetics, the lip shape is used as one of the primary factors in the categorization of vowels (rounded or not), and consonants (bilabial, labio-dental etc.). Lip shape also determines certain voice quality sub-factors. In fact, the shape of the vocal tract can be determined from the impulse response at the lips [53].

8. **Size and position of the larynx**: Laryngeal musculature controls the vocal fold movements, and is crucial to voice production. It is redundant to discuss its effect on voice in detail here, since the subject is also covered in Chap. 2. The physical *position* of the larynx in the vocal tract is implicated in voice quality terms such as *raised-larynx voice* and *lowered-larynx voice*. These refer to the raising and lowering of the larynx resulting in raising and lowering of all formants (indeed the entire spectrum) [54] respectively. Obviously, these changes in *larynx height* also influence voice pitch [55]. Other structures in the larynx [56] also cause changes in various voice characteristics, such as the intensity of sounds produced.

9. **Size of the epilaryngeal region**: The epilarynx couples the vocal folds to the supra-laryngeal vocal tract, and plays an important role in the mechanics of voice production [57]. It lies just above the larynx, comprising the *epilaryngeal region*. The glottal flow waveform (shown in Chap. 2) exhibits a skew towards the right in time. Part of the reason for this is that as soon as the glottis opens, and the vocal tract fills with air, the air volume exerts a resistance to further incoming air. This mass-like

behavior of air delays the increase in glottal flow. This is called vocal tract *inertance*. This air column inertance within the vocal tract has been shown to directly affect the resonant properties of voice [58]. Inertance is significantly related to the size of the lower vocal tract, especially of the epilarygeal region (its diameter, narrowness etc.) [51].

10. **Physical size of the vocal folds**: Since the vocal folds are three dimensional structures, their length, breadth and thickness, all play a role in voice production. In Chap. 2 we discussed their intricate role in the process of phonation. Their movements control the pressure variations across the glottis. During phonation, the cyclic opening and closing of the vocal folds creates specific asymmetries in the aerodynamic variables across the glottis, which in turn cause specific patterns of self-sustained vocal fold oscillations [59]. The exact nature of the oscillations—i.e. the excitation signal, is also dependent on the patterns of eigenmode synchronization during phonation, which are in turn dependent on the physical size and structure of the vocal folds, in addition to other factors discussed in Chap. 2.

The physical size of the vocal folds also determines the relative proportions of the histological layers of the vocal folds. This distribution significantly affects the mechanical properties of the vocal folds [60] during phonation. Of these layers, the lamina propria is especially significant. It is composed of three types of substances: (a) collagenous fibers that lie parallel to each other, tightly connected along their longitudinal direction and loosely connected in a transverse direction, (b) elastic fibers that form a network, running in every direction, and (c) a ground substance that has a gelatinous property. The effects of variations in these manifest in many ways in the voice signal, affecting $F0$, all formants, harmonics, energy and other measurable entities in the voice spectrum. These have been discussed in Chap. 2.

11. **Physical size of the lungs, rib cage and abdomen**: The dimensions of the structures below the larynx, pertaining to the individual's respiratory system play a significant influence on voice production. Lung volume affects many characteristics of voice, e.g. voicing onset time [61], the nature of the excitation signal during phonation [62] etc. As the lung volume decreases, sub-glottal pressures decrease during voice production. This decreases the dynamic range (the peak-to-peak amplitude) of the voice signal, usually as a response to reduce glottal leakage. Natural variations in lung volume thus have a direct effect on the acoustic intensity of the voice produced in the vocal tract.

In addition to the size of the lungs, the volume displacement of the rib cage and abdomen, i.e., the volume of air displaced during speech production, depends on their size, and that of the diaphragm (among other factors) [63, 64]. These affect voice by affecting respiration (rate and volume), and can also cause a shift in the position of the larynx [65], affecting voice characteristics.

The factors discussed above are all parts of the vocal tract. More broadly, studies have shown that the vocal tract shape in an overall sense plays an important role in being able to produce different registers, to tune formants [66] and so on.

An important point to note regarding the relationship of voice to physical parameters is that there are aspects of the voice that can change instantaneously in response to dynamic changes in some parameters. On the other hand, there are aspects of the voice that cannot change over a person's lifetime. The vocal tract provides a good example of both. We know that the shape of the vocal tract can be changed dynamically using our articulators, and in fact this is the enabling mechanism whereby speech is produced. However, the overall dimensions of the vocal tract are fixed in adult humans. Some of their effects cannot be masked or changed. This applies, for instance, to the fourth and fifth formants, $F4$ and $F5$. They have been found to be very dependent on the length of the vocal tract and the configuration of the deep pharynx. In adult males, $F4$ and $F5$ lie in the range of 2500–4000 and 3000–5000 Hz respectively (for vowels). There are hard to change by articulation, and hard to change significantly by any natural means.

3.2 Physiological Parameters

Physiological parameters are those that relate to *functional* (in a biochemical/biological sense) aspects of the living body. Multiple physiological parameters of humans influence their voice. Of these, we discuss some key ones below.

1. **Age**: Medically, age is considered to be a physiological parameter, since its cause and effects are largely physiological. The changes caused by age to voice are marked enough that there is a medical term for it: *presbyphonia*. Both perceptual and causal aspects of presbyphonia have been extensively investigated, e.g. [67–69].

Some age-related changes to voice are a result of natural growth from childhood to adulthood. Biologically, the vocal tract and its components that are key to producing sound, namely the vocal folds in the larynx, change continuously with age as a person transitions from childhood to adulthood. Consequently, during that period voice changes with age. In the pre-adolescent years, this change is accelerated, and marked changes in voice characteristics are seen with smaller age differentials than in adulthood. The natural length of the vocal folds is a factor in deciding its mechanics, which contribute (among other things) to $F0$ [70–72]—the pitch of a person's voice is higher when the vocal folds are thinner or shorter, and lower when the vocal folds are thicker or longer. The vocal folds are short at birth and grow through childhood and early adulthood, as does the length of the vocal tract in its entirety. The growth rate is different for males and females. The pitches of children's voices are thus much higher than those of adults. The relative position of the larynx within the vocal tract also influences pitch, which is a function of neck length. This positioning also changes continually with age until puberty, influencing pitch simultaneously.

Generally, the relation between the pitch $F0$, the length of the vocal fold L, the longitudinal stress on the vocal fold σ and the tissue density μ can be expressed by the following equation:

$$F0 = \frac{1}{2L}\sqrt{\frac{\sigma}{\mu}} \tag{3.5}$$

Here L is the *total* length of the vocal fold, a small fraction of which (at the base) does not actually vibrate to produce sound. The part that is free to vibrate is about 2 mm in length in both males and females at birth. Thereafter, L grows at approximately 0.7 mm per year for males and 0.4 mm per year for females. The growth stops in early adulthood (early twenties).

In Eq. 3.5, the longitudinal stress on the vocal fold σ and the tissue density μ refer in part to the response of the vocal fold epithelium and superficial layer of the lamina propria under tension [73], and these are two parameters among the many that are also affected by age. Age causes changes to the cellular physiology of vocal folds [74–77], by affecting the cellular and molecular composition of the vocal fold layers, which affect its mechanical and aerodynamic properties. The longitudinal axis of the vocal folds being one that is parallel to the elongation of the vocal folds, the precise longitudinal tension, transverse tension and shear modulus (measured with a rotation around it), all affect $F0$ and other voice characteristics significantly [60, 78, 79]. Changes in the mechanics of the vocal folds as a result of changes in tissue composition in turn affect the phonation threshold pressure, and all of these cumulatively affect phonation and voice quality [80].

It is no surprise, therefore, that we perceive the effects of age by perceiving changes in voice quality. There have been numerous studies that have shown correlations between voice quality and age, e.g. [81, 82]. Other signal characteristics that have also been positively correlated positively with age include pitch, the formant positions, formant bandwidths, energy distribution in the long-term spectra etc. Figure 3.4 shows the long-term average spectra (LTAS) for the word "anyone" spoken by Woody Allen (a Hollywood producer/director/actor), in two different public interviews conducted twenty years apart. The upper (black) contour in the figure corresponds to the case when he was young, and the lower (dark gray) corresponds to the same word spoken 20 years later. We note that the energy in the mid-frequency regions has subsided. This is characteristic of advancing age, and is due to the physiological changes in tissue composition mentioned earlier.

From early adulthood until the age of about 55 years, voice pitch remains relatively constant, and then changes begin to happen again as the tissue structure within the vocal tract begins to undergo deteriorating changes [83]. These changes are different in different people, depending on their musculo-skeletal structure and composition, and their state of health. In some people, the larynx undergoes neurological changes [84, 85], and the voice develops tremors. In some, the vocal folds thicken and the pitch lowers, while in others they atrophy or shorten, and the pitch is raised. In some men, the mucosa becomes thicker and more viscous, leading to slurred enunciation of speech sounds. All of these changes are however continuous and can often be related to the pitch of the person at or close to the beginning or end of steady-state for voice, i.e., adulthood and middle-age. The possible changes in a person's voice with the progression of age can therefore be predicted.

2. **Physical and mental exhaustion**: Voice was documented to be linked to the level of exhaustion of the speaker as early as 1940 [86], and the relation between the two ("tired voice," "she sounded tired" …) has been mentioned in the general literature

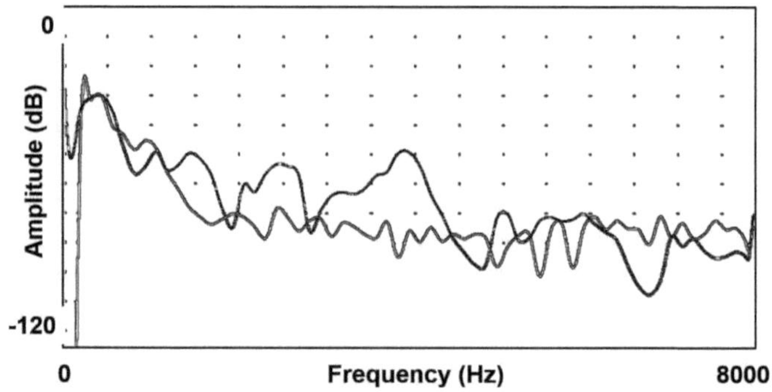

Long-term spectral energy distribution for the word "anyone" uttered by Woody Allen in public interviews given 20 years apart. The upper contour corresponds to the word spoken.

Fig. 3.4 Changes in spectral characteristics of voice with age

for centuries. Physical and mental exhaustion elevate **blood pressure** and **heart rate**. Voice pitch has been found to change non-linearly with both kinds of exhaustion, increasing in plateaus with increasing loads of each kind [87]. This may either be directly due to exhaustion-related physiological changes that affect the muscles of the vocal tract, or indirectly due to blood pressure or heart rate changes.

The former is plausible since studies have related electromyographic activity of laryngeal muscles to sub-glottal air pressure, and thereby to the fundamental frequency of vibration of the vocal folds (pitch) [88]. The latter is *also* plausible, since the effects of blood pressure on the nervous system are well known. All muscles in the larynx are innervated in a complex fashion. The TA muscle in the larynx, for example, is innervated by the recurrent laryngeal nerve and a second source of motor innervation called the "communication nerve" that also provides sensory innervation to the sub-glottal area and the cricoarytenoid joint [89]. These, like all other nerves in the body, respond to blood pressure changes. Direct studies relating blood pressure to voice characteristics, however have not been performed (or at least are not easily found at the time of writing this chapter). The indirect relationship of voice to blood pressure, indicated by many studies, makes it an ideal candidate for feature engineering based profiling, discussed in Chap. 7.

3. **Heart rate**: Heart rate has been found to be highly correlated with the Hayre stress parameters H_1 and G_1 which are computed from the voice signal and based on its spectral properties [90].

The fact that heart-rate and voice are correlated, and the former can be deduced from the latter has been empirically shown in studies, e.g. [91], in which, in a speaker-dependent setting, the heart rate and skin conductance (another physiological parameter) could be determined with a correlation coefficient of 0.861/0.960 and mean absolute error of 8.1 BPM/88.2 μMhO respectively. In the same study, pulse-rate classification as high/low could be done with over 80% accuracy. Interestingly, the study also found that both heart-rate and skin conductance could also be identified from breathing periods.

Some other physiological parameters that have been observed to affect voice are **sleep deprivation** [92], **normal body rhythms** and other changes caused by hormonal fluctuations, e.g. menstrual cycles [10, 11, 93] (discussed earlier in this chapter) etc.

3.3 Demographic Parameters

1. **Nativity and geographical origin**: Speech is as much a learned process as a natural, bio-mechanical one. Early childhood exposure to language, and learning to speak a language, cause the formation of tenacious speaking habits. These habits determine how a person co-ordinates and moves their articulators during speech, and what sounds they are able to form, or even perceive. The speaking habits of early childhood remain in a person's speech throughout their life, unless they make extreme efforts to "unlearn" or mask them (as in voice changes done in drama and performances). These are often not completely successful, and work best as long as the person remains alert to the required changes in speaking patterns.

These habits translate to specific patterns of signal characteristics which encode information about the speaker's nativity and geographical origin. The latter can often be deduced through their accent or dialect, which in turn can be gauged using the analysis of $F1$–$F2$ formant spaces of vowels [94, 95]. There is a vast body of literature on deducing a speaker's accent, dialect, and thereby their nativity and geographical origin from voice using other signal characteristics such as nasalence, prosody etc., but the analysis of $F1$–$F2$ spaces remains the most effective strategy.

2. **Level of education**: Often, we can tell a person's level of education from their speech. This is largely cued by the vocabulary they use, the depth of thought and understanding they express on specific subjects (not all, though) and other pragmatic information in their speech. Voice reflects the level education, through the level of control and deliberation exerted in enunciating sounds. This link is however tenuous and not quantified. There has been some sparse work in gauging how the level of education is codified in voice, but the subject largely remains to be explored from a profiling standpoint.

3. **Ethnicity/Race**: A question that immediately arises in the context of demographics and voice is—does race influence voice? Earlier studies have found that this is so, e.g. [6, 96, 97], but, as mentioned earlier in this chapter, there is a general fallacy among the populace that there exist distinct human races. This is not true. According to [98, 99], races can be defined in two ways: one refers to "*highly genetically differentiated populations with sharp geographical boundaries*," and the second to "*distinct evolutionary lineages within a species*." The following is further argued by [99] (quote:)

- *By either definition, races do not exist in humans but do exist in chimpanzees.*
- *Adaptive traits such as skin color do not define races and are often discordant with one another.*
- *Humans populations are interwoven by genetic interchanges; there is no tree of populations.*

Although the above viewpoint has been contested by some [100], it is also strongly supported by genetic evidence [101]. Currently, "racial" segregations are based on skull types. There have been several racial maps produced based on skull types, two noteworthy ones being that of Huxley [102], which gave an estimate of the distribution of the three races (people of three skull types)—caucasoid, mongoloid and negroid across the world. In another exposition, published in 1962, Coon segregated mankind into five races based on skull type [103], which was later strongly disputed [104].

Anthropological classifications based on skeletal structure, especially the skull structure, can be related to acoustical differences in voices by looking at the manner in which the dimensions of vocal structures change with associated phenotypic variations. Ethnic differences do exist in human bone geometry, and there have been several studies that support this, e.g. [105]. Many studies have considered differences caused by such factors (outside the voice production structure), e.g. [106]. Of the numerous studies that have shown that positive correlations exist between race and voice, most have defaulted to correlating skeletal structure, as differentiated above, to voice. For example, in one study [107] that asked listeners to identify race when presented with voices of Chinese Americans, Korean Americans, European Americans, Latinos, and African Americans, listeners could identify race from voice with a better than random accuracy, but were unable to distinguish between Chinese and Korean Americans (both of who share very similar Mongoloid skeletal structures). In another study [108], for example, researchers noted that the voice pitch of Caucasian males and females, on average, is around 117 and 193 Hz respectively, while for Africans, it is 110 and 217 Hz for males and females respectively. But in such studies, race is differentiated based on skin color, which is a fallacy as mentioned above. Arguably, the results would not be similar if the populations were differentiated by skull type (if this was not normalized for in the study). South Africa is a particular case in point, where the term "Africans" could refer to both light-skinned and dark-skinned people. Figure 3.5 shows "Caucasians" of different ethnicities across the world. They all share the same skull and skeletal type.

The gist of the discussion above is that through the lens of the mixed studies, we do find that different skeletal structures lead to different patterns of phonation and exhibit characteristic spectral structures, with differences extending into almost every part and measurement of the spectrum of voice.

Fig. 3.5 The Caucasians of the world (graphics: Arsalan Khan, Deviant Art)

3.4 Medical Parameters

Diseases and other medical conditions influence voice by influencing the bio-mechanical and neurological processes that play a role in speech production. Depending on the condition, these influences manifest differently in the different sounds produced by the vocal tract. Many studies, for example, have shown that certain medical conditions have clear biomarkers in vowel sounds, others have focused on other categories of sounds as being most relevant etc.

The relationship between the medical condition of a person, and their voice is in fact a deeply researched topic, with thousands of articles and studies in the published literature. Some good overall accounts exist, such as [84, 109], which can serve as starting points, and talk about many different medical reasons for voice dysfunction (or deviation of the voice from the canonical "normal" voice). The very same findings can provide valuable information for profiling. A small, random and *very* sparse sampling of the literature is given below to give an idea of how wide-ranging the effects of medical conditions are on voice.

1. **General state of health**: Within controlled age groups, it has been found that people in good medical health have significantly higher vowel phonation durations, lower jitter and shimmer, and larger phonation ranges than people who are not healthy [82]. Of course, it has also been noted that older people are less healthy on an average, but age-related changes to voice (as discussed earlier in this chapter) are caused by physiological factors other than simply changes in physical health status.

2. **Autoimmune disorders**: Patients with **Sjögren's syndrome** [110] exhibit voice, speech and swallowing abnormalities, among other things. These are thought to be associated with xerosis and neurological abnormalities that are secondary to the disease. In the study reported in [110], the voice pitch for the vowel /aa/ and the formants $F1$–$F5$ was charted for changes in structure, and for anti-resonances in the frequency bands 0–4000, 4001–10,000 and 10,001–14,000 Hz. In *Type I* syndrome, it was noted that there was an absence of harmonics in the high frequencies (above 4 kHz) and noise in formant regions. In *Type II*, the symptoms of Type I persisted, in addition to the absence of harmonics between $F1$ and $F2$, and in *Type III*, the symptoms of Type II were augmented by the disappearance of $F2$. For *Type IV*, more changes in voice characteristics were noted.

AIDS (Auto Immune Deficiency Syndrome) has been noted to cause voice changes such as hoarseness and xerostomia.

3. **Neurological disorders**: **Amyotrophic lateral sclerosis (ALS)**, also known as **Motor neuron disease (MND)** or **Lou Gehrig's disease** is characterized (from the perspective of voice) by increased instability in phonation, and reduced range (or limits) of phonation as the disease progresses [111, 112]. Some observations with the progress of the disease are an increase in amplitude and frequency variation, reflected in higher jitter and shimmer values, reduced harmonics-to-noise ratio, and reduced maximum vowel duration.
   Parkinson's disease has been shown to have voice biomarkers in many studies, e.g. [113–116]. It affects laryngeal muscle control (which has a direct effect on vocal cord movements) [117]. Pitch, jitter and shimmer have also been found to be significantly affected. Studies have shown that the voices of Parkinson's patients have significantly lower vocal SPL than control subjects.
   In **Dementia** [118, 119], the periodicity of the energy envelope in the modulation spectrum is affected, and the envelope shows marked aperiodicity in comparison to healthy subjects. Many linguistic features are also affected, such as the number of pauses and their duration in continuous speech, number of errors made etc. The studies cited here further cross-reference other studies that have related voice to impairment of cognitive function, specially in the elderly population. In the case of **Alzheimer's** [120], in addition to semantic issues with language production, some key vocal characteristics that are affected are syllabic perseveration, increased shouting and inappropriate laughter—these are vocal features rather than linguistic, since they are language agnostic.
   Other neurological conditions that have been shown to have biomarkers in voice are **epilepsy** [121], **impaired motor co-ordination** [122], etc.
4. **Endocrine system disorders**: Many endocrine system issues have been shown to have voice biomarkers. In **hypothyroidism** [123, 124], for example, studies have found that there is a positive correlation between TSH (Thyroid Stimulating Hormone) concentration and variation in $F0$, and also a prevalence of different voice disorders caused by dryness in larynx and pharynx, dyspnea etc. In addition to $F0$, turbulence and phonation characteristics of the voice are affected.
   Disturbances in the functioning of the **thyroid** gland [125], **ovaries**, **gonads**, **adrenal**, **parathyroid**, **pineal** and **pituitary** glands can cause hormone-induced dysphonia.
5. **Medications, toxins, substance abuse**: Many **medical drugs**, e.g. [126, 127], intoxicants such as **alcohol** [128], **recreational drugs**, **exposure to chemicals and toxins** [129] etc. cause changes in voice.
   As an example of medication-related voice changes, we consider **birth control pills** [84, 109]. These pills are often ovulation inhibitors and change the hormonal levels in the body, which in turn alters the range and quality of voice. In some women, voice changes are observed within a few months of therapy. These pills enhance and cause the same vocal dysfunctions that occur in the normal pre-

menstrual period for women. Studies have shown (especially on singers) that this (pre-menstrual) period is characterized by decreased vocal efficiency, inability to reach higher vocal notes or registers, voice fatigue, hoarseness and muffling of voice. Sub-mucosal haemorrhages are also common in the pre-menstrual period, and these symptoms can extend well into the period of taking birth control pills. Note that even before scientific studies revealed this, there was a time when singers in European opera houses were given grace period (time off singing) in their pre-menstrual and shortly-after periods because they were thought to not be singing to their best capacity during that time. **Androgens** given to women to treat endometriosis etc. also cause unsteadiness of the voice, rapid changes of timbre, and lowering of fundamental frequency. These changes can remain permanent.

In **alcohol intoxication**, changes have been observed in the harmonic-to-noise ratio, the distribution of $F0$, and changes in the long term average spectra (LTAS). The ratio of $F1$ and $F2$ especially also varies with, and marks the states, where the blood alcohol level is high. **Opioid**-induced respiratory depression (and in extreme cases, glottis closure leading to stopped respiration) is a well known effect of opioids such as Codeine, Hydrocodone, Morphine, Oxycodone, Hydromorphone, Fentanyl etc. It is also common in recreational drugs such as **heroin**, **cocaine** etc. A direct effect of changes in respiratory pattern is evident on phonation cycles, which are dependent on a critical balance of air pressure across the glottis—i.e., in the supra- and sub-glottal regions. At finer levels, drugs affect muscle movements, cause tissue constrictions, and cause changes in articulator positioning and movements within the vocal tract. Changes in muscle movements also affect the control on vocal cords. All of these cause different changes in voice.

6. **Diseases**: Diseases of the respiratory system such as **lung cancer**, **lower respiratory tract infections** etc. are known to cause changes in voice. Problems in the vicinity of the vocal tract, such as **sinusitis**, **tonsillitis** etc. significantly affect voice. **Systemic illnesses** such as **anemia**, **Lyme disease**, **mononucleosis**, **chronic fatigue syndrome**, or other diseases associated with malaise and weakness impair the ability of vocal musculature to recover from heavy or long-term use (such as singing or prolonged conversation), alter mucosal secretions in the vocal tract, and reflect in the quality of the voice.

   Conditions such as **cerebral palsy** [130], even **gastroesophageal reflux** and **dehydration** [131, 132] cause changes in voice.

7. **Other medically relevant conditions**: **Physical abnormalities** that affect the vocal tract, **speech and vocal disabilities** that are congenital or acquired, and many other generic medical conditions affect the voice. Voice and neonatal issues have been linked [133]. Voice changes during **pregnancy** are medically known as *Laryngopathia gravidarum*, and can become permanent.

   Trauma, Lesions and injuries [134], **Medical procedures** such as chemotherapy and radiation therapy [135], many surgical procedures [136, 137] etc. have also been observed to affect the voice.

8. **Psychological conditions**: Many psychological conditions affect the voice, and have been discussed elsewhere in this chapter. Some examples are **Asperger's syndrome** [138], **Autism** [139], **Anxiety** and so on. Although anxiety is different from stress, the symptoms are much the same. Stress (even that induced by lying in specific situations) causes changes in voice patterns at a very fine levels—by altering the stiffness of the vocal folds, and by affecting laryngeal muscle movements.

9. **Psychiatric illnesses**: Almost all psychiatric illnesses are likely to have biomarkers in voice since speech and voice are both mediated intricately by, and affected by, activities in multiple centers of the brain, as discussed in Chap. 2. Examples of psychiatric conditions that have been correlated with voice changes are **depression** [140, 141], **schizophrenia** [142], **suicidal risk** [143–145] etc.
   The changes noted for each type of illness can range from one to several characteristics of the voice signal. For example, in the case of depression, breath patterns change, silences are inappropriate in the linguistic context, syllabic stress is reduced, short rushes or bursts of speech are noted, voice quality is found to be harsher and breathier, voice is quieter in general (lower SPL)—the loudness-decay is faster, prosodic anomalies are noted even by human listeners (interpreted by humans a "reduced vitality"), and so on. The lowered variation in loudness gives an impression of reduced respiratory drive. In addition to all of these, there is a tendency to trail off at ends of utterances (before breath intake), pitch inflection and stress patterns are found to settle into a sing-song repetitious drone etc. Similarly, studies have measured many aspects of voice in the context of other psychiatric illnesses (not mentioned here).
   Under controlled conditions, **suicidal risk** classification through voice can reportedly achieve good (>80%) accuracies with automated methods that simply use speech-silence ratio as a transition parameter between normal and suicidal states [143]. This is also true for depression detection.

10. **Genetic disorders**: Many genetic disorders affect voice, each in different characteristic ways, such as **Down's syndrome** [146], **Ehlers–Danlos syndrome** [147], **Prader–Willi syndrome** [148] etc.

3.5 Psychological Parameters

Personality and Emotions are the two broad categories of parameters that have been widely studied in the context of voice. These are rather complex entities and cannot be categorically mentioned in the profiling context without establishing some background for discussion. The goal of this section is to provide such a basis. The aspects of voice that are affected by psychological factors mostly fall within the purview of voice quality, which is discussed in detail in Chap. 6.

3.5.1 Personality

Quoted directly from the information released publicly by the American Psychological Association, and according to the Encyclopedia of Psychology: "*Personality refers to individual differences in characteristic patterns of thinking, feeling and behaving. The study of personality focuses on two broad areas: One is understanding individual differences in particular personality characteristics, such as sociability or irritability. The other is understanding how the various parts of a person come together as a whole.*"

3.5.1.1 The Five Factor Model of Personality

In 1990, Goldberg propounded the **five-factor model** (also known as the **big-five model**) for personality [149], which is now widely accepted in the scientific community. This model has been supported by many later studies, and is thought to be *human universal*, i.e. to generalize across all cultures [150–152].

In the five factor model (FFM), five factors comprise personality in its entirety. These are:

1. **Agreeableness**: This is equated to the qualities of being good-tempered, helpful, friendly, compassionate, cooperative, as opposed to being challenging, suspicious, antagonistic, detached, competitive, and untrustworthy. Too much of this trait is viewed as submissiveness.
2. **Conscientiousness**: This is equated to being efficient, organized, dependable, disciplined, dutiful, striving towards achievement, planned etc., as opposed to being easy-going, careless and capricious. Too much of this trait can be perceived as stubbornness and obsession. Too little can be perceived as lack of reliability and sloppiness, though sometimes also as flexibility and spontaneity.
3. **Extraversion**: This trait is characterized by energy, positive emotions, surgency, assertiveness, sociability with the tendency to seek stimulation in the company of others, and talkativeness. High extraversion is often perceived as attention-seeking and domineering. In contrast, low extraversion is perceived as a reserved, reflective personality, which can be further perceived as aloof or self-absorbed. Extraverted people tend to be more dominant in social settings, as opposed to introverted people who may be shy and reserved in such settings.
4. **Neuroticism**: Neuroticism as a trait reflects a person's propensity to psychological stress, or the tendency to allow the effects of personal experiences to manifest easily as anger, anxiety, depression, and vulnerability. Such people are considered to have low emotional constancy, low impulse control and low stability. To

put the latter in perspective, people who are perceived as highly stable are perceived as calm, and often uninspiring and unconcerned. In contrast, people who have low stability are perceived to have a reactive and excitable personality—on one hand they are considered to be dynamic, on the other hand unstable or insecure. Individuals with higher levels of neuroticism also tend to be in worse psychological health.

5. **Openness to experience**: Openness implies a readiness to experience emotion, adventure, and unusual ideas. This trait is characterized by curiosity, attraction to novelty, high intellectual curiosity and imagination, and often appreciation of intellectual and artistic pursuits. People high in this trait prefer to be independent, engage in a variety of activities and do not remain confined to strict routine. High openness can however also be perceived as unpredictability or lack of focus: people with this trait are also likely to engage in risky behavior such as drug abuse, and seek fulfillment through intense, euphoric experiences. Conversely, those with low openness seek to gain fulfillment through perseverance and deliberation, and are characterized as "pragmatic and data-driven," sometimes even perceived to be dogmatic and closed-minded.

The traits in FFM are further broken down into multiple sub-factors in a widely accepted NEO model [153] that is used in clinical practice, as a checklist for judging potential job success, for job profiling, to generate quantitative personality scores [154] etc. This model is simple, and provides six *facets* to each personality factor in the FFM as shown in Table 3.1.

Personality is thought to remain relatively stable over adulthood [155] and varies across genders [156], but can be moderated by circumstances, trauma, experience and other (usually extreme or intense) factors. Personality factors have been known to regulate other psychological traits such as emotion [157] and mood [158]. They have been extensively linked to behavioral factors [159] as well. Surprisingly, personality factors are also thought to be an inherited [160–163]. In addition, studies have shown that different traits of personality are inherited to different degrees [164].

3.5.1.2 Personality and Common Mental Disorders

The five-factor model of personality is now widely used in gauging various mental disorders, and risks associated with them, e.g. drug consumption risk [165]. Figure 3.6 (reproduced from [165]), shows the average deviation of personality factors of heroin users from the population mean.

In this context, it is instructive to know what common Mental Disorders (CMDs) are, and how they are categorized. In current schools of thought, there are three categories of mental disorders that are seen in general populations. These are

- **Depressive disorders**: These include disorders like major depressive disorder (MDD), dysthymic disorder etc.
- **Anxiety disorders**: These include disorders like generalized anxiety disorder (GAD), post-traumatic stress disorder (PTSD), panic disorder, phobias etc.

Fig. 3.6 Personality factor trends for heroin users

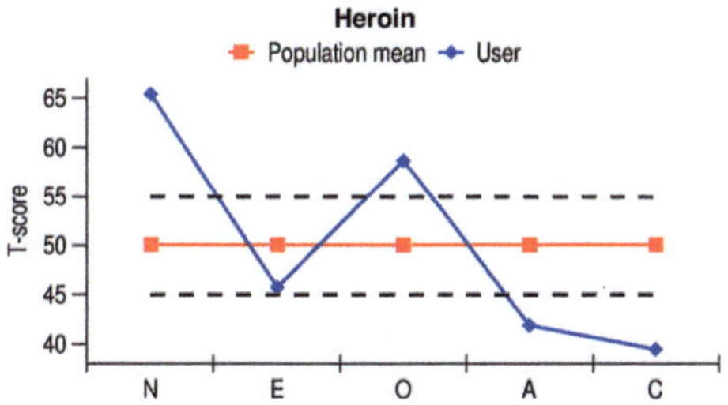

Deviation of personality factors from population mean values for heroin users. O,C,E,A,N (OCEAN) stand for Openness, Conscientiousness, Extravertness, Agreeability and Neuroticism respectively.

Table 3.1 Personality factors and their facets according to the NEO-P1-3 [154] model

Factor	Facets
Agreeableness (A)	A1: Trust, A2: Straightforwardness, A3: Altruism, A4: Compliance, A5: Modesty, A6: Tender-mindedness
Conscientiousness (C)	C1: Competence, C2: Order, C3: Dutifulness, C4: Achievement-striving, C5: Self-discipline, C6: Deliberation
Extraversion (E)	E1: Warmth, E2: Gregariousness, E3: Assertiveness, E4: Activity, E5: Excitement-seeking, E6: Positive emotions
Neuroticism (N)	N1: Anxiety, N2: Angry hostility, N3: Depression, N4: Self-consciousness, N5: Impulsiveness, N6: Vulnerability
Openness (O)	O1: Fantasy, O2: Aesthetics, O3: Feelings, O4: Actions, O5: Ideas, O6: Values

- **Substance use disorders (SUD)**: These include addictions to different drug types, such as *opiates* (painkillers that cause drowsiness, and intense euphoria—feeling of well-being, happiness, joy etc.), *stimulants* (chemicals that stimulate the brain and nervous system and cause intense focus and achievement, such as drugs that treat ADHD), *depressants* (substances that cause drowsiness and reduce anxiety), and *hallucinogens* (chemicals that cause hallucinations). Opiates include substances such as heroin, opium, codeine and some medical painkillers. Stimulants include cocaine, amphetamines such as methylphenidate, ritalin etc. Depressants include alcohol, barbiturates, benzodiazepines, chloral hydrate, paraldehyde etc. Hallucinogens include LSD, Marijuana (cannabis, or hashish), mescaline, psilocybin ("mushrooms"), PCP (phencyclidine or "angel dust") etc.

The risk of developing certain CMDs has been positively linked to specific personality factors, such as neuroticism [166, 167]. These include anxiety, depression, substance abuse, psychosis, schizophrenia etc. [168]. Other studies have found that low conscientiousness significantly increases the risk of CMDs [169] such as panic disorders, phobias and SUD. Different personality factors have also been correlated to increased or decreased mortality. There is in a fact a vast literature on the subject of CMDs and personality factors in existence today.

Currently, there are five key models that prevail simultaneously (there is no established "best" model), that attempt to relate personality and CMDs. Often, more than

Table 3.2 Examples of emotions, affects and related terms

Category	Exemplars
Emotion	Anger (and all emotions named on the Plutchik model)
Cognitive state	Doubt
Trait	Competitive (and all personality factors named in the NEO model)
Behavior	Cry
Attitude	Skepticism
Feeling	Pleasure
Mood	Pensive

one of these may be used to explain a person's observed mental state. These models [169–171], and their key contentions are as follows:

- **The Vulnerability/Risk Model**: Pre-existing personality factors cause or enhance the causal factors for CMDs. Of the five factors, neuroticism has been linked to the greatest degree of vulnerability to CMDs.
- **The Pathoplasty Model**: Pre-morbid personality factors influence the expression, course, severity, and/or treatment response of CMDs [172]. A depressed person with low levels of constraint is likely to become more suicidal.
- **The Common Cause Model**: Personality factors are predictive of CMDs (even if there is no causal relationship between the two) since both have shared genetic and environmental determinants.
- **The Spectrum Model**: CMDs are part of the spectrum of personality factors—they are simply a display of extremes of normal personality function.
- **The Scar Model**: Episodes of a CMD 'scar' an individual—changing their personality factors in significant ways.

Of interest to us in the context of profiling are the studies that have linked various personality factors and their sub-factors to different characteristics of the voice signal, as well as to CMDs and their different associations with personality factors. This forms a circular chain of relationships that can be exploited for profiling. In the paragraphs below we give sparse overview of these studies.

3.5.1.3 Voice Characteristics and Personality Factors

Studies that establish correlations between voice and different personality traits date back nearly a century, if not more. In a multitude of studies by Fay and Middleton, and some other scientists, the different traits that were investigated and found positively correlated to voice early in the last century included **intelligence** (c. 1939) [173], **Spranger personality types** (c. 1939) [174], **sincerity** (lies) (c. 1941, 1953) [175, 176], **general personality** (c. 1931) [177], **leadership** (c. 1943) [175], and **sociability** (c. 1941) [178], **introversion** (c. 1942) [179], among others. Since then, there have been thousands of studies carried out in this context, and the understanding today is sufficient to build automated systems for profiling humans for personality traits (*and* for recognition of CMDs) at least at a basic level.

In more recent times, voice has been measurably positively linked to the perception of personality factors and sub-factors such as **benevolence** [180], **competence** [181], **overall personality type** [182–184], **intellectual capacity** [185] etc.

The key characteristics that have been most related to personality are voice quality (this was in fact one of the earliest investigated entity [186]), its sub-components, speaking rate, prosodic components such as pitch, energy and formants, and myriad others. For example, a study conducted on schizophrenics and non-schizophrenics found that the two states or conditions could be discriminated based on voice quality alone, and that voice quality could also be correlated with (in these populations) specific impressions of a speaker's physical characteristics and demeanor [187]. In an early exemplary study [188], speakers were trained to emulate seven different sub-qualities in their voice: breathy, tense, thin, flat, throaty, nasal, and orotund, combined with three different speaking rates: slow, normal and fast, and pitch variations between normal, more, and less than normal. The study conclusively showed that depending on different combinations of these, human perceptions of personality traits were different for the same persons. The traits perceived had high reliability scores, ranging from 0.94 to 0.64, and the interesting list of traits included the following: feminine-masculine, young-old, enthusiastic-apathetic, energetic-lazy, good looking-ugly, cooperative-uncooperative, unemotional-emotional, talkative-quiet, intelligent-stupid, interesting-uninteresting, mature-immature, polite-boorish, educated-uneducated, convincing-unconvincing, well adjusted-neurotic, sensitive-insensitive, sense of humor-no sense of humor, jovial-morose kind-cruel, romantic-unromantic, tall-short, sophisticated-naive, active-passive, proud-humble, sexy-sexless, orderly-disorderly, careful-careless, artistic-inartistic, strong willed-weak willed, idealistic-realistic, conventional-unconventional, brave-cowardly, rich-poor, sincere-insincere, skinny-fat, self respecting-servile, healthy-sickly, law abiding-criminal, honest-dishonest and extraverted-introverted. These are listed here for a reason: presumably, with sufficient advances in profiling technology, *all* of these may be found deducible from voice.

Another example of a characteristic that has been widely linked to personality traits is speaking rate [189–191]. In one study [189], for example, voices of six subjects, representing various social and educational backgrounds, were presented at different rates. Perceptions of competence increased linearly with speaking rate, perceptions of benevolence peaked at normal speaking rate, going down for other speaking rates (slow or fast) rapidly.

Yet another example of a characteristic of relevance in this context is prosody. In a study [192] which investigated the effects of pitch, formants, energy and speaking rate on (the automatic attribution of) personality traits in the big-five inventory, it was found that these prosodic features can predict certain personality parameters with up to 75% accuracy. These correlate well with those perceived by humans even if they do not understand the language spoken.

Automated profiling for personality, however is likely to be less straightforward, since even at the level of human perception, personality is an extremely subjective characteristic. To support the subjectivity argument, we refer to studies that have shown that factors other than voice quality that are more subjective in nature affect personality judgment, e.g regional dialect [193], gender of judge [194], intelligibility of speech [191]—this changes with accent, and unfortunately is linked to the perception of the speaker's competence and competence-related traits, such as leadership qualities, hard-working, well-educated etc.

The cues for personality are strong enough in voice that people with voice disorders (such as functional dysphonia, vocal nodules, spasmodic dysphonia, unilateral vocal fold paralysis etc.) are sometimes adversely affected [195] in how other people perceive their personalities.

3.5.2 Emotions

The cues that allow for the perception of emotions are not completely encoded in linguistic syntax. Studies have shown that they are more ubiquitously encoded in the voice signal. Emotions affect the voice production mechanism. The laryngeal nerves that feed the muscles of the larynx have connections to the vagus nerve, among others. The vagus nerve has been highly implicated in the body's response to emotion. There are many other reasons why emotions are expected to have measurable effects on voice, but to better understand those, it is important to understand the nature of emotions in a biological sense.

What is an emotion? To quote from [196], "*Therefore, in the light of the psycho-evolutionary structural theory of emotion, the following definition is proposed. An emotion is an inferred complex sequence of reactions to stimulus, and includes cognitive evaluations, subjective changes, autonomic and neural arousal, impulses to action, and behavior designed to have an effect upon the stimulus that initiated the complex sequence.*" Other researchers often define emotion by its associations to other related entities, e.g. [197]: "*Emotion is a psycho-physiological process triggered by conscious and/or unconscious perception of an object or situation and is often associated with mood, temperament, personality and disposition, and motivation.*" A study in fact compiled 92 different definitions of emotion in an attempt to find a consensual one [198].

In simpler terms, generally, emotions can be said to be mental states that are brief in duration and influence our behavior. They are a result of, and cause changes in, physical and physiological states of a person [199]. Loosely, they are a range of feelings that are experienced by humans on a continuum of intensity, according to some researchers [200]. Physiologically, they are related to the state of the nervous system, and affect many functions of the body as well.

3.5.2.1 A Compendium of Emotions

Although there is no scientific consensus on the precise definition of emotion, there have nevertheless been many attempts to categorize emotion and to build an ontology of emotions.

At the outset, emotions must be distinguished from what are recognized as *affective* states (pleasure, pain etc), motivational states (hunger or curiosity), moods, dispositions and traits. They are however thought to be related to or even facets of these [201]. The classification of emotions is thus complex and ambiguous. There are two views that apply to the categorizations of emotions. In one, emotions are considered to be discrete, measurable, and physiologically distinct entities on their own. In the second, emotions are characterized as groupings along different dimensions of *basic* emotions, and (possibly) influenced by entities such as the ones mentioned above.

The first view was propounded and supported by the studies of Paul Ekman and others, based on which he proposed that there are six basic emotions: anger, disgust, fear, happiness, sadness and surprise [202]. For the most part, the key characteristics that make them distinct and make them fit all criteria in the first viewpoint are that (a) they evoke distinct physiological reactions, (b) conversely, they are also resultant from them, and (c) they are universal across cultures and recognized by people through associated facial and body expressions without prior learning (e.g. learning based on advanced mechanisms such as observation of emotional enactments).

The second point of view led to the development of more detailed models of emotions. One of them, by Robert Plutchik [203] has been widely accepted. Figure 3.7 shows the model he proposed, drawn slightly differently using concentric octagons for ease of representation ("web of emotions"), although note that the clear linear boundary between emotions in this figure misrepresents Plutchik's contentions. In reality, and also according to his theory and model, there is no abrupt transition between any two adjacent emotions, and that is possibly the reason for the manner in which he represented his "wheel of emotions," or the circumplex model of emotions, that propounds that emotions are distributed in a two-dimensional circular space, containing arousal and valence dimensions (discussed later in this section) [204].

Plutchik's circumplex model is an elegant one, wherein the eight basic emotions could be seen as more intense versions of others that could in turn be grouped into very plausible categories of synergistics and opposites, according to *valence* (positive or negative), e.g joy versus sadness, ecstasy versus grief etc. In Fig. 3.7, valence is represented along the vertical axis. The horizontal axis represents *arousal*, or degree of intensity, diminishing in both directions from the center of the figure. Thus apprehension can be seen as a milder version of terror, annoyance as a milder version of rage etc. The entire space of emotions can be traversed and, guided by the valence and arousal values, can be seen to cluster into focal points that are identified as other emotions that are named on the web in Fig. 3.7, clustered at the centroid of each cell.

Note that there have been many other theories of basic emotions and their variations, some of which are noted below. For the purpose of this book, we will follow the Plutchik model.

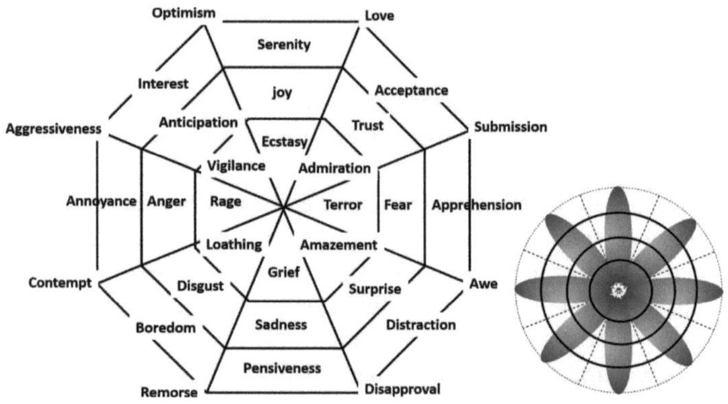

Left: Depiction of Plutchik's model for emotions as a *Web of emotions* **Right:** Layout of the original Plutchik model, the emotion labels appear in regions of one-to-one correspondence in both depictions.

Fig. 3.7 Plutchik's model for emotions

Computational analysis of emotions was initially largely confined to *affective computing* [205]. Affective computing involves using visual and auditory sensors to capture human data (facial expressions, gestures, postures, voice etc.) to automatically gauge a person's *affective state*. Affective states, however, are not completely coincident with emotional states. In fact, while we discuss affect, we must at the same time differentiate between emotions, feelings, mood and affect. Feelings are subjective to an individual, and refer to how a person feels as they experience any emotion. Mood refers to what translates into behavior that can be perceived by others (or described as such by the individual) and is correlated to how a person feels. There is no precise definition for mood. Affect, in psychology, describes observable behavior that represents the expression of a subjectively experienced feeling state (emotion). Examples of *types* of affect are euthymic, irritable, constricted, blunted, flat, inappropriate, labile etc. However, examples of affect span a broad range of entities. Some examples of affect are the same as emotions: happiness, sadness, fear, joy etc. Others are not emotions, but closer to emotional *state* or mood of the person, such as bored, relaxed, elated, stressed. While often the term is used interchangeably with emotion, affect does not factor in theories of basic emotion. Affect is more of a *derivative* of those theories—more of a concept used in neuroscience and computing. Part of the reason may be that it was introduced as a derivative of other factors at a much later stage in the order of chronological developments in human psychology.

Another factor to consider in this context are *dimensions* of emotions. It is now common practice to describe affective states, emotions etc. in a three dimensional space of *valence*, *arousal* and *dominance*. Valence is the positivity or negativity of an emotion (or related entity). Arousal is the level of activation, or intensity. Dominance (along with tension and kinetics) relates to the psychosomatic aspects of

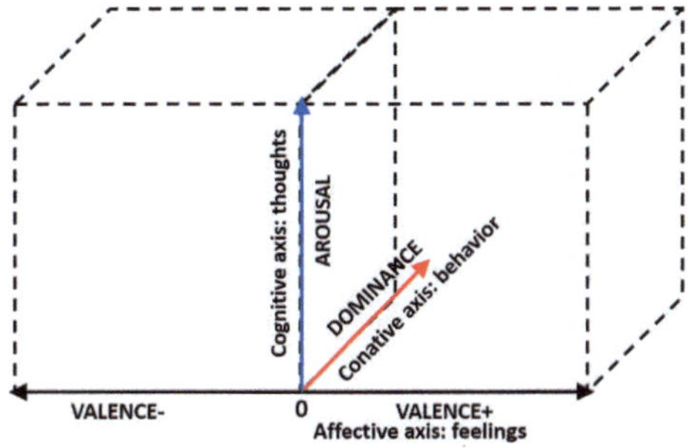

Fig. 3.8 Dimensions of emotions: dominance, valence and arousal

emotions. In a very loose sense, dominance is the level to which an emotion prevails over other competing entities that affect humans at the same time (behaviorally, physiologically, psychologically or in other respects) These dimensions are shown in Fig. 3.8, fashioned after the figure suggested in [206]. The definitions/descriptions of arousal, valence and dominance are as in [207].

Plutchik's model is just one of many models of emotion that have been proposed, and the theories of emotion mentioned in this section are only a couple of the many theories proposed—as an example of those that were left out, we mention the Hourglass model of emotions [208].

3.5.2.2 Emotions and Voice Characteristics

Voice is significantly involved both in the simulation [209] and expression of emotion [210], and also in the perception of emotion. Studies have shown that listeners' rating of emotions from voice are sixfold better than chance [211]. Research on the relationship between voice and emotions in fact spans over eighty years.

Emotions affect the physiology of a person. The effects also extend to the speech production mechanism, and the nervous system, including the brain. The process of phonation is affected by emotions [212]. Multiple studies in the literature support this. To quote directly from one [213]: "*Emotional state affects the physiological mechanism involved in phonation.*"

Why do emotions cause changes in speech, specifically? Why do emotions affect the physiological mechanism involved in phonation? A speculative answer may be found in the manner in which they change the levels of neurotransmitters in the brain. In 2012, based on multiple studies, Lövheim proposed that there seemed to be a direct relation between the levels of the neurotransmitters dopamine, noradrenaline and serotonin, and the eight basic emotions represented by the model proposed by Sivan Tomkins. The relationships were represented as a *cube of emotions*, as shown in Fig. 3.9. This model suggests, for example, that enjoyment or joy are associated with high dopamine and serotonin levels, and very low noradrenaline levels. The reason why this model is critical in understanding why voice, or phonation, is affected, is that these neurotransmitters have been linked to the speech production process in

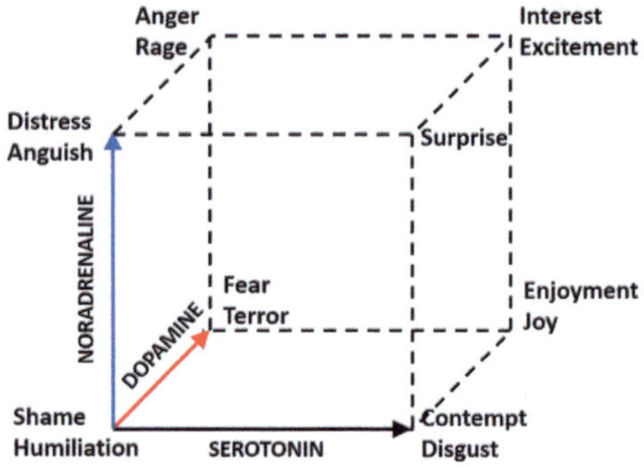

Fig. 3.9 Lövheim's cube model depicting the relationship between the neurotransmitters noradrenaline, dopamine and serotonin and eight basic emotions

myriad studies in the literature. In the light of these studies, a very strong relationship exists between voice and neurotransmitter levels, and given this model, between neurotransmitter levels and emotions. It is no surprise then that other studies have found strong correlations between various voice characteristics and emotions.

Not all researchers are in agreement with this, though. Some studies are not supportive of the relation between voice (especially voice quality) and strong emotions, but these studies are themselves not quite realistic, and have not ruled out links between emotion and voice. For example, in one study [214], the researchers synthesized seven different voice qualities: harsh, tense, modal, breathy, whispery, creaky and lax-creaky using a formant synthesizer. These were then presented to listeners to gauge perceived affect or emotions. The listeners did not perceive *strong* emotions, though they did positively perceive emotions. This may be due to the fact that synthesized voice quality is not really realistic in all aspects. The positive results in this study indicated that breathiness was associated with anger and happiness, sadness was associated with increased resonance etc. There have been many studies that relate different entities in the Plutchik range of emotions to characteristics of the voice signal. Most show that variations in the same underlying voice characteristics are correlated with different emotions e.g. common-cause variations in joy, anger, sadness, fear are named in [215].

The relationship between the *perception* of emotions, and voice quality, has been investigated by many studies, and is strongly supported by them, e.g. [216]. Different emotional states also bring about significant changes in the *style* of speaking. While there are no clear measures of style, it is generally thought that style is encoded largely in speech prosody. Ergo, emotions must also be strongly encoded in prosody [217].

In addition, how the voice *conveys* emotions is influenced by other correlates, such as gender and age. Studies have shown that female voices convey emotions more reliably than males [218] and that females encode emotions better, and are much better than males at decoding emotions in voices [219].

3.6 Behavioral Parameters

The Merriam-Webster dictionary defines behavior as *"a: the manner of conducting oneself, b: anything that an organism does involving action and response to stimulation, and c: the response of an individual, group, or species to its environment (e.g. They are studying the behavior of elephants in the wild")*.

Behavior can be categorized in many ways depending on the underlying criteria, e.g. patient behavior, child behavior, customer behavior etc. At the core, these categorizations are based on behavioral *traits*, many of which have been identified as distinct in the scientific literature. Examples of such traits include anxiety, impulsiveness [220], drive, motivation, ambition, energy, tenacity, integrity, honesty, self-confidence, cognitive ability, charisma, creativity, flexibility etc. and their emergents such as leadership [221]. There is no standardized list of traits though, and different researchers adhere to different lists. Behavioral traits have been found to be inherently different in males and females. For example, some studies have indicated that females have greater potential for creativity in scientific and other fields [222].

Due to the complexity and depth of the topic and the ambiguities therein, it is beyond the scope of this book to go deeper into behavioral theories and comprehensive lists of behavioral traits for humans. However, when relevant, the term "behavioral trait" will be used synonymously with a "behavioral parameter" that has been studied in the context of voice, and for that reason, can be potentially profiled.

3.6.1 Why is Voice Influenced by Behavior?

A speculative answer to the question of why voice has been found to correlate with behavior may lie in the known fact that behavior is influenced by the state of the nervous [223, 224] and endocrine [225, 226] systems in humans. From an evolutionary viewpoint, there is consensus that the complexity of an organism's nervous system is indicative of its behavioral responses and capacity to control them, or to adjust them by learning new behavioral responses [227, 228]. In other studies, voice has been conclusively shown to be influenced by changes in both the nervous and endocrine systems (discussed earlier in this chapter). The relation between the two (voice and behavior) is thus apparent through the commonality of these causes. There are many other reasons why positive correlations between voice and behavior are not surprising. For example, voice and behavior are both thought to have a genetic basis at their core. There have been many studies in support of this [229, 230].

In another set of examples, while on one hand voice has been linked to myriad aspects of a person's physical and mental health, on the other hand behaviors have been linked to multiple physical and mental disorders that affect the health of an individual. In other words, behavior affects health, and health affects voice. For instance, several disorders have been linked to the tendency to internalize or externalize behavior. People who, due to fear or inordinate levels of perceived "self-control," do not exhibit natural behavioral impulses, even when they are socially acceptable, are said to *internalize* their behavior. Disorders that correlate to internalizing behavior include depression, apathy, non-suicidal self-injury, substance abuse, anxiety disorders, certain phobias, panic disorders, eating disorders, obsessive-compulsive disorders etc. In contrast, people who externalize behavior to excessive degrees develop the personality traits of aggression, and are thought to be non-compliant and socially disruptive. Correlations have been shown to exist between externalizing behavior and conditions like bipolar disorder, attention deficit hyperactivity disorder (ADHD), oppositional defiant disorder (ODD) etc. In several instances, voice has been found to respond to these disorders.

In the light of all these considerations, it is reasonable to expect that specific types of voice, or voice characteristics, may correlate with behavior. In addition, *correlates of behavior* [231], such as diet [232] and exercise habits, smoking and drinking habits, sleep patterns etc. are also likely to be amenable to inference from voice, and some studies have indeed found correlations between voice and these entities.

Studies that have related voice to behavioral parameters include parameters like perception of **dominance** [233], **dynamism** [234–236], **leadership** [237, 238], **public/private behavior**, **voting behavior** [239], **sexual behavior** [240], **sexual orientation** [241], **normal and abnormal behavior** [242], **opposite sex bias** [15, 243], etc. For the most part, voice characteristics that have been correlated to these parameters relate to voice quality and its sub-components, discussed in Chap. 6.

Fundamental frequency plays a key role in this the perception of leadership and dominance. People with lower $F0$ have been found to be more likely to obtain leadership positions. This also gives precedence to men over women [237]. Lower voice pitch in both females and males has been linearly, and positively correlated with dominance [244]. Low-pitched voices lead to a perception of competence and strength [245], two highly desirable characteristics of leaders.

3.7 Sociological Parameters

Voice has been linked to sociological parameters from very early times, e.g. to **occupation** (c. 1940) [246]. One of the more consistently studied sociological parameters in the context of voice is **social status**. Even in smaller settings such as within an office, a study found that supervisors could place their employees' status in the hierarchy by their voice [247]. Managerial voice affective states have been correlated with **success in business** [248]. **Work environment** [249, 250] and **quality of life** [251] are other examples of sociological parameters linked to voice.

Voice, in social settings, is not easily controlled and gives away more social cues than facial expressions do [252–254]. For example, it has been found to be revealing in the case of **deception** in social settings. Deception, however, is one out of a set of parameters that *somewhat* belong to the sociological category, but in reality span multiple categories. Deceptive behavior can both be seen as a behavioral trait, and as a sociological (sociopathic) parameter.

Income or **socio-economic status** [255], and profession have also both been correlated to voice in other studies. In the former case, the correlations are explained based on "*voluntary long-term muscular settings of the larynx and vocal tract*" [256] that are epigenetic developments in humans as a result of their environment (social and physical). Other reasons have also been mentioned in these reports.

3.8 Environmental Parameters

The environment affects the voice signal in two ways: by affecting the speaker in a way that the voice production mechanism is affected, and by modifying the voice signal *after* it is generated (e.g. acoustic effects on the sound recoding, and instrumentation/transmission related changes).

In this section, we discuss the effect of the environment on the voice as generated by the human, rather than on the voice signal after it is produced.

The physical environment of the speaker directly causes changes in the voice through the effects of multiple factors such as humidity, pressure, directly inhaled atmospheric toxins etc. on the human body. Many of these effects have been investigated in the context of space research, environmental pollution, environmental phenomena, disasters etc.

The speaker's physical environment also exerts *indirect* influences on voice. This follows a chain of relationships, which begin with the effect of the environment on the speaker's psychology and behavior (quote from [257]: "*environmental stimuli are linked to behavioral responses by the primary emotional responses of arousal, pleasure, and dominance*"). As we have discussed in earlier sections in this chapter, psychological and behavioral responses can affect the physiological and mental processes within the speaker's body, which in turn can cause changes to the vocal mechanism.

Some key environmental parameters and their effects on voice are mentioned below.

1. **Geomagnetic activity**: The term "geomagnetic activity" refers to geomagnetic storms, or solar storms. Such a storm happens when a solar wind—a stream of charged particles from the sun's corona, comprising electrons, protons and alpha particles with thermal energy between 1.5 and 10 keV—impinges the earth's magnetosphere, causing increased electric field strengths and plasma movements inside it. This results in an increase in electric current in the magnetosphere and ionosphere. According to a recent study carried out in the context of space research, this effect of geomagnetic activity causes changes to physiological parameters of humans. Notably, the study [258] found that arterial blood pressure of the study group increased significantly (as much as 9%) with increase in geomagnetic activity, regardless of gender. Increased blood pressure has conse-

quences on the voice, and correlations have been found between voice and blood pressure as part of aging studies [82]. Conversely, vocal behavior can increase blood pressure and cardiovascular reactivity [259]. The links between voice and blood pressure, however, need to be further explored.

2. **Atmospheric pollution**: It is well known that substances that cause atmospheric pollution are toxic to humans at different levels. The effect of many extend to causing changes in voice [260], although the number of studies that have explored such cause-effect relationships is relatively small. Direct effects are caused by injury to the larynx, parts of the vocal tract and to the respiratory system (chronic injuries to the airway, pulmonary dysfunction, constrictions of the trachea or bronchi etc.) [261, 262]. Indirect effects of neurological injury may also extend to the larynx and vocal tract, and cause changes in voice. When pollutants cause sudden physiologic reactions such a coughing, it may result in vocal fold injury, tearing, haemorrhage, inflammation, and damage to the tissue of the vocal folds, muscles and ligaments. The resultant changes in voice manifest as changes in voice quality, such as increased breathiness, roughness, hoarseness, etc. They also result in changes in the acoustic intensity and other characteristics of the voice signal. Often, in such cases, voice reflects vocal fatigue. Smoke inhalation during a fire can cause many such effects expeditiously.
3. **Noise pollution**: Noise pollution causes voice abnormalities. By association, the sound-emitting objects in a speaker's environment can cause elevated levels of noise, and influence the way a person speaks. This is a well known phenomenon, called the Lombard effect [263, 264], wherein the voice characteristics are changed instinctively by humans to be intelligible above the noise in the environment. If the noise is confined to certain frequency bands, it is likely that a speaker will try to "evade" the band by altering their voice characteristics. Noise pollution also affects voice indirectly by affecting human hearing (and at sufficiently high and sustained levels can cause damage to the body or ears). Damage caused to the ear hampers the speaker's acoustic feedback mechanism which the brain uses to adjust the quality of speech during production, which can cause changes and disabilities in this process. Other effects of noise are linked to the body. Infrasound, for example, is not heard by humans and does not damage human hearing, but is nevertheless felt by the body and can cause serious sickness if sustained long enough, the changes in voice occur due to the secondary effects of the sickness induced by infrasound.
4. **Mobile and static physical placement**: It is often easy to discern that a speaker is on a moving platform while speaking, such as in a moving car, a boat, on a bicycle, a treadmill, etc. There are two ways in which motion is implicated in bringing about changes in voice characteristics:
   One is by affecting the feedback mechanism between voice production and the brain, mediated by the ear. The mechanism is hampered by issues with balancing and changes in sound characteristics that are heard and fed back to the brain as a result of the motion.

The second is through the instabilities in voice production caused by the moving platform: in this case, presumably the Coanda effect [265] comes into play. The effect causes asymmetric airflow in the glottis and results in a difference in the driving force on the left and right vocal folds, even if they have *quote: "left-right symmetric prephonatory shape and tissue properties."* As a consequence, pressure differences across the left and right vocal folds induce displacement asymmetry. This effect is well explained in various models of vocal fold kinematics, e.g. [266], that can also account for the additional affects of asymmetry-inducing physical displacements, related to motion.

5. **Clutter/Neatness**: clutter and other descriptives for the speaker's environment (neat, messy, clean, dirty, filthy, sloppy, etc.) affect a person psychologically [267], and cause secondary effects on voice. Counseling rooms, for example, are designed based on the desirable attributes of spaciousness, natural light, pleasing decor etc., since these are observed to have a calming effect on the patient. The effect of these choices on voice have also been observed—the voice takes on a "calm" quality (the opposite in some environments, wherein it sounds "agitated"). Evidence of such links, though, is still tenuous for lack of appropriate studies.

6. **Atmospheric conditions (pressure, moisture, relative humidity)**: When the mouth (or nasal passages) and glottis are open, the alveolar pressure equals atmospheric pressure [268]. As the atmospheric pressure changes, so does the lung pressure [269] (to facilitate respiration). This can cause changes in the phonation patterns of the vocal folds, since their self-sustained vibration is critically dependent on air pressures in the sub-glottal and supra-glottal regions. Voice changes with increased barometric pressure have been noted in many studies [270]. Similarly, changes in relative humidity can cause changes in voice [271–274].
   In the context of professional voice artists who travel by air, several factors that are different in an aircraft cabin environment have been thought to cause changes to voice [275]. These include ambient noise levels, cabin temperature, pressurization, humidity, and ozone. Outside the cabin environment, these can also potentially cause changes to voice, when present in deviant amounts.

7. **Pleasant and unpleasant odors**: Surprisingly, the fundamental frequency of voice changes with odors [276]. For an individual, it is higher in the presence of pleasant odors, and the opposite is true for bad odors. This may have implications for people who wear pleasant perfumes—their voice pitch is slightly higher, and especially in the case of women, more attractive, according to the studies mentioned earlier in this chapter.

We close this chapter on this pleasant and fragrant note. In the next three chapters, we will discuss the manifestation of all of these influences on the voice signal, in terms of its information content and quality.

References

1. Massaro, D. W. (1998). *Perceiving talking faces: From speech perception to a behavioral principle* (Vol. 1). Cambridge, Massachusetts: MIT Press.
2. Rendall, D., Vokey, J. R., & Nemeth, C. (2007). Lifting the curtain on the Wizard of Oz: Biased voice-based impressions of speaker size. *Journal of Experimental Psychology: Human Perception and Performance*, *33*(5), 1208.
3. Titze, I. R. (1989). Physiologic and acoustic differences between male and female voices. *The Journal of the Acoustical Society of America*, *85*(4), 1699–1707.
4. Fitch, W. T., & Giedd, J. (1999). Morphology and development of the human vocal tract: A study using magnetic resonance imaging. *The Journal of the Acoustical Society of America*, *106*(3), 1511–1522.
5. Sergeant, D. C., & Welch, G. F. (2009). Gender differences in long-term average spectra of children's singing voices. *Journal of Voice*, *23*(3), 319–336.
6. Ryalls, J., Zipprer, A., & Baldauff, P. (1997). A preliminary investigation of the effects of gender and race on voice onset time. *Journal of Speech, Language, and Hearing Research*, *40*(3), 642–645.
7. Swartz, B. L. (1992). Gender difference in voice onset time. *Perceptual and Motor Skills*, *75*(3), 983–992.
8. Sundberg, J., & Sataloff, R. T. (2005). *Vocal tract resonance* (pp. 275–292). San Diego, California: Plural Publishing.
9. Oh, E. (2011). Effects of speaker gender on voice onset time in Korean stops. *Journal of Phonetics*, *39*(1), 59–67.
10. Abitbol, J., Abitbol, P., & Abitbol, B. (1999). Sex hormones and the female voice. *Journal of Voice*, *13*(3), 424–446.
11. Pipitone, R. N., & Gallup, G. G., Jr. (2008). Women's voice attractiveness varies across the menstrual cycle. *Evolution and Human Behavior*, *29*(4), 268–274.
12. Chae, S. W., Choi, G., Kang, H. J., Choi, J. O., & Jin, S. M. (2001). Clinical analysis of voice change as a parameter of premenstrual syndrome. *Journal of Voice*, *15*(2), 278–283.
13. Mulac, A., & Giles, H. (1996). 'Your're only as old as you sound': Perceived vocal age and social meanings. *Health Communication*, *8*(3), 199–215.
14. Dabbs, J. M., Jr., & Mallinger, A. (1999). High testosterone levels predict low voice pitch among men. *Personality and Individual Differences*, *27*(4), 801–804.
15. Collins, S. A. (2000). Men's voices and women's choices. *Animal Behaviour*, *60*(6), 773–780.
16. Butler, J. E., Hammond, T. H., & Gray, S. D. (2001). Gender-related differences of hyaluronic acid distribution in the human vocal fold. *The Laryngoscope*, *111*(5), 907–911.
17. Hammond, T. H., Zhou, R., Hammond, E. H., Pawlak, A., & Gray, S. D. (1997). The intermediate layer: A morphologic study of the elastin and hyaluronic acid constituents of normal human vocal folds. *Journal of Voice*, *11*(1), 59–66.
18. Jekosch, U. (2006). *Voice and speech quality perception: Assessment and evaluation*. Berlin: Springer Science & Business Media.
19. Erlik, Y., Meldrum, D. R., & Judd, H. L. (1982). Estrogen levels in postmenopausal women with hot flashes. *Obstetrics and Gynecology*, *59*(4), 403–407.
20. Huang, Z., Hankinson, S. E., Colditz, G. A., Stampfer, M. J., Hunter, D. J., Manson, J. E., et al. (1997). Dual effects of weight and weight gain on breast cancer risk. *JAMA*, *278*(17), 1407–1411.
21. Shi, H., & Clegg, D. J. (2009). Sex differences in the regulation of body weight. *Physiology & Behavior*, *97*(2), 199–204.
22. Schneider, G., Kirschner, M. A., Berkowitz, R., & Ertel, N. H. (1979). Increased estrogen production in obese men. *The Journal of Clinical Endocrinology & Metabolism*, *48*(4), 633–638.
23. Fant, G. (2012). *Acoustic theory of speech production: With calculations based on X-ray studies of Russian articulations* (Vol. 2). Berlin: Walter de Gruyter.

24. Thornhill, R., & Grammer, K. (1999). The body and face of woman: One ornament that signals quality? *Evolution and Human Behavior*, *20*(2), 105–120.
25. Collins, S. A., & Missing, C. (2003). Vocal and visual attractiveness are related in women. *Animal Behaviour*, *65*(5), 997–1004.
26. Abend, P., Pflüger, L. S., Koppensteiner, M., Coquerelle, M., & Grammer, K. (2015). The sound of female shape: A redundant signal of vocal and facial attractiveness. *Evolution and Human Behavior*, *36*(3), 174–181.
27. Valentova, J. V., Varella, M. A. C., Havlíček, J., & Kleisner, K. (2017). Positive association between vocal and facial attractiveness in women but not in men: A cross-cultural study. *Behavioural Processes*, *135*, 95–100.
28. Feinberg, D. R., Jones, B. C., DeBruine, L. M., Moore, F. R., Smith, M. J. L., Cornwell, R. E., et al. (2005). The voice and face of woman: One ornament that signals quality? *Evolution and Human Behavior*, *26*(5), 398–408.
29. Feinberg, D. R., Jones, B. C., Little, A. C., Burt, D. M., & Perrett, D. I. (2005). Manipulations of fundamental and formant frequencies influence the attractiveness of human male voices. *Animal Behaviour*, *69*(3), 561–568.
30. González, J. (2003). Estimation of speakers' weight and height from speech: A re-analysis of data from multiple studies by Lass and colleagues. *Perceptual and Motor Skills*, *96*(1), 297–304.
31. Lass, N. J., Phillips, J. K., & Bruchey, C. A. (1980). The effect of filtered speech on speaker height and weight identification. *Journal of Phonetics*, *8*(1), 91–100.
32. Mporas, I., & Ganchev, T. (2009). Estimation of unknown speaker's height from speech. *International Journal of Speech Technology*, *12*(4), 149–160.
33. González, J. (2004). Formant frequencies and body size of speaker: A weak relationship in adult humans. *Journal of Phonetics*, *32*(2), 277–287.
34. Evans, S., Neave, N., & Wakelin, D. (2006). Relationships between vocal characteristics and body size and shape in human males: An evolutionary explanation for a deep male voice. *Biological Psychology*, *72*(2), 160–163.
35. Fay, P. J., & Middleton, W. C. (1940). Judgment of Kretschmerian body types from the voice as transmitted over a public address system. *The Journal of Social Psychology*, *12*(1), 151–162.
36. Pough, R. (1949). *Audubon Land Bird Guide*. New York City, New York: Doubleday Publishing Company.
37. Worthy, M. (1999). *Eye color: A key to human and animal behavior*. Bloomington: iUniverse.
38. Cartmill, M. (1998). The status of the race concept in physical anthropology. *American Anthropologist*, *100*(3), 651–660.
39. Wolpoff, M. H., & Caspari, R. (1997). *Race and Human Evolution*. New York City, New York: Simon & Schuster.
40. Walton, J. H., & Orlikoff, R. F. (1994). Speaker race identification from acoustic cues in the vocal signal. *Journal of Speech, Language, and Hearing Research*, *37*(4), 738–745.
41. Dang, J., Honda, K., & Suzuki, H. (1994). Morphological and acoustical analysis of the nasal and the paranasal cavities. *The Journal of the Acoustical Society of America*, *96*(4), 2088–2100.
42. Bloothooft, G., Bringmann, E., Van Cappellen, M., Van Luipen, J. B., & Thomassen, K. P. (1992). Acoustics and perception of overtone singing. *The Journal of the Acoustical Society of America*, *92*(4), 1827–1836.
43. Glenn, J. W., & Kleiner, N. (1968). Speaker identification based on nasal phonation. *The Journal of the Acoustical Society of America*, *43*(2), 368–372.
44. Mayo, R., Floyd, L. A., Warren, D. W., Dalston, R. M., & Mayo, C. M. (1996). Nasalance and nasal area values: Cross-racial study. *The Cleft Palate-Craniofacial Journal*, *33*(2), 143–149.
45. Beck, J. M. (1997). Organic variation of the vocal apparatus. In W. J. Hardcastle & J. Laver (Eds.), *The handbook of phonetic sciences* (pp. 256–297). Oxford, UK: Blackwell Publishers; Malden, Massachusetts.
46. Sundberg, J. (1974). Articulatory interpretation of the "singing formant". *The Journal of the Acoustical Society of America*, *55*(4), 838–844.

47. Sundberg, J., & Rossing, T. D. (1990). The science of singing voice. *The Journal of the Acoustical Society of America*, *87*(1), 462–463.
48. Vurma, A., & Ross, J. (2003). The perception of 'forward' and 'backward placement' of the singing voice. *Logopedics Phoniatrics Vocology*, *28*(1), 19–28.
49. Kingston, J., Macmillan, N. A., Dickey, L. W., Thorburn, R., & Bartels, C. (1997). Integrality in the perception of tongue root position and voice quality in vowels. *The Journal of the Acoustical Society of America*, *101*(3), 1696–1709.
50. Sundberg, J. (1977). The acoustics of the singing voice. *Scientific American*, *236*(3), 82–91.
51. Titze, I. R., & Story, B. H. (1997). Acoustic interactions of the voice source with the lower vocal tract. *The Journal of the Acoustical Society of America*, *101*(4), 2234–2243.
52. Lofqvist, A., & Gracco, V. L. (1997). Lip and jaw kinematics in bilabial stop consonant production. *Journal of Speech, Language, and Hearing Research*, *40*(4), 877–893.
53. Sondhi, M. M., & Gopinath, B. (1971). Determination of vocal-tract shape from impulse response at the lips. *The Journal of the Acoustical Society of America*, *49*(6B), 1867–1873.
54. Laver, J. (1980). *The phonetic description of voice quality* (Vol. 31, pp. 1–186). Cambridge studies in linguistics. London: Cambridge University Press.
55. Esling, J. H., Heap, L. M., Snell, R. C., & Dickson, B. C. (1994). Analysis of pitch dependence of pharyngeal, faucal, and larynx-height voice quality settings. In *Proceedings of the Third International Conference on Spoken Language Processing (ICSLP)* (pp. 1475–1478). Japan: Yokohama.
56. Hirano, M., & Bless, D. M. (1993). *Videostroboscopic examination of the larynx* (pp. 23–24). San Diego, California: Singular Publishing Group.
57. Moisik, S. R. (2013). The epilarynx in speech. Doctoral Dissertation, Department of Linguistics, University of Victoria, Canada.
58. Titze, I. R. (2001). Acoustic interpretation of resonant voice. *Journal of Voice*, *15*(4), 519–528.
59. Thomson, S. L., Mongeau, L., & Frankel, S. H. (2005). Aerodynamic transfer of energy to the vocal folds. *The Journal of the Acoustical Society of America*, *118*(3), 1689–1700.
60. Hirano, M., Kakita, Y., Ohmaru, K., & Kurita, S. (1982). Structure and mechanical properties of the vocal fold. In *Speech and language* (Vol. 7, pp. 271–297). Amsterdam: Elsevier. ScienceDirect.com
61. Hoit, J. D., Solomon, N. P., & Hixon, T. J. (1993). Effect of lung volume on voice onset time (VOT). *Journal of Speech, Language, and Hearing Research*, *36*(3), 516–520.
62. Iwarsson, J., Thomasson, M., & Sundberg, J. (1998). Effects of lung volume on the glottal voice source. *Journal of Voice*, *12*(4), 424–433.
63. Hixon, T. J., Goldman, M. D., & Mead, J. (1973). Kinematics of the chest wall during speech production: Volume displacements of the rib cage, abdomen, and lung. *Journal of Speech, Language, and Hearing Research*, *16*(1), 78–115.
64. Hixon, T. J., Watson, P. J., Harris, F. P., & Pearl, N. B. (1988). Relative volume changes of the rib cage and abdomen during prephonatory chest wall posturing. *Journal of Voice*, *2*(1), 13–19.
65. Iwarsson, J. (2001). Effects of inhalatory abdominal wall movement on vertical laryngeal position during phonation. *Journal of Voice*, *15*(3), 384–394.
66. Tom, K., Titze, I. R., Hoffman, E. A., & Story, B. H. (2001). Three-dimensional vocal tract imaging and formant structure: Varying vocal register, pitch, and loudness. *The Journal of the Acoustical Society of America*, *109*(2), 742–747.
67. Kendall, K. (2007). Presbyphonia: A review. *Current Opinion in Otolaryngology & Head and Neck Surgery*, *15*(3), 137–140.
68. Martins, R. H. G., Gonçalvez, T. M., Pessin, A. B. B., & Branco, A. (2014). Aging voice: Presbyphonia. *Aging Clinical and Experimental Research*, *26*(1), 1–5.
69. Martins, R. H. G., Benito Pessin, A. B., Nassib, D. J., Branco, A., Rodrigues, S. A., & Matheus, S. M. M. (2015). Aging voice and the laryngeal muscle atrophy. *The Laryngoscope*, *125*(11), 2518–2521.
70. Harries, M., Hawkins, S., Hacking, J., & Hughes, I. (1998). Changes in the male voice at puberty: Vocal fold length and its relationship to the fundamental frequency of the voice. *The Journal of Laryngology & Otology*, *112*(5), 451–454.

71. Schuberth, S., Hoppe, U., Döllinger, M., Lohscheller, J., & Eysholdt, U. (2002). High-precision measurement of the vocal fold length and vibratory amplitudes. *The Laryngoscope*, *112*(6), 1043–1049.
72. Hollien, H., Brown, W. S., Jr., & Hollien, K. (1971). Vocal fold length associated with modal, falsetto and varying intensity phonations. *Folia Phoniatrica et Logopaedica*, *23*(1), 66–78.
73. Zhang, K., Siegmund, T., & Chan, R. W. (2006). A constitutive model of the human vocal fold cover for fundamental frequency regulation. *The Journal of the Acoustical Society of America*, *119*(2), 1050–1062.
74. Gray, S. D. (2000). Cellular physiology of the vocal folds. *Otolaryngologic Clinics of North America*, *33*(4), 679–697.
75. Sato, K., & Hirano, M. (1997). Age-related changes of elastic fibers in the superficial layer of the lamina propria of vocal folds. *Annals of Otology, Rhinology & Laryngology*, *106*(1), 44–48.
76. Sato, K., Hirano, M., & Nakashima, T. (2002). Age-related changes of collagenous fibers in the human vocal fold mucosa. *Annals of Otology, Rhinology & Laryngology*, *111*(1), 15–20.
77. Hirano, M., Kurita, S., & Sakaguchi, S. (1989). Ageing of the vibratory tissue of human vocal folds. *Acta Oto-Laryngologica*, *107*(5–6), 428–433.
78. Titze, I. R. (1988). The physics of small-amplitude oscillation of the vocal folds. *The Journal of the Acoustical Society of America*, *83*(4), 1536–1552.
79. Story, B. H., & Titze, I. R. (1995). Voice simulation with a body-cover model of the vocal folds. *The Journal of the Acoustical Society of America*, *97*(2), 1249–1260.
80. Chan, R. W., & Titze, I. R. (2006). Dependence of phonation threshold pressure on vocal tract acoustics and vocal fold tissue mechanics. *The Journal of the Acoustical Society of America*, *119*(4), 2351–2362.
81. Sundberg, J., Thörnvik, M. N., & Söderström, A. M. (1998). Age and voice quality in professional singers. *Logopedics Phoniatrics Vocology*, *23*(4), 169–176.
82. Ramig, L. A., & Ringel, R. L. (1983). Effects of physiological aging on selected acoustic characteristics of voice. *Journal of Speech, Language, and Hearing Research*, *26*(1), 22–30.
83. Linville, S. E. (1996). The sound of senescence. *Journal of Voice*, *10*(2), 190–200.
84. Sataloff, R. T. (2006). *Vocal health and pedagogy, volume II: Advanced assessment and practice* (Vol. 2). San Diego, California: Plural Publishing.
85. Brown, W. S., Jr., Morris, R. J., Hollien, H., & Howell, E. (1991). Speaking fundamental frequency characteristics as a function of age and professional singing. *Journal of Voice*, *5*(4), 310–315.
86. Fay, P. J., & Middleton, W. C. (1940). The ability to judge the rested or tired condition of a speaker from his voice as transmitted over a public address system. *Journal of Applied Psychology*, *24*(5), 645.
87. Johannes, B., Wittels, P., Enne, R., Eisinger, G., Castro, C. A., Thomas, J. L., et al. (2007). Non-linear function model of voice pitch dependency on physical and mental load. *European Journal of Applied Physiology*, *101*(3), 267–276.
88. Atkinson, J. E. (1978). Correlation analysis of the physiological factors controlling fundamental voice frequency. *The Journal of the Acoustical Society of America*, *63*(1), 211–222.
89. Wu, B. L., Sanders, I., Mu, L., & Biller, H. F. (1994). The human communicating nerve: An extension of the external superior laryngeal nerve that innervates the vocal cord. *Archives of Otolaryngology - Head & Neck Surgery*, *120*(12), 1321–1328.
90. Hayre, H. S., & Holland, J. C. (1980). Cross-correlation of voice and heart rate as stress measures. *Applied Acoustics*, *13*(1), 57–62.
91. Schuller, B., Friedmann, F., & Eyben, F. (2013). Automatic recognition of physiological parameters in the human voice: Heart rate and skin conductance. In *Proceedings of the International Conference on Acoustics, Speech and Signal Processing (ICASSP)* (pp. 7219–7223). Vancouver, Canada: IEEE.
92. Harrison, Y., & Horne, J. A. (1997). Sleep deprivation affects speech. *Sleep*, *20*(10), 871–877.
93. Wadnerkar, M. B., Cowell, P. E., & Whiteside, S. P. (2006). Speech across the menstrual cycle: A replication and extension study. *Neuroscience Letters*, *408*(1), 21–24.

94. Hillenbrand, J., Getty, L. A., Clark, M. J., & Wheeler, K. (1995). Acoustic characteristics of American English vowels. *The Journal of the Acoustical Society of America, 97*(5), 3099–3111.
95. Jacewicz, E., Fox, R. A., & Salmons, J. (2007). Vowel space areas across dialects and gender. In *Proceedings of the International Congress of Phonetic Sciences* (Vol. 16, pp. 1465–1468).
96. Xue, S. A., & Fucci, D. (2000). Effects of race and sex on acoustic features of voice analysis. *Perceptual and Motor Skills, 91*(3), 951–958.
97. Andrianopoulos, M. V., Darrow, K., & Chen, J. (2001). Multimodal standardization of voice among four multicultural populations formant structures. *Journal of Voice, 15*(1), 61–77.
98. Templeton, A. R. (1998). Human races: A genetic and evolutionary perspective. *American Anthropologist, 100*(3), 632–650.
99. Templeton, A. R. (2013). Biological races in humans. *Studies in History and Philosophy of Science Part C: Studies in History and Philosophy of Biological and Biomedical Sciences, 44*(3), 262–271.
100. Pigliucci, M., & Kaplan, J. (2003). On the concept of biological race and its applicability to humans. *Philosophy of Science, 70*(5), 1161–1172.
101. Barbujani, G., Magagni, A., Minch, E., & Cavalli-Sforza, L. L. (1997). An apportionment of human DNA diversity. *Proceedings of the National Academy of Sciences, 94*(9), 4516–4519.
102. Huxley, T. H. (1870). On the geographical distribution of the chief modifications of mankind. *The Journal of the Ethnological Society of London (1869–1870), 2*(4), 404–412.
103. Coon, C. S. (1962). *The origin of races*. New York: Alfred A. Knopf Publishers.
104. Jackson, J. P., (2001). "In Ways Unacademical": The Reception of Carleton S. Coon's The Origin of Races. *Journal of the History of Biology, 34*(2), 247–285.
105. Zengin, A., Pye, S. R., Cook, M. J., Adams, J. E., Wu, F. C. W., O'Neill, T. W., et al. (2016). Ethnic differences in bone geometry between White, Black and South Asian men in the UK. *Bone, 91*, 180–185.
106. Fry, F. J., & Barger, J. E. (1978). Acoustical properties of the human skull. *The Journal of the Acoustical Society of America, 63*(5), 1576–1590.
107. Newman, M., & Wu, A. (2011). "Do you sound Asian when you speak English?" Racial identification and voice in Chinese and Korean Americans' English. *American Speech, 86*(2), 152–178.
108. Hudson, A. I., & Holbrook, A. (1982). Fundamental frequency characteristics of young black adults: Spontaneous speaking and oral reading. *Journal of Speech, Language, and Hearing Research, 25*(1), 25–28.
109. Sataloff, R. T., & Hawkshaw, M. J. (2006). Common medical diagnoses and treatments in patients with voice disorders. In J. S. Rubin, R. T. Sataloff, & G. S. Korovin (Eds.), *Diagnosis and treatment of voice disorders* (pp. 331–347). San Diego, California: Plural Publishing.
110. Allec, L. D. R., López, X. H., Porras, J. B. A., Ramos, R. V., del Valle, J. C. P., & García, Á. I. P. (2011). Alterations in voice, speech and swallowing in patients with Sjögren's syndrome. *Acta Otorrinolaringologica (English Edition), 62*(4), 255–264.
111. Ramig, L. O., Scherer, R. C., Klasner, E. R., Titze, I. R., & Horii, Y. (1990). Acoustic analysis of voice in amyotrophic lateral sclerosis: A longitudinal case study. *Journal of Speech and Hearing Disorders, 55*(1), 2–14.
112. Carrow, E., Rivera, V., Mauldin, M., & Shamblin, L. (1974). Deviant speech characteristics in motor neuron disease. *Archives of Otolaryngology, 100*(3), 212–218.
113. Fox, C. M., & Ramig, L. O. (1997). Vocal sound pressure level and self-perception of speech and voice in men and women with idiopathic Parkinson disease. *American Journal of Speech-Language Pathology, 6*(2), 85–94.
114. Ho, A. K., Iansek, R., Marigliani, C., Bradshaw, J. L., & Gates, S. (1999). Speech impairment in a large sample of patients with Parkinson's disease. *Behavioural Neurology, 11*(3), 131–137.
115. Lieberman, P., Kako, E., Friedman, J., Tajchman, G., Feldman, L. S., & Jiminez, E. B. (1992). Speech production, syntax comprehension, and cognitive deficits in Parkinson's disease. *Brain and Language, 43*(2), 169–189.

116. Critchley, E. M. (1981). Speech disorders of Parkinsonism: A review. *Journal of Neurology, Neurosurgery & Psychiatry*, *44*(9), 751–758.
117. Perez, K. S., Ramig, L. O., Smith, M. E., & Dromey, C. (1996). The Parkinson larynx: Tremor and videostroboscopic findings. *Journal of Voice*, *10*(4), 354–361.
118. Satt, A., Sorin, A., Toledo-Ronen, O., Barkan, O., Kompatsiaris, I., Kokonozi, A., et al. (2013). Evaluation of speech-based protocol for detection of early-stage dementia. In *Proceedings of Annual Conference of the International Speech Communication Association (INTERSPEECH), Lyon, France* (pp. 1692–1696).
119. Reilly, J., Rodriguez, A. D., Lamy, M., & Neils-Strunjas, J. (2010). Cognition, language, and clinical pathological features of non-Alzheimer's dementias: An overview. *Journal of Communication Disorders*, *43*(5), 438–452.
120. Appell, J., Kertesz, A., & Fisman, M. (1982). A study of language functioning in Alzheimer patients. *Brain and Language*, *17*(1), 73–91.
121. Hara, K., Ohta, K., Miyajima, M., Hara, M., Iino, H., Matsuda, A., et al. (2012). Mismatch negativity for speech sounds in temporal lobe epilepsy. *Epilepsy and Behavior*, *23*(3), 335–341.
122. Amorosa, H., von Benda, U., Dames, M., & Schäferskϋpper, P. (1986). Deficits in fine motor coordination in children with unintelligible speech. *European Archives of Psychiatry and Neurological Sciences*, *236*(1), 26–30.
123. Mohammadzadeh, A., Heydari, E., & Azizi, F. (2011). Speech impairment in primary hypothyroidism. *Journal of Endocrinological Investigation*, *34*(6), 431–433.
124. Ahmed, R. G. Congenital hypothyroidism, developmental skills, and voice-speech disorders. *ARC Journal of Neuroscience*, *3*(2), 1–5.
125. Verma, A., & Hasan, T. (2009). Thyroid disorders in older people. *Reviews in Clinical Gerontology*, *19*(3), 193–206.
126. Baker, K. K., Ramig, L. O., Johnson, A. B., & Freed, C. R. (1997). Preliminary voice and speech analysis following fetal dopamine transplants in 5 individuals with Parkinson disease. *Journal of Speech, Language, and Hearing Research*, *40*(3), 615–626.
127. Martin, F. G. (1988). Drugs and vocal function. *Journal of Voice*, *2*(4), 338–344.
128. Klingholz, F., Penning, R., & Liebhardt, E. (1988). Recognition of low-level alcohol intoxication from speech signal. *The Journal of the Acoustical Society of America*, *84*(3), 929–935.
129. Heydari, F., & Ghanei, M. (2011). Effects of exposure to sulfur mustard on speech aerodynamics. *Journal of Communication Disorders*, *44*(3), 331–335.
130. Farmer, A. (1980). Voice onset time production in cerebral palsied speakers. *Folia Phoniatrica et Logopaedica*, *32*(4), 267–273.
131. Verdolini, K., Min, Y., Titze, I. R., Lemke, J., Brown, K., Van Mersbergen, M., et al. (2002). Biological mechanisms underlying voice changes due to dehydration. *Journal of Speech, Language, and Hearing Research*, *45*(2), 268–281.
132. Solomon, N. P., & DiMattia, M. S. (2000). Effects of a vocally fatiguing task and systemic hydration on phonation threshold pressure. *Journal of Voice*, *14*(3), 341–362.
133. Jennische, M., & Sedin, G. (1998). Speech and language skills in children who required neonatal intensive care. I. Spontaneous speech at 6.5 years of age. *Acta Paediatrica*, *87*(6), 654–666.
134. Botez, M. I., & Wertheim, N. (1959). Expressive aphasia and amusia following right frontal lesion in a right-handed man. *Brain*, *82*(2), 186–202.
135. Jacobi, I., van der Molen, L., Huiskens, H., Van Rossum, M. A., & Hilgers, F. J. (2010). Voice and speech outcomes of chemoradiation for advanced head and neck cancer: A systematic review. *European Archives of Oto-Rhino-Laryngology*, *267*(10), 1495–1505.
136. Mora, R., Jankowska, B., Mora, F., Crippa, B., Dellepiane, M., & Salami, A. (2009). Effects of tonsillectomy on speech and voice. *Journal of Voice*, *23*(5), 614–618.
137. Shi, X. H., Chen, N., Xing, S. Z., Yuan, Y., & Yao, W. P. (2008). Study on spectrum features of speech before and after repair in cleft palate patients. *Stomatology*, *28*(2), 65–69.
138. Scharfstein, L. A., Beidel, D. C., Sims, V. K., & Finnell, L. R. (2011). Social skills deficits and vocal characteristics of children with social phobia or Asperger's disorder: A comparative study. *Journal of Abnormal Child Psychology*, *39*(6), 865–875.

139. Rutherford, M. D., Baron-Cohen, S., & Wheelwright, S. (2002). Reading the mind in the voice: A study with normal adults and adults with Asperger syndrome and high functioning autism. *Journal of Autism and Developmental Disorders*, *32*(3), 189–194.
140. Darby, J. K., & Hollien, H. (1977). Vocal and speech patterns of depressive patients. *Folia Phoniatrica et Logopaedica*, *29*(4), 279–291.
141. Darby, J. K., Simmons, N., & Berger, P. A. (1984). Speech and voice parameters of depression: A pilot study. *Journal of Communication Disorders*, *17*(2), 75–85.
142. Vercammen, A., de Haan, E. H., & Aleman, A. (2008). Hearing a voice in the noise: Auditory hallucinations and speech perception. *Psychological Medicine*, *38*(8), 1177–1184.
143. Hashim, N. W., Wilkes, M., Salomon, R., & Meggs, J. (2012). Analysis of timing pattern of speech as possible indicator for near-term suicidal risk and depression in male patients. *International Proceedings of Computer Science and Information Technology*, *58*, 6.
144. Palinkas-Sanches, E., Sanches, M., Ferrari, M. C. C., Oliveira, G., & Behlau, M. (2010). Vocal analysis of suicidal movie characters. *Revista Brasileira de Psiquiatria*, *32*(4), 409–415.
145. Firestone, R. W. (1988). *Voice therapy: A psychotherapeutic approach to self-destructive behavior*. New York: Human Sciences Press.
146. Dodd, B. J. (1972). Comparison of babbling patterns in normal and Down'syndrome infants. *Journal of Intellectual Disability Research*, *16*(12), 35–40.
147. Rimmer, J., Giddings, C. E. B., Cavalli, L., & Hartley, B. E. J. (2008). Dysphonia - a rare early symptom of Ehlers-Danlos syndrome? *International Journal of Pediatric Otorhinolaryngology*, *72*(12), 1889–1892.
148. Akefeldt, A., Åkefeldt, B., & Gillberg, C. (1997). Voice, speech and language characteristics of children with Prader-Willi syndrome. *Journal of Intellectual Disability Research*, *41*(4), 302–311.
149. Goldberg, L. R. (1990). "An alternative description of personality": The big-five factor structure. *Journal of Personality and Social Psychology*, *59*(6), 1216.
150. McCrae, R. R., & John, O. P. (1992). An introduction to the five-factor model and its applications. *Journal of Personality*, *60*(2), 175–215.
151. McCrae, R. R., & Costa, P. T., Jr. (1997). Personality trait structure as a human universal. *American Psychologist*, *52*(5), 509.
152. McCrae, R. R., & Costa, P. T., Jr. (1999). A five-factor theory of personality. *Handbook of Personality: Theory and Research*, *2*(1999), 139–153.
153. Costa, P. T., & McCrae, R. R. (1992). Normal personality assessment in clinical practice: The NEO personality inventory. *Psychological Assessment*, *4*(1), 5.
154. Costa, P. T., & McCrae, R. R. (2010). *The NEO personality inventory: 3*. Odessa, Florida: Psychological Assessment Resources.
155. McCrae, R. R., & Costa, P. T. (2003). *Personality in adulthood: A five-factor theory perspective*. New York City, New York: Guilford Press.
156. Costa, P. T., Jr., Terracciano, A., & McCrae, R. R. (2001). Gender differences in personality traits across cultures: Robust and surprising findings. *Journal of Personality and Social Psychology*, *81*(2), 322.
157. Kokkonen, M., & Pulkkinen, L. E. A. (2001). Extraversion and neuroticism as antecedents of emotion regulation and dysregulation in adulthood. *European Journal of Personality*, *15*(6), 407–424.
158. Kokkonen, M., & Pulkkinen, L. (2001). Examination of the paths between personality, current mood, its evaluation, and emotion regulation. *European Journal of Personality*, *15*(2), 83–104.
159. Salgado, J. F. (2002). The Big Five personality dimensions and counterproductive behaviors. *International Journal of Selection and Assessment*, *10*(1–2), 117–125.
160. Benjamin, J., Ebstein, R. P., & Belmaker, R. H. (1997). Personality genetics. *The Israel Journal of Psychiatry and Related Sciences*, *34*(4), 270–280.
161. Buss, A. H., & Plomin, R. (2014). *Temperament (PLE: Emotion): Early developing personality traits*. London: Psychology Press, Taylor & Francis Group.
162. Thapar, A., & McGuffin, P. (1993). Is personality disorder inherited? An overview of the evidence. *Journal of Psychopathology and Behavioral Assessment*, *15*(4), 325–345.

163. Plomin, R., Caspi, A., Pervin, L. A., & John, O. P. (1990). Behavioral genetics and personality. *Handbook of Personality: Theory and Research*, *2*, 251–276.
164. Bouchard, T. J., Jr., & McGue, M. (2003). Genetic and environmental influences on human psychological differences. *Journal of Neurobiology*, *54*(1), 4–45.
165. Fehrman, E., Muhammad, A. K., Mirkes, E.M., Egan, V., & Gorban, A. N. (2017). In F. Palumbo, A. Montanari, & M. Vichi (Eds.), *Data science* (pp. 231–242). Studies in classification, data analysis, and knowledge organization. Cham: Springer.
166. Khan, A. A., Jacobson, K. C., Gardner, C. O., Prescott, C. A., & Kendler, K. S. (2005). Personality and comorbidity of common psychiatric disorders. *The British Journal of Psychiatry*, *186*(3), 190–196.
167. Cuijpers, P., Smit, F., Penninx, B. W., de Graaf, R., ten Have, M., & Beekman, A. T. (2010). Economic costs of neuroticism: A population-based study. *Archives of General Psychiatry*, *67*(10), 1086–1093.
168. Jeronimus, B. F., Kotov, R., Riese, H., & Ormel, J. (2016). Neuroticism's prospective association with mental disorders halves after adjustment for baseline symptoms and psychiatric history, but the adjusted association hardly decays with time: A meta-analysis on 59 longitudinal/prospective studies with 443 313 participants. *Psychological Medicine*, *46*(14), 2883–2906.
169. Kotov, R., Gamez, W., Schmidt, F., & Watson, D. (2010). Linking "big" personality traits to anxiety, depressive, and substance use disorders: A meta-analysis. *Psychological Bulletin*, *136*(5), 768.
170. Millon, T., Krueger, R. F., & Simonsen, E. (Eds.). (2011). *Contemporary directions in psychopathology: Scientific foundations of the DSM-V and ICD-11*. New York City, New York: Guilford Press.
171. Widiger, T. A. (2011). Personality and psychopathology. *World Psychiatry*, *10*(2), 103–106.
172. De Bolle, M., Beyers, W., De Clercq, B., & De Fruyt, F. (2012). General personality and psychopathology in referred and nonreferred children and adolescents: An investigation of continuity, pathoplasty, and complication models. *Journal of Abnormal Psychology*, *121*(4), 958.
173. Michael, W., & Crawford, C. C. (1927). An experiment in judging intelligence by the voice. *Journal of Educational Psychology*, *18*(2), 107.
174. Fay, P. J., & Middleton, W. C. (1939). Judgment of Spranger personality types from the voice as transmitted over a public address system. *Journal of Personality*, *8*(2), 144–155.
175. Fay, P. J., & Middleton, W. C. (1943). Judgment of leadership from the transmitted voice. *The Journal of Social Psychology*, *17*(1), 99–102.
176. Hildreth, R. A. (1953). *An experimental study of audiences' ability to distinguish between sincere and insincere speeches*. Doctoral dissertation, University of Southern California, USA.
177. Pear, T. H. (1931). *Voice and personality*. CRC Press, UK: Chapman and Hall.
178. Fay, P. J., & Middleton, W. C. (1941). The ability to judge sociability from the voice as transmitted over a public address system. *The Journal of Social Psychology*, *13*(2), 303–309.
179. Fay, P. J., & Middleton, W. C. (1942). Judgment of introversion from the transcribed voice. *Quarterly Journal of Speech*, *28*(2), 226–228.
180. Ray, G. B. (1986). Vocally cued personality prototypes: An implicit personality theory approach. *Communications Monographs*, *53*(3), 266–276.
181. Thakerar, J. N., & Giles, H. (1981). They are - so they spoke: Noncontent speech stereotypes. *Language & Communication*, *1*(2–3), 255–261.
182. Brown, B. L., Strong, W. J., & Rencher, A. C. (1975). Acoustic determinants of perceptions of personality from speech. *International Journal of the Sociology of Language*, *1975*(6), 11–32.
183. Warner, R. M., & Sugarman, D. B. (1986). Attributions of personality based on physical appearance, speech, and handwriting. *Journal of Personality and Social Psychology*, *50*(4), 792.
184. Allport, G. W., & Cantril, H. (1934). Judging personality from voice. *The Journal of Social Psychology*, *5*(1), 37–55.

185. Schroeder, J., & Epley, N. (2015). The sound of intellect: Speech reveals a thoughtful mind, increasing a job candidate's appeal. *Psychological Science*, *26*(6), 877–891.
186. Paves, L. A. (1965). *A study to determine relationships between voice qualities and attributed traits of personality*. Doctoral dissertation, School of Education, New York University, USA.
187. Markel, N. N., Meisels, M., & Houck, J. E. (1964). Judging personality from voice quality. *The Journal of Abnormal and Social Psychology*, *69*(4), 458.
188. Addington, D. W. (1968). The relationship of selected vocal characteristics to personality perception. *Speech Monographs*, *35*(4).
189. Smith, B. L., Brown, B. L., Strong, W. J., & Rencher, A. C. (1975). Effects of speech rate on personality perception. *Language and Speech*, *18*(2), 145–152.
190. Brown, B. L. (1980). Effects of speech rate on personality attributions and competency evaluations. In H. Giles, W. P. Robinson, & P. Smith (Eds.), *Language: Social psychological perspectives* (pp. 294–300). Oxford: Pergamon.
191. Llurda, E. (2000). Effects of intelligibility and speaking rate on judgements of non-native speakers' personalities. *IRAL-International Review of Applied Linguistics in Language Teaching*, *38*(3–4), 289–300.
192. Mohammadi, G., Vinciarelli, A., & Mortillaro, M. (2010). The voice of personality: Mapping nonverbal vocal behavior into trait attributions. In *Proceedings of the 2nd International Workshop on Social Signal Processing* (pp. 17–20). ACM.
193. Markel, N. N., Eisler, R. M., & Reese, H. W. (1967). Judging personality from dialect. *Journal of Verbal Learning and Verbal Behavior*, *6*(1), 33–35.
194. Markel, N. N., & Roblin, G. L. (1965). The effect of content and sex-of-judge on judgments of personality from voice. *International Journal of Social Psychiatry*, *11*(4), 295–300.
195. Roy, N., Bless, D. M., & Heisey, D. (2000). Personality and voice disorders: A multitrait-multidisorder analysis. *Journal of Voice*, *14*(4), 521–548.
196. Plutchik, R. (1984). Emotions: A general psychoevolutionary theory. In K. R. Scherer, & P. L. Ekman (Eds.), *Approaches to emotion* (pp. 197–219). New Jersey: Lawrence Erlbaum Associates.
197. Koelstra, S., Muhl, C., Soleymani, M., Lee, J. S., Yazdani, A., Ebrahimi, T., et al. (2012). DEAP: A database for emotion analysis; using physiological signals. *IEEE Transactions on Affective Computing*, *3*(1), 18–31.
198. Kleinginna, P. R., & Kleinginna, A. M. (1981). A categorized list of emotion definitions, with suggestions for a consensual definition. *Motivation and Emotion*, *5*(4), 345–379.
199. Fox, E. (2008). *Emotion science: Cognitive and neuroscientific approaches to understanding human emotions*. Basingstoke, UK: Palgrave Macmillan.
200. Graham, M. C., Priddy, L., & Graham, S. (2014). *Facts of life: Ten issues of contentment*. Denver, Colorado: Outskirts Press.
201. Schwarz, N. (1990). Feelings as information: Informational and motivational functions of affective states. In E. T. Higgins & R. M. Sorrentino (Eds.) Handbook of motivation and cognition: Foundations of social behavior (Vol. 2, pp. 527–561). New York: Guilford Press.
202. Thamm, R. A. (2006). The classification of emotions. *Handbook of the sociology of emotions* (pp. 11–37). Boston, Massachusetts: Springer.
203. Plutchik, R. (2001). The nature of emotions: Human emotions have deep evolutionary roots, a fact that may explain their complexity and provide tools for clinical practice. *American Scientist*, *89*(4), 344–350.
204. Plutchik, R. (1997). The circumplex as a general model of the structure of emotions and personality. In R. Plutchik & H. R. Conte (Eds.), *Circumplex models of personality and emotions* (pp. 17–45). Washington: American Psychological Association.
205. Picard, R. W. (1995). *Affective computing*. MIT Media Laboratory Perceptual Computing Section Technical Report No. 321.
206. Bakker, I., van der Voordt, T., Vink, P., & de Boon, J. (2014). Pleasure, arousal, dominance: Mehrabian and Russell revisited. *Current Psychology*, *33*(3), 405–421.
207. Russell, J. A., & Mehrabian, A. (1977). Evidence for a three-factor theory of emotions. *Journal of Research in Personality*, *11*(3), 273–294.

208. Cambria, E., Livingstone, A., & Hussain, A. (2012). The hourglass of emotions. In A. Esposito, A. M. Esposito, A. Vinciarelli, R. Hoffmann, & V. C. Müller (Eds.), *Cognitive behavioural systems* (pp. 144–157). Berlin, Heidelberg: Springer.
209. Murray, I. R., & Arnott, J. L. (1993). Toward the simulation of emotion in synthetic speech: A review of the literature on human vocal emotion. *The Journal of the Acoustical Society of America, 93*(2), 1097–1108.
210. Ohala, J. J. (1996). Ethological theory and the expression of emotion in the voice. In *Proceedings of the Fourth International Conference on Spoken Language Processing (ICSLP)* (Vol. 3, pp. 1812–1815). Philadelphia, PA, USA: IEEE.
211. Pittam, J., & Scherer, K. R. (1993). *Vocal expression and communication of emotion*. New York: Guilford Press.
212. Cummings, K. E., & Clements, M. A. (1995). Analysis of the glottal excitation of emotionally styled and stressed speech. *The Journal of the Acoustical Society of America, 98*(1), 88–98.
213. Hugh-Munier, M., Scherer, K. R., Lehmann, W., & Scherer, U. (1997). Coping strategies, personality, and voice quality in patients with vocal fold nodules and polyps. *Journal of Voice, 11*(4), 452–461.
214. Gobl, C., & Chasaide, A. N. (2003). The role of voice quality in communicating emotion, mood and attitude. *Speech Communication, 40*(1–2), 189–212.
215. Sobin, C., & Alpert, M. (1999). Emotion in speech: The acoustic attributes of fear, anger, sadness, and joy. *Journal of Psycholinguistic Research, 28*(4), 347–365.
216. Laukkanen, A. M., Vilkman, E., Alku, P., & Oksanen, H. (1997). On the perception of emotions in speech: The role of voice quality. *Logopedics Phoniatrics Vocology, 22*(4), 157–168.
217. Frick, R. W. (1985). Communicating emotion: The role of prosodic features. *Psychological Bulletin, 97*(3), 412.
218. Mehrabian, A. (2017). *Nonverbal communication*. Abingdon, UK: Routledge.
219. Zuckerman, M., Lipets, M. S., Koivumaki, J. H., & Rosenthal, R. (1975). Encoding and decoding nonverbal cues of emotion. *Journal of Personality and Social Psychology, 32*(6), 1068.
220. Hariri, A. R. (2009). The neurobiology of individual differences in complex behavioral traits. *Annual Review of Neuroscience, 32*, 225–247.
221. Derue, D. S., Nahrgang, J. D., Wellman, N. E. D., & Humphrey, S. E. (2011). Trait and behavioral theories of leadership: An integration and meta-analytic test of their relative validity. *Personnel Psychology, 64*(1), 7–52.
222. Garai, J. E., & Scheinfeld, A. (1968). Sex differences in mental and behavioral traits. *Genetic Psychology Monographs, 77*, 169–299.
223. Kandel, E. R., Schwartz, J. H., & Jessell, T. M. (Eds.). (1995). *Essentials of neural science and behavior* (p. 5). Norwalk, Connecticut: Appleton & Lange.
224. Kandel, E. R., Schwartz, J. H., & Jessell, T. M. (Eds.). (2000). *Principles of neural science* (Vol. 4, pp. 1227–1246). New York: McGraw-hill.
225. Nelson, R. J. (2011). *An introduction to behavioral endocrinology*. Sunderland, Massachusetts: Sinauer Associates.
226. Neave, N. (2007). *Hormones and behaviour: A psychological approach*. Cambridge, UK: Cambridge University Press.
227. Chiel, H. J., & Beer, R. D. (1997). The brain has a body: Adaptive behavior emerges from interactions of nervous system, body and environment. *Trends in Neurosciences, 20*(12), 553–557.
228. Gottlieb, G. (1976). The roles of experience in the development of behavior and the nervous system. *Studies on the development of behavior and the nervous system* (Vol. 3, pp. 25–54). Amsterdam: Elsevier. ScienceDirect.com.
229. Plomin, R. (1990). *Nature and nurture: An introduction to human behavioral genetics*. Pacific Grove, California: Thomson Brooks/Cole Publishing Co.
230. Sataloff, R. T. (1995). Genetics of the voice. *Journal of Voice, 9*(1), 16–19.
231. Stroebe, W. (2011). *Social psychology and health*. UK: McGraw-Hill Education.

232. Pivik, R. T., Andres, A., & Badger, T. M. (2012). Effects of diet on early stage cortical perception and discrimination of syllables differing in voice-onset time: A longitudinal ERP study in 3 and 6 month old infants. *Brain and Language*, *120*(1), 27–41.
233. Eisenberg, P., & Zalowitz, E. (1938). Judging expressive movement: III. Judgments of dominance-feeling from phonograph records of voice. *Journal of Applied Psychology*, *22*(6), 620.
234. Burgoon, J. K., Birk, T., & Pfau, M. (1990). Nonverbal behaviors, persuasion, and credibility. *Human Communication Research*, *17*(1), 140–169.
235. Puts, D. A., Hodges, C. R., Cárdenas, R. A., & Gaulin, S. J. (2007). Men's voices as dominance signals: Vocal fundamental and formant frequencies influence dominance attributions among men. *Evolution and Human Behavior*, *28*(5), 340–344.
236. Puts, D. A., Gaulin, S. J., & Verdolini, K. (2006). Dominance and the evolution of sexual dimorphism in human voice pitch. *Evolution and Human Behavior*, *27*(4), 283–296.
237. Klofstad, C. A., Anderson, R. C., & Peters, S. (2012). Sounds like a winner: Voice pitch influences perception of leadership capacity in both men and women. *Proceedings of the Royal Society of London B: Biological Sciences* rspb20120311.
238. Anderson, R. C., & Klofstad, C. A. (2012). Preference for leaders with masculine voices holds in the case of feminine leadership roles. *PloS One*, *7*(12), e51216.
239. Tigue, C. C., Borak, D. J., O'Connor, J. J., Schandl, C., & Feinberg, D. R. (2012). Voice pitch influences voting behavior. *Evolution and Human Behavior*, *33*(3), 210–216.
240. Hughes, S. M., Dispenza, F., & Gallup, G. G. (2004). Ratings of voice attractiveness predict sexual behavior and body configuration. *Evolution and Human Behavior*, *25*(5), 295–304.
241. Munson, B., McDonald, E. C., DeBoe, N. L., & White, A. R. (2006). The acoustic and perceptual bases of judgments of women and men's sexual orientation from read speech. *Journal of Phonetics*, *34*(2), 202–240.
242. Luria, A. R. (2014). *The role of speech in the regulation of normal and abnormal behavior*. Oxford, UK: Liveright.
243. Jones, B. C., Feinberg, D. R., DeBruine, L. M., Little, A. C., & Vukovic, J. (2010). A domain-specific opposite-sex bias in human preferences for manipulated voice pitch. *Animal Behaviour*, *79*(1), 57–62.
244. Borkowska, B., & Pawlowski, B. (2011). Female voice frequency in the context of dominance and attractiveness perception. *Animal Behaviour*, *82*(1), 55–59.
245. Klofstad, C. A., Anderson, R. C., & Nowicki, S. (2015). Perceptions of competence, strength, and age influence voters to select leaders with lower-pitched voices. *PloS One*, *10*(8), e0133779.
246. Fay, P. J., & Middleton, W. C. (1939). Judgment of occupation from the voice as transmitted over a public address system and over a radio. *Journal of Applied Psychology*, *23*(5), 586.
247. Howell, T. M., Harrison, D. A., Burris, E. R., & Detert, J. R. (2015). Who gets credit for input? Demographic and structural status cues in voice recognition. *Journal of Applied Psychology*, *100*(6), 1765.
248. Mayew, W. J., & Venkatachalam, M. (2012). The power of voice: Managerial affective states and future firm performance. *The Journal of Finance*, *67*(1), 1–43.
249. Ohlsson, A. C. (1988). *Voice and work environment: Towards an ecology of vocal behaviour*. Doctoral dissertation. Department of Logopedics and Phoniatrics, Gothenburg University, Sweden.
250. Ohlsson, A. C., Järvholm, B., Löfqvist, A., Näslund, P. E., & Stenborg, R. (1987). Vocal behaviour in welders - a preliminary study. *Folia Phoniatrica et Logopaedica*, *39*(2), 98–103.
251. Hogikyan, N. D., & Sethuraman, G. (1999). Validation of an instrument to measure voice-related quality of life (V-RQOL). *Journal of Voice*, *13*(4), 557–569.
252. Ekman, P., Friesen, W. V., & Scherer, K. R. (1976). Body movement and voice pitch in deceptive interaction. *Semiotica*, *16*(1), 23–28.
253. Ekman, P., O'Sullivan, M., Friesen, W. V., & Scherer, K. R. (1991). Invited article: Face, voice, and body in detecting deceit. *Journal of Nonverbal Behavior*, *15*(2), 125–135.

254. Zuckerman, M., DePaulo, B. M., & Rosenthal, R. (1981). Verbal and nonverbal communication of deception. *Advances in experimental social psychology* (Vol. 14, pp. 1–59). Cambridge, Massachusetts: Academic Press.
255. Esling, J. (1978). The identification of features of voice quality in social groups. *Journal of the International Phonetic Association*, *8*(1–2), 18–23.
256. Laver, J. (1970). *The synthesis of components in voice quality* (p. 575). Academia Publishing House of the Czechoslovak Academy of Sciences.
257. Mehrabian, A., & Russell, J. A. (1974). *An approach to environmental psychology*. Cambridge, Massachusetts: MIT Press.
258. Dimitrova, S. (2006). Relationship between human physiological parameters and geomagnetic variations of solar origin. *Advances in Space Research*, *37*(6), 1251–1257.
259. Siegman, A. W., Anderson, R. A., & Berger, T. (1990). The angry voice: Its effect on the experience of anger and cardiovascular reactivity. *Psychosomatic Medicine*, *52*(6), 631–643.
260. Sataloff, R. T. (2001). Essay the impact of pollution on the voice. *Voice and Speech Review*, *2*(1), 247–253.
261. Sataloff, R. T. (1992). The impact of pollution on the voice. *Otolaryngology - Head and Neck Surgery*, *106*(6), 701–705.
262. Sataloff, R. T. (2006). Pollution and its effect on the voice. *Vocal health and pedagogy, volume II: Advanced assessment and practice* (Vol. 2, p. 133). San Diego: Plural Publishing.
263. Junqua, J. C. (1993). The Lombard reflex and its role on human listeners and automatic speech recognizers. *The Journal of the Acoustical Society of America*, *93*(1), 510–524.
264. Brumm, H., & Zollinger, S. A. (2011). The evolution of the Lombard effect: 100 years of psychoacoustic research. *Behaviour*, *148*(11–13), 1173–1198.
265. Erath, B. D., & Plesniak, M. W. (2006). The occurrence of the Coanda effect in pulsatile flow through static models of the human vocal folds. *The Journal of the Acoustical Society of America*, *120*(2), 1000–1011.
266. Tao, C., Zhang, Y., Hottinger, D. G., & Jiang, J. J. (2007). Asymmetric airflow and vibration induced by the Coanda effect in a symmetric model of the vocal folds. *The Journal of the Acoustical Society of America*, *122*(4), 2270–2278.
267. Farshchi, M., & Fisher, N. (1997). The emotional content of the physical space. In *Proceedings of the Royal Institution of Chartered Surveyors (RICS) COBRA Conference, Portsmouth, New Hampshire, USA* (pp. 10–12).
268. Bouhuys, A., Mead, J., Proctor, D. F., & Stevens, K. N. (1968). Pressure-flow events during singing. *Annals of the New York Academy of Sciences*, *155*(1), 165–176.
269. Lucero, J. C., & Koenig, L. L. (2007). On the relation between the phonation threshold lung pressure and the oscillation frequency of the vocal folds. *The Journal of the Acoustical Society of America*, *121*(6), 3280–3283.
270. Hill, L. E., & Greenwood, M. (1906). The influence of increased barometric pressure on man. *Proceedings of the Royal Society of London B*, *77*(519), 442–453.
271. Vilkman, E., Lauri, E. R., Alku, P., Sala, E., & Sihvo, M. (1998). Ergonomic conditions and voice. *Logopedics Phoniatrics Vocology*, *23*(1), 11–19.
272. Hemler, R. J., Wieneke, G. H., & Dejonckere, P. H. (1997). The effect of relative humidity of inhaled air onacoustic parameters of voice in normal subjects. *Journal of Voice*, *11*(3), 295–300.
273. Hemler, R. J., Wieneke, G. H., Lebacq, J., & Dejonckere, P. H. (2001). Laryngeal mucosa elasticity and viscosity in high and low relative air humidity. *European Archives of Oto-Rhino-Laryngology*, *258*(3), 125–129.
274. Leydon, C., Sivasankar, M., Falciglia, D. L., Atkins, C., & Fisher, K. V. (2009). Vocal fold surface hydration: A review. *Journal of Voice*, *23*(6), 658–665.
275. Feder, R. J. (1984). *The professional voice and airline flight*. Los Angeles, California: Sage Publications.
276. Millot, J. L., & Brand, G. (2001). Effects of pleasant and unpleasant ambient odors on human voice pitch. *Neuroscience Letters*, *297*(1), 61–63.

Chapter 4
The Voice Signal and Its Information Content—1

The voice signal, like all sounds, is a pressure wave. The actions of the speaker's vocal tract result in continuous variations of pressure in the air surrounding the speaker's mouth. These pressure waves radiate outward from the speaker's mouth and are sensed by the listener's ear. The information in the voice signal is encoded in these time variations of air pressure. Any computer-based analysis of voice must first convert these variations into a sequence of numbers that the computer can operate upon. This requires *transduction* of the pressure wave into a sequence of numbers in a manner that assuredly retains most of the information in it with minimal distortion. From the perspective of the computer, this sequence of numbers now represents the voice signal. We refer to the sequence of numbers representing the voice signal as a "digital" signal, and the process of converting the pressure wave into it as "digitization." Subsequent computational procedures must be performed on this digitized signal in order to derive information from it. The sequence of procedures followed for computer analysis of sounds is illustrated in Fig. 4.1a.

In this chapter we discuss how a voice signal is digitized in a manner that its information content is preserved, and the nature and analysis of that information. All of this falls under the purview of signal processing. The discussion herein is concise, and primarily relates to the topics relevant to the analysis of voice. For a more detailed study of digital signal processing, readers are directed to the dozens of excellent books available on the subject, e.g. [1–4].

An important note on notations: The notations used in this chapter and in the rest of this book are not standard in the context of signal processing. In this chapter, the index t for example, denotes time in the case of both continuous-time and discrete-time signals, to explicitly bring across the fact the index represents temporal sequences (which is true for voice), while most standard text books will use a symbol other than t, e.g. n, to represent discrete time. Also, while most conventional text books on signal processing represent discrete-time frequency as a value between 0 and 2π, in this book, frequencies for discrete-time signals are represented as values between 0 and 1, where 1 represents the *sampling frequency* with which the continuous-time signal has been discretized. *Thus the maximum frequency in any*

R. Singh, *Profiling Humans from their Voice*,
https://doi.org/10.1007/978-981-13-8403-5_4

signal is scaled to 1, and the minimum is 0. As long as we remember this extremely important point, the notation in this chapter will be easy to follow. Such a representation not only simplifies explanations from a notational perspective, it is also more intuitive in a setting where comprehending frequency in the physical world is important. Negative frequencies are avoided, since for discrete-time signals, a negative frequency f is indistinguishable from the positive frequency $1 - f$.

Similarly, since this book is intended for a wide range of readers, the explanations provided favor intuitive understanding while maintaining brevity. In a small number of cases this may come at the cost of conceptual exactness from a mathematical perspective. References to more in-depth articles in the literature are provided in those cases.

4.1 Capturing the Voice Signal

The process of "capturing" the pressure wave that represents voice on the computer requires digitizing it. Digitization must be performed such that the resulting sequence of numbers fully and faithfully represents the information in the original pressure wave. The only true criterion to verify that this sequence does indeed retain the information is that we must be able to recreate the original pressure wave *exactly* from it. This is illustrated in Fig. 4.1b. However, this criterion is often relaxed to state that it is sufficient to be able to recreate a pressure wave that is *perceived* by listeners to be similar to the original sound.

4.1.1 Analog Transduction

The first step to digitization is to convert the continuously-varying pressure wave into a continuously-varying analog electrical signal, e.g. a time-varying voltage or current. Such conversion could be done with perfect fidelity by an *ideal* microphone. Note that the term "analog" here specifically refers to a signal that is *continuous-valued* in *continuous-time*.

A microphone generally comprises a membrane or similarly pliable surface that moves in response to the time-varying air pressure incident on it. The motion of the surface results in time-varying changes in electrical properties of the microphones, such as its resistance, capacitance or inductance, which in turn result in continuous variations of voltage or current in attached circuitry.

The time-varying electrical signal may also be converted back to a pressure wave through an ideal *speaker* (short for *loudspeaker*). In a speaker, the electrical signal is used to *move* a physical surface, e.g. by continuously varying the magnetic field in a coil that causes a membrane to move [5]. The motion of the membrane results in pressure waves in the surrounding air, that are heard as sound.

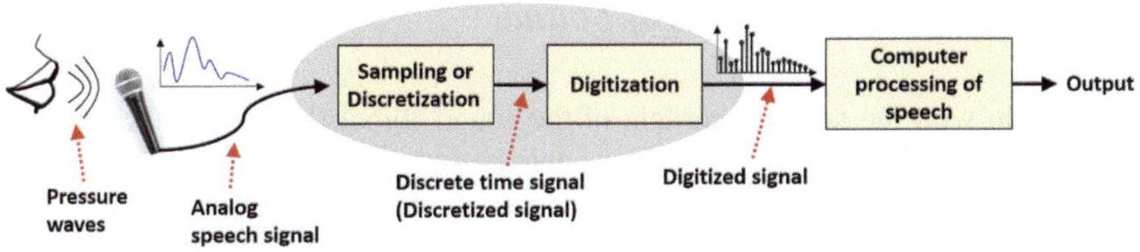

(a) The overall process of capturing voice for computer processing.

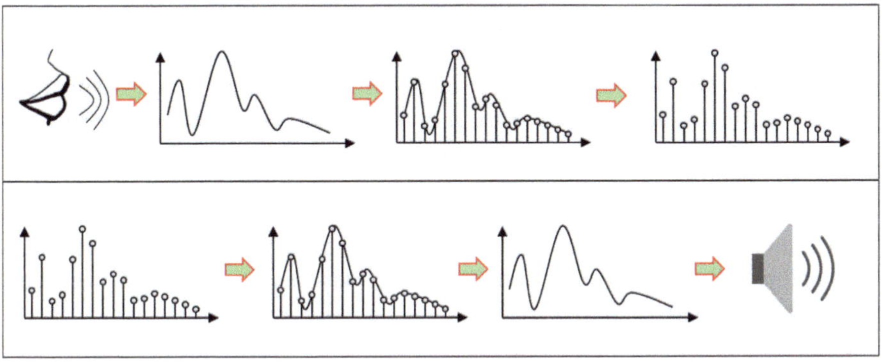

(b) **Top panel:** Sound pressure waves are transduced to an analog electrical signal, which is converted to a sequence of numbers by sensing it at periodic intervals of time. The values are stored as the digital signal. **Bottom panel:** To reproduce the original sound from the digital signal the sequence of numbers is used to construct an analog electrical signal that moves a membrane to create pressure waves. The process of digitization is perfect if the input pressure wave and the reconstructed pressure wave are the same.

Fig. 4.1 The process of digitization

If the transduction is perfect, the pressure wave recorded by the microphone and that produced by the speaker would be perfect facsimiles of each other. The time-varying electrical signal produced by the ideal microphone would trace the time variations of the sensed pressure wave exactly. The ideal speaker, in turn, would exactly recreate the time-varying electrical signal as a pressure wave. In practice, both microphones and speakers are imperfect.

Needless to say, for the purpose of *analyzing* voice, we require the microphone to be as close to ideal as possible. For the purpose of this chapter and book, we assume perfect transduction of the pressure wave into an electrical signal, since imperfections in the transduction process are generally indistinguishable from the effect of noise (external sound sources that are concurrent with voice, causing distortions in the signal) or other imperfections caused by downstream processing. Thus, from our perspective, conversion or recreation of this signal is equivalent to conversion or recreation of the original pressure wave itself. In our discussions below, references to the sound "signal" may be interpreted either as the electrical signal derived from perfect transduction of a pressure wave, or the pressure wave itself. The reference should be generally clear from the context.

4.1.2 *The Spectral Content of Sound*

Let us begin with the *spectral* content of the sound signal. The spectral content of sound can be best understood through an analogy with the physics of an ideal tuning fork. When struck, a tuning fork ideally produces a pure tone—a perfectly sinusoidal pressure wave resulting from the oscillations of the fork at its resonant frequency. The amplitude of the tone depends on the energy with which the tuning fork is struck. Conversely, when the tuning fork is subjected to a sound tone of its resonant frequency, it vibrates. The amplitude of its vibration depends on the sound *intensity* of the incident tone.

In principle, it is possible to impinge a tone on a tuning fork, and sense its vibrations (e.g. as a perfectly transduced electrical current). One can then recreate the tone perfectly at a different location by "exciting" a clone of the original tuning fork by the sensed vibrations of the original.

If we had a *bank* of tuning forks, each of a different resonant frequency as in Fig. 4.2, and if all of them were excited simultaneously, the resulting sound would be an additive mix of the tones produced by the individual forks. The actual signal produced would depend on three factors:

- the *frequencies* of the tuning forks
- the *amplitudes* with which each of them is excited
- the *synchrony* between the forks, or equivalently, what *phase* in the oscillatory cycle the forks are at the "initial" instant of observation.

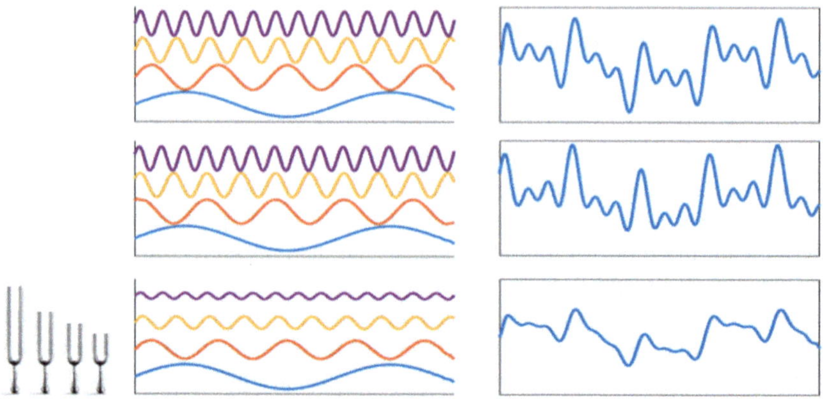

The signal produced by four ideal tuning forks of different resonant frequencies. In each row, the left panel shows the outputs of the individual tuning forks, while the right panel shows the sum signal. In the top row all tuning forks have identical amplitudes and start at the same time. In the second they have identical amplitudes but start asynchronously. In the third, they are synchronous, but have different amplitudes.

Fig. 4.2 Resolution of sound into its spectral components by tuning forks

Figure 4.2 illustrates the effect of varying the amplitudes and synchrony/phase of the tuning forks.

Conversely, any sound that is exactly a mixture of tones at the resonant frequencies of the tuning forks would excite the forks. If we were to measure the amplitudes and synchrony with which the forks vibrate, we could recreate the sound exactly by simply exciting an identical bank of tuning forks with the corresponding amplitudes and synchrony.

The key point here is that this procedure can *only* recreate sounds that are exactly the sums of the resonant tones of the tuning forks in our bank. If the original signal included other components comprising tones at other frequencies, those would not be registered. The recreated sounds would only include those components of the original signal that correspond to the tones of the forks in our bank; the rest would be absent.

For more complex signals, we would therefore require larger banks of forks. For the purpose of analysis, consider (completely unrealistic) tuning forks that, when excited, continue to vibrate forever. The summed signal produced by the bank of forks will be periodic, with a period that is the least common multiple of their respective individual periods. In order to "record" a signal that never repeats, we would require infinite tuning forks. We will not discuss the mathematical connotations of infinity here, or the difference between summation and integration that results from it, since computational analysis of discrete-time signals (explained later in this chapter), deals with discrete sets of frequencies.

We say that a sound *contains* a discrete set of frequencies (or frequency *components*) if it can be perfectly recreated by a bank of ideal tuning forks with those resonant frequencies. We refer to these frequency components and their amplitudes as the **spectral** content of the signal.

Note that the analogy to tuning forks drawn above is somewhat naïve, in that real tuning forks do not quite behave in the idealized manner described above. They do not produce pure tones, nor respond exactly and only to a single tone, nor do they continue to ring forever uniformly. They also do not capture the synchrony between different tones exactly as suggested. The analogy above nevertheless remains valid for the discussion in this chapter. Also, with reference to Chap. 2, the analogy is particularly appropriate since the basilar membrane and hair cells in the human ear behave very similarly to tuning forks.

4.1.3 Digitizing the Analog Signal

We return to the problem of digitizing audio signals. The outcome of the analog transduction described in Sect. 4.1.1 must be converted to a sequence of numbers on the computer. We must take into consideration the following three issues when doing so:

1. Actual sound signals are *continuous*, i.e. they are present at *every* instant of time, and there are uncountably infinite such instants. The finite set of numbers on the computer must represent all infinite instants of the original sound.
2. Most real-life sounds include an infinity of frequencies, i.e., using the tuning fork analogy above, they would need an infinity of tuning forks to characterize perfectly. Once again, we are required to represent this entire infinity with the finite set of numbers we use to represent the sound on a computer.
3. Sound signals are *continuous valued*, i.e. the signal can take *any* value within the limits of the physical (or electrical) medium within which they occur. Digital computers, on the other hand, store numbers with finite precision, which are generally unable to take more than 2^K values, where K is the number of bits used to represent a number.

We deal with each of these issues in order below.

4.1.3.1 Discretization, Sampling and Sampling Frequency

We first consider the problem of converting a *continuous-time* (i.e. a continuously varying) signal into a discrete set of numbers. We refer to this process as *discretization*, and the resulting sequence of numbers as a *discrete-time* signal.

The simplest mechanism for discretizing a continuous-time signal is to capture its value at discrete instants of time (as illustrated in Fig. 4.1). The process of taking "snapshots" of the signal at discrete instants of time is called "sampling," and the instants at which the snapshots are taken are the "sampling instants." The measurements we obtain at these discrete instants must, however, be such that they are sufficient to "fill in" the rest of the values of the signal between the measured instants unambiguously, and accurately.

To enable unambiguous reconstruction, we must make a number of assumptions. We assume that the sampling instants are equally spaced in time,[1] and that this spacing ΔT is known. We will refer to this spacing as the "sampling period." The number of samples captured per second is known as the *sampling frequency*, and is given by $f_{samp} = \frac{1}{\Delta T}$. Its unit of measurement is *samples per second.*

This assumption of uniformly-spaced sampling instants immediately imposes some constraints on the types of signals we can represent. A sampling frequency of f_{samp} cannot distinguish between tones of frequency f_{samp} and Nf_{samp}, where N is any integer. In other words, given two tones for frequency $N_1 f_{samp}$ and $N_2 f_{samp}$ for any two integers N_1 and N_2, the discretized samples we would obtain using a sampling frequency f_{samp} would be indistinguishable. Also, more generally, we cannot distinguish between tones of frequency $\frac{f_{samp}}{2} - \delta f$ and $\frac{f_{samp}}{2} + \delta f$ for any δf. This is illustrated in Fig. 4.3. Further information and proof of this can be found in [6, 7].

[1]While non-uniform sample spacing is also possible, it is not general practice in sampling audio signals.

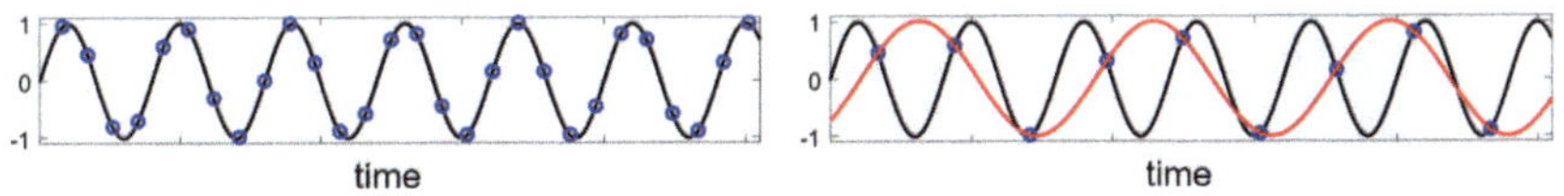

The panels show the discretization of a sinusoidal analog signal. The blue circles show the discrete samples captured. In both panels, the black curve shows several cycles of a continuous -time sinusoid with frequency $f = 400$ Hz. **Left:** If the signal is sampled more than 800 samples per second ($\frac{f_{samp}}{2} > f$), there is a unique way of interpolating between the samples which reconstructs the original sinusoid. **Right:** If the signal is sampled at less than 800 samples per second ($\frac{f_{samp}}{2} < f$), then interpolation results in a different frequency f' that is lesser than $\frac{f_{samp}}{2}$ by the same amount that f is above it. Here, interpolation between samples obtained at 600 samples/sec results in a frequency of 200Hz (in red), rather than 400Hz.

Fig. 4.3 Correct and incorrect sampling

To resolve this ambiguity, we must agree that the *reconstruction* process will only include frequency components up to $\frac{f_{samp}}{2}$. In turn, if we require perfect reconstruction of the originally discretized signal, we also require that the original signal itself must have no components at frequencies[2] greater than $\frac{f_{samp}}{2}$. Alternately stated, if we wish to discretize a signal with frequency components up to f_{max} Hertz (Hz, henceforth), we must sample it at at least $f_{samp} = 2f_{max}$ samples per second.

For a signal with a highest frequency component at f_{max} Hz, $2f_{max}$ is called the *Nyquist* sampling rate for the signal [6]. If a signal is sampled at less than the Nyquist rate, this results in an effect called *aliasing*—signal components at $\frac{f_{samp}}{2} + \delta f$ in the original signal manifest as components at $\frac{f_{samp}}{2} - \delta f$ in the sampled signal.

Figure 4.4 shows the effect on sampling below the Nyquist rate on the reconstruction of a signal. The signal represented here is a single tone whose frequency varies with time. This is represented as a two-dimensional image called a spectrogram, where the horizontal axis represents time, the vertical axis represents frequency, and the brightness of the image at any time, at any particular frequency, represents the energy in that frequency component at that instant of time.

In the example shown, the original signal sweeps linearly through frequencies from 0 Hz to 20 kHz. The signal is discretized at three different sampling rates (one for each panel), and then reconverted into an audio signal from those frequencies. Figure 4.4a shows the spectrograms of the recomposed signals. The spectrogram of the original signal sampled at 48 kHz[3] is shown in the first inset in Fig. 4.4a. The reconstructed signal faithfully reproduces the original. As we see in the subsequent insets, when the sampling rate falls below twice the current frequency of the original

[2]We can also obtain unambiguous and perfect reconstruction of signals with frequency components greater than half the sampling frequency, provided the signal is bandlimited, the limits on the frequencies of the signal are known, and the sampling frequency is carefully chosen to enable such reconstruction. Audio signals do not generally satisfy these conditions, however.

[3]Note that although "Hz" is used here as a unit for sampling rate, in its strictest definition "Hz" refers to *cycles-per-second* with reference to waves of any kind.

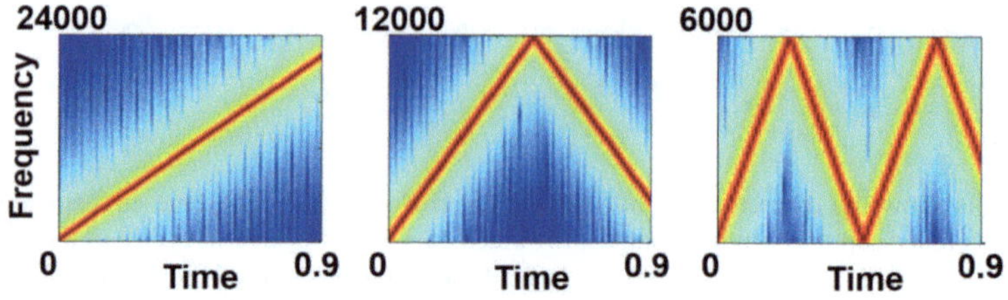

(a) Spectrogram of a chirp signal with frequencies sweeping from 0 Hz to 20,000 Hz in 1 second. **Left:**Sampled at 48 kHz, **Middle:** Sampled at 24 kHz and **Right:** Sampled at 12 kHz.

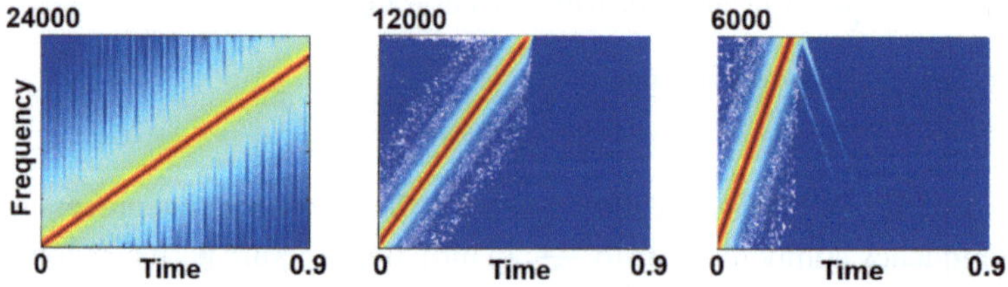

(b) The examples from above when the signals have been filtered by an anti-aliasing filter prior to sampling. Note that frequencies above half the sampling frequency are now eliminated and their aliased "ghosts" do not reappear at lower frequencies.

Fig. 4.4 Aliasing and the effect of an anti-aliasing filter

signal, not only does the reconstructed signal not represent the original signal, it actually recomposes as an entirely different signal with a different set of frequencies than those present in the original signal. Frequencies above half the sampling rate "fold over" and appear as other frequency components below half the sampling rate. Thus, sampling below the Nyquist rate not only *loses* information in the frequency components above half the sampling rate, it also *distorts* the frequency components *below* it. Thus, it becomes important to sample at the Nyquist rate or higher, in order to retain the fidelity in the signal.

4.1.3.2 Discretizing the Voice Signal

Real-life audio signals (including voice), however, do not have an actual uppermost frequency. While they may be inaudible, there is no true upper limit on the frequency components present in the signal. Thus, sampling at *any* frequency will result in a distortion of *all* frequency components of the signal. Not only will it result in the loss of the frequency components *above* half the sampling rate, it will also cause misrepresentation of the frequency components below it.

In order to avoid such distortion, it is customary to first *filter* the continuous-time signal using an analog *anti-aliasing* filter before sampling, in order to eliminate all frequency components above half the sampling frequency, to prevent them from "folding over" and distorting lower-frequency components. The effect of an anti-aliasing filter is illustrated in Fig. 4.4b.

The human ear can perceive frequencies within (approximately) the range 20 Hz–20 kHz, depending on the listener's age, with the higher end of frequencies audible primarily to younger people. The most prominent frequency components in normally produced *speech* sounds usually lie in the range of 50–6800 Hz. Commercially available hardware for audio capture is usually designed to retain the most important frequency components of the speech signal, but audio capture is also constrained by other logistics such as restrictions of the hardware, desired fidelity, storage or transmission capacity, and the natural bandwidth of the signal. Consequently, commercial hardware typically supports several standard sampling rates, e.g. 8, 16, 11.025, or 44.1 kHz. In most practical applications, a sampling rate of 16 kHz captures nearly all the relevant components of the signal. However, if the signal has been recorded over a telephone channel, a sampling rate of 8 kHz is assumed to suffice, since the telephone channel generally filters out frequencies above 4 kHz.

4.1.3.3 Digitizing the Discretized Signal

Discretization of the signal is only half the story in the conversion of audio signals to computer-analyzable series of numbers. As mentioned earlier, the actual audio signal is *continuous-valued.* Digital computers, however, can only record and store a discrete set of numbers. A device which stores numbers using an N-bit representation cannot store more than 2^N unique values for samples in the signal.

This means that each continuous-valued number must be *quantized* into one of the 2^N values. *Uniform quantization* maps uniform ranges of the analog signal into corresponding integers according the relation $x = \lfloor \frac{s}{\Delta s} \rfloor$, where x is the *digitized* value stored, s is the original analog signal, Δs is quantization step size, and $\lfloor . \rfloor$ represents a flooring operation. Figure 4.5a shows the mapping effect of a uniform quantizer. The digitized value x is directly proportional (to within the *quantization error* introduced by flooring) to the original signal, and may be directly used in lieu of the original

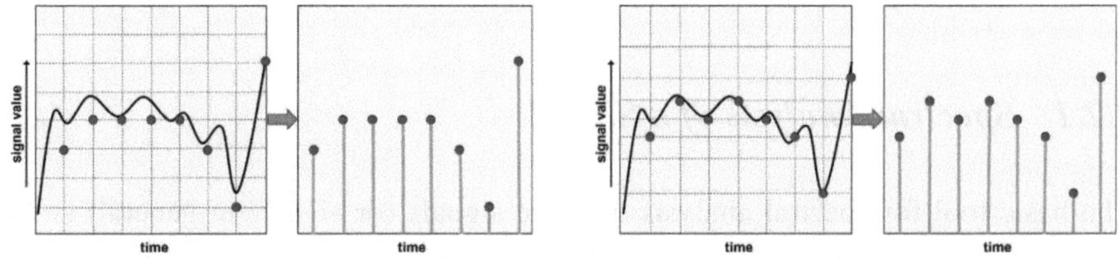

(a) Uniform quantization. Grid lines are uniformly spaced. Analog signal values are floored to their nearest quantized level.

(b) Nonuniform quantization. Grid lines are non-uniformly spaced. The resulting quantized values are different from uniform quantization.

Fig. 4.5 Uniform and non-uniform quantization

signal for further analysis, i.e. we may assume that $\hat{s} = x$ represents the actual signal at each time, and perform our analyses on this value. For sufficiently large N (bits per sample), the largest error $0.5\Delta s$ is sufficiently small to not have a significant influence on further analyses of the signal. For effective analysis of speech signals $N = 16$ bits is generally considered to be sufficient.

The sample values of the speech signal generally tend to lie largely within a small range around their mean value (with respect to the dynamic range of values that are usually available to represent them on a digital device). For speech signals specifically, *non-uniform* quantization attempts to achieve greater fidelity in representation by having smaller quantization steps closer to the mean value of the signal, and larger ones farther away. Quantization is now performed as $x = \lfloor g(s) \rfloor$, where $g(.)$ is a squashing function that compresses the dynamic range of the signal non-linearly. As a result, most of the time, when the signal lies close to its mean, the error introduced by quantization is small, although it comes at the cost of larger error when the signal wanders far away from its mean. Note that this is not the primary reason why non-uniform quantization is done for audio signals in general—the reasons come from psychoacoustics, but we will not discuss them here.

Figure 4.5b shows the mapping function used by a typical non-uniform quantizer.

For non-uniformly quantized signals, the digitized value x does not directly represent the actual signal. Instead, an approximation of the signal must be derived as $\hat{s} = g^{-1}(x)$. The inverse mapping is usually stored in a mapping table. The two most popular standards for non-uniform quantization are μ-law and A-law quantization [8], both of which use only 8 bits to represent the signal, but achieve the same average error as a 12-bit uniform quantizer.

4.2 Analyzing the Digital Signal

The digitized speech signal is only a sequence of numbers that by itself conveys little information. In order to extract information from it, it must first be decomposed into more detailed characterizations, from which information may be extracted.

The most common decompositions employed in the analysis of voice signals utilize *spectral* characterizations of the signals, which attempt to decompose them into their frequency components.

4.2.1 Spectral Analysis of a Signal

The basic tool for spectral analysis of voice signals (or signals in general) is the Fourier transform. The Fourier transform decomposes the signal into the sum of a set of sinusoids (or, more precisely, complex exponentials), representing the frequency components in it.

4.2.1.1 The Discrete-Time Spectrum

Let $s[t]$ represent a digitized audio signal. The sample index t represents time, but only takes integer values since the signal is discretized (i.e. it is a *discrete-time* signal). $s[t]$ specifically represents the tth sample of the signal, although we will also sometimes use it to refer to the signal in general.

The *discrete time Fourier transform* (DTFT) characterizes the signal as a weighted combination of complex exponentials.

$$s[t] = \int_0^1 S(f)e^{i2\pi ft}df \tag{4.1}$$

where $i = \sqrt{-1}$ and f represents *discrete-time frequency*. $S(f) = |S(f)|\exp(i\phi_f)$ is a complex-valued coefficient whose magnitude $|S(f)|$ specifies the contribution of the frequency component at frequency f and ϕ_f is the initial *phase* of that component. Coefficients $S(f),\ \forall f \in [0, 1]$ are referred to as the DTFT coefficients (or simply as the DTFT) of the signal.

Note that the discrete-time frequency f is assumed to lie between 0 and 1. This is because for discretized signals (with integer t) higher frequencies cannot be distinguished from frequencies in the range [0, 1], since $\exp(2i\pi ft) = \exp(2i\pi(f + l)t)$ for any integer l. Since discrete-time signals are generally derived by sampling continuous-time signals, it is convenient to think of discrete-time frequencies in terms of their corresponding continuous-time frequencies. Thus, f represents the *fraction of the sampling frequency*. A frequency of 1 for the discrete-time signal corresponds to the sampling frequency f_{samp} in continuous time.

The DTFT coefficients themselves can be computed as

$$S(f) = \sum_{-\infty}^{\infty} s[t]\exp(-i2\pi ft) \tag{4.2}$$

Note that the decomposition is expressed in terms of the complex exponential $\exp(2i\pi ft)$ rather than pure tones (Sines and Cosines). The two are related through Euler's formula: $\exp(2i\pi ft) = \cos(2\pi ft) + i\sin(2\pi ft)$, or alternatively, cos $(2\pi ft) = 0.5(\exp(2i\pi ft) + \exp(2i\pi(1 - f)t)$. The complex exponential also confers on us the benefit that we are now able to decompose *complex-valued* signals.

The DTFT of the signal is, in some sense, the "true" spectrum of the signal that computes the exact contribution of every frequency component to the signal.

An aperiodic signal may have an infinity of frequency components within this range, i.e. $S(f)$ can have non-zero values for every real f from 0 to 1, and a complete characterization of the signal will require specification of all of these values. For real valued signals, the spectrum at a frequency f will be the complex conjugate of its spectrum at $1 - f$, i.e. $S(f) = conj(S(1 - f))$, hence, it is sufficient to specify spectral values for $f \in [0, \frac{1}{2}]$. This ties back to the fact that a discrete-time signal

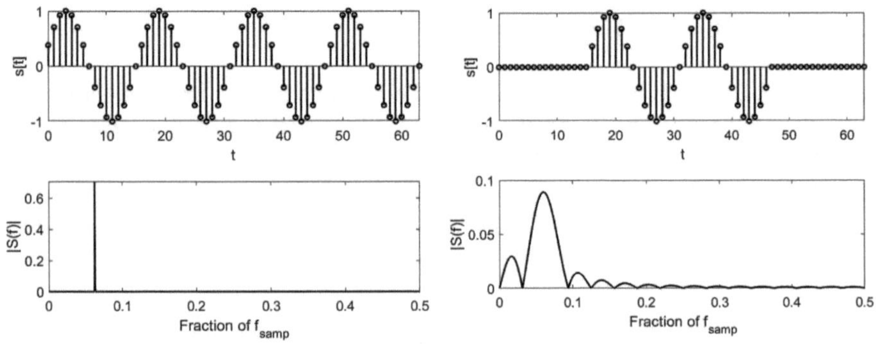

(a) Infinite-duration discrete-time sine and its spectrum.

(b) A signal comprising exactly one cycle of a sine wave, and its spectrum.

Fig. 4.6 Effect of truncating a sine on its spectrum

obtained by sampling a continuous-time signal at a rate f_{samp} can only represent frequencies up to $\frac{f_{samp}}{2}$ in the continuous-time signal.

4.2.1.2 Windowing

Computation of the DTFT requires the knowledge of all samples of the signal from $t = -\infty$ to $t = \infty$ (as evidenced from Eq. 4.2). In practice, all signals (especially all voice signals) we work with are of finite duration. The natural assumption that these signals are in fact zero-valued outside this duration can lead to some unfortunate consequences, as illustrated by Fig. 4.6. The spectrum of the perfect tone should ideally show just a single frequency component corresponding to the frequency of the tone; however, when the tone is truncated in duration, the spectrum obtained has a rippled structure with spurious peaks, called "sidelobes," spread over the entire spectrum.

In order to reduce the spurious structure in the computed spectrum, it is customary to apply a "window" function to the signal prior to computing its spectrum, which tapers the ends of the signal. The window function has the effect of minimizing the sidelobes in the spectrum at the cost of "smearing" or "smoothing" the spectrum itself, and the degree of smoothing depends on the actual window function applied. In fact, not applying any window may itself be viewed as a case of applying a "rectangular" window. Figure 4.7 shows the spectra computed for a tone, applying a variety of different window functions. Each of them provides a different degree of tradeoff between the level of sidelobes and the smearing of the actual peaks in the spectrum.

The choice of window function is often a key issue in the analysis of voice signals. In situations where the aim is to derive a good estimate of the overall "envelope" of the spectrum of the signal, a window function that maximally suppresses spurious

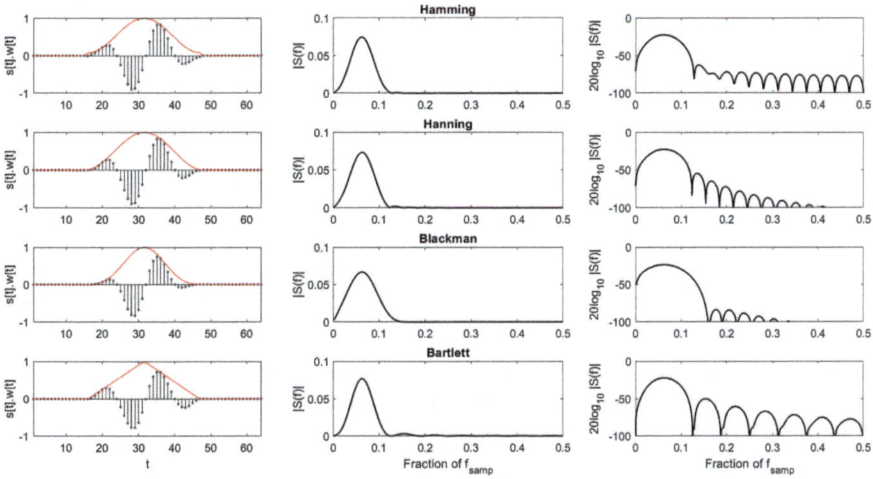

Windowed spectra computed for the partial sine from Figure 4.6(b). **Left column:** Windowed signal in the time domain. The windows are shown in red. **Middle column:** Magnitude spectrum of the windowed signal. **Right column:** The spectrum of the windowed signal shown on a dB scale.

Fig. 4.7 Effect of different windows on the signal spectrum

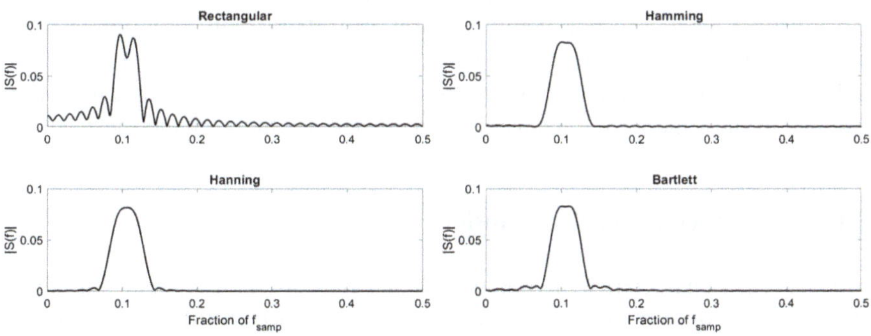

(a) The spectrum of a 64-sample sequence comprising the sum of two tones of 2250 and 2500 Hz respectively, sampled at 16000 Hz, computed using different window functions.

Fig. 4.8 The effect of windowing on frequency resolution

sidelobes is desirable. On the other hand, when the goal is to resolve closely-spaced high-energy spectral components, then tapering the signal may be less desirable. Figure 4.8 shows this through examples of the spectra of a finite length of a signal comprising a sum of tones. As can be seen, only the rectangular window is able to resolve the frequencies in this example while the tapered windows cannot, although this comes at the expense of significant spurious sidelobes.

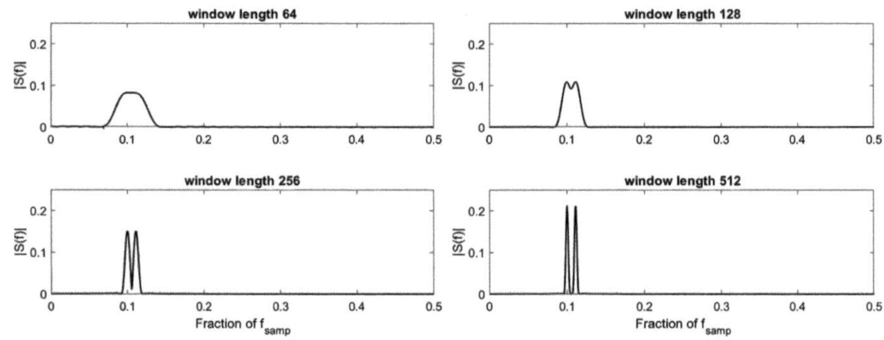

The Hamming-windowed spectrum of the sum of two tones of 2250 and 2500 Hz, sampled at 16000 Hz for analysis windows of different lengths. The frequency components and their actual energies are better resolved for longer windows.

Fig. 4.9 The effect of window length on frequency resolution

The *width* of the window also affects the resolution of the spectrum. Wider windows permit greater resolution between close frequencies for all window functions. This is also seen in Fig. 4.9. For an M-sample segment of a signal sampled at sampling rate f_{samp}, two tones must be at least $\frac{f_{samp}}{M}$ apart in frequency for them to be resolvable as two distinct tones, and not merely spectral artifacts resulting from finite signal length, if no windowing is performed (i.e. using a rectangular window). If tapering windows are employed, the resolution decreases to $\frac{2f_{samp}}{M}$. In either case, increasing the length of the window M increases the frequency resolution of the spectrum.

4.2.1.3 The Discrete Fourier Transform

In order to fully characterize the spectrum of a signal, we must compute all infinite spectral values for every frequency between 0 and 1 (or 0 and 0.5 for real-valued signals). Computing and manipulating all of these is generally infeasible. However, for finite-duration signals there is a computational approximation that is possible. A finite-duration signal may be viewed as a single period of an infinitely long periodic signal (Fig. 4.10). A discrete-time periodic signal with period M can be expressed entirely as a sum of a discrete set of M frequency components with frequencies k/M, $k = 1 \dots M$ [1]. Consequently, for a signal of length M, we get the following relation:

$$s[t] = \frac{1}{M} \sum_{k=0}^{M-1} S[k] e^{\frac{i2\pi kt}{M}} \tag{4.3}$$

where $e^{\frac{i2\pi kt}{M}}$ is a complex exponential with period k. $S[k]$, $k = 1 \cdots K$ are known as the *discrete Fourier transform* (DFT) of the signal. The DFT values can be computed in turn as

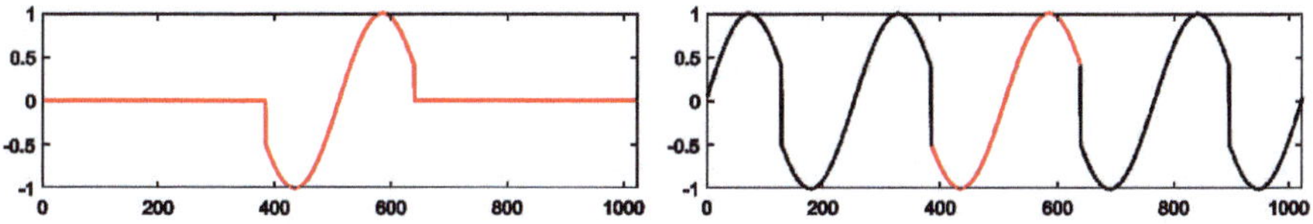

Left: A finite-length segment of a perfect sinusoid. **Right**: This is viewed as one period (in red) of an infinite-length periodic signal. The periodic signal that is analyzed is not sinusoidal, even though the original finite signal is from a sinusoid.

Fig. 4.10 A finite-length signal viewed as one period of a periodic signal

$$S[k] = \sum_{t=0}^{M-1} s[t]e^{-\frac{i2\pi kt}{M}} \tag{4.4}$$

The DFT computes DTFT values at a discrete set of frequencies, $f = \frac{k}{M}$, $k = 0 \ldots M - 1$; however, these are sufficient to recover the original signal. DTFT values at other frequencies may be obtained by padding the signal with zeros prior to computing the DFT. In fact, it is customary to zero-pad signals to a length that is a power of 2, in order to be able to utilize the Fast Fourier Transform (FFT), an algorithm that efficiently computes DFTs of signals of lengths that are powers of two.[4]

For a real signal, $S[k] = conj(S[M - k])$. As a result, only $S[k],\ k = 0 \cdots \frac{M}{2}$ are required to fully describe the spectrum of the signal. $S[k]$ represents the spectral component of the continuous-time counterpart of $s[t]$ at $\frac{kf_{samp}}{M}$ Hz.

4.2.1.4 Magnitude and Power Spectra

Both the DTFT and the DFT of a signal consist of complex-valued parameters characterizing its frequency composition. The *magnitude spectrum* of a signal $s[t]$ is simply the magnitude of its DTFT or DFT coefficients, namely $|S(f)|$ or $|S[k]|$.

The *energy spectrum* of the signal is the square of the magnitude spectrum, i.e. $|S[f]|^2$ (or $|S(k)|^2$).

The term "energy" here is derived from analogy to the underlying analog electrical signal that was sampled (to produce $s[t]$), which in turn, characterizes a physical process (pressure wave) that does have energy in a very physical sense. Assuming that the pressure wave is ideally transduced, and that the digitization of the resulting electrical signal is also perfect, the value $energy(s) = \sum_{t=-\infty}^{\infty} s[t]^2$ computed from the digitized signal is proportional to the energy in the physical sound wave. The energy spectrum value $|S(f)|^2$ is in turn proportional to the energy in the corresponding frequency component of the original pressure wave.

However, the term "energy" is not always appropriate in this context. The energy required to generate a sound—a pressure wave—of infinite duration is infinite. In

[4]In fact FFTs exist for signals of any length. Power-of-2 FFTs are preferred due to their simplicity.

such signals, it is more appropriate to speak of signal *power*, i.e. energy per unit time. This principle also transfers to the domain of digital signals.

If the signal $s[t]$ is a finite segment of an otherwise infinite-duration signal of unbounded energy, it is more appropriate to view the spectrum $S(f)$ (or the DFT $S[k]$) as representing the *instantaneous* characteristics of the (infinite-duration) signal in the brief window of time from which the finite-length $s[t]$ was extracted. In such a situation, the squared magnitude spectrum $|S(f)|^2$ (or $|S[k]|^2$) is more appropriately defined as the *power spectrum* of the signal.

To be more didactic, the DTFT $S(f)$ does not actually represent the specific frequency component at frequency f—there being an infinity of frequencies in the range [0, 1], attributing non-zero power to individual frequencies would amount to attributing infinite power to the signal. Rather, $S(f)$ is better interpreted as representing the power in an infinitesimal *band* of frequencies around f. Consequently, $|S(f)|^2$ is better explained as the *power spectral density* (PSD) of the signal (thus the integral of $|S(f)|^2$ over f is the power in the signal). To be even more didactic, $|S(f)|^2$ is an *estimate* of the PSD of the infinitely long signal from which the finite-length segment $s[t]$ was extracted. The reader is referred to textbooks on stochastic processes for more information on the topic, e.g. [9].

4.2.1.5 Autocorrelation and Power Spectra

The *self-similarity* of a signal $s[t]$ is quantified by its *autocorrelation function*, or ACF, which is itself a time-series. The ACF of a signal $s[t]$ at lag τ, denoted as $R_{ss}[\tau]$, is the *expected* value of the product $s[t]\bar{s}[t+\tau]$, i.e.

$$R_{ss}[\tau] = E[s[t]\bar{s}[t+\tau]] \tag{4.5}$$

where $\bar{s}[t] = conj(s[t])$, the complex conjugate of $s[t]$ and $E[.]$ represents the statistical expectation operator.

To interpret the ACF, consider a simple sinusoid with period T. For the sinusoid, $R_{ss}[\tau]$ too will be a perfect sinusoid with period T; specifically $R_{ss}[0] = R_{ss}[kT]$ for any integer k, indicating that the signal repeats itself perfectly with period T. This is illustrated in Fig. 4.11a. The true significance of the *statistical expectation* of self-similarity emerges when the signal is not perfectly periodic, but rather has an underlying periodic structure embedded in noise. The ACF can extract the underlying periodicities even for such signals, as illustrated by the example in Fig. 4.11b.

In principle, the ACF computes the expected value of $s[t]\bar{s}[t+\tau]$ over *all* time. In practice, signals are of finite duration. Also, they tend to be *non-stationary*—i.e., their characteristics change with time. Hence the ACF is typically estimated from relatively short segments of time, such that the estimate quantifies the self-similarity of the signal within that segment.

The autocorrelation of a *finite length* signal of length N may be estimated as

$$R_{ss}[\tau] = \frac{1}{N-\tau}\sum_{n=0}^{N-1} s[t]\bar{s}[t+\tau] \quad \tau = 0, 1, 2, \ldots, N-1 \tag{4.6}$$

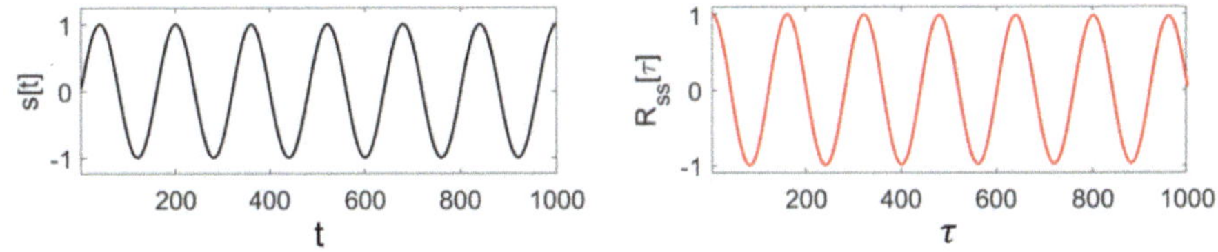

(a) A sinusoid (left) and its autocorrelation (right). The ACF has been normalized such that the value at $\tau = 0$ is 1.

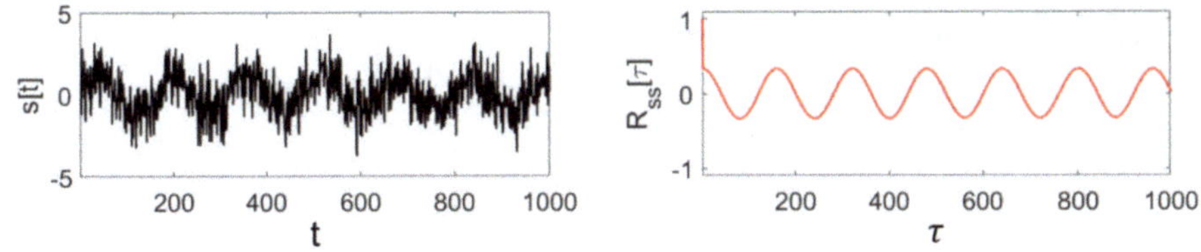

(b) The same sinusoid embedded in high noise (left) and its autocorrelation (right).

Fig. 4.11 Statistical self-similarities captured by the autocorrelation function

where N is the number of samples in the time series $s[t]$, τ is the lag, and it is assumed that $s[t] = 0, \ t \geq N$. The estimator in Eq. 4.6 is only one of several variants proposed in the literature, which differ based on whether or not the signal is windowed prior to computation of the ACF, and in the normalizing denominator $N - \tau$ (which may be set to N), among other possibilities.

The ACF also relates to the power spectral density of the signal, which, in principle, is the Fourier transform of the ACF [4].

4.2.1.6 Filtering

The most basic operation performed on any signal is that of *filtering*. A filter $h()$ is any operation on a signal $s[t]$ to produce a modified signal $y[t]$. The most commonly used filters are causal *linear* filters which take the form

$$y[t] = \sum_{k=0}^{K} a_k s[t-k] + \sum_{l=1}^{L} b_l y[t-l] \tag{4.7}$$

The effect of the filter is to modify the spectrum of the signal by enhancing some frequency components while suppressing others. Figure 4.12 shows a typical example of a signal whose spectrum has been modified by filtering. In principle, filters can modify the spectra in arbitrary ways. The most basic filtering operations however are **low-pass** filtering, which suppresses high frequencies with respect to lower ones, **band-pass** filtering, which enhances a band of frequencies while suppressing all frequencies on either side, and **high-pass** filtering, which suppresses low frequency components with respect to the higher frequencies. Figure 4.12 shows instances of all three types of filters.

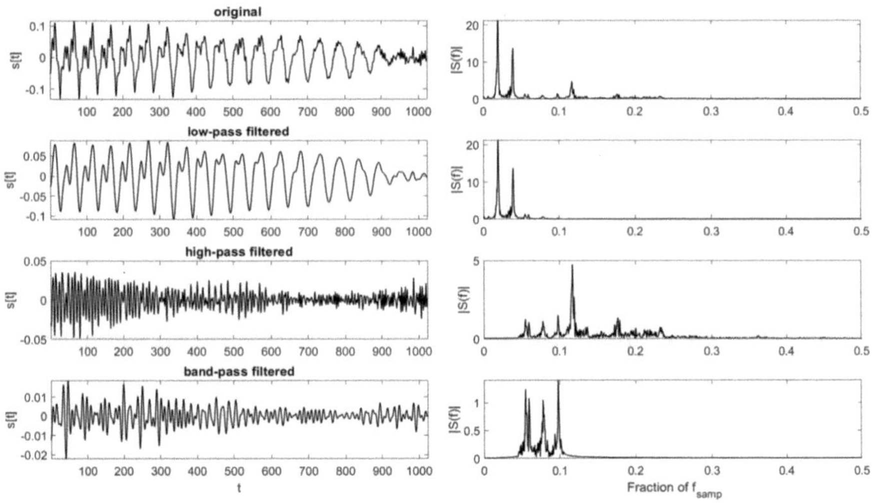

Top row: A segment of a speech signal and its spectrum **Second row:** The same signal and its spectrum after it has been filtered by a low-pass filter **Third row:** The same as the second row for a high-pass filter **Fourth row:** The same as the second row for a band-pass filter.

Fig. 4.12 Effect of filtering on the signal and spectrum

The actual effect of the filter on the spectrum of the signal can be inferred from the *impulse response* of the filter. The impulse response is the signal that is produced by the filter when the input is just a single impulse at time $t = 0$, and can generally be determined from the filter Eq. 4.7 by setting $s[0] = 1,\ s[t] = 0$ for $t \neq 0$.

Let $h[t]$ be the impulse response of the filter. It can be shown that for any signal $s[t]$

$$y[t] = \sum_{k=-\infty}^{\infty} s[k]h[t-k]$$

The above operation is known as a "convolution," and is represented as

$$y[t] = s[t] \otimes h[t]$$

.

Let $H(f)$ be the DTFT of $h[t]$. $H(f)$ is often called the "spectral response" of the filter. The effect of the filter on a signal $s[t]$ with DTFT $S(f)$ is to modify it to

$$Y(f) = H(f)S(f)$$

i.e. the spectral response multiplies the spectrum of the signal to yield the DTFT of the filtered signal, $Y(f)$. Figures 4.13 and 4.12 illustrate these through examples.

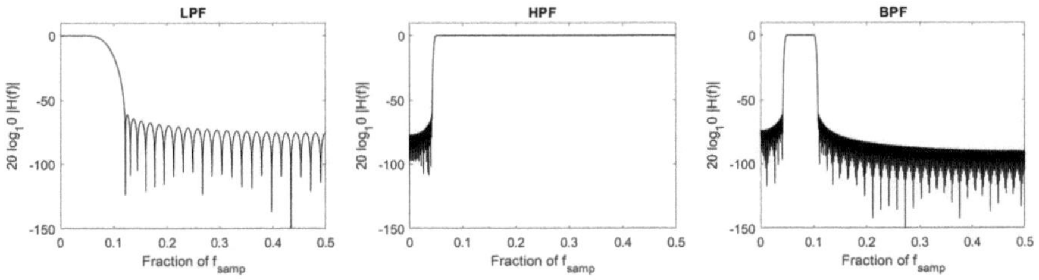

(a) Filter responses of the low-pass (LPF), high-pass (HPF) and band-pass (BPF) filters that modify the spectrum of the example in Figure 4.12 to produce the filtered versions of the signal.

Fig. 4.13 Filter responses of some filters

4.2.2 Analytic Models for Signal Analysis

4.2.2.1 Poles and Zeros

Linear filters are often expressed in terms of the "z-transform." The z transform of any signal $s[t]$ is given by $S(z) = \sum_{t=-\infty}^{\infty} s[t]z^{-t}$, where $z = re^{i\theta}$ is a complex number. While the intricate details of z transforms are beyond the scope of this chapter (there are many textbooks on the subject e.g. [1, 4]), it is nevertheless useful to cover some basics. The discussion below and in the rest of this chapter assumes *causal* filters where the output at any time only depends on current and past inputs and past outputs.

Consider the linear filtering operation $\mathbf{y} = h(\mathbf{s})$ (where $\mathbf{y} = y[t],\ t = -\infty, \ldots, \infty$ and $\mathbf{s} = s[t],\ t = -\infty, \ldots, \infty$ represent entire signals), given by

$$y[t] = \sum_{k=0}^{K} a_k s[t-k] + \sum_{l=1}^{L} b_l y[t-l] \tag{4.8}$$

The z transforms of $y[t]$ and $s[t]$ are related by $Y(z) = S(z)H(z)$, where $H(z)$ represents the *transfer function* of the filter and can be shown to be the following polynomial:

$$H(z) = \frac{\sum_{k=0}^{K} a_k z^{-k}}{1 - \sum_{l=1}^{L} b_l z^{-l}} \tag{4.9}$$

The spectral response of the filter $H(f)$ can be computed by setting $r = 1$ and $f = \frac{\theta}{2\pi}$.

Since the numerator and denominator of $H(z)$ are polynomials, we can rewrite it as

$$H(z) = \frac{z^{L-K} \prod_{k=1}^{K} (z - z_k)}{\prod_{l=1}^{L} (z - p_l)} \tag{4.10}$$

The z_k terms are the *zeros* of the filter. For any $z_k = r_k e^{i\theta_k}$, $\frac{\theta_k}{2\pi}$ are the frequencies that are suppressed by the filter. r_k quantifies the degree of suppression. When $r_k = 1$, the corresponding frequency component is entirely eliminated by the filter.

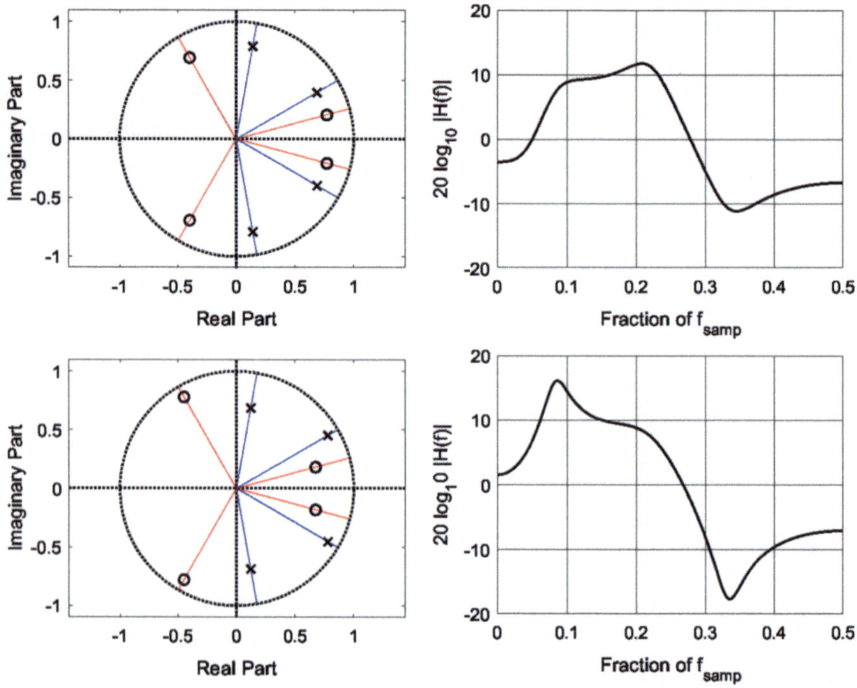

Pole-zero plot for two filters. Crosses represent poles and circles represent zeros. The plots in the lower panel show a filter with poles and zeros (indicated by circles and crosses respectively), at exactly the same frequency (angle) as in the upper panel, but their distances r_k and q_l from origin have been changed. While the positions of the peaks and valleys in the frequency response remain relatively unchanged (right panels), their actual values change.

Fig. 4.14 Pole-zero plots for a filter and its frequency response

The p_l terms are the *poles* of the filter. For any $p_l = q_l e^{\phi_l}$, $\frac{\phi_l}{2\pi}$ are the frequencies that are *boosted* by the filter. The filter may be viewed as a resonant system, and the poles correspond to the resonances of the filter and $\frac{\phi_l}{2\pi}$ are its resonant frequencies. The degree to which the resonant frequencies are boosted depend on q_l. A q_l value of 0 implies that the frequency is unmodified. As q_l increases to 1 the degree to which the corresponding frequency is enhanced increases rapidly. At $q_l = 1$ or more, the corresponding frequency component increases with time and the signal "blows up", i.e. the filter becomes unstable.

The spectral response of the filter also shows clear valleys and peaks at the frequencies corresponding to the zeros and poles of the filter. This is illustrated in Fig. 4.14.

4.2.2.2 Parametric Estimation

Parametric estimation methods model the signal as the outcome of a linear filter operating on an "excitation." By appropriately characterizing the filter and its excitation to best explain the signal, the signal itself may be characterized.

The most commonly used parametric model is the *auto-regressive* or AR model, given by

$$s[t] = \sum_{k=1}^{P} a_k s[t-k] + e[t]. \tag{4.11}$$

The signal $s[t]$ is modeled as the output of a recurrent ("auto-regressive") linear filter that has been excited by an input $e[t]$. The parameters $a_k,\ \ k = 1 \dots P$ are known as the AR parameters. The order P of the filter is also known as the AR order. $e[t]$ is the excitation signal to the AR filter.

The transfer function of the linear filter in Eq. 4.11 is given by

$$H(z) = \frac{1}{1 - \sum_{k=1}^{P} a_k z^{-k}} \tag{4.12}$$

As is evident by comparing this to Eq. 4.10, this is an *all-pole* system, with the sole zero at the origin ($z = 0$). In the Fourier domain we can write the relationship of Eq. 4.11 as

$$|S(f)|^2 = |H(f)|^2 |E(f)|^2. \tag{4.13}$$

$H(f)$ can be computed from $H(z)$ by setting $z = exp(i2\pi f)$.

If $e[t]$ is assumed to be "white noise," i.e. a signal whose power spectral density is constant across all frequencies, then $E(f)$ is flat.[5] In this case $S(f) \propto H(f)$, i.e. the spectrum of the signal is identical (within a scaling term) to the spectral response of the filter $H(z)$. It becomes possible to analyze the signal by analyzing the filter $H(z)$ instead. The poles of $H(z)$ determine the resonant frequencies of the filter; consequently, the angles and magnitudes of the poles are indicative of the frequencies and magnitudes of the spectral peaks of $s[t]$.

Given a signal $s[t]$, the AR model parameters for the signal can be estimated through least-squares estimation. The estimation procedure assumes that the autoregressive component of Eq. 4.11, $\sum_k a_k s[t-k]$, is a *predictor* for $s[t]$, and that $e[t]$ is a prediction error whose energy must be minimized. This leads to the following estimator for the model parameters:
indentThe prediction error $Err(a_1, \dots, a_P)$ is defined as

$$Err(a_1, \dots, a_P) = \sum_t \left(s[t] - \sum_k a_k s[t-k] \right)^2 \tag{4.14}$$

[5]The term "white" is analogous to the definition of white light, which has equal contribution from all wavelengths.

and the actual parameter values are estimated to minimize this error as

$$\{a_i,\ i = 1 \ldots P\} = \arg\min_{\{\hat{a}_i\}} Err(\hat{a}_1, \ldots, \hat{a}_P) \tag{4.15}$$

The least-squares minimization of Eq. 4.15 implicitly assumes that $e[t]$ is Gaussian and the estimated AR parameters are used to derive the spectrum of the signal $s[t]$ itself.

In practice, the signal $s[t]$ is tapered using a window function, typically a Hamming window, before the predictor parameters are estimated. The manner in which the bounds of the summation in Eq. 4.14 are defined, and the actual minimization algorithm used for Eq. 4.15, lead to a variety of different estimation procedures.

- Explicitly solving the quadratic minimization problem of Eq. 4.15 by differentiating it and equating the derivative to zero results in a set of simultaneous equations that can be solved for the AR parameters. Different settings of the limits of summation in Eq. 4.14 result in different solutions, including the well-known autocorrelation and covariance methods [10]. These solutions generally fall under the umbrella of **Yule–Walker estimators**.
- **Burg estimators** utilize the fact that Eq. 4.11 also provides a mechanism for predicting $s[t-P]$ from $s[t-k],\ k = 0, \ldots, P-1$. This *backward* prediction is combined with the forward predictor of Eq. 4.11 to estimate the AR parameters. Burg associated these with maximum entropy spectral estimation [11].
- **Maximum likelihood estimators** assume that the excitation $e[t]$ is a white noise produced by a Gaussian process, and estimate the parameters using a maximum likelihood approach. This involves a nonlinear optimization and is more complex than the first two methods [12].

The outcome of the estimation process is a set of AR parameters $a_k,\ k = 1 \ldots P$. Substituting these values into Eq. 4.12, we can now estimate the spectrum of $s[t]$ as $S(f) \propto H(z)|_{z=\exp(i2\pi f)}$. Figure 4.15 shows a typical spectrum estimated using the AR model.

The parameters can now also be used to recover the excitation $e[t]$ as

$$e[t] = s[t] - \sum_{k=1}^{P} a_k s[t-k]. \tag{4.16}$$

Interestingly, although the AR model itself is motivated by assuming that $e[t]$ is white, in practice the estimation procedures can be used even if $e[t]$ is impulsive, such as an impulse train. This makes it particularly useful for the extraction of vocal tract excitations and vocal tract parameters of speech (discussed in Chap. 5).

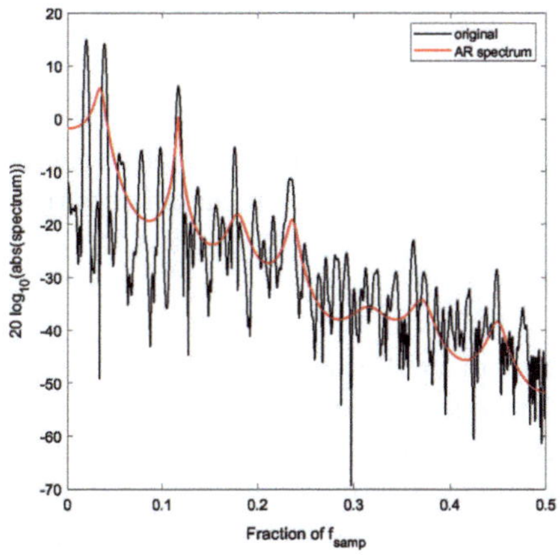

Left: The black curve shows the DTFT of a segment of speech. The red curve shows the spectrum estimated using an AR model. Note how the AR model highlights spectral peaks. **Right:** DTFT of a hamming-windowed 400-sample segment of speech, and its cepstrally smoothed variant.

Fig. 4.15 Comparison of spectral estimates obtained from an AR model, and through cepstral smoothing

4.2.2.3 Homomorphic Processing and Liftering

Homomorphic processing operates on the *logarithm* of signal spectra and their derivatives to analyze the signal. Let $S(f)$ be the spectrum of the signal (here $S(f)$ is being generically used to represent the complex spectrum, its magnitude, or its power; while we use the DTFT notation, in practice, the DFT is used). Homomorphic processing of the signal is typically performed as $H(\log(S(f)))$, where H is any linear filtering operation.

In the context of the speech signal, the most commonly used homomorphic representation is the *cepstrum*, which is computed from the DFT of the speech signal. Representing the sequence of DFT coefficients from a K-sample speech signal as $\mathbf{S} = [S[0], S[1], \ldots, S[K-1]]$, the cepstrum is generally computed as

$$\mathbf{C}_s = IDCT(\log(abs(\mathbf{S}))) \tag{4.17}$$

$abs(\mathbf{S})$ represents the sequence of component-wise absolute values of $\mathbf{S}$, and $\mathbf{C} = [C[0], C[1], \ldots, C[K-1]]$ is also a sequence. "IDCT" is the inverse discrete cosine transform, which is derived from the Fourier transform and guarantees real outputs if the input is real [13, 14].

One feature of the cepstrum that we often find useful is that due to the fact that the logarithm of a product is the sum of logs, the cepstrum $\mathbf{C}_y$ of a filtered signal $y[t] = s[t] \otimes h[t]$ is simply the sum of the cepstrum of the input $\mathbf{C}_s$ and that of the impulse response of the filter $\mathbf{C}_h$, i.e. $\mathbf{C}_y = \mathbf{C}_s + \mathbf{C}_h$.

Homomorphic operations may now be performed on the cepstrum. The most common operation is truncation, which sets higher-order cepstral terms to 0, i.e.

$$H(C[k]) = \begin{cases} C[k], & \text{if } k \leq L \\ 0 & \text{otherwise} \end{cases}$$

where $L + 1$ is the number of coefficients retained. The effect of truncation can be visualized by recomputing the modified log spectrum from the processed cepstrum:

$$log(\tilde{\mathbf{S}}) = DCT(H(\mathbf{C})_s)$$

Figure 4.15 compares the modified log spectrum of a 400-sample segment of speech, whose cepstrum has been truncated to only 20 values, to the log of the magnitude of its DTFT. As we can see, truncation of the cepstra results in a smoothing of the log spectrum, retaining only the envelope, but losing the detail.

Other forms of commonly used homomorphic processing for speech includes RASTA, which applies a band-pass filter to the log spectrum of the signal (treating the log spectrum as a signal itself) [15]. The RASTA filter typically takes the form

$$T(z) = 0.1\frac{2 + z^{-1} - z^{-3} - 2z^{-4}}{z^{-4}(0.91 - 0.98z^{-1})}$$

RASTA filtering has been found to be useful for computer speech recognition, but less so for voice profiling.

4.3 Analyzing Time-Varying Signals

The spectral analyses we have considered so far assume that the signal's spectral composition stays stationary for its entire duration. The frequency content of real-life signals such as speech varies continuously with time. In fact, it is the time-varying nature of the speech signal that enables it to carry information; a stationary speech signal would be uninformative.

Spectral analyses of time-varying signals must ideally derive representations from them that show their *instantaneous* spectral composition and how it varies with time. Such representations are often called "time-frequency" representations or "spectrographic" representations. Visual depictions of such representations are generally called spectrograms.

Spectrographic representations are fundamental to the analysis of speech signals. Much of the science (and art) of deriving representations of speech signals deals with *how* to derive such representations, and what additional information may be derived from them. The sections below delve briefly into this topic.

4.3.1 Short-Time Fourier Transform

The Short-time Fourier transform derives time-frequency representations of time-varying signals through repeated application of the Fourier transform to short segments of the signal. The actual computation is performed by "sliding" an analysis window over the signal, and computing the spectrum at each position of the window. In principle, this results in the following computation:

$$S(f,t) = \sum_{n=-\infty}^{\infty} s[n]w[t-n]e^{-i2\pi fn}$$

where $S(f,t)$ represents the contribution of frequency f at time t to signal $s[t]$. The actual value obtained depends on the window $w[t]$, which is typically a Hamming or Hanning window of some width M.

In practice, we compute the STFT using the DFT, rather than the DTFT. We also do not compute it for every instant t; instead it is sufficient to compute it every r samples, where $r \leq 0.5M$, and M is the size of the window. Thus the STFT is computed as

$$S[k,l] = e^{\frac{-i2\pi klr}{M}} \sum_{n=0}^{M-1} s[n+lr]w[n]e^{\frac{-i2\pi kn}{M}} \tag{4.18}$$

It is useful to be familiar with the terminology employed in time-frequency analyses of speech, such as the STFT. Each windowed segment of $s[t]$ is called a *frame*. The shift r between frames is often referred to as the "hop" or "skip." The STFT itself is computed as a sequence of complex spectral vectors, one per frame, and is often represented as a complex matrix. For visual depictions (i.e. spectrograms), as well as for actual analysis of speech, we usually work with the *magnitude* of the STFT components, i.e. $|S[k,l]|$.

There is a natural tradeoff between time resolution and frequency resolution in the computation of the STFT. As explained in Sect. 4.2.1.2, the *frequency resolution* of the spectrum—the separation in frequency required between significant frequency components for them to be resolved as being separate—depends on the window width M. Increasing the window results in greater frequency resolution for $S(f,t)$. However, *temporal* changes of signal characteristics within the width of the window cannot be resolved. On the other hand narrower windows permit tracking faster time changes in the signal, but at the cost of frequency resolution. The general time-frequency tradeoff is given by $\Delta t \Delta f \geq \frac{1}{2}$, where Δt is the smallest spacing in time between changes in the spectrum where the change can be resolved, and Δf is the smallest difference in frequency that can be resolved.

Figure 4.16 shows the time-frequency tradeoff on spectrograms of speech. The two panels show spectrograms obtained with windows of two different widths for a typical speech signal. The wider-window spectrogram shows fine spectral details in the signal, including the pitch frequency and its harmonics. Spectrograms computed

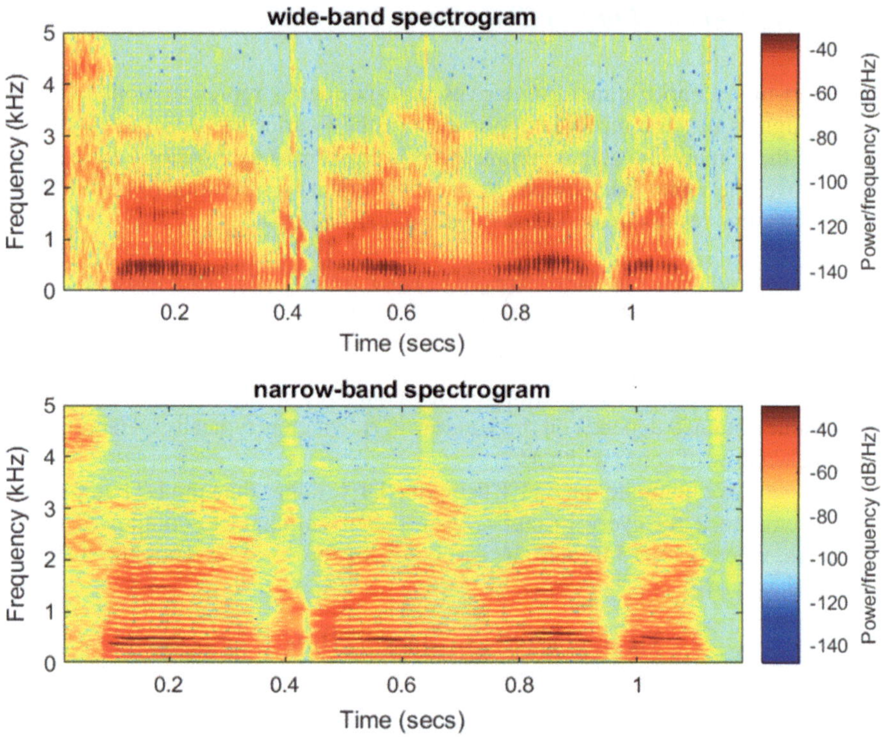

Upper panel: "wide-band" spectrogram of speech, computed with short 8 ms analysis windows. Lower panel: "narrow-band" spectrogram of speech, computed with wide, 64 ms analysis windows.

Fig. 4.16 Wideband and narrowband spectrograms of a speech signal

with wide windows (typically 40 ms or more) show such fine spectral detail and are called "narrow-band" spectrograms. The narrower-window spectrogram, on the other hand, shows coarse spectral structure including formants, which nevertheless informs about the signal. Such spectrograms computed with narrow windows (typically 10 ms or less) are referred to as "wide-band" spectrograms.

4.3.2 Shortcomings of the STFT

The short-time Fourier transform is, in fact, a computational approximation of a *filter bank*, comprising a bank of (approximately) band-pass filters, each centered at frequency $\frac{k}{M}$. STFT computation comprises filtering the signal $s[t]$ with each of the filters in the bank and capturing samples (taken with a sampling period of r) from the envelope of the output of the filters. Each filter in the bank has an impulse response comprising the actual window function $w[t]$ modulated by a complex exponential $\exp(i2\pi\frac{kt}{M})$, such that its frequency response is now centered at $\frac{k}{M}$. As a result, all filters have impulse responses of identical width—that of the window $w[t]$. The time and frequency resolution of the STFT is hence identical at all frequencies.

On the other hand, in most real-life signals the low-frequency components with their larger periods change more slowly with time than high-frequency components with their much shorter periods. As a result, higher frequency components require greater time resolution than lower frequencies. Conversely, a difference of δ_f between two frequencies f_1 and f_2 expands to $N\delta_f$ between their Nth harmonics. Distinguishing between the two frequency components and their harmonics only requires a frequency resolution of $N\delta_f$ in the vicinity of the Nth harmonic, whereas it requires the much finer resolution of δ_f at the fundamental frequencies f_1 and f_2. In general, real-life signals require lower frequency resolution at high frequencies than at low frequencies. Thus, the fixed time and frequency resolutions at all bands that the STFT provides is suboptimal from both aspects.

A second issue with the STFT relates to the suboptimality of the representation of fundamentals and harmonics alluded to above. Since the frequency axis is linear, harmonics of different fundamental frequencies become increasingly farther apart (with respect to each other's counterparts) in frequency, with increasing frequency.

The influence of this is most evident on harmonic signals—signals whose spectral structure consists primarily of spectral peaks at harmonics of a fundamental frequency (i.e. $S(f)$ is negligible for $f \neq kf_*$, where f_* is the fundamental frequency, and k is an integer). Consider two harmonic signals $s_1[t]$ and $s_2[t]$ which have different fundamental frequencies f_1 and f_2, but otherwise have identical structure over the set of harmonics such that $S_1(kf_1) = S_2(kf_2)$. In spite of their perfect coupling, their STFTs will be entirely different. To illustrate, consider two signals with fundamental frequencies of 100 and 150 Hz respectively (a difference of 50 Hz). Their fourth harmonics will respectively be at 400 and 600 Hz (a difference of 200 Hz), while their sixth harmonics are at 600 and 900 Hz (a difference of 300 Hz). Thus, a simple operation such as increasing the pitch of a note on a musical instrument can result in a dramatic change in the spectrogram of the signal.

4.3.3 Filter Banks and Filter-Bank Analysis

An alternative approach to the STFT (which implicitly analyzes the signal with a filter bank) that potentially addresses one or both of the above drawbacks is to utilize explicit filter-bank analysis to derive time-frequency representations. In FB analysis, the signal is explicitly filtered by a bank of band-pass filters, and the outputs of the filters are sampled periodically to obtain the time-frequency representation of the signal. Figure 4.17 illustrates this approach.

Mathematically, filter-bank analysis computes the time-frequency representation of a signal $s[t]$ as follows. Let $h_k[t],\ k = 1 \ldots B$ represent the impulse responses of the filters in the bank, and let B be the number of filters in the bank. The output of the kth fiter is computed as

$$s_k[t] = s[t] \otimes h_k[t]. \tag{4.19}$$

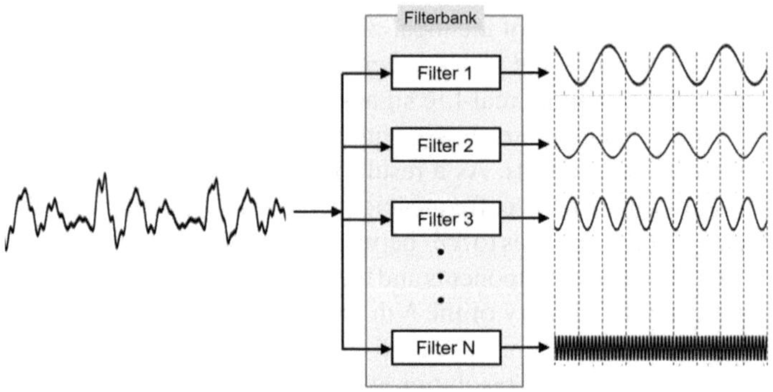

A bank of N filters operates on the incoming signal to produce N filtered signals. The output signals are sampled at regular intervals of time. The dotted lines show sampling instants.

Fig. 4.17 Depiction of a filterbank

The time-frequency representation is typically derived either by sampling the filtered signal $s_k[t]$ directly every r samples as

$$S[k, l] = s_k[lr] \tag{4.20}$$

or by sampling the *Hilbert envelope* of the $s_k[t]$ at $t = lr$. The extraction of Hilbert envelopes is discussed in Chap. 6. In the derived representation, the output of each filter may be viewed as representative of the signal within the bank of frequencies it passes, or more specifically, its center frequency or frequency of peak response.

Traditional filter-bank analysis prefers filters $h_k[t]$ that are designed such that the original signal $s[t]$ can be reconstructed from $S[k, l]$. Each of the filters is itself generally implemented as a simple IIR or FIR filter. For speech analysis, however, we usually do not impose this requirement, and a variety of filter banks, such as Mel-frequency filter banks [16], ERB filter banks [17] and Gammatone filter banks [18] are used. We discuss these in Chap. 5. Below we describe two non-traditional filter-banks that are potentially useful for analyzing speech for profiling.

4.3.3.1 Constant-Q Transforms

One particular type of filter-bank analysis that addresses both drawbacks of the STFT is the constant-Q transform [19].

The constant-Q transform is a variant of the conventional STFT analysis, which also employs complex-exponential modulated window functions as filter responses. However there are two key modifications.

1. The analysis window used with each filter of the filterbank varies with the filter, to ensure that the number of cycles of the center frequency of any filter that occur within the analysis window is constant. This ensures that the frequency resolution at any frequency is proportional to the frequency itself. Thus in this transform,

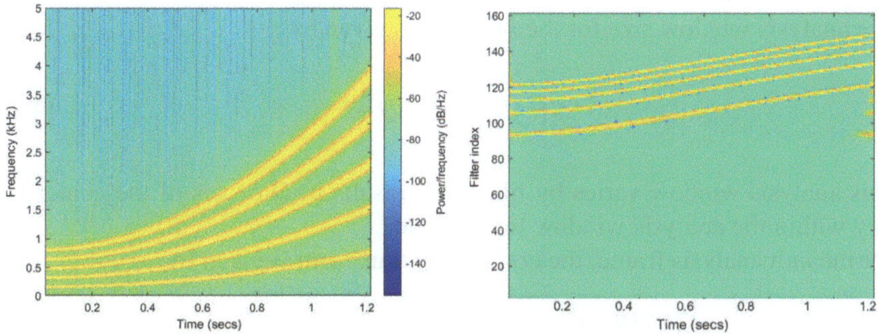

(a) **Left:** The STFT of a quadratic chirp signal and four of its harmonics. **Right:** The constant-Q transform of the same signal. In the STFT, the spacing between harmonics increases with fundamental frequency. In the constant-Q transform, the inter-harmonic spacing remains independent of the fundamental frequency.

Fig. 4.18 Comparing the STFT and the constant-Q transform

the windows for filters with lower center frequencies are wider, and the widths decrease with increasing center frequency.

2. The center frequencies of all filters are spaced on a logarithmic scale. On the logarithmic frequency scale, the frequency resolution at all frequencies is constant. An example is shown in Fig. 4.18, using the base-2 logarithm.

The term “constant-Q” is itself derived from the manner in which the filters are designed. The *Q-factor* of a filter is the ratio of the center-frequency of a filter to its bandwidth. It is a measure of how selective the filter is with respect to its center frequency. Higher-Q filters are more selective, but generally harder to design. In a constant-Q transform, the logarithmically-spaced filters in the filter bank are designed such that all of them have identical Q factors.

Like the STFT, the actual implementation of the constant-Q analysis depends on the number of filters and the width of the analysis window (frame width) chosen. The most popular version, as originally proposed in [19], uses an $1/24$th octave bank of filters (useful for music signal processing applications). In this version, there are exactly 24 filters between any frequency f and its harmonic $2f$.

The center frequency of kth filter is

$$f_k = \left(2^{1/24}\right)^k f_{min} \tag{4.21}$$

where f varies from f_{min} to a frequency chosen to be below 0.5 (corresponding to the continuous-time frequency $f_{samp}/2$). The minimum frequency f_{min} can be chosen as the lowest frequency for which information is desired.

Assuming that the kth spectral component is the center frequency of the kth filter, the Q for the filter is given by $Q_k = f_k/\delta f_k$, where δf_k is the bandwidth of the filter. For a *constant-Q* filter, we have $Q_1 = Q_2 = \cdots Q_K = Q$, where K is the number of filters in the bank.

The analysis window size for the kth filter is given by

$$N[k] = \frac{1}{\delta f_k} = Q\frac{1}{f_k} \tag{4.22}$$

This analysis window varies by filter, and results in Q cycles of the center frequency within the analysis window for any filter.

For the mth analysis frame, the constant-Q transform is given by

$$S_{cqt}[k, m] = \frac{1}{N[k]} \sum_{n=0}^{N[k]-1} W[k, n]s[n + mr]exp(-i2\pi Qn/N[k]). \tag{4.23}$$

where r is the hop between frames. The window function $W[k, n]$ is generally chosen to be the Hamming window.

Equation 4.21 used for the computation of Q as described earlier uses powers of the 24th root of 2. The number of frequencies that would be resolved by the filterbank between any f and $2f$ is therefore 24. This is independent of f, unlike the STFT. In *any* implementation of the constant-Q transform, the number of frequencies between any f and kf depends only on k. Thus a shift in pitch simply becomes a linear and constant shift of all harmonics in the constant-Q representation.

The constant-Q transform is especially effective for representing signals with harmonic structure. It is particularly useful for musical signals, as it produces equal frequency resolution for every octave. It is also useful for analysis of speech, which too generally has harmonic structure. The logarithmic spacing of the filters also ensures that the spacing between any two harmonic peaks is no longer dependent on the fundamental frequency and only depends on the ordinality of the harmonic. Any change in the fundamental frequency of a signal (and thereby its harmonic structure) is reflected simply as a linear shift of the constant-Q spectral representation without any change in the overall pattern. For instance, if $S[k, t]$ is a constant-Q feature obtained from signal derived by exciting a vocal tract with a pitch frequency f_0, the feature for the signal obtained when the same vocal tract is excited by a pitch f_0' is simply $S[k + \lceil \log(f_0'/f_0) \rceil, t]$. This rather simple translation model is particularly effective in modeling pitch variations, among other advantages.

4.3.3.2 Wavelet Transform

The STFT decomposes signals into weighted sums of pure sinusoids (embodied as complex exponentials). While sinusoids are highly localized in frequency, they are infinite in duration. As a result, while the STFT is able to identify the frequency components in a signal, it cannot localize *where* they occur. Any attempts at localizing the frequencies in time by truncating the complex exponential bases results in a smearing of their spectrum, and a loss of localization in frequency.

The constant-Q transform attempts to remedy this through frequency-appropriate time windows leading to better temporal localization, however the transform is not invertible and is lossy.

The *wavelet transform* is a filter-bank based signal representation that addresses both drawbacks. Unlike the STFT or CQT which employ complex exponentials as their bases for representing signals, the wavelet transform is composed from *wavelets*. A wavelet is a time-limited square-integrable[6] signal $\psi(t)$ that is localized in both time and frequency. $\psi(t)$ can be used to generate an entire family of functions $\psi_{a,b}(t) = \frac{1}{\sqrt{a}}\psi(\frac{t-b}{a})$ [20]. Each $\psi_{a,b}(t)$ is a version of $\psi(t)$ that has been scaled in time by a factor a, and translated to begin at time b. $\psi(t)$ is often called the *mother wavelet* to the family $\psi_{a,b}(t)$.

The actual choice of the mother wavelet depends on a variety of factors, including desired structural attributes, computational considerations and application domain. A large number of wavelets have been proposed in the literature. Some examples are Daubechies, Biorthogonal, Coiflets, Symlets, Morlet, Mexican Hat, Meyer etc. Figure 4.19a shows a *complex* wavelet—the Gaussian Hilbert wavelet bases, as well as a variety of other real-valued wavelets. For speech signals the Morlet wavelet is generally found to be the most useful.

Although the scale factor a and the translation b may take on any real value, it is customary (for a variety of reasons including computational efficiency) to restrict a to integer powers of a real number, typically 2, and b to integers. This gives rise to the following family of wavelets

$$\psi_{j,k}(t) = \frac{1}{\sqrt{2^j}}\psi\left(\frac{t-2^jk}{2^j}\right), \tag{4.24}$$

where j and k can take any integer value. j is referred to as the *scale* of the wavelet. The wavelet transform based on the wavelet family described by Eq. 4.24 is known as the *discrete* wavelet transform (DWT), where the word "discrete" refers to the fact that the scale and translation factors j and k can only take discrete integer values (and not to discrete time).

By itself, $\psi(t)$ has a band-pass spectrum—a filter with $\psi(t)$ as its impulse response would be approximately band-pass. As a result, both the center frequency and the bandwidth of $\psi_{j,k}(t)$ are scaled by a factor 2^j with respect to $\psi(x)$. Thus the set $\{\psi_{j,k}(t),\ j,k \in \mathcal{Z}\}$ represents the impulse responses of a set of constant-Q band-pass filters.

The mother wavelet $\psi(t)$ (and every member of its family) is also paired with a corresponding low-pass function $\phi(t)$, called the *scaling function*, which complements its spectral characteristics. A filter with $\phi(t)$ as its impulse response would have approximately low-pass spectral response, and permit the lower frequencies that $\psi(t)$ cuts off. Figure 4.20 shows an example of a wavelet and its scaling function. The scaling function $\phi(t)$ can in fact be derived from the wavelet and vice versa [21].

---

[6]By "square integrable" we mean that $\int_{-\infty}^{\infty}\psi^2(t)dt$ is finite.

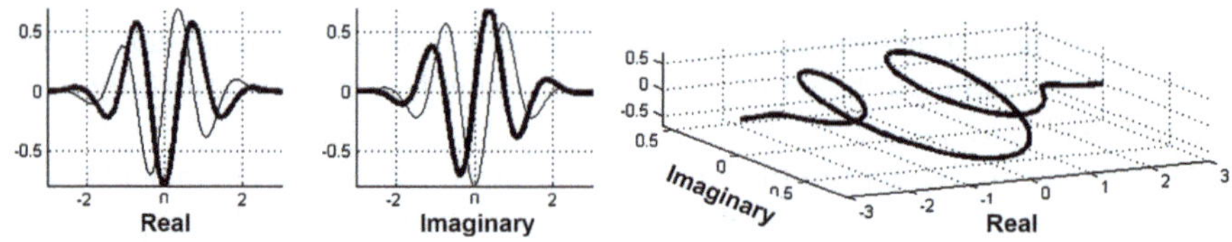

(a) A pair of orthogonal wavelet basis functions: the ones shown are complex Gaussian-Hilbert bases.

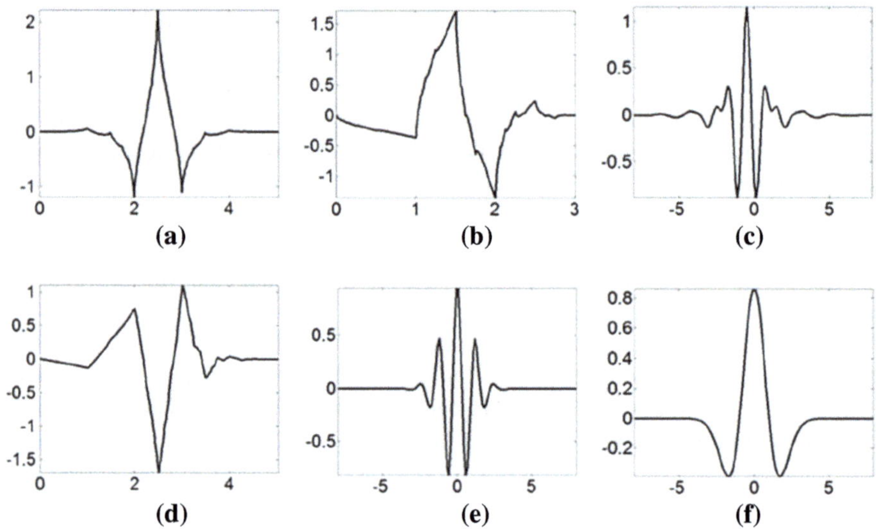

(b) Some other types of wavelets: (a) Coif1 wavelet (b) db2 wavelet (c) Meyer-wavelet (d) Sym3 wavelet (e) Morlet wavelet (f) Mexican hat wavelet

Fig. 4.19 Some examples of different types of wavelets

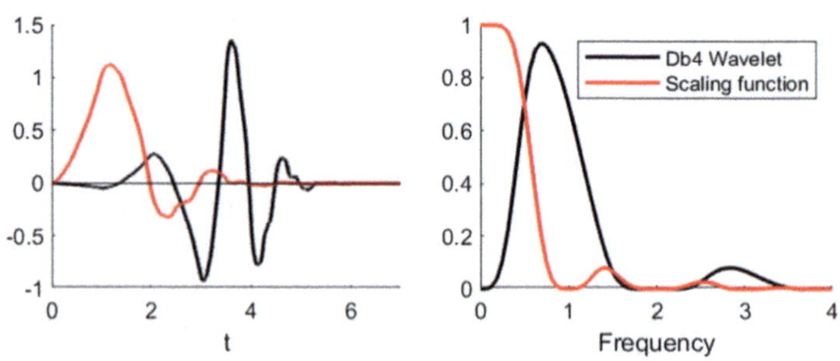

Left: A Daubechies wavelet and its scaling function. **Right**: The spectra of the wavelet and its scaling function.

Fig. 4.20 A Daubechies wavelet and its scaling function

Together, the wavelet $\psi(t)$ and its scaling function $\phi(t)$ are used to compute a pair of complementary finite-impulse-response filters $h[t]$ and $g[t]$, where $h[t]$ is derived from the wavelet $\psi(t)$ and $g[t]$ represents $\phi(t)$ [21], and $h[t]$ and $g[t]$ are *quadrature mirror* images of one another (i.e. their z transforms are related as $H(z) = G(-z)$ and $h[t] = (-1)^t g[t]$). $h[t]$ has an approximately high-pass filter response, truncating lower frequencies, while $g[t]$ largely truncates frequencies from the upper half of the frequency axis and passes lower frequencies.

To transform any signal $s[t]$, it is simultaneously filtered by both $h[t]$ and $g[t]$. Since the output of each of them now only has half the frequencies in the original signal (due to the high-pass/low-pass nature of the filters), their outputs can be *downsampled* by a factor of 2, by simply deleting every second sample. Thus the outputs of the filters are obtained as

$$d_0[t] = (s[t] \otimes h[t]) \downarrow_2 \tag{4.25}$$

$$a_0[t] = (s[t] \otimes g[t]) \downarrow_2 \tag{4.26}$$

where $\downarrow_2$ represents the operation of downsampling by a factor of 2. $d_0[t]$ and $a_0[t]$ are known as *detail* and *approximation* coefficients, referring to the fact that $d_0[t]$ is derived from the high-pass filter $h[t]$ and captures the details of the signal, while $a_0[t]$, being low-pass filtered, is a smoothed version of it.

The quadrature mirror relationship between $h[t]$ and $g[t]$ implies that the original signal $s[t]$ can be reconstructed from the downsampled outputs $d_0[t]$ and $a_0[t]$ by upsampling them again (by introducing a 0 between every two samples), refiltering the upsampled signals with $h[t]$ and $g[t]$ respectively and adding the outputs. Representing the process of upsampling by 2 (introducing a zero between every two samples) by $\uparrow_2$, we can write

$$s[t] = (d_0[t] \uparrow_2) \otimes h[t] + (a_0[t] \uparrow_2) \otimes g[t] \tag{4.27}$$

Thus the filters are jointly invertible to recover the original signal. This is illustrated in the upper panel of Fig. 4.21.

The frequency components of low-pass signal $a_0[t]$ now also span the entire frequency range (0, 1) as a result of the downsampling. $h[t]$ and $g[t]$ can now be applied to it, and the outputs downsampled yet again. This recursion can continue indefinitely, or, in the case of a finite-length input $s[t]$, until the low-pass version of the signal contains only one sample.

The entire computation can thus be performed as

$$d_k[t] = (a_{k-1}[t] \otimes h[t]) \downarrow_2 \tag{4.28}$$

$$a_k[t] = (a_{k-1}[t] \otimes g[t]) \downarrow_2 \tag{4.29}$$

where k is the level of the recursion. The overall process is illustrated by the lower panel of Fig. 4.21. The set of resultant detail and approximation coefficients

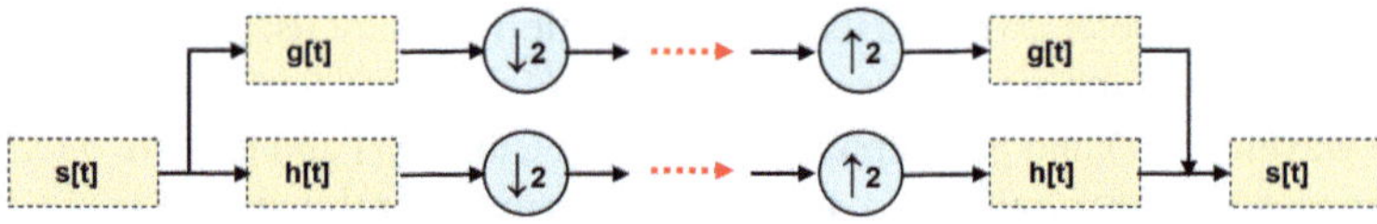

The down sampled outputs of $h[t]$ and $g[t]$ can be upsampled back to the original sampling rate, refiltered by $h[t]$ and $g[t]$ respectively, and added to recover the original signal $s[t]$.

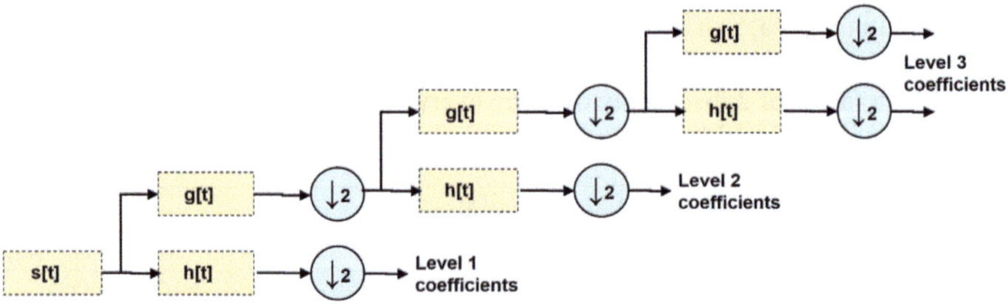

A complete DWT filter bank has several levels. In each level, both the sequences of detail coefficients output by the high-pass filter $h[t]$ and approximation coefficients output by the low-pass filter $g[t]$ are downsampled. The resulting sequence of approximation coefficients is passed up to the next level.

Fig. 4.21 The discrete wavelet transform filterbank

$\{d_k[t], a_k[t]; k = 1 \ldots K\}$, where K is the depth of the recursion, are known as the *wavelet coefficients* of the signal.

The original signal $s[t]$ can be reconstructed from its wavelet coefficients by recursive application of Eq. 4.27.

The detail and approximation coefficients provide a hierarchical decomposition of $s[t]$ that accurately detects and localizes significant events in a speech signal in a manner that is not possible using the STFT. Figure 4.22 highlights the differences between the STFT and the wavelet transform, and Fig. 4.23 shows an example of the type of detailed decomposition that can be obtained using wavelets.

There are many excellent articles in the literature that give further details about the wavelet transform e.g. [21–23]. DWT is widely used in time-series analysis of various kinds, e.g. [24, 25] and image processing. It is easy to extend this transform into multiple dimensions. For example, a wavelet transform of a 2-D image is a row-wise or column-wise 1-D wavelet transform of the image matrix, which is then stacked appropriately into a matrix of the same size as the image. The use of wavelets for profiling and biometric analysis of speech, however, remains underexplored at the time of writing of this chapter.

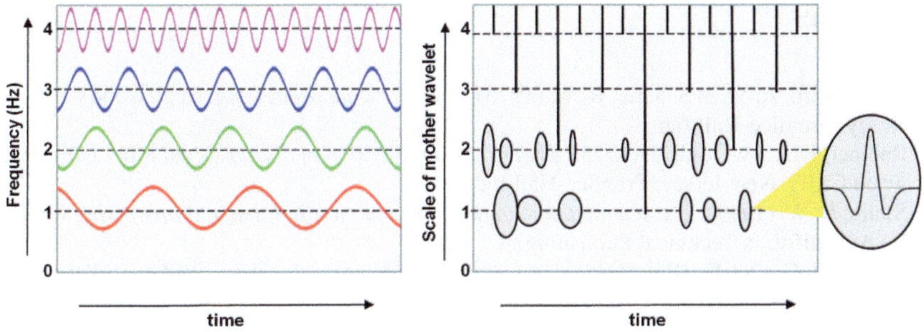

The axes and waves shown in this figure are only symbolic. **Left panel**: The Fourier transform resolves the signal into its frequency components in terms of bases of infinite duration and as a superposition of infinite bases. The STFT samples the frequency axis into a finite number of frequencies as shown. **Right panel**: The Wavelet transform resolves the signal in terms of a finite set of finite length bases (wavelets). The mother wavelet is compressed or dilated to different extents at different levels, and the transform gives the locations of these compressed or dilated versions of the mother wavelet in the signal. At each level, shown by each decimated sequence of bins, the wavelet in terms of which the signal is resolved for that bin may be scaled to different magnitudes. The time-location of the wavelet within a bin is its translation parameter. The figure shows circles at various levels instead of actual wavelets for visual clarity. The inset shows an example wavelet.

Fig. 4.22 A schematic illustration of the difference between Fourier and wavelet transforms

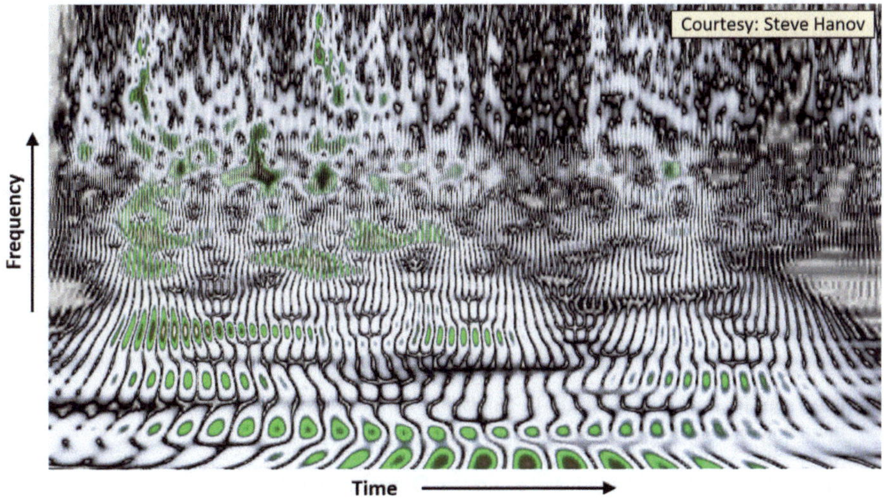

Fig. 4.23 Wavelet transform of the sound of a paper crumpling

References

1. Oppenheim, A. V., & Schafer, R. W. (1975). *Digital signal processing*. Englewood Cliffs, New Jersey: Prentice-Hall Inc.
2. Rabiner, L. R., & Gold, B. (1975). *Theory and application of digital signal processing*. Englewood Cliffs, New Jersey: Prentice-Hall Inc.
3. Smith, S. W. (1999). *The scientist and engineer's guide to digital signal processing* (2nd ed.). USA: California Technical Publishing.
4. Proakis, J. G. (2001). *Digital signal processing: Principles, algorithms and applications*. Pearson Education India.
5. Wente, E. C., & Thuras, A. L. (1934). Loud speakers and microphones. *Bell System Technical Journal, 13*(2), 259–277.
6. Landau, H. J. (1967). Sampling, data transmission, and the Nyquist rate. *Proceedings of the IEEE, 55*(10), 1701–1706.
7. Vaidyanathan, P. P. (2001). Generalizations of the sampling theorem: Seven decades after Nyquist. *IEEE Transactions on Circuits and Systems I: Fundamental Theory and Applications, 48*(9), 1094–1109.
8. Jayant, N. S., & Noll, P. (1984). *Digital coding of waveforms: Principles and applications to speech and video*. Englewood Cliffs, New Jersey: Prentice-Hall Inc.
9. Ross, S. M., Kelly, J. J., Sullivan, R. J., Perry, W. J., Mercer, D., Davis, R. M., et al. (1996). *Stochastic processes* (Vol. 2). New York: Wiley.
10. Marple, S. L., Jr. (1989). *Digital spectral analysis: With applications*. Prentice-Hall series in signal processing. Englewood Cliffs, New Jersey: Prentice-Hall Inc.
11. Burg, J. P. (1975). *Maximum entropy spectral analysis*. Doctoral dissertation, Stanford University, Palo Alto, California.
12. Yao, Q., & Brockwell, P. J. (2006). Gaussian maximum likelihood estimation for ARMA models. I. Time series. *Journal of Time Series Analysis, 27*(6), 857–875.
13. Ahmed, N., Natarajan, T., & Rao, K. R. (1974). Discrete cosine transform. *IEEE Transactions on Computers, 100*(1), 90–93.
14. Rao, K. R., & Yip, P. (2014). *Discrete cosine transform: Algorithms, advantages, applications*. Cambridge: Academic Press.
15. Hermansky, H., & Morgan, N. (1994). RASTA processing of speech. *IEEE Transactions on Speech and Audio Processing, 2*(4), 578–589.
16. Davis, S., & Mermelstein, P. (1980). Comparison of parametric representations for monosyllabic word recognition in continuously spoken sentences. *IEEE Transactions on Acoustics, Speech, and Signal Processing, 28*(4), 357–366.
17. Moore, B. C., & Glasberg, B. R. (1983). Suggested formulae for calculating auditory-filter bandwidths and excitation patterns. *The Journal of the Acoustical Society of America, 74*(3), 750–753.
18. Patterson, R. D., Nimmo-Smith, I., Holdsworth, J., & Rice, P. (1987). An efficient auditory filterbank based on the gammatone function. In *The Meeting of the IOC Speech Group on Auditory Modelling at RSRE* (Vol. 2(7)).
19. Brown, J. C. (1991). Calculation of a constant Q spectral transform. *The Journal of the Acoustical Society of America, 89*(1), 425–434.
20. De Moortel, I., Munday, S. A., & Hood, A. W. (2004). Wavelet analysis: The effect of varying basic wavelet parameters. *Solar Physics, 222*(2), 203–228.
21. Daubechies, I. (1990). The wavelet transform, time-frequency localization and signal analysis. *IEEE Transactions on Information Theory, 36*(5), 961–1005.
22. Antonini, M., Barlaud, M., Mathieu, P., & Daubechies, I. (1992). Image coding using wavelet transform. *IEEE Transactions on Image Processing, 1*(2), 205–220.
23. Chui, C. K. (1992). *An introduction to wavelets; Wavelets: A tutorial in theory and applications*. Wavelet analysis and its applications (Vols. 1, 2). San Diego, California; London, UK: Academic Press Inc.; Harcourt Brace Jovanovich Publishers.

24. Ocak, H. (2009). Automatic detection of epileptic seizures in EEG using discrete wavelet transform and approximate entropy. *Expert Systems with Applications*, *36*(2), 2027–2036.
25. Nason, G. P., & Silverman, B. W. (1995). The stationary wavelet transform and some statistical applications. In R. R. Coifman & D. L. Donoho (Eds.), *Wavelets and statistics* (pp. 281–299). Lecture notes in statistics. New York: Springer.

Chapter 5
The Voice Signal and Its Information Content—2

Information in the voice signal is embedded in both its time progression and in its spectral content, i.e. in its *time domain* and *spectrographic domain* respectively. Within these domains, information relevant to profiling may be present in the patterns exhibited by specific characteristics of the voice signal. The signal may however also have characteristics that are not evident in these domains, and must be searched for in other (derivative) mathematical domains where the relevant patterns become more tangible for measurement and analysis. This, however, is the subject of *feature discovery*—a subject that is discussed in Part II of this book. A third domain that reflects the information in the voice signal is that of physical or abstract models that simulate or explain the voice signal and the processes that generate it. We will refer to this as the *model domain*. Regardless of the domain, we refer to any information extracted from the signal as a **feature** that can be used for profiling. The process of feature computation is often referred to as "parametrization," but we will not use this term here, since it is confusable with the term "parameter" in the context of profiling. The two terms are not related.

This chapter is divided into six sections. In the first section we discuss the key steps involved in the process of "preparing" the voice signal for information extraction. The rest of the sections deal with different characterizations of the voice signal, focusing more on those in time, spectrographic and model domains. The list of topics discussed under each section is not exhaustive, and only includes some selected ones that are likely to be very useful for profiling.

5.1 Pre-processing a Voice Signal for Analysis

Continuous speech—comprising spoken utterances—is not a stream of seamless vocalizations. In this respect, a voice recording is rarely that of truly "continuous" speech. More typically, it contains segments of speech activity against a backdrop of "background" acoustics, which may be just silence, or some subset of other non-speech sounds in the environment. As a very first step, therefore, it is necessary

R. Singh, *Profiling Humans from their Voice*,
https://doi.org/10.1007/978-981-13-8403-5_5

to identify and isolate the speech from the background, and perhaps remove any interfering noise overlaid on the speech. We begin with the latter, generally referred to as *denoising*.

5.1.1 Denoising

A variety of speech denoising algorithms have been proposed in the literature. Denoising algorithms obtain a statistical estimate of the noise, which they then subtract or filter out of the noisy speech. However, the noise estimates are not perfect. Removing noise based on these estimates usually (a) leaves a little of the real noise behind, and (b) *removes some speech, or distorts it.*

Consequently, the *perceptual quality* of the signal may improve, but the *intelligibility* often decreases. The algorithms that implicitly remove noise have similar issues. In the context of most algorithms, it is difficult to strike an acceptable tradeoff between removing corrupting noise and retaining intelligibility.

For profiling, in most contexts, denoising algorithms may neither be needed nor advisable, since they are likely to remove information pertinent to profiling, or leave behind artifacts that may obfuscate the information needed for profiling. For profiling *in the presence of noise*, however, the spectral characteristics of noise must be studied and factored out of the profiling decisions by either profiling from those aspects of the signal that are not "corrupted" by noise, or by using basic representations and transformations that are unaffected by noise.

We do not discuss noise compensation procedures explicitly in this chapter. A few special cases where denoising may be useful in the context of profiling are discussed in Chap. 8.

5.1.2 Voice Activity Detection

Voice activity detection (VAD) is the process of distinguishing voice from the background acoustics in a recording. Most popular VAD approaches are based on the following observations:

- The *onset* of speech is generally dramatic, marked by a sudden production of air pressure as the person begins to speak. This, in turn, is characterized by a sudden increase in the instantaneous energy in the speech signal.
- Even after the onset, the overall energy in the speech signal remains higher than that in the background. The termination of speech is marked by a drop in the energy of the recorded signal.

Figure 5.1 illustrates this for a typical speech recording.

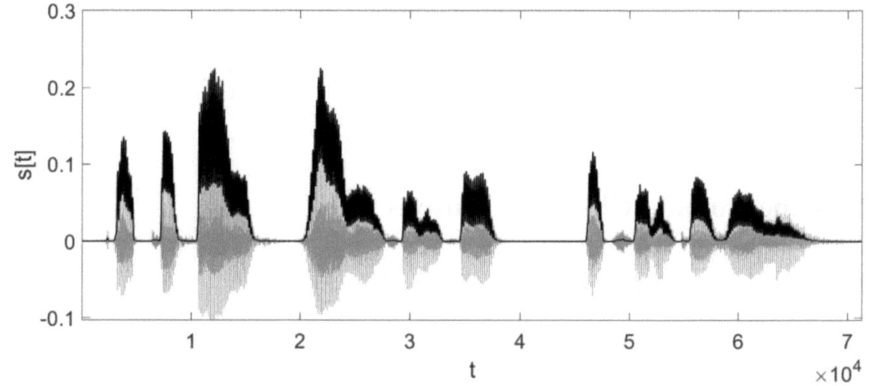

A signal (grey) and its energy contour (black). Energy values are computed over windows of length 16ms. Every onset of speech is indicated by a sudden increase in signal energy.

Fig. 5.1 Energy contour of a speech signal

A typical VAD algorithm based on these observations may do the following:

1. Maintain instantaneous energy thresholds that must be exceeded to indicate the presence of speech. In practical applications, two thresholds are maintained: a relatively-high *onset* threshold T_{onset} that the energy in the signal must exceed (with respect to the background) to indicate the onset of speech, and a lower *offset* threshold T_{offset} that the energy must fall below to indicate that the speech has ended.
2. Compute the energy of the signal as described in the section below (Sect. 5.2). The first few frames of the recording are assumed to contain no speech, and the initial estimate of the background energy level B is set to be the average of the energy in the first few frames.
3. Flag subsequent frames as "speech" or "not speech" using the following logic:

if $!IS\_SPEECH$ **then**
  if $E[l] < T_{onset}B$ **then**
    $B = \alpha B + (1 - \alpha)E_l$
  else
    $IS\_SPEECH = 1$
  end if
else
  if $E_l < T_{offset}B$ **then**
    $IS\_SPEECH = 0$
  end if
end if

Here α is a smoothing factor, typically set to 0.5, and IS_SPEECH is a flag output by the VAD algorithm that indicates whether the corresponding frame of the recording is within a speech segment or not, which is initialized to 0.

The VAD output is often refined through post-processing. An example of a post-processing step is where adjacent speech segments that are very close (separated by less than 100 ms, for instance) are merged into longer segments.

5.2 Time Domain Characterizations

5.2.1 *Energy and Power in the Signal*

Let $x[n]$ be a *square integrable* signal, i.e. a signal for which $\sum_{n=-\infty}^{\infty} x^2[n] < \infty$. The energy in the signal is defined as

$$E = \sum_{n=-\infty}^{\infty} x^2[n] \tag{5.1}$$

Energy is only defined for square integrable signals, such as finite-duration transient signals.

For more generic signals that are not square integrable, such as signals that are the outcome of random processes and speech signals, it is more appropriate to speak of *power*—energy per unit time. Unlike energy, the power in a signal can vary with time. A typical estimate of signal power for a signal $s[t]$ is given by

$$P[t] = \frac{1}{C} \sum_{k=-\infty}^{\infty} |s[k]w[t-k]|^2 \tag{5.2}$$

where $w[n]$ is a *window* function, $C = \sum_{t=0}^{N-1} |w[t]|^2$, and N is the length of the window. Typically a Hamming window is used.

In practice we do not compute the power at every instant. Instead, it is computed over *frames* of the signal that are N samples wide (typically 10 ms), where the skip R between adjacent frames is typically 5ms. Consequently, it is more appropriate to speak of the *energy in a frame*, rather than the power.

The energy in the lth frame of the signal is estimated as

$$E[l] = \sum_{t=0}^{N-1} (s[t+lR]w[t])^2 \tag{5.3}$$

5.2.2 *Energy Contours*

The energy contour of a speech signal is simply the sequence of energy values $E[l]$, expressed directly as energy or as the logarithm of the energy.

The energy contour of the signal carries information about both the spoken content of the signal and about the *speaker*—their affect, emotional and psychological state, health, and a variety of other factors. Typical statistical characterizations of the energy contour include the mean, standard deviation, median, as well as spectral characterizations of the time series signal $E[l]$.

5.2.3 Zero and Mean Crossing Rates

A *zero crossing* occurs when the signal value crosses from one side of the mean value of the signal to the other. Two widely-used signal features are the *zero-crossing rate* (ZCR), and the *mean crossing rate* (MCR). ZCR is merely the number of times successive samples change their sign per unit time in the signal. For a signal $s[.]$, the zero crossing rate ZCR_x is computed over a segment of T samples as

$$\text{ZCR} = \frac{1}{T}\sum_{n=0}^{T-2} \frac{|s[t]||s[t+1]| - s[t]s[t+1]}{2|s[t]||s[t+1]|} \tag{5.4}$$

The MCR is the ZCR of a mean-normalized signal. A sinusoid has $\text{ZCR} = 2f$ where f is its frequency. White noise has a high ZCR. ZCR and MCR are however easily affected by external noise, and therefore more of use in clean recordings.

ZCR is related to the fundamental frequency in voice signals, but also carries information about the spoken content of the signal, and the speaker.

5.3 Spectrographic Domain Characterizations

5.3.1 The Spectrogram

The most basic spectrographic representation of the speech signal $s[t]$ is its short-time Fourier transform $S[k,t]$ (explained in Sect. 4.3.1). A number of representations are derived from it. The **magnitude spectrogram** $|S[k,t]|$ comprises the magnitudes of the complex STFT coefficients. The **power spectrogram** comprises the set of squared magnitude values $|S[k,t]|^2$. The **log magnitude spectrogram** is the logarithm of the magnitude spectrogram, given by $\log |S[k,t]|$. The **phase spectrogram** comprises the matrix of phase values derived from the complex STFT, $\angle S[k,t]$.

A particularly useful type of spectrogram for the analysis of voiced speech is the **constant-Q spectrogram** described in Sect. 4.3.3.1. Figure 5.2 shows examples of different types of spectrograms.

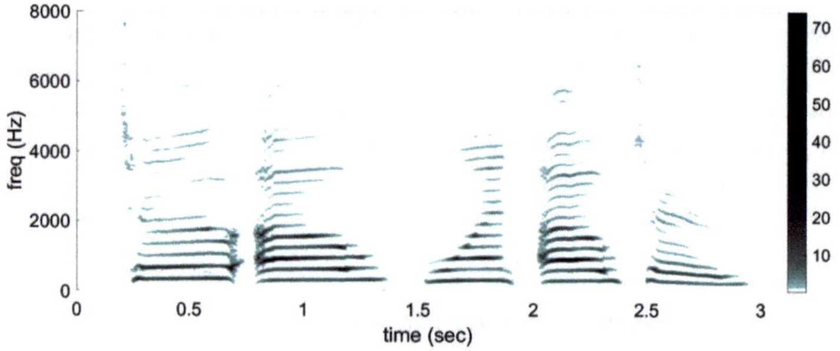

(a) Magnitude spectrogram

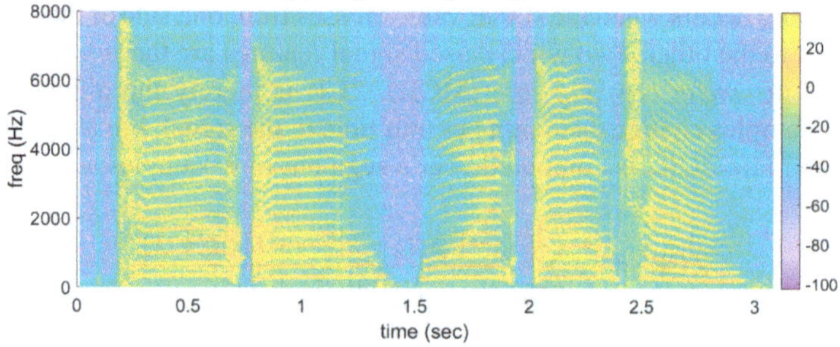

(b) Log (magnitude) spectrogram

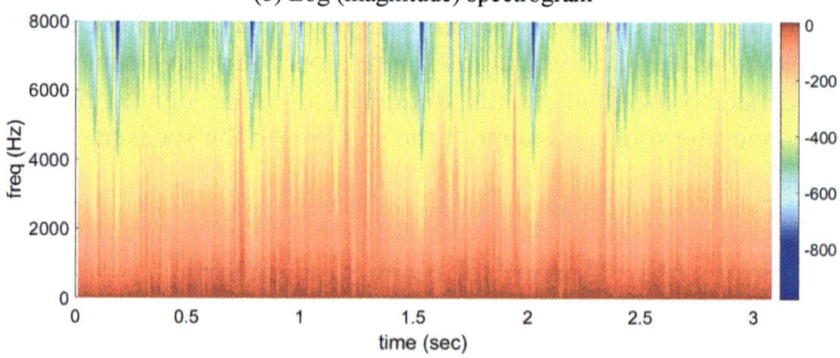

(c) Phase spectrogram

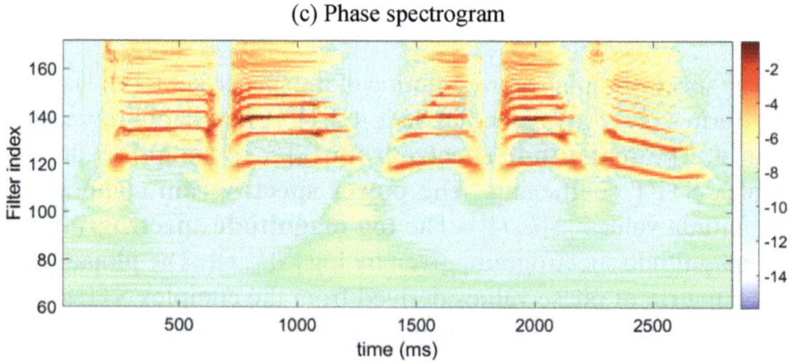

(d) Constant-Q spectrogram

Fig. 5.2 Magnitude, power, phase and constant-Q spectrograms of the same utterance

5.3.1.1 Spectral Entropy

The *spectral entropy* is merely the Shannon entropy calculated on the *power spectrum* viewed as a distribution (called a *spectral mass function*). For the measurement of spectral entropy, the area under this distribution must be normalized to 1. This is done as follows:

Given the spectrum $S[f]$ for a signal, the normalized power is given by

$$p_{S[f]} = \frac{|S[f]|^2}{\sum_f |S[f]|^2} \tag{5.5}$$

The spectral entropy is then

$$S_{entropy} = -\sum_f p_{S[f]} log p_{S[f]} \tag{5.6}$$

In practice, when performing spectrographic analysis of speech, we compute the spectral entropy for each frame of the signal to obtain a sequence of spectral entropy values that correspond to the complete signal. These values carry information about the content of speech, and relate to various aspects of the voice production process.

5.3.2 The Reassigned Spectrogram

Although a signal sampled at N samples per second has all frequencies up to $N/2$ Hz, the STFT only includes L distinct frequency components, where L is the length of the STFT window in terms of number of samples. Thus each STFT value $S[k, t]$ represents a time-frequency *bin* comprising the frequency range $f \in (\frac{k-0.5}{L}, \frac{k+0.5}{L})$, in the time range $t' \in (t - 0.5, t + 0.5)$. The STFT attributes all spectral components in the cell to $[k, t]$, effectively quantizing both time and frequency. This quantization results in loss of precision in identifying the actual frequencies that contribute to the signal, and the precise times at which they occur.

The method of *reassignment* or *remapping* a spectrogram attempts to correct the STFT placements of time and frequency by identifying the most appropriate time and frequency in each time-frequency bin to assign the energy of that bin to. In other words, the method of reassignment maps the spectral estimates in the spectrogram to time-frequency co-ordinates that are nearer to the true "support" of the analyzed signal. The resultant spectrogram is called the **reassigned spectrogram**.

Reassignment was introduced by Kodera et al. [1, 2] as the "modified moving window method" (MMWM). The method of reassignment uses the phase components of the STFT to correct the time-frequency co-ordinates of the points on the spectrogram [3, 4]. This is based on the fact that the rate of change of *phase* of a sinusoid is identical to its frequency. The actual *instantaneous frequency* within any frequency bin of the STFT (assuming there is only one dominant frequency component in it) at any time can hence be calculated as the *time derivative* of the complex phase within the bin, at that time. A similar relationship holds for the time axis as well, where the most appropriate time instant to assign to the bin can be computed from the *frequency derivative*.

Let $(\phi[f,t], \tau[f,t])$ be the reassigned frequency and time locations of the STFT cell (f,t) (we use the index $f = \frac{k}{L}$ in this section to emphasize the fact that we are reassigning frequencies and time). The matrix of reassigned frequency values $\phi[f,t]$, $\forall f$, $\forall t$ are collectively called the *channelized instantaneous frequency* (CIF) of the signal [5]. The matrix of *corrections* to obtain reassigned time values $\tau[f,t]$ are called *local group delay* (LGD), i.e. $\tau[f,t] = t + LGD[f,t]$.

For the digital signal these can be calculated by the finite difference method [1], or its equivalent—Nelson's method [5, 6] as follows. We first compute the spectral "cross-surface" terms C and L as

$$C[f,t] = STFT\left[f, t+\frac{\varepsilon}{2}\right] STFT^*\left[f, t-\frac{\varepsilon}{2}\right] \tag{5.7}$$

$$L[f,t] = STFT\left[f+\frac{\varepsilon}{2}, t\right] STFT^*\left[f-\frac{\varepsilon}{2}, t\right] \tag{5.8}$$

C and L are both complex. ε is usually set to 1.0 (i.e. the equivalent of a single time sample for C and a single frequency index for L). The CIF and LGD are computed as

$$CIF[f,t] = \tfrac{1}{2\pi\varepsilon}\angle C[f,t] \tag{5.9}$$

$$LGD[f,t] = \frac{-1}{2\pi\varepsilon}\angle L[f,t] \tag{5.10}$$

and the reassigned time-frequency location for $S[f,t]$ is now $(CIF[f,t], t + LGD[f,t])$.

This correction makes the spectrographic patterns much sharper than their original STFT counterparts, and enables far better resolution of frequencies and temporal phenomena. An example is shown in Fig. 5.3.

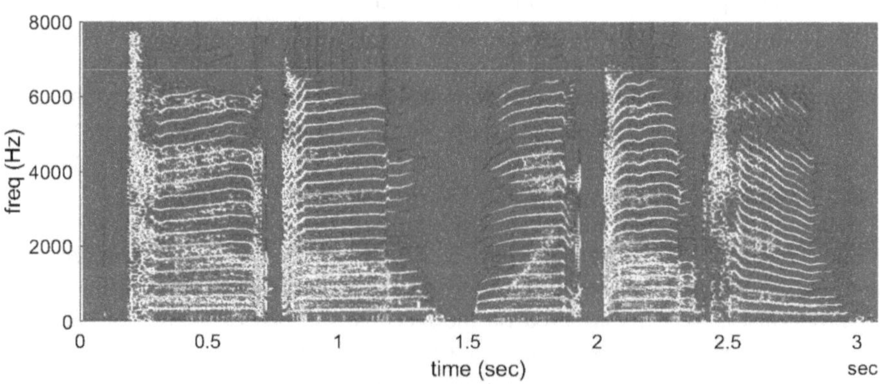

(a) The reassigned spectrogram for the signal in Figure 5.2

Fig. 5.3 Time-frequency correction by reassignment—an example

In practice, the actual outcome depends on both the type and analysis width of the window used by the STFT.

5.3.3 *Auditory Representations*

Other, more voice-appropriate spectrographic representations are based on the knowledge of human peripheral auditory processing. As explained in Chap. 2, the principal mechanism for conveying the perception of sound to the brain is through the resonant movements of the basilar membrane, which are sensed by hair cells and transmitted to the brain. The conventional STFT-based representations described in Sect. 5.3.1 decompose the signal into a set of equally spaced frequencies. However the response of human hearing is not uniform across all frequencies. Lower frequencies are better resolved by the human ear, and higher frequencies are more coarsely differentiated.

Auditory representations attempt to derive spectrographic characterizations by modeling the frequency sensitivity of the ear closely. They do so in a couple of ways:

In the first, instead of having uniformly spaced frequency bins in the decomposition, they are spaced according to the frequency sensitivity of the human ear. In other words, they are resolved more finely at lower frequencies (greater resolution in frequency) and coarsely at higher frequencies. This is typically done using a perceptual *warping* function $p(f)$, which *warps* the frequency scale. The warping function is designed such that if for any two pairs of frequencies (f_1, f_2) and (f_3, f_4), $p(f_1) - p(f_2) = p(f_3) - p(f_4)$, then the *perceived* separation between f_1 and f_2 is approximately the same as the perceived separation of f_3 and f_4. Section 2.5.2 lists a number of different warping functions that have been proposed in the literature.

The second approach is based on the observation that the STFT may be viewed as the sampled response of a bank of filters with identically-shaped frequency responses differing only in their center frequency. In auditory representations, instead of simply utilizing the standard window-based filters of the STFT, the filters themselves are modeled according to the response of the (basilar membrane and) hair cells in the ear. The representations described below utilize different combinations of these two approaches for emulating auditory responses.

5.3.3.1 The Mel-frequency Spectrogram and Cepstra

Mel-frequency spectral analysis models the frequency sensitivity of the human ear by the *Mel* function.

$$mel(f) = 2595 \log_{10}\left(1 + \frac{f}{700}\right) \tag{5.11}$$

The Mel function represents a nonlinear warping of the frequency axis to represent human sensitivity to frequencies. Figure 5.4 shows the $mel(f)$ curve.

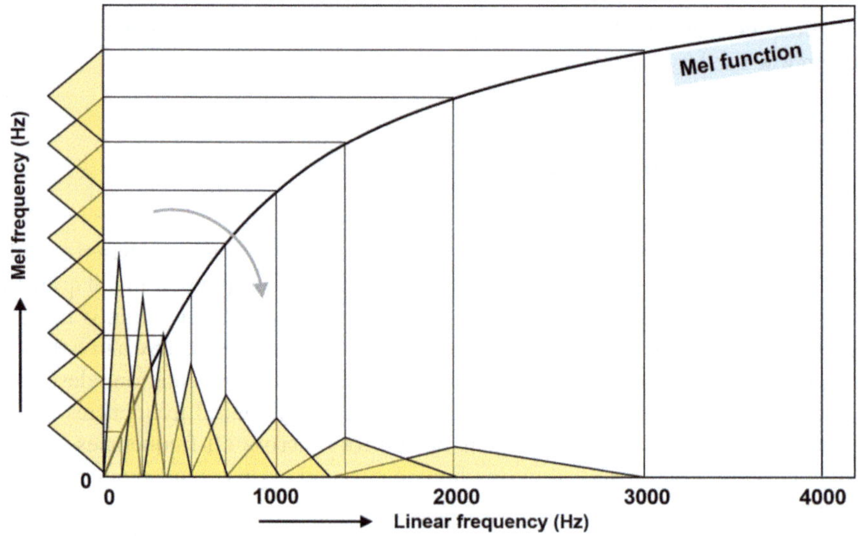

The curve shows the relation of perceptual frequencies (vertical axis) to physical frequency (horizontal axis), as specified by the Mel function. The bank of uniformly spaced triangles with 50% overlap set against the vertical axis represents the Mel-frequency filter bank, plotted against perceptual frequency. The bank of non-uniformly spaced triangles on the horizontal frequency axis shows the corresponding filter bank over the linear (physical) frequency axis. The figure is only illustrative and shows far fewer filters than would be used in a practical implementation.

Fig. 5.4 Mel warping and the Mel filter bank

The spectrographic analysis is performed using a bank of triangular overlapping filters with center frequencies spaced uniformly on the warped frequency axis. We refer to this as the *Mel-frequency filter bank* (or *Mel filter bank*). Figure 5.4 also shows the frequency response of Mel filter bank against the Mel-frequency axis. The triangular shape of the filters in this filter bank is a crude approximation to the actual frequency response of the basilar membrane (shown in Chap. 2, Fig. 2.23) at any location, which peaks at a particular frequency, but also responds to a *band* of frequencies around it.

On the linear frequency scale, the Mel filters translate approximately to a bank of non-equilaterally triangular filters whose *peak* response frequencies are spaced increasingly farther apart with increasing frequency. Figure 5.4 also shows the inverse frequency warping of the Mel filterbank to the linear frequency axis and the resulting filters. A spectrogram that is filtered using the Mel-frequency filter bank is called the *Mel-spectrogram*.

In practice, the computation of Mel spectrograms builds over the conventional STFT. The STFT is typically computed using frames of 20–25 ms duration, with a

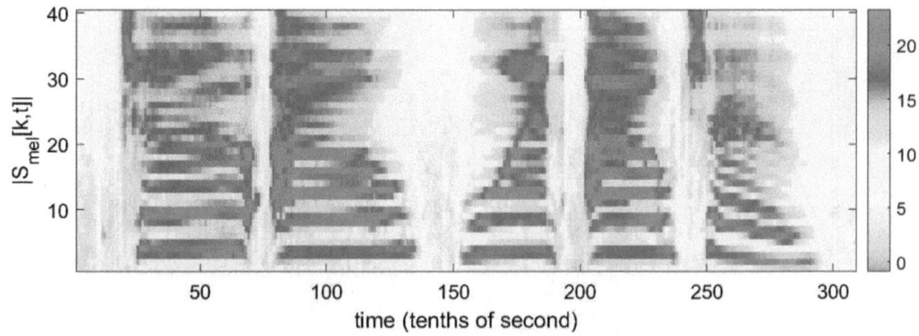

(a) Mel spectrogram of the signal in Figure 5.3

Fig. 5.5 Mel spectrogram

10 ms skip between frames (giving us a *frame rate* of 100 frames per second). A Hamming window is generally used for the STFT.

The actual imposition of the Mel filter bank on the signal is simulated by impressing the triangular Mel filters on the individual spectral vectors of the SFFT, as shown in Fig. 5.4. The coefficients of the Mel spectrogram are computed as

$$S_{mel}[k, t] = \sum_{l=0}^{L-1} m_k[l] |S[l, t]|^2 \tag{5.12}$$

Here L is the number of frequency components of the original STFT $S[k, t]$, and $m_k[l]$ is the lth triangular filter in the mel filter bank. $S_{mel}[k, t]$ is the kth coefficient of the Mel spectral vector for the tth frame. Each Mel spectral vector has as many frequency components as filters in the Mel filter bank—more filters provide greater frequency resolution; however increased (and sometimes unnecessary) resolution at higher frequencies comes at the price of noisy representations at lower frequencies where the filters may now become too narrow to capture reliable information. Typical analyses use 40–64 filters covering the frequency range 100–7000 Hz. Figure 5.5a shows an example of a Mel spectrogram. Note the warping of the frequency axis.

Mel spectral vectors are often further transformed into *Mel cepstra*. To do so, first the log of the Mel filtered values is computed, then a discrete cosine transform DCT (IDCT), is performed on each resulting *log Mel spectral* vector. The IDCT of a K-component log-spectral vector returns a K-component cepstral vector (see Fig. 5.6b). The higher-order terms of the Mel cepstral vector capture often irrelevant spectral detail. In most speech processing applications, the cepstral vector is truncated to retain the first thirteen coefficients—an empirically determined choice that has been found to be optimal for speech recognition, specifically. For biometric and forensic analysis, it may be advantageous to retain at least the first twenty cepstral coefficients. The resulting truncated vector is called a "Mel-frequency cepstral coefficient" (MFCC) vector. The operations above are repeated for each frame. The spectrogram as a whole is thus transformed into a sequence of MFCC vectors.

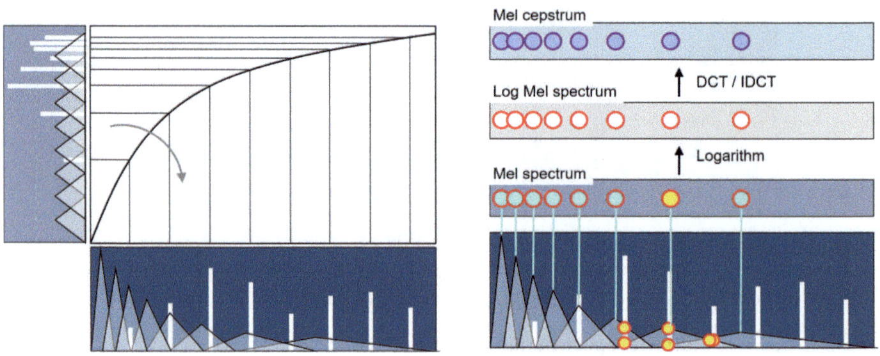

(a) **Spectrum warping:** Filtering the linear axis with warped Mel filters.

(b) Further processing of the Mel spectra. All weighted spectral values for each filter are added (integrated) giving one value per filter.

Fig. 5.6 Warping the Mel filter bank and further processing for computation of Mel cepstra

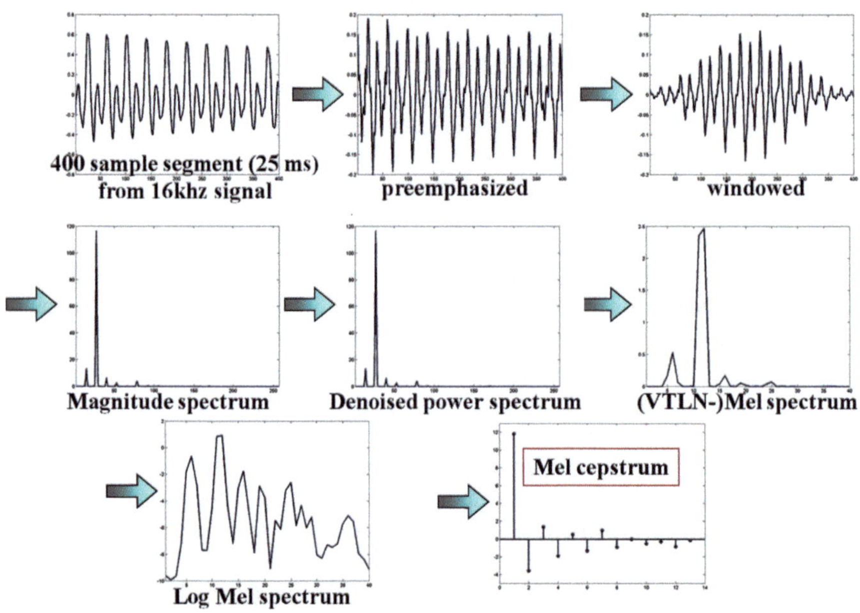

The process shown includes denoising and speaker normalization via *vocal tract length normalization* (VTLN). These steps are important for ASR systems, but not recommended for profiling, since VTLN normalizes speaker differences and denoising introduces artifacts.

Fig. 5.7 The complete mfc computation process

The MFCC computation process is schematically illustrated in Fig. 5.7.

The extraction of Mel cepstra as explained above is just one of the many variants of perceptually frequency-warped spectrographic representations that may be derived from the voice signal. Alternate perceptually warped spectrograms may use the Bark warping function or a log-linear approximation to it instead of the Mel curve, to derive the perceptual filter bank. For the purposes of analysis, the distinction is minor. The constant-Q transform of Sect. 4.3.3.1 may in fact also be viewed as an instance of a perceptually warped spectral representation, where the perceptual warping has been modeled by $\log(f)$.

Other variants of perceptually motivated features are derived by modifying the finer details of the processing described above. For example, *perceptual linear prediction* (PLP) features use *equivalent rectangular bandwidth* (ERB) filters instead of Mel filters, cube-root compression rather than a logarithmic one, and cepstral features derived from an auto-regressive model fitted to the warped and compressed spectrum [7, 8]. Other types of cepstra, such as *power normalized cepstra* (PNC) [9], are designed for noise tolerance and robustness to distortions introduced by reverberation etc. These are not described here.

5.3.3.2 Auditory Filter Bank

The term "auditory filter bank" is generally used to refer to any filter bank analysis where the filters are modeled after the human auditory system. In that sense, the Mel-frequency filter bank is an auditory filter bank, as is any other filter bank based on popular models of the frequency response of the auditory system, such as the Bark scale or the ERB scale.

More commonly, however, the term "auditory filter bank" is used to refer to a *gammatone* filter bank. A gammatone filter has an impulse response that is the product of a gamma distribution and sinusoidal tone and is given by

$$h_f[t] = at^{n-1}e^{-2\pi bt}\cos(2\pi ft + \phi) \tag{5.13}$$

where f is the center frequency of the filter, ϕ is the phase of the carrier, a is the amplitude, n is the filter's order, and b is the filter's bandwidth. The filter models the response of the auditory filters in the peripheral auditory system [10].

A gammatone filter bank is a bank of gammatone filters. The parameters of the filters are generally set based on *equivalent rectangular bandwidth* (ERB). The ERB in turn is based on the observation that at moderate sound levels the bandwidth of human auditory filters centered at frequency f is approximately given by

$$ERB(f) = 24.7(0.00437f + 1) \tag{5.14}$$

This is the "equivalent rectangular bandwidth" at f. In fact, Eq. 5.14 leads to the ERB scale, which we recall is given by

$$ERBs(f) = 21.4\log_{10}(0.00437f + 1) \tag{5.15}$$

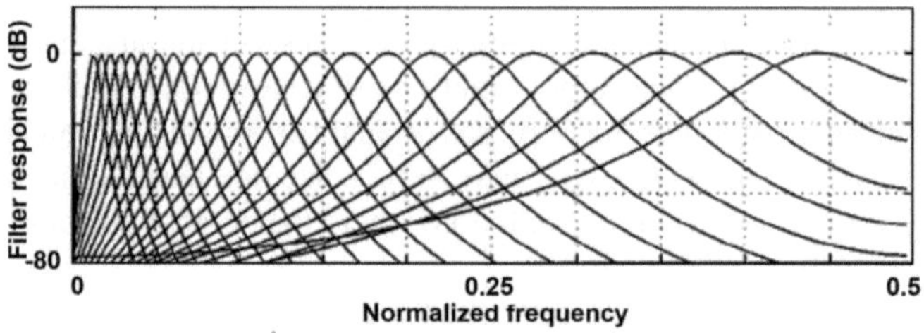

The spectral response of a 23-filter Gammatone filter bank. The characteristic frequencies of the filters are uniformly spaced on the ERB scale.

Fig. 5.8 The Gammatone filter bank

In the typical setting, the center frequencies of the gammatone filters in the filter bank are distributed uniformly along the ERB scaled frequency axis. The filter order n is typically set to 4, and the bandwidth of the filter b is set to $b = 1.019ERB(f)$. Figure 5.8 illustrates a gammatone filter bank.

Conventional implementation of auditory filters (including the Mel filter bank) are only a crude approximation to the actual frequency response of the auditory system. As we have seen in Chap. 2 and Fig. 2.24b, the response of the and hair cells is not triangular at any frequency. Also, the cochlea has $\approx$*3000* hair cells in aggregate across all hair bundles, while we use only a small number of filters in typical auditory filter bank representations—about 40 in most applications. Greater resolution obtained using larger filter banks with narrower filters is often more effective in profiling applications.

5.3.4 The Correlogram

The correlogram is a three-dimensional characterization of the speech signal that captures the time-varying self-similarities and periodicities of the speech signal *within individual frequency bands*.

The signal $s[t]$ is first filtered by a filter bank. Typically an auditory filter bank such as a Mel filter bank or a Gammatone filter bank is used. Alternate variants may simply use the STFT as a filter bank. The output of the filter bank is a bank of outputs, $s_f[t],\ f = f_1, \ldots, f_L$, where $s_{f_l}[t]$ is the output of the lth filter, f_l is the frequency of peak response of the filter, and L is the number of filters.

The correlogram computes the autocorrelation function (ACF) of the outputs of each of the filters over a sliding window, to obtain the following triple-indexed value

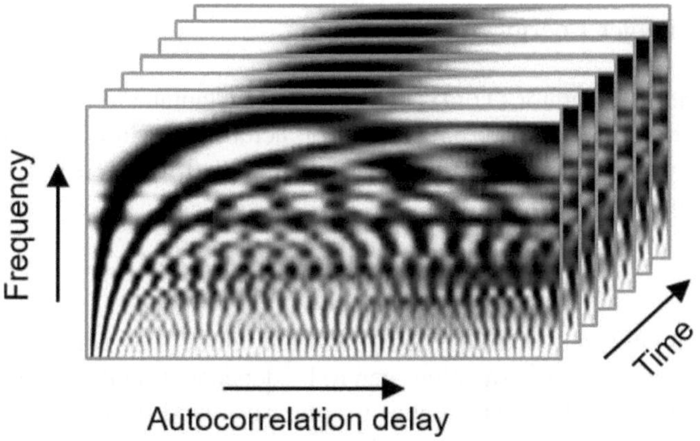

A small time-section of a correlogram of a voice signal, computed from the complex STFT

Fig. 5.9 Example of a correlogram

$$R[f,t,\tau] = \frac{1}{K-\tau} \sum_{k=0}^{K-1-\tau} s_f[t+k]w[k]s_f[t+k+\tau]w[k+\tau]$$
$$\tau = 0, 1, \ldots, K-1; \quad f = f_1, \ldots, f_L \qquad (5.16)$$

$R[f, t, \tau]$ is the autocorrelation for lag τ, at time t, in the fth frequency band (i.e., the output of the filter with peak response frequency f), $w[k]$ is a window function, and K is the length of the analysis window. Figure 5.9 shows an example of a correlogram derived from the STFT of a voice signal.

The correlogram $R[f, t, \tau]$ is often visualized as a video, where the nth frame of the video is the two-dimensional image $\hat{R}[f, \tau] = R[f, n, \tau]$.

For profiling, analysis of the sequence of 2-D autocorrelation matrices that comprise the correlogram using image and video processing techniques provides cues to a number of forensic-related factors, including the recording environment of the speech signal.

5.3.5 Modulation Features

Modulation features capture systematic patterns of variation of signal amplitude and frequency within frequency bands. Below we describe some methods of extracting such information.

5.3.5.1 The Teager Energy Operator

The Teager Energy Operator (TEO) [11] tracks the instantaneous energy in a signal by modeling the process that produces it. The actual formula derived is obtained from a second-order mass-on-a-string approximation to the process [12], and results in the following formula

$$\psi(x[n]) = x^2[n] - x[n-1]x[n+1] \tag{5.17}$$

The TEO for a pure tone of frequency f and amplitude A is computed using the above formula as $4\pi^2 A^2 f^2$—a value seen to be dependent on the frequency of the tone. The TEO of a falling exponential is similarly found to be 0. TEO is in fact generally used in most AM-FM demodulation procedures, e.g. [13].

The TEO also performs crude amplitude demodulation of signals. It is more commonly used, however, as key component of more sophisticated demodulation schemes, such as the one described in the next section.

5.3.5.2 Amplitude and Frequency Modulations

Consider the following signal

$$s[t] = (1 + a[t])cos(2\pi(f_c + f[t])t)$$

This signal is ostensibly a cosine, but its amplitude $(1 + a[t])$ itself varies with time. Its frequency $(f_c + f[t])$ too is time-varying about a center f_c. Figure 5.10 shows an instance of such a signal. The signal is thus both *amplitude modulated* by $a[t]$ and *frequency modulated* by $f[t]$. The frequency f_c, representing the frequency of the signal in the absence of amplitude or frequency modulation (i.e when $a[t] = 0$ and $f[t] = 0$) is called the carrier frequency. The top panel of Fig. 5.10 shows an example of an amplitude and frequency modulated signal.

In general, amplitude modulation (AM) refers to the time-variation of the amplitude of a band-limited signal within a frequency band. Frequency modulation (FM) refers to the time-variation of the frequency itself. Estimation of the AM ($a[t]$ in the above example) and FM ($f[t]$ in the above example) from the signal is called *demodulation*.

A band-pass filtered speech signal filtered by a frequency with center frequency f_c too demonstrates AM and FM characteristics. AM and FM features may be extracted from it by demodulating the filtered signal. Of the many algorithms available for demodulation, a widely used one for speech signals is the Discrete Energy Separation Algorithm (DESA) [14] algorithm. It separates a signal into its instantaneous (sample level) AM and FM components (thus also yielding accurate formant measurements [15]).

For any bandlimited signal $s[t]$, the FM component $\Omega[t]$ and the magnitude of its AM component $a[t]$ can be derived as follows:

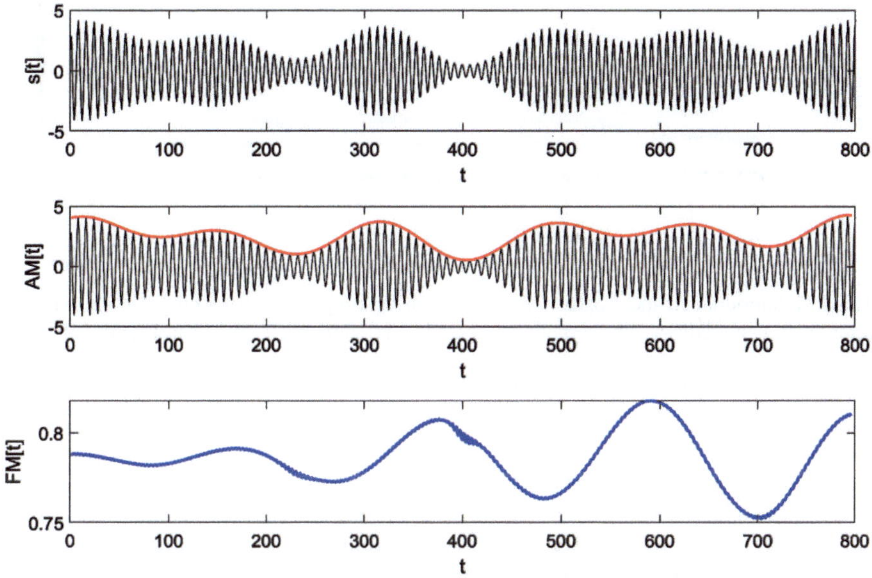

Top: A narrow-band signal with both amplitude and frequency modulation. Middle: the demodulated AM obtained using DESA (red curve). **Bottom:** The demodulated FM obtained using DESA.

Fig. 5.10 AM and FM demodulation

$$\Omega[t] \approx cos^{-1}\left(1 - \frac{\psi(s[t])+\psi(s[N-1])}{4\psi(s[t])}\right) \tag{5.18}$$

$$|a[t]| \approx \sqrt{\frac{\psi(s[t])}{1-cos(\Omega[t])^2}} \tag{5.19}$$

If we assume that the narrowband/instantaneous FM signal is approximately equal to the center frequency f_c for any of the filters with sufficiently bandlimited filters, i.e. for any filter with center frequency f_c the FM signal $\Omega[t] \approx f_c$, the instantaneous AM signal at the filter is:

$$a[t] \approx \sqrt{\frac{|s^2[t] - s[t-1]s[t+1]|}{f_c^2}} \tag{5.20}$$

The lower two panels of Fig. 5.10 show demodulated AM and FM singals obtained with DESA.

Note that although DESA is an efficient and effective algorithm, other methods too are often used to demodulate the frequency bands. Hilbert transforms are frequently used when only AM demodulation is required [16]. In [17] pole-zero models are used to effect the demodulation. Others such as [18, 19] use a combination of half-wave rectification and low-pass filtering to extract the AM signal.

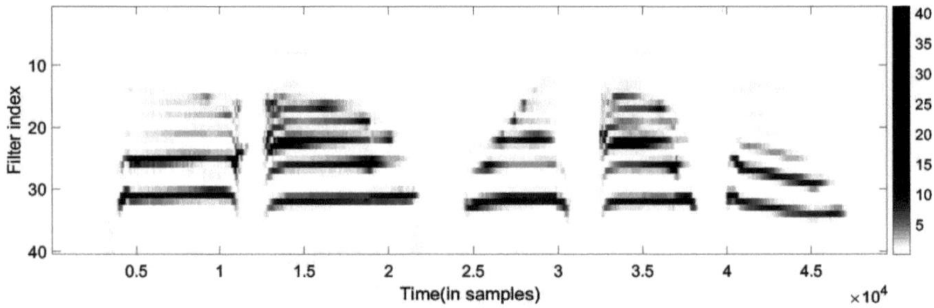

The AM demodulated spectrogram obtained from the output of a 40-component gammatone filter bank, using DESA. The signal represented is the same as the one from Figure 5.2.

Fig. 5.11 AM demodulated spectrogram

In practical implementations of demodulation, the signal is filtered with a filter bank such as a Gammatone filter bank with upper and lower cutoff frequencies 200 Hz and 7500 Hz. The output of each filter is is demodulated—AM demodulation extracts AM features $A[f, t]$ (representing the tth AM demodulated sample in frequency band f), while FM demodulation extracts FM features $F[f, t]$ (representing the tth FM demodulated sample in frequency band f). Figure 5.11 shows examples of AM features represented in the form of a spectrogram (the FM demodulation does not exhibit visually meaningful patterns in this example and is not shown). The output may be used as is, or more frequently it is downsampled for downstream processing to extract additional features as outlined below in Sect. 5.3.5.4.

5.3.5.3 2-D Modulations: The Gabor Filter

A Gabor filter is defined in the spatial domain as the modulation of a sinusoidal plane wave by a Gaussian shaped kernel function.

These filters are very good for detecting edges and textures in images, the characterizations of which can be strong features for many machine learning-based applications of images and sound (represented in visual forms).

A two-dimensional Gabor filter $g(x, y; \Theta)$ with parameters Θ is given by

$$g(x, y; \Theta) = s(x, y; \Theta_s)w(x, y; \Theta_w) \tag{5.21}$$

where $s(x, y; \Theta_s)$ is a complex exponential called the *carrier*, and $w(x, y; \Theta_w)$ is a Gaussian function known as the envelope. Θ_s and Θ_w are the parameters of the carrier and the envelope respectively and together compose Θ, the parameters of the Gabor itself.

The complex sinusoid is given by

$$s(x, y; \Theta_s) = \exp\left(i\, 2\pi F_0(x \cos\psi + y \sin\psi) + \phi\right) \tag{5.22}$$

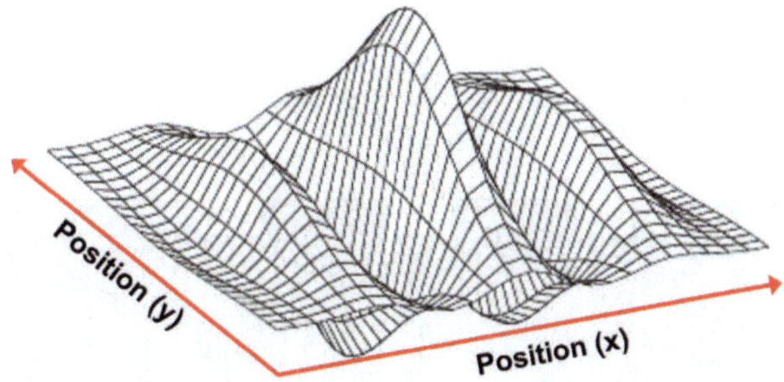

Fig. 5.12 A Gabor filter in the spatial domain

where F_0 is the magnitude of the exponential, ψ is its direction, and ϕ represents its phase. Thus $\Theta_s = \{F_0, \psi, \phi\}$.

The Gaussian envelope is given by

$$w(x, y; \Theta_w) = K\ exp\left(-\frac{x_\theta^2 + a^2 y_\theta^2}{2\sigma^2}\right) \tag{5.23}$$

where σ is the standard deviation of the Gaussian and a is its *aspect ratio*. (x_θ, y_θ) is a *rotation* of (x, y) by θ. The relation between the two is given by

$$\begin{aligned} x_\theta &= x \cos\theta + y \sin\theta \\ y_\theta &= -x \sin\theta + y \cos\theta \end{aligned} \tag{5.24}$$

θ is the *orientation* of the Gaussian. Thus, the parameters of the Gaussian are given by $\theta_w = \{\sigma, a, \theta\}$.

In the special case where the exponential and the Gaussian have the same orientation, we can also write the exponential as

$$s(x, y; \Theta_s) = \exp\left(i2\pi F_0 x_\theta + \phi\right) \tag{5.25}$$

Here F_0 is the spatial frequency of the exponential and $F_0\sigma$ is the spatial bandwidth of the filter. Gabor functions may also be parametrized by $\lambda = \frac{1}{F_0}$, rather than F_0. In this case λ is the *wavelength* of the Gabor function.

Figure 5.12 shows a typical Gabor filter.

The *power spectrum* of the Gabor filter is the squared magnitude of its 2-D Fourier transform [16]. It characterizes how the Gabor filter responds to spatial frequencies along x and y. Figure 5.13 shows the Gabor function in spatial and power spectral domains.

A two-dimensional input $S(x, y)$ is filtered by a Gabor filter $g(x, y; \Theta)$ (where we have collectively represented the parameters of the filter by Θ) through two-dimensional convolution as follows

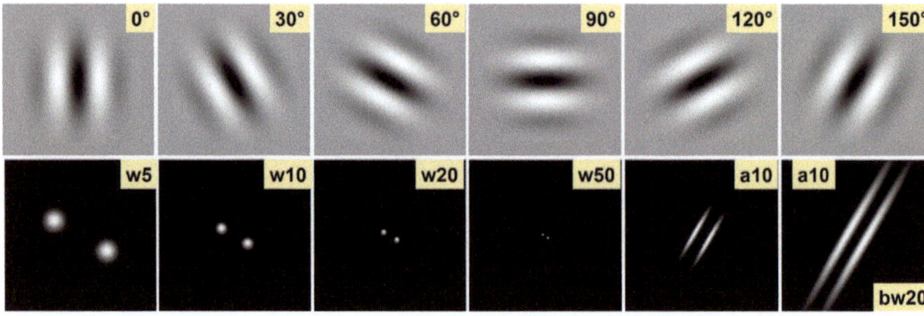

Top panel: The Gabor function in the spatial domain with six different orientations. The wavelength in all cases is 150. The aspect ratio and bandwidths are all unity. **Bottom panel:** Power spectrum of the Gabor function with 150° orientation, and different wavelengths, aspect ratios and bandwidths. The phase offset is 0 in all cases. ***In sequence***: wavelengths of 5, 10, 20 and 50, with unity aspect ratio and bandwidths; wavelength 20, aspect ratio 10 and unity bandwidth; wavelength 20, aspect ratio 10, and bandwidth 20.

Fig. 5.13 Gabor function in spatial and power spectral domains

$$Y_\Theta(x, y) = \sum_{x'=0}^{K_x} \sum_{y'=0}^{K_y} S(x - x', y - y')g(x', y'; \Theta) \tag{5.26}$$

In practice, a bank of filters with parameters $\Theta_1, \ldots, \Theta_K$ is employed, to obtain a collection of outputs $Y_{\Theta_1}, \ldots, Y_{\Theta_K}$. These outputs may either be used separately, or added into a single combined representation for further processing.

Figure 5.14 shows an example of the application of a bank of Gabor filters to a speech spectrogram. Here, the outputs of a bank of filters with different orientations have been added to result in a single output representation.

More details about extracting these features are given in [20] and many other articles.

5.3.5.4 Modulation Spectra

Modulation spectra are features derived (typically) from the temporal envelopes obtained through AM demodulation of the bandpass-filtered components of speech signals. Typically, the modulation spectrum is obtained by passing the speech signal through bandpass auditory filters, computing the envelopes of the filter outputs, and passing these envelopes through a second set of parallel bandpass modulation filters with center frequencies between 2 and 16 Hz. As a result, the modulation spectrum is a joint function of the center frequencies of the initial peripheral auditory filters, which span the range of useful speech frequencies, and the center frequencies of the modulation filters.

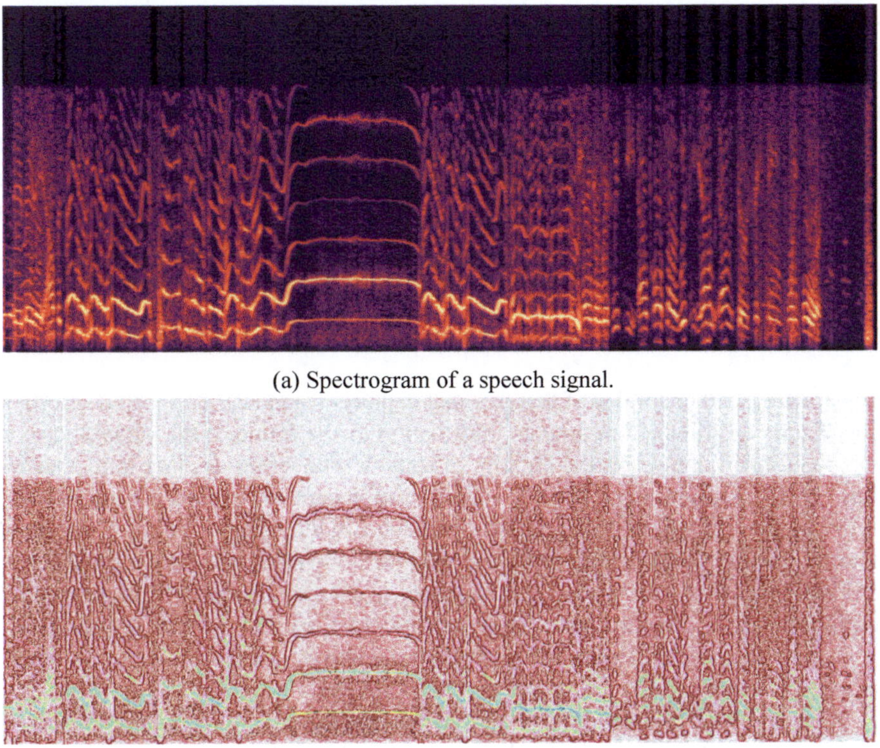

(a) Spectrogram of a speech signal.

(b) The spectrogram above filtered using a bank of Gabor filters.

Fig. 5.14 Gabor filters used on a spectrogram

Modulation spectra are useful in reducing the effect of background noise, since they pass the temporal modulation patterns of speech. Noise components usually exhibit frequencies of amplitude modulation outside this range, and are suppressed as a result. In speech signals, the greatest amount of temporal modulation has been observed at modulation frequencies of approximately 6 Hz [21]. Low-pass filtering the envelopes of the outputs of each channel reduces the variability introduced by background noise.

Modulation spectra play a role in perception of sound, and influence human judgments made from speech. In the context of perception, studies have shown that human brainstems are sensitive to specific modulation frequencies of amplitude-modulated signals [22, 23], and modulation spectra approximate these and other psychoacoustic findings, e.g. [24].

The actual computation of modulation spectra can be done in a variety of ways, based on the application. Examples include compute a Fourier transform of the envelopes in each band, to compute the spectral composition of the AM signals as a preliminary step to denoising the signal [25], and computing the average crosscorrelation between the AM signals in different bands, to obtain features to analyze and synthesize natural sounds [26]. Yet other methods, developed to improve the robustness of ASR systems, include downsampling the envelopes and computing the Fourier transforms of the resulting signals in each of the bands [19], and using a DCT to compute features from the downsampled spectrum [28]. From the perspective of profiling, however, these representations are lossy. It is more profitable to derive modulation spectral features using automated feature extraction techniques such as those listed in Chap. 7 directly on the AM and FM demodulation spectrograms.

5.3.5.5 Secondary Features That Capture Temporal Variations

Cepstra are good examples of features that capture instantaneous spectral information. Each cepstral vector captures information in the frame of the signal that it is computed from. However, the identity of sounds (phonemes or other units that we may consider) resides not just in the snapshots of instantaneous information, but also in the manner in which this information changes with time, i.e., in their temporal patterns. Features that capture these temporal variations are called temporal features.

Two of the most widely used temporal features get their identities from the spectral features they are computed from. These are *velocity features*—which capture the rate of change of spectral feature values with time, and *acceleration features*—which capture the rate at which the velocity features change.

In most speech processing applications that use cepstra, velocity features corresponding to any frame are obtained by computing the difference between the corresponding feature value for the next frame and the value for the previous frame. This is usually a component-wise difference, and thus the dimension of a velocity vector is the same as that of the original features. Acceleration features for a given frame are computed as the difference between the velocity feature for the next frame and that of the previous frame. Their dimensionality is the same as that of the velocity features. For any cepstral vector $\mathbf{C}$ corresponding to frame t, velocity and acceleration features are given by:

$$\Delta\mathbf{C}(t) = \mathbf{C}(t+\tau) - \mathbf{C}(t-\tau) \tag{5.27}$$

$$\Delta\Delta\mathbf{C}(t) = \Delta\mathbf{C}(t+\tau) - \Delta\mathbf{C}(t-\tau) \tag{5.28}$$

where $\Delta\mathbf{C}(t)$ is the velocity vector corresponding to frame t and $\Delta\Delta\mathbf{C}(t)$ is the acceleration vector. These are also referred to as *delta* and *double-delta* features respectively. In practical applications such as ASR systems, the spectral feature for a frame, and the velocity and acceleration features for the frame are generally concatenated and used as a single feature representation for the frame.

5.4 Model Domain Characterizations

As discussed in Chap. 2, the sounds we hear are the glottal pressure waves (and sounds created by other sources such as turbulence) modified by the resonances of the vocal tract. Models of the vocal tract attempt to analytically explain and derive the characteristics of both—the sound sources and the resonances of the vocal tract. These then form the basis for many features that can be used for profiling. We describe some key characteristics of these models in the sections below.

5.4.1 The Source-Filter Model

As discussed in Sect. 2.3.3.2, the source-filter model for the human voice production mechanism comprises two parts. The first is a "source" that represents signals that originate at the glottis. The second is a linear filter, representing the upper vocal tract beyond the glottis (which we refer to as the "vocal tract" in this section). The source generates "excitation signals" that are input to the filter. The filter operates on the excitation to output the actual signal. Practically, the excitation represents the glottal airflow. The output of the filter is the actual speech that results from the modulation of the glottal airflow by the vocal tract. Figure 2.9 illustrates the model.

The source filter-model is motivated by the fact that the physics of the vocal tract can be largely explained by linear differential equations. The linear-system model also results in a mathematically tractable framework that enables us to approximately separate out the actual glottal excitation and also compute the resonances of the vocal tract from the speech signal, as we see below.

5.4.2 The Filter: Modeling the Vocal Tract

The vocal tract can be viewed as an acoustic tube into which air pressure waves from the glottal excitation are "pushed" at one end and emerge from the other as speech. In Sect. 2.3.3.6 we briefly discussed the physics of air flow through the vocal tract and explained how it may be mathematically approximated as a sequence of uniform lossless-tubes, as shown in Fig. 2.14 of Chap. 2.

The "filter" component of the source-filter model characterizes the vocal tract as a discrete-time version of this lossless-tube model.

Figure 5.15 illustrates the discrete-time model for the vocal tract. For the greatest resolution in the model, the lengths of the individual segments must be the smallest that can be resolved by our computation. For the discretized signal, this corresponds to segments whose lengths are equal to the distance traveled by sound in half a sampling instant. The length of the vocal tract of a typical adult Caucasian male is 17 cm. Assuming the velocity of sound to be 340 m/s and a sampling frequency

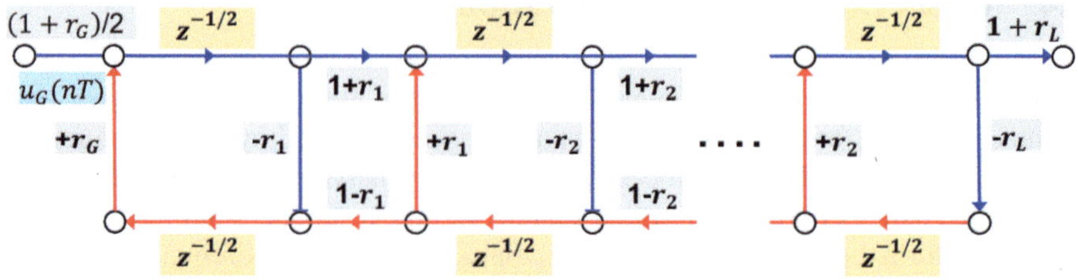

Discrete-time equivalent of the lossless tube model of the vocal tract. Each section represents one of the tube segments. The r_is are their reflection coefficients.

Fig. 5.15 Discrete-time model of vocal tract

of 16000 samples/second, each section of the lossless-tube model would be about approximately 1 cm long, and the entire model would require 16 tube segments. Female vocal tract lengths are typically shorter (around 15 cm) and would require fewer segments in the model.

Under the specific condition where the length of each section equals the distance traveled by sound in half a sampling instant and the glottis is closed, the transfer function of entire model, or the **vocal-tract transfer function**, is given by the discrete-time analog of Eq. 2.5 of Sect. 2.3.3.4:

$$T(z) = \frac{G}{1 - \sum_{k=1}^{N} a_k z^- k} \tag{5.29}$$

where G is a gain term and N is the number of tube sections in the model. The vocal tract transfer function is a filter that modifies or modulates the speech spectrum to produce its characteristic shape [29]. Representing the discrete-time version of the excitation to the vocal tract as $e[t]$ and the actual speech signal as $s[t]$, Eq. 5.29 represents the relation

$$s[t] = \sum_{k=1}^{N} a_k s[t-k] + Ge[t] \tag{5.30}$$

Under the lossless-tube assumptions, the denominator of the transfer function $T(z)$, given by $D(z) = 1 - \sum_{k=1}^{N} a_k z^- k$, can be derived incrementally over the sections of the model from the reflection coefficients using the recursion [29].

$$D_0(z) = 1 \tag{5.31}$$

$$D_k(z) = D_{k-1}(z) - r_k z^{-k} D_{k-1}(z^{-1}) \tag{5.32}$$

$$D(z) = D_N(z) \tag{5.33}$$

where $D_k(z)$ is the denominator coefficient of the transfer function of the first k segments.

Equation 5.29 is seen to be identical to the autoregressive model described in Sect. 4.2.2.2. Thus, if we were to derive AR parameter estimates for a speech segment from the closed-glottis period of the speech signal, they may be expected to correspond to the parameters of $D(z)$. Specifically, Eq. 5.33 is very similar to relations obtained in Burg's method for estimating AR models. In particular, if the AR parameters are derived using the Levinson Durbin algorithm [30, 31], the reflection coefficients that are obtained as part of the update correspond to the reflection coefficients of the lossless-tube vocal tract model. As explained in Sect. 2.3.3.4 the reflection coefficients are, in turn, related to the ratios of the areas of adjacent tube segments of the model. As a consequence, the estimation of AR parameters also potentially leads to the estimation of these area ratios.

All of the above methods for estimating the parameters of the AR model for any segment of speech $s[t]$ are based on minimization of the squared error: $E = \sum_t (s[t] - \sum_k a_k s[t-k])^2$. An alternate approach, referred to as "Discrete All Pole" (DAP) estimation [32], instead minimizes the *Itakura Saito* distortion [33] between the signal and the filter response. While somewhat less simple than squared-error minimization, DAP has been found to result in more accurate estimates of AR parameters for signals such as speech and is often used instead of squared-error-based AR techniques in the context of speech signal analysis.

The AR model of Eq. 5.29 thus facilitates both, the estimation of the vocal tract configuration from a speech signal (from the reflection coefficients) and the estimation of the resonant characteristics of any configuration of the vocal tract by expressing its structure as the concentric, lossless tube model. The model may further be extended to also consider the anti-resonances of the nasal tract for nasalized sounds [34] although most analyses do not utilize this extension.

The model of Eq. 5.29 however only represents the upper vocal tract—the region of the vocal tract above the glottis, which is considered as a filter by the source-filter model—when it is estimated from speech recorded during the *closed-glottis* phase of the glottal excitation. When the glottis is open, the estimated parameters represent the influences of not only the upper vocal tract, but also the *subglottal* regions below the glottis, and the resulting filter is thus an unreliable characterization of the vocal tract filter.

We discuss the challenges of estimating the AR model parameters accurately under the circumstances in Sect. 5.5.3.2, as part of the discussion on recovering the glottal excitation signal.

5.4.3 *The Vocal Excitation: The Source*

In Chap. 2 we have discussed the generation of the glottal signal that excites the vocal tract. The primary source for this excitation is the air that is expelled from the lungs. For phonated (voiced) sounds, the glottis repeatedly opens and closes, letting the air out as semi-periodic (or quasi-periodic) pulses. For unvoiced sounds the glottis stays open and the air rushes through. The resultant air turbulence then comprises

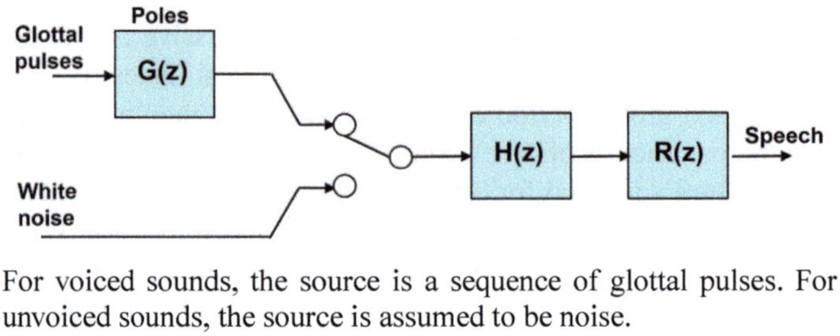

Fig. 5.16 Source assumptions in the concatenated tube model for different types of sounds

the excitation signal for the vocal tract. For profiling purposes, it is extremely useful to extract or estimate the vocal excitation source for further analysis.

For voiced speech, the excitation signal has detailed and consistent structure that can be identified and quantified. For unvoiced speech, however, the excitation can be ill-defined and it may be less effective to try to isolate and analyze it. Indeed, most variants of the source-filter model simply model the source for unvoiced speech as white noise. Figure 5.16 illustrates the source filter model.

In the following section we discuss how the excitation signal for voiced sounds may be extracted from the speech signal and how its key features may be estimated.

5.4.3.1 Extracting the Excitation Signal: Estimation of the Glottal Pulse Waveform

For voiced signals, the excitation primarily comprises a noisy semi-periodic sequence of glottal pulses. Figure 5.17 shows a typical glottal-flow waveform and indicates the key timing characteristics of glottal flow, including the *Glottal Closure Instant* (GCI), which is the instant at which the glottis closes, the *Glottal Opening Instant* (GOI), which is the instant at which the glottis opens, the peak flow instant (GP), and the glottal cycle time $T0$, which is the time between two consecutive glottal closures. The glottal cycle time is also the *pitch period*. Various timing-based measures may be derived from the glottal flow from knowledge of these attributes.

Nearly all methods of extracting the vocal excitation from a voice signal do so through *inverse filtering*. As discussed in Sect. 5.4.2, the upper vocal tract is modeled as a linear filter that filters the excitation signal following Eqs. 5.29 and 5.30. The excitation signal $e[t]$ can be recovered by passing the speech signal through an *inverse* of this filter, i.e. one that undoes the effect of the vocal-tract filter.

$$e[t] = s[t] - \sum_{k=1}^{N} a_k s[t-k] \tag{5.34}$$

(where the gain term G in Eq. 5.30 has been subsumed into $e[t]$ itself).

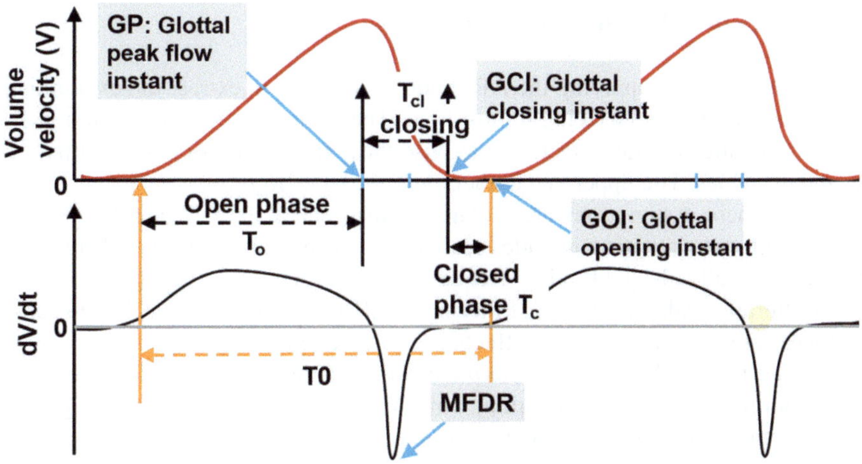

Top: Glottal flow waveform. The glottal closure instants (GCI), glottal peak flow instants (GP) and glottal opening instants (GOI) are marked. Also marked are the periods of the various phases of the glottal cycle, including the pitch period T0. **Bottom:** The derivative of the glottal flow. It usually exhibits a sharp negative peak just before the GCI.

Fig. 5.17 Glottal flow and its derivative

The parameters of Eq. 5.29 can however only be accurately measured during the *closed glottis* period of the vocal tract excitation, i.e. the period between a GCI and the subsequent GOI, since this is the only phase of voice production during which the poles of the all-pole filter of Eq. 5.29 represent only the resonances of the upper vocal tract. During other phases, it is hard to separate the resonances of the upper vocal tract from those of subglottal regions.

Ideally, therefore, the AR parameters must be estimated only from the portions of the speech signal that occur during glottal closure, but must subsequently be used in the inversion Eq. 5.34 to derive the excitation signal over the entire glottal cycle [35–37]. This requires knowledge of the GCI and GOI instants and the pitch period. However, identifying these timing cues in the glottal waveform requires knowledge of the excitation $e[t]$ itself, resulting in a gridlock.

Early methods to deal with the problem used auxiliary readings through devices such as electro-glotto graphs (EGG) to identify GCI and GOI [38–40]. Later methods added a pneumotachograph mask, which captures the volume velocity of the air directly at the mouth [41], to eliminate the effect of lip radiation and open-air transmission. More generally however, auxiliary measurements may not be available and the estimation must be done from the signal itself.

A number of solutions have been proposed in the literature in an attempt to resolve the problem and derive good estimates of $e[t]$. Below we outline three approaches, which approach the problem in different ways. Note that all of them benefit from knowledge of the pitch. We address the issue of pitch extraction later in Sect. 5.5.1.

5.4.3.2 Iterative Adaptive Inverse Filtering

The key challenge in applying the AR model effectively to extract the excitation is that the assumptions underlying it must hold. The z-transform of the excitation $e[t]$ must not include any poles itself—otherwise these too would become incorporated into the a_k values. The upper vocal tract too must have no zeros.

Both of these assumptions are violated in the general case. When the glottis is open, the glottal excitation includes resonances of the sub-glottal vocal tract. Lip radiation typically also adds a low-frequency anti-resonance to the signal.

The "Iterative Adaptive Inverse Filtering" (IAIF) algorithm [42] resolves the problem by explicitly modeling these effects in an iterative framework.

It models the subglottal resonances manifesting in the excitation as a low-order AR filter. It compensates for lip radiation, which can be modeled as a simple high-pass filter with the transfer function $R(z) = 1 - \alpha z^{-1}$, using an integrator

$$\hat{s}[t] = s[t] + \alpha\hat{s}[t-1] \tag{5.35}$$

where the parameter α is typically set in the range [0.95, 0.99].

The rest of the process is shown in Fig. 5.18.

The signal is first high-pass filtered. This is an anticipatory step, intended to compensate for the later integration of the signal. Subsequently, a single-pole ($N = 1$) AR filter is estimated from the signal. This order-1 AR is assumed to capture the resonances in the glottal waveform. The speech is inverse filtered by this filter. This inverse filtered signal is assumed to no longer include glottal resonances and a higher-order ($N = 10$) AR model is derived from it. This AR model nominally now only models resonances of the upper vocal tract.

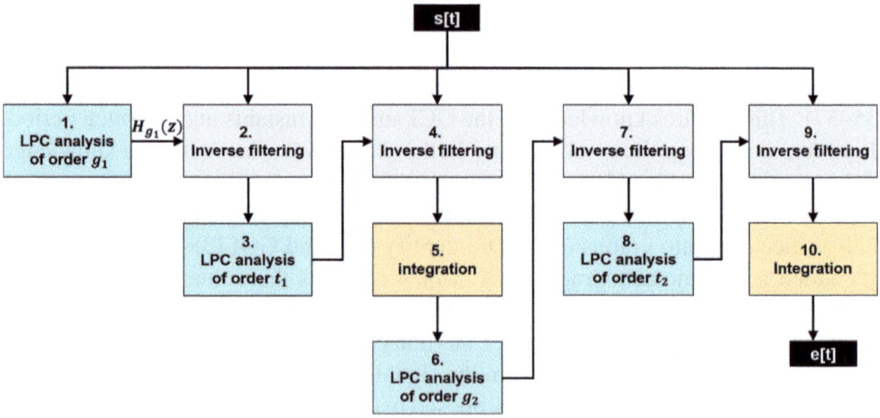

Fig. 5.18 The IAIF method for computing the excitation signal

The speech signal is inverse filtered by this second AR model, and integrated to eliminate lip radiation. The resulting signal supposedly now only contains the glottal excitation.

The entire process is iterated a second time to refine it. The inverse filtered signal from the first stage is first modeled by an order-4 AR model, to capture the resonances in it. The original (high-pass filtered) speech signal is inverse filtered by this model to obtain a second estimate of a signal free of glottal and sub-glottal resonances. A higher-order ($N = 11$) AR model is estimated from the inverse-filtered signal. This AR model is now assumed to be a better and more accurate model of the upper vocal tract. The original speech signal is inverse filtered by it, and finally integrated to remove effects of lip radiation, to obtain the final estimate of the excitation signal $e[t]$.

In the initial version of the method [40], the actual glottal events were not considered in the of the glottal pulse. However, the process was subsequently upgraded to first identify individual pitch pulses from the output of the IAIF, followed by a recomputation of the IAIF, except that the analysis frames are now synchronized to only include one glottal pulse each. This "Pitch Synchronous" IAIF (or PSIAIF) is still among the most accurate mechanisms for estimating vocal tract excitation through glottal inverse filtering. In practice, PSIAIF does not require performing IAIF twice such as the IAIF described above, provided the glottal events that enable isolation of individual glottal pulses can be identified by other means (e.g. the methods discussed in the next section).

5.4.3.3 Extracting the Excitation Signal by Homomorphic Processing

An alternate approach to extracting the excitation signal is through *homomorphic processing*. The cepstrum of the speech signal can be shown to be the sum of the cepstra of the excitation and that of the vocal-tract filter. The effect of the filter can be canceled by the following simple operation on the cepstra of the speech signal [43].

As a first step the *complex cepstrum* of the speech signal is computed as follows

$$C_s = IDFT(\log|S| + i\angle S) \tag{5.36}$$

where S is the complex spectrum of the signal $s[t]$ and $\angle S$ is its *unwrapped* phase [44, 45]. The complex cepstrum of the excitation can now be computed as

$$C_e[i] = \begin{cases} 0.5C_s[1], & \text{if } i = 1; \\ 0, & \text{if } 2 \le i \le \frac{N}{2} \\ C_s[i], & \text{if } \frac{N}{2} < i \le N \end{cases} \tag{5.37}$$

where N is the length of the analysis frame. The vocal excitation signal can subsequently be computed as

$$e = IDFT(\exp(DFT(C_e))) \tag{5.38}$$

The actual extraction is most effective if the analysis frame size is approximately equal to the duration of a glottal pulse, and is most effective when the analysis is pitch synchronous.

5.4.3.4 Extracting Glottal Timing Features

The methods in the previous two sections focused on recovering the glottal excitation signal $e[t]$ from the signal, without explicitly requiring knowledge of the glottal timing events.

The methods described in this section do the reverse—they attempts to derive glottal timing events without explicitly attempting accurate extraction of the excitation signal. The key timing events in the glottal pulse waveform are the glottal closure instants (GCIs), the glottal opening instants (GOIs) and the instant of peak flow (GP). Knowledge of these instants enables us to derive various profiling-relevant features. The importance of identifying these events is well recognized and there is a large body of literature on the topic e.g. [46, 47].

Early methods attempted to estimate the events, specifically the GCI from the "residual" signals ($e[t]$) from low-order AR models, by finding instants of peak energy in them [48, 49]. Since the AR analysis itself doesn't consider the timing of glottal events, the derived residual is noisy and an unreliable indicator of glottal events, and the estimates derived from them tend to be unreliable too.

The more reliable approach is based on the fact that the derivative of the glottal signal has a negative spike just before the the GCI (see Fig. 5.17) and abruptly becomes 0 at the GCI. Thus, a *time reassignment* of the energy in any analysis window of the AR residual approximately assigns it to the GCI. For analysis windows centered exactly at the GCI, the reassignment shift is 0.

This leads to the following algorithm. As a first step conventional AR analysis is performed on the speech and the residual is extracted.[1] Subsequently, at each instant t the averaged group delay $\tau[t]$ at that instant (see Sect. 5.3.2) is computed as follows. First, the windowed sequence $s_r[t]$ positioned at time instant r and its time-scaled variant $\tilde{s}_r[t]$ are computed as:

$$s_r[t] = s[t + r]w[t], \quad t = 0 \cdots N - 1 \tag{5.39}$$

$$\tilde{s}_r[t] = ts_r[t] \tag{5.40}$$

where N is the width of the window $w[t]$. Let $S_r = DFT(s_r)$ and $\tilde{S}_r = DFT(\tilde{s}_r)$. The *group delay* in the kth frequency band of the DFT is given by:

[1] More sophisticated methods for excitation extraction such as IAIF can be expected to produce improved results.

$$\tau_r[k] = real\left(\frac{\tilde{S}_r[k]}{S_r[k]}\right) \tag{5.41}$$

and the averaged group delay at r is obtained as

$$d[r] = \frac{\sum_{k=0}^{N-1} \beta_k \tau_r[k]}{\sum_{l=0}^{N-1} \beta_l} \tag{5.42}$$

where β_k is the weight of the kth frequency band in estimating the group delay. Common values of β_k are $\beta_k = 1$ and $\beta_k = |S_r[k]|^2$.

Note that $\tau_r[k]$ is actually the time-reassignment shift for the kth frequency band, and $d[r]$ is the average reassignment across all bands. $d[r]$ is often called the "phase-slope" function.

A plot of $d[r]$ as a function of r for different window sizes (with a Hamming window) is shown for a pulse train in Fig. 5.19. We see that the locations where $d[r]$ cross zero with a negative slope match the locations of the pulses, but this is only

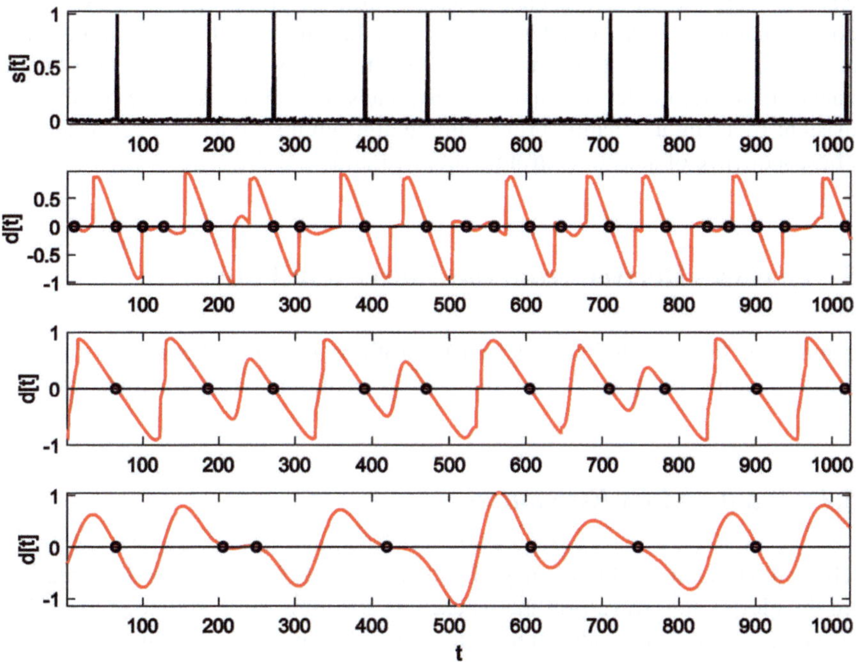

Plots of $d[t]$ computed using different windows. **Top:** Original signal. It consists of impulses in white noise. Average impulse spacing is 100 samples. **Second:** $d[t]$ computed with window size 64. **Third:** $d[t]$ computed with window size 128. **Fourth:** $d[t]$ computed with window size 256. In all cases, the circles indicate the estimated location of the impulses. The best result is obtained with window sizes comparable to inter-pulse spacing.

Fig. 5.19 Phase-slope function

reliable if the analysis window size is approximately equal to the inter-pulse spacing. The identical logic now applies to the detection of GCIs—the instances where $d[r]$ computed from the AR residual crosses zero with a negative slope are candidate GCI locations [50]. The most reliable estimates are obtained if the analysis window is approximately equal to a pitch period, and selecting the optimal window size can be critical.

In practice, additional postprocessing is required to eliminate spurious GCI estimates, and to insert good ones in locations where they are missing. One popular approach utilizes heuristics to *propose* additional candidate GCI locations, followed by a dynamic programming algorithm (called DYPSA) that selects the optimal sequence of GCI candidates based on known properties of speech, such as its quasi-stationarity [51].

The approaches we have considered so far focus on identifying glottal *closure* instants—GCIs. Glottal *opening* instants are far more difficult to identify, as their characteristics are not as distinctive as those of GCIs. On the other hand, GCIs are far more critical to the identification of glottal pulse cycles, and some laxness may be considered acceptable in identifying GOIs. A GOI is identified as the instant where the energy in the AR residual is least—once the glottis opens, the AR model becomes unable to track the sudden change and the energy in the residual increases. Thus a common heuristic to identifying GOIs is to simply identify the lowest-energy instants in the AR model residual. The estimates may be made more reliable and consistent by using a dynamic programming algorithm such as DYPSA to identify the best sequence of selections from a set of candidate GOIs.

5.5 Features Derived from the Source and Filter

A number of features that are relevant to profiling may now be derived from the characterizations of the vocal tract filter and its excitation. We list them below.

5.5.1 Pitch and Its Estimation

Pitch (or the fundamental frequency of phonation, or $F0$) is one of the key measures from the glottal flow waveform that is used as a feature and correlate of many bio-relevant parameters. The pitch frequency $F0$ is the inverse of the pitch period $T0$, which is the time lag between two glottal cycles in the excitation (i.e. $F0 = \frac{1}{T0}$). The problem of estimating the pitch frequency $F0$ is identical to that of estimating the pitch period $T0$.

We have defined and discussed pitch in multiple contexts so far. In this section we discuss its direct estimation from the voice *signal*, rather than from estimates of the glottal flow waveform. Over the past several decades, a variety of techniques have been proposed for the estimation of pitch from the voice signal. Good reviews of these are available, e.g. [52]. Of the many available methods, in the paragraphs below we discuss the key steps involved in two of the most exemplary and effective approaches to pitch estimation—the autocorrelation method, and the filter bank-based method. We also briefly comment on the currently popular data-driven empirical methods that provide some advantages over purely signal-processing based methods.

5.5.1.1 The Autocorrelation Method for Pitch Extraction

Voiced speech signals are produced by the excitation of the vocal tract by a quasi-periodic sequence of glottal pulses. As a result, the signals too have a quasi-periodic structure, where almost the exact same waveform segment repeats. The period of repetition of the signal is the period between two consecutive glottal pulses, i.e. the pitch period. Hence, in order to determine the pitch of a voiced signal, it is sufficient to determine the period with which the speech signal itself repeats.

We have previously seen in Sect. 4.2.1.5 that the autocorrelation function captures self-similarity and periodicities in a signal. The autocorrelation method for pitch extraction exploits this property.

The autocorrelation method segments the speech signal $s[t]$ into analysis frames, typically of length 20–30 ms. The length is chosen to include at least one complete glottal cycle, but not more than 2–3 cycles (since the pitch can change in that time). Within each frame, it computes the autocorrelation, which we recall is given by

$$R_{ss}[\tau] = \frac{1}{N-\tau} \sum_{t=0}^{N-1-\tau} s[t]s[t+\tau] \tag{5.43}$$

where $s[t]$ is the signal within the frame, and N is the length of the frame. The pitch period within the analysis frame is computed as

$$\tau_{pitch} = \arg \max_{\tau \in [\tau_{min}, \tau_{max}]} R_{ss}[\tau] \tag{5.44}$$

where τ_{min} and τ_{max} specify the range of values that the pitch period is expected to lie in. Figure 5.20 shows a segment of speech and its autocorrelation. We note a clear peak in the autocorrelation at the lag τ corresponding to the period with which the speech signal repeats, i.e. its pitch.

Practical implementations of pitch estimation may include several modifications over the simple algorithm described above. In general pitch estimation is not performed in unvoiced frames. Unvoiced frames are indicated by excessive zero crossings, or absurd pitch period estimates.

The signal may be half-wave rectified (by setting all values below 0 to 0) prior to computing the autocorrelation for the purpose of pitch estimation, to increase the prominence of the peak. A unique pitch value may not be assigned to each frame; instead a set of candidate pitch values may be assigned based on multiple peaks, or based on integer fractions or multiples of the identified peak lag that also fall between

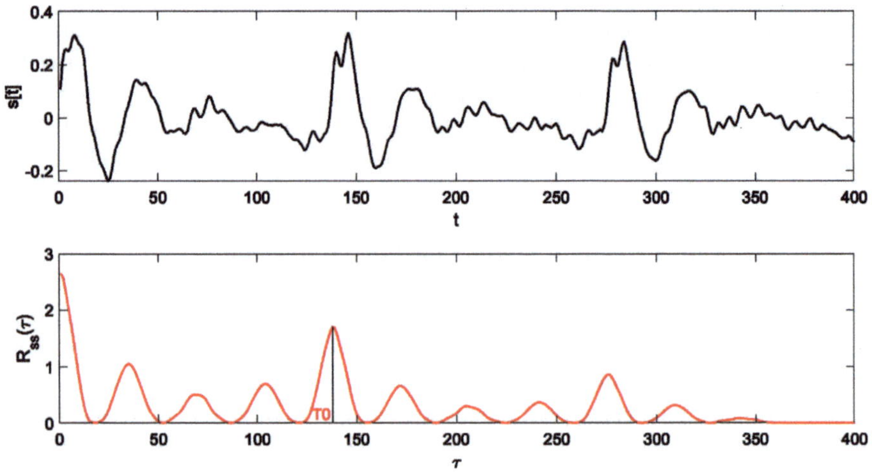

A voiced signal (top) and its autocorrelation (bottom). The ACF is computed after half-wave rectifying the signal. A distinct peak is seen at the pitch period.

Fig. 5.20 Finding the pitch in the ACF

τ_{min} and τ_{max}, and the overall sequence of pitch values may subsequently be derived using a dynamic programming algorithm that uses typical pitch value ranges, and the fact that pitch values do not generally change dramatically between adjacent frames, as constraints.

The manner of *framing* of the signal, i.e. the segmentation of the signal into frames for pitch estimation may also vary. If the primary purpose of pitch estimation is to derive a pitch track, then the hop between adjacent frames can be fixed. On the other hand, if the purpose of pitch extraction is to facilitate pitch-synchronous estimation of other parameters, such as the excitation signal (for IAIF) or glottal timing instants, then the hop between one frame and the next must be equal to the pitch period for that frame.

5.5.1.2 The Filter Bank Based Method for Pitch Extraction

Filter-bank-based methods for pitch extraction separate the signal $s[t]$ into multiple frequency bands using a filter bank such as a gammatone filter bank. For each frame of the input signal $s[t]$, the autocorrelation method is then used to derive an estimate of the pitch from the corresponding segments of the filter bank outputs in each of the frequency bands. This results in as many pitch estimates as there are filters in the bank.

Subsequently, the actual pitch estimate for the analysis frame is obtained through a simple vote across all frequency bands. Voting is typically performed by composing a histogram of pitch estimates and choosing the most populous bin. If the histogram has no distinct peak (with a bin that includes at the estimates from at least 25% of the frequency bands), the frame is declared as unvoiced [53].

Filter-bank-based estimates tend to be more robust than simple estimates from the original signal, particularly in noisy conditions, or on data that have undergone filtering, such as by a telephone channel.

In practice, all of the methods above result in occasional spurious pitch estimates, as well as some amount of jitter in the estimates. As a result, the estimates are often smoothed, e.g. by low-pass filtering the sequence of pitch values, to obtain the final estimates.

5.5.1.3 Empirical Methods for Pitch Extraction

Empirical methods use data-driven (or example-driven) learning mechanisms for mapping a voice signal to its pitch contour. A pitch estimator (a machine learning or AI mechanism) is first shown examples of voice-signal, paired with estimated pitch. In the training phase it learns the relationship between the two, using its specific abstract internal representations for the mappings. Once these are learned, the architecture and parameters of the estimator are fixed, and the estimator may be used for estimating the pitch of new voice signals. Myriad mechanisms can be devised to implement this basic empirical learning approach. At the time of writing this book, some effective ones are based on deep neural networks. We do not discuss these here, and refer the reader to in-depth expositions in the literature instead e.g. [54–56].

5.5.2 Timing Features from the Excitation Signal

A number of different features may be derived from the glottal pulse signal. Of the many possible, a few widely used ones are mentioned below. We do not describe the related algorithms here.

5.5.2.1 Properties Related to the Physics of Phonation

Due to the nature of their source and mechanics, glottal pulses can be related to entities such as transglottal flow, subglottal pressure and intensity [57] which are in turn used to estimate laryngeal resistance, efficiency and power. The laryngeal resistance, especially, is a measure of the resistance to the initiation of phonation. It is used to estimate the *phonation threshold pressure* [58–60], or the "vocal tract inertia," which is considered to be the cause of *voicing onset time* (the time taken for the vocal folds to move from a state of rest to a state of phonation). It is also indicative of the viscoelastic properties of the mucosa.

5.5.2.2 Timing Measurements

Timing measures quantify the proportions of the durations of the various phases of the glottal opening and closing cycle. As a reminder, we recall that the movement of the glottis during phonation has three distinct phases: a *closed* phase, when the glottis is fully closed, an *opening* phase, when it begins to open up, and a *closing* phase, when it reverses direction and begins to close. We have already seen how GCI and GOI may be determined. GP can be obtained from the excitation signal $e[t]$. The durations of the closed, opening and closing phases of a glottal pulse, which we represent as T_c, T_o and T_{cl} respectively, can be obtained from them as $T_c = GOI - GCI, T_o = GP - GOI$, and $T_{cl} = GCI - GP$ respectively.

The following measures are defined based on these:

- **Open Quotient (OQ)**: This is the fraction of the total duration during which the glottis is *not* closed (i.e., it is open to any degree) and is given by

$$OQ = \frac{T_o + T_{cl}}{T0} \tag{5.45}$$

- **Closing Quotient (CIQ)**: This is the fraction of the total pitch period during which the glottis is closing:

$$CIQ = \frac{T_{cl}}{T0} \tag{5.46}$$

- **Speed Quotient (SQ)**: This relates to how fast the glottis opens to how quickly it closes, and is given by:

$$SQ = \frac{T_o}{T_{cl}} \tag{5.47}$$

The various timing quotients have been found relate to both the affect of the speaker and their physical and health characteristics, sometimes in non-obvious ways. For instance, OQ has been observed to decrease with age in females, while it increases with increasing age for men.

5.5.2.3 Amplitude Based Measures

These are based on the amplitude relationships of the signal. The most widely used amplitude parameters are minimum flow or "DC offset", which is the residual air flow during glottal closure, the "AC-flow," which is the peak glottal signal, and the peak amplitude of the derivative, d_{mim}. These have been used in the quantification of the glottal source [61]. The normalized amplitude quotient (NAQ), which is the ratio of AC and d_{min}, is seen to provide a more robust estimate of the CIQ.

Both AC and d_{min} can be computed from the excitation signal $e[t]$. In particular, d_{min} occurs close to the GCI.

5.5.2.4 Mathematical-Model Based Features

Features based on mathematical models apply functional fits to the glottal flow, and characterize the signal by the parameters of the model. For instance, Klatt [62] used a third-order model fit to the glottal pulse during its open phase: $e[t] = at^2 - bt^3$, where a and b are fit to get the desired OQ for the signal. A number of other such models have been proposed [63–66].

Perhaps the most popular model is the *Liliencrantz Fant* (LF) model [67]. This models the flow *derivative* in two segments. The first segment models the flow derivative from the glottal opening (GOI) until the instant of the peak excitation (GP) η, with an exponential/sinusoidal function:

$$E(t) = E_0 e^{\alpha t} sin(\omega_g t) \tag{5.48}$$

For the second segment, the flow derivative is modeled from the instant of the main excitation to the instant of (full) closure (GCI) with exponential functions as follows:

$$E(t) = \frac{-EE}{\varepsilon T_a} e^{-\varepsilon(t-\eta) - e^{-\varepsilon T_b}} \tag{5.49}$$

In addition to critical time instants GOI and GP (or η) the seven additional parameters in the equations above are:

1. E_0: a scaling factor used to obtain the right balance between the opening and closing phases
2. α: the rate of amplitude increase
3. ω_g: the sine frequency
4. EE: the amplitude of the negative peak in the derivative of the glottal signal
5. ε: the inverse of the time constant of the exponential function that fits the return phase
6. T_a: the effective duration of the return phase
7. T_b: the actual duration of the return phase.

In practice, the LF model can be computed using T0 and four additional parameters through an iterative algorithm that requires the integral of the pulse over the glottal cycle to be zero [68].

Other time-domain models estimate the glottal signal by explicitly modeling the physics of voice production [69, 70]. Unlike the LF model which attempts to fit to the *signal*, these methods search for the physical parameters of the system (*e.g.* vocal fold mass and stiffness) that best explain the observed glottal pulse.

5.5.2.5 Frequency-Domain Characterizations of the Glottal Pulse

Timing-based quantifications of the glottal pulse are relatively crude. Speakers have significant control over vocal excitation, which they can change by controlling the tension of laryngeal muscles and their respiratory effort. This can change the shape, periodicity and timing patterns of the glottal pulses. Other involuntary and pathological conditions too can change their structure.

More detailed descriptions of the glottal pulse are obtained in the frequency domain. Variations in the glottal pulse structure can change the shape, periodicity and the decay with frequency of the spectrum of the excitation. This is illustrated in Fig. 5.21.

Frequency-domain characterizations of the glottal signal can be computed using either the conventional DFT, or using auto-regressive models (c.f. Sect. 4.2.2.2). The most effective analysis is *pitch synchronous*, where individual analysis frames span exactly one pitch period. The analysis spans an integral number of pitch pulses. Quantifications include the overall shape of the spectrum, spectral decay, and skewness. For instance, the skewness is often quantified by the *alpha ratio*, which is the ratio between the spectral energies below and above a certain frequency (typically 1.0 kHz) [71].

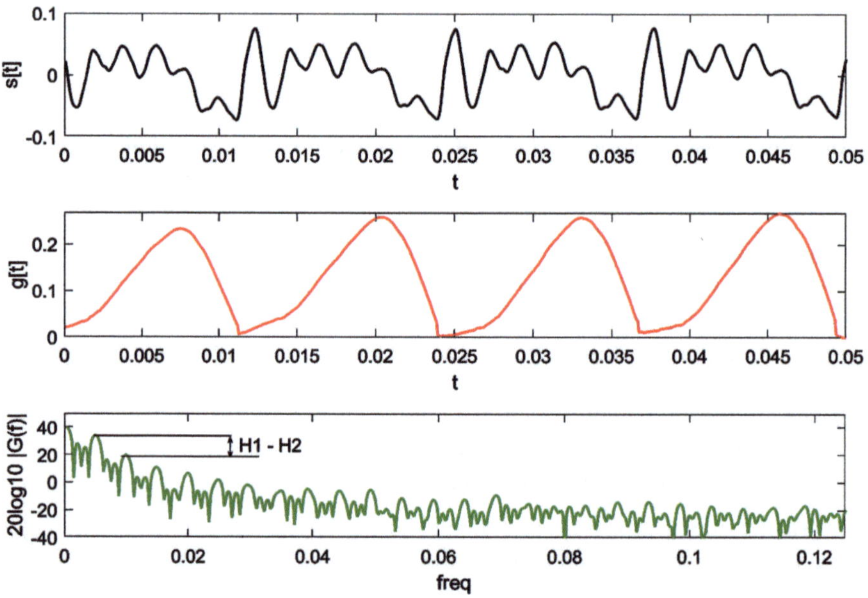

Top: A segment of voiced speech. **Middle:** Its glottal waveform. **Bottom:** The spectrum of the glottal waveform, and its H1-H2 feature.

Fig. 5.21 Frequency domain representation of glottal waveform

The **spectral decay** of the glottal source is often quantified using the level of the fundamental frequency F0, and its harmonics. This can be done in multiple ways. For example, a quotient called *harmonic richness factor* (HRF), is defined as the ratio between the sum of the amplitudes of harmonics above the fundamental and the amplitude of the fundamental frequency (pitch) in the glottal pulse signal [72].

Another measure is the *H1–H2 feature*, which is defined as the difference between the amplitudes of the fundamental and the second harmonic of the glottal pulse signal [73].

A third measure is the *parabolic spectral parameter* (PSP), which matches a second-order polynomial to the flow spectrum computed over a single glottal cycle [74]. The PSP has been shown to be able to differentiate phonation types of varying spectral slopes effectively. Other metrics such as the harmonic-to-noise ratio (HNR) compute the ratio of the energies in the harmonic and non-harmonic components of the spectrum (e.g., [75]). The HNR (also discussed in Chap. 6), is useful for the analysis of disordered voices, in which the glottal excitation is typically affected in different ways.

A number of studies have shown that characterizations of the vocal-tract excitation are of high relevance to voice profiling, not surprisingly since they relate directly to the bio-physical mechanisms that produce them. The excitation, or the glottal flow waveform and its various measures, have been linked to emotions [76–81], age [82–85], vocal fatigue [86], mental problems such as depression and suicidal risk [87], psychological stress [88], intoxication [89], various pathologies [90–94], and myriad other bio-relevant factors. An exhaustive list is out of this chapter's scope. Analysis of vocal tract excitations is also a key aspect of the analysis of singing voices, e.g [95, 96].

5.5.3 Formants and Their Estimation

As discussed in Chap. 2 and earlier sections of this chapter, the primary behavior of the vocal tract "filter" is to produce combinations of audible resonances in response to the excitation. These resonances, or "formants," not only carry information about the underlying sound, but also about the speaker. Informative formant characteristics include their peak frequencies, their bandwidths, their Q-values that quantify the relationship of the peak frequency of a formant to its bandwidth, and the relative arrangement of the formants, among others, and these are all relevant to profiling. As such, the subject of accurate estimation of formants from the voice signal is an important one.

There are many approaches for measuring or estimating formants. In this section we discuss two of these briefly. In each case, the actual estimation begins with a *framing* of the speech recording in a manner akin to that used to derive spectrographic representations. The signal is first segmented into frames, typically 20–40 ms wide, where adjacent frames may overlap by as much as 75%. Formants (or candidate formants) are estimated from each frame. In simple estimation schemes the formants are estimated individually from each frame. More sophisticated approaches estimate *candidate* formants in each frame, and select the actual formants from these candidates using a dynamic programming algorithm that applies constraints on the choice of formant values. The latter is termed a "delayed-decision" estimator for reasons that are explained later in this section.

5.5.3.1 Cepstrum Based Estimation of Formants

Cepstrum-based estimation of formants was introduced in 1970 [97]. As we recall from Sect. 5.3.3.1, the cepstrum **C** of a signal is the IDCT of the log its magnitude spectrum $|S|$:

$$\mathbf{C} = IDCT(log(abs(\mathbf{S}))) \tag{5.50}$$

Here **C** and **S** represent *sequences* of cepstral and spectral components respectively, and $abs(.)$ represents a component-wise operation that computes the absolute value of the components of **S**.

In Chap. 4, we discussed how performing the DCT of the *truncated* cepstrum of a signal results in a smoothed estimate of the spectrum of the signal. Formants typically manifest as peaks in this smoothed spectrum, although not every peak is a formant. Thus the peaks in the spectrum may be viewed as *candidates* for formants. Figure 5.22 illustrates this with an example.

To detect formants, a cepstrally smoothed estimate of the spectrum is obtained for every frame of the speech signal. A peak-picking algorithm is then employed to locate the peaks in the cepstrally smoothed spectrum of the signal [98, 99]. In simple estimation methods, the first three peaks in the smoothed spectrum are identified as

Fig. 5.22 Deriving formants and their bandwidths from cepstrally smoothed spectra

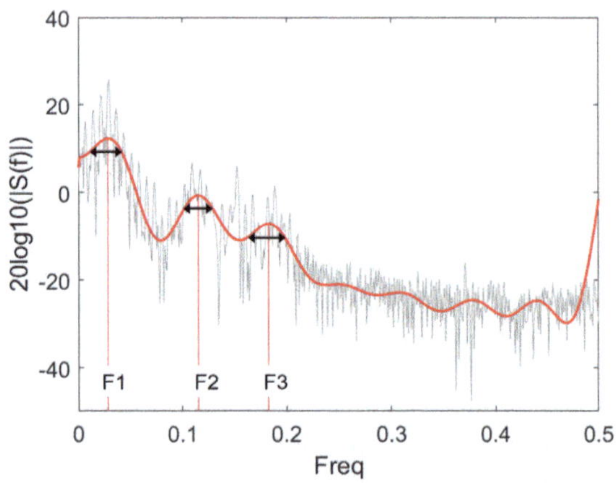

Cepstrally-smoothed spectrum for segment of speech. The red lines indicate frequencies at which the first three formants are located. The bandwidths of the formants are shown by the double-headed arrows.

the first three formant frequencies. If a delayed-decision dynamic programming algorithm is to be used many more peaks are picked [100, 101], and their corresponding frequencies are chosen as the candidate formant frequencies for the frame.

5.5.3.2 AR Model Based Estimator

The AR model for the vocal tract, when computed over a glottis-closed phase of the signal, results in the model of the vocal-tract filter given by Eq. 5.29, which can be rewritten as

$$T(z) = \frac{Gz^k}{\prod_{k=1}^{N}(z - p_k)} \tag{5.51}$$

where p_k are the poles (resonances) of the filter. As can be seen, this is an *all pole* filter. The poles ideally capture the resonances of the vocal-tract filter. An Nth order AR model has N poles. Since the filter coefficients are all real, the poles occur in complex conjugate pairs. Thus an Nth order all-pole filter captures $N/2$ resonances, each of which is a candidate formant frequency [102].

The poles of the AR model can be obtained using any polynomial root-finding algorithm [103]. Since roots occur in complex conjugate pairs, only roots with positive imaginary part are retained and their corresponding angles are determined by:

$$\angle p_k = \sum tan^{-1}\frac{Im(p_k)}{Re(p_k)} \tag{5.52}$$

$\angle p_k$ gives us the resonant frequency for the pole p_k for the analysis frame. The N poles provide us $N/2$ resonant frequencies. Format values may be directly chosen from these (e.g. as the lowest three resonances), or derived from all $N/2$ roots by a subsequent dynamic program.

The AR order must be carefully chosen, particularly if formants are directly estimated from the frames of speech with no subsequent dynamic program for selection of formants from a candidate set. The optimal AR order is a function of the (approximate) vocal tract length of the speaker and the sampling frequency. An AR order of 10 is sufficient for male speakers if the first four formant frequencies are directly estimated. Female speakers require a lower AR order—increasing the order may result in spurious peaks in the spectrum. For speech sampled at 16 kHz, for male speakers with a vocal tract length of approximately 17 cm, an AR order of 16 assumes tube segment lengths of approximately 1.06 cm, which corresponds to a half-sample delay introduced into the signal by each segment. Higher orders result in shorter segments that cannot be resolved. For female speakers with shorter vocal tracts, the highest resolvable order is 14.

5.5.3.3 Formant Estimation and the Principle of Delayed Decision

Ideally, formant analysis must be done differently for different sound classes. The all-pole model of Eq. 5.29 assumes that the vocal tract has only resonances, and no anti-resonances. Some classes of sounds however have extra resonance-antiresonance pairs—vocal tract configurations that include both a resonance-causing part and an *anti-resonance* causing part. These manifest as *pole-zero pairs* in the transfer function of the vocal tract. Sounds for which the phenomenon happens include fricatives, nasals and liquids. These pole-zero pairs are not arbitrary, though. They have different characteristic frequencies for each sound. For example, for retroflex sounds such as /R/, and rhotic vowels, the pole-zero pair is in the range of the third formant, with the pole at a lower frequency and the zero often canceling out the third formant. For a non-retroflex sound, a similar spectrographic pattern must be interpreted as a sudden drop in frequency of the third formant, with the corresponding tongue motion to lengthen the cavity associated with the third formant; however this rule does not apply to the retroflexed sound.

While the loss of information from pole-zero cancellation cannot be reversed, we can nevertheless prevent erasures or completely inaccurate estimates of formants through the application of appropriate continuity and positional constraints, through dynamic programming. This is an embodiment of the principle of *delayed decision*, where the decision regarding formant positions at any instant of time is taken only after the entire signal has been evaluated.

The dynamic programming solution attempts to *select* the sequence of candidate formants from each frame, to compose the most plausible sequence of formant frequencies. The algorithm considers several constraints. For each formant, the algorithm considers candidates in the general frequency range within which the formant under consideration is generally found. It considers the fact that formant positions and bandwidths do not generally change abruptly between adjacent frames (except at abrupt speech events such as the termination or onset of speech or voicing). In addition, it considers the fact that in the case where AR models are used to obtain candidate formants, reflection coefficients and area ratios (which can be derived from reflection coefficients) do not change abruptly across adjacent frames. Other constraints too may be applied. The dynamic programming algorithm must also consider the fact that formants may disappear entirely, e.g. when canceled by an anti-resonance, or when the speech energy goes down to zero.

The output of the process of delayed decision dynamic programming is a sequence (or *trajectory*) of formant position values. Separate trajectories are obtained for each of the formants F1, F2, F3, etc. Figure 5.23 shows an example of tracked formants for a speech segment.

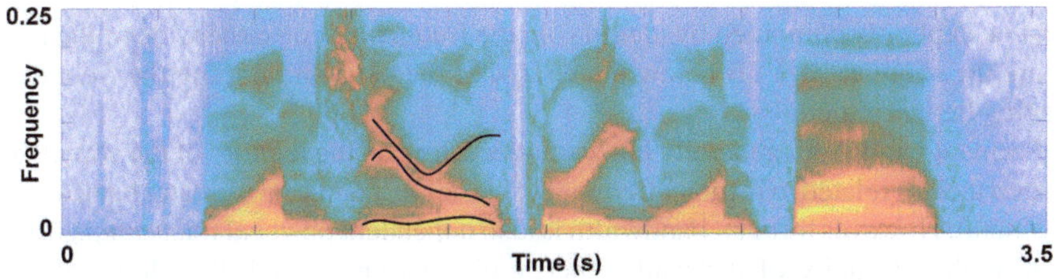

Spectrogram for a speech recording. The estimates of the first three formants are overlaid on a short segment.

Fig. 5.23 Tracking formants

5.5.4 Formant-Based Features

5.5.4.1 Formant Frequency and Magnitude

The outcome of formant estimation is the set of formant frequencies at each time. Not all formants may be found at every time. F1 is generally more observable than F2, and F2 in turn is more easily measured than F3. The higher formants F4, F5 and F6 are increasingly hard to detect and measure.

Both the formant frequencies and *the relationship* between the frequencies, such as their differences and ratios, carry information and can be derived from the estimated formant frequencies. The spectral magnitudes at the formants, the relation between these values for different formants etc. also carry information and can be obtained from the spectral magnitude values at these frequencies.

5.5.4.2 Formant Bandwidth

The spectral energy that peaks at formant frequencies also rolls off as the frequency moves away from the formant. The formant *bandwidth* is defined as the spread of frequencies around the formant within which the energy remains within 3 dB (a factor of 0.5) of the formant energy. Figure 5.22 also shows the bandwidths of the identified formants.

For a transfer function with a *single* pole, the bandwidth of the corresponding resonance is given by

$$BW = -\frac{ln(|p|)f_s}{\pi} \tag{5.53}$$

where p is the sole pole of the system. For a system with multiple poles, Eq. 5.53 still gives the 3 dB bandwidth of the resonance, provided the poles are sufficiently distant from each other in frequency. More realistically, however, Eq. 5.53 does not apply to an AR filter with order greater than 1, and the bandwidth of the resonances must be explicitly computed by identifying the frequencies on either side of the peak where the response is 3 dB below the energy at the peak and finding the frequency spacing between them.

Formant bandwidth generally has a smaller effect on the identity of the sound, but is nevertheless related to formant frequency: higher formants have greater bandwidths in general. However formant bandwidth is also dependent on vocal tract *composition*, such as the elasticity of the walls, energy dissipation through the glottis etc., in addition to other configuration-related characteristics [104].

Not surprisingly then, formant bandwidth too has speaker-specific characteristics. In particular, damping effects due to the nature of the tissue of the vocal tract, and damping due to energy dissipation from the glottis and nasal pathways are related to the size and coupling of these passages with the main vocal cavity. These effects may be expected to manifest directly in a speaker's voice in a manner that they cannot entirely control.

5.5.4.3 Formant-Q

A vocal tract configuration that generates a specific formant may be viewed as a filter with a particular resonant frequency. The Q-factor of any filter is defined as the ratio of the peak frequency of the filter to its bandwidth. In the case of formants, the formant Q is the ratio of the formant frequency to the formant bandwidth.

There is an inherent relationship between the formant frequency and the formant bandwidth. In general, as the formant frequency increases, so does its bandwidth. However, the actual Q of any formant depends not only on the frequency of the formant, but also on the characteristics of the vocal tract of the speaker.

5.5.4.4 Formant Dispersion

Formant dispersion is defined as the average spacing between formants. Formant dispersion, being dependent only on the spacing between formants and not on the absolute position of the formants themselves, has been suggested as being more characteristic of the speaker's vocal tract length than the formant positions themselves [105]. Once again, as in the other measurements, this translates to speaker dependence of formant dispersion. Consequently, we may expect the distribution of formant dispersion values to have speaker-specific characteristics.

We note here that the common definition of formant dispersion, given by $D = \frac{(F_2-F_1)+(F_3-F_2)+\cdots+(F_n-F_{n-1})}{n-1}$, simply collapses to $\frac{F_n-F_1}{n-1}$, i.e. the average distance between formants, which loses the distinction between the individual spacings. In order to better reflect the effect of individual formant spacings, a more informative dispersion measure is the *geometric* mean of the formant spacings, i.e. $D = \sqrt[n-1]{\prod_i F_i - F_{i-1}}$.

5.6 Style and Content Based Characterizations

In this extremely brief closing section of this chapter, we mention an important category of features that have not been discussed so far: those based on an analysis of the *style* of spoken utterances. The features we have discussed thus far are based on the *structure* of the speech signal. Style based characterizations are those that require the delimitation of speech in a pragmatic or syntactic sense with reference to its content—e.g. an "utterance," or "syllable," or "segment" of speech. An utterance for this purpose, for instance, may be defined as a full sentence by some definition in the grammar of the underlying language which is spoken.

Stylistic and content-based features are widely used in studies related to the perception of emotion, stress and other human factors, e.g. [106–108]. Most of these, however, comprise secondary trend-computations over temporal contours of signal-level features such as pitch, energy, entropy, duration etc. They are often derived through various macro-level analyses of these signal-level features, in conjunction with the content of the signal. We do not discuss these features in detail here. Only a few widely used ones are mentioned below:

- **Duration based (segmental) cues**: These include the absolute length (in time) of phonemes, words, utterances and other syntactic and semantic segments of language represented in the speech signal, and measures of variations of these in continuous speech.
- **Prosody and rhythm**: Prosody captures the pitch and energy contours of the waveform, typically over utterances, but not always so. It measures the speaker's emphasis and intonation styles, which relate to changes in the envelope of the time-domain waveform, and changes in the fundamental frequency over time. Rhythm captures the periodicity of these variations.
- **Lexical features**: Lexical features relate to idiosyncratic long-term patterns in a speaker's use of linguistic units, such as specific pronunciations and contractions of words, word-sequences, phrases etc.
- **Speaking rate**: This is a general term that includes various characterizations of the rate at which a speaker enunciates the content of speech. Specific characterizations include measurements such as syllable rate (number of syllables uttered per unit time), phoneme rate (number of phonemes uttered per unit time), vowel rate (number of vowels uttered per unit time) etc.

References

1. Kodera, K., De Villedary, C., & Gendrin, R. (1976). A new method for the numerical analysis of non-stationary signals. *Physics of the Earth and Planetary Interiors*, *12*(2–3), 142–150.
2. Kodera, K., Gendrin, R., & Villedary, C. D. (1978). Analysis of time-varying signals with small BT values. *IEEE Transactions on Acoustics, Speech, and Signal Processing*, *26*(1), 64–76.

3. Flandrin, P., Auger, F., & Chassande-Mottin, E. (2003). Time-frequency reassignment: From principles to algorithms. *Applications in Time-Frequency Signal Processing*, *5*(179–203), 102.
4. Auger, F., & Flandrin, P. (1995). Improving the readability of time-frequency and time-scale representations by the reassignment method. *IEEE Transactions on Signal Processing*, *43*(5), 1068–1089.
5. Nelson, D. J. (2001). Cross-spectral methods for processing speech. *The Journal of the Acoustical Society of America*, *110*(5), 2575–2592.
6. Nelson, D. (1993). Special purpose correlation functions for improved signal detection and parameter estimation. In *Proceedings of the IEEE International Conference on Acoustics, Speech, and Signal Processing (ICASSP), Minneapolis, Minnesota, USA* (Vol. 4, pp. 73–76).
7. Hermansky, H., Hanson, B., & Wakita, H. (1985). Perceptually based linear predictive analysis of speech. In *Proceedings of the IEEE International Conference on Acoustics, Speech, and Signal Processing (ICASSP), Tampa, Florida, USA* (Vol. 10, pp. 509–512).
8. Hermansky, H. (1990). Perceptual linear predictive (PLP) analysis of speech. *The Journal of the Acoustical Society of America*, *87*(4), 1738–1752.
9. Kim, C., & Stern, R. M. (2016). Power-normalized cepstral coefficients (PNCC) for robust speech recognition. *IEEE/ACM Transactions on Audio, Speech and Language Processing (TASLP)*, *24*(7), 1315–1329.
10. Darling, A. M. (1991). Properties and implementation of the gammatone filter: A tutorial. A report. Department of Phonetics and Linguistics, University College London (pp. 43–61).
11. Teager, H. (1980). Some observations on oral air flow during phonation. *IEEE Transactions on Acoustics, Speech, and Signal Processing*, *28*(5), 599–601.
12. Kaiser, J. F. (1993). Some useful properties of Teager's energy operators. In *Proceedings of the IEEE International Conference on Acoustics, Speech, and Signal Processing (ICASSP), Minneapolis, Minnesota, USA* (Vol. 3, pp. 149–152).
13. Kvedalen, E. (2003). Signal processing using the Teager Energy Operator and other nonlinear operators. Norway, Department of Informatics: Thesis for the Candidatus Scientiarum degree at the University of Oslo.
14. Maragos, P., Kaiser, J. F., & Quatieri, T. F. (1993). Energy separation in signal modulations with application to speech analysis. *IEEE Transactions on Signal Processing*, *41*(10), 3024–3051.
15. Jabloun, F., Cetin, A. E., & Erzin, E. (1999). Teager energy based feature parameters for speech recognition in car noise. *IEEE Signal Processing Letters*, *6*(10), 259–261.
16. Oppenheim, A. V., & Schafer, R. W. (1975). *Digital signal processing*. Englewood Cliffs, New Jersey: Prentice-Hall Inc.
17. Kumaresan, R., & Rao, A. (1999). Model-based approach to envelope and positive instantaneous frequency estimation of signals with speech applications. *The Journal of the Acoustical Society of America*, *105*(3), 1912–1924.
18. Kingsbury, B. E., Morgan, N., & Greenberg, S. (1998). Robust speech recognition using the modulation spectrogram. *Speech Communication*, *25*(1–3), 117–132.
19. Gallun, F., & Souza, P. (2008). Exploring the role of the modulation spectrum in phoneme recognition. *Ear and Hearing*, *29*(5), 800.
20. Meyer, B. T., Ravuri, S. V., Schädler, M. R., & Morgan, N. (2011). Comparing different flavors of spectro-temporal features for ASR. *Twelfth Annual Conference of the International Speech Communication Association (INTERSPEECH)* (pp. 1269–1272). Italy: Florence.
21. Tchorz, J., & Kollmeier, B. (1999). A model of auditory perception as front end for automatic speech recognition. *The Journal of the Acoustical Society of America*, *106*(4), 2040–2050.
22. Viemeister, N. F. (1979). Temporal modulation transfer functions based upon modulation thresholds. *The Journal of the Acoustical Society of America*, *66*(5), 1364–1380.
23. Yost, W. A., & Moore, M. J. (1987). Temporal changes in a complex spectral profile. *The Journal of the Acoustical Society of America*, *81*(6), 1896–1905.
24. Joris, P. X., Schreiner, C. E., & Rees, A. (2004). Neural processing of amplitude-modulated sounds. *Physiological Reviews*, *84*(2), 541–577.

25. Kollmeier, B., & Koch, R. (1994). Speech enhancement based on physiological and psychoacoustical models of modulation perception and binaural interaction. *The Journal of the Acoustical Society of America*, *95*(3), 1593–1602.
26. Singh, N. C., & Theunissen, F. E. (2003). Modulation spectra of natural sounds and ethological theories of auditory processing. *The Journal of the Acoustical Society of America*, *114*(6), 3394–3411.
27. Tyagi, V. (2011). Fepstrum features: Design and application to conversational speech recognition. IBM Research Report (p. 11009).
28. Atal, B. S., & Hanauer, S. L. (1971). Speech analysis and synthesis by linear prediction of the speech wave. *The Journal of the Acoustical Society of America*, *50*(2B), 637–655.
29. Rabiner, L. R., & Schafer, R. W. (1978). *Digital processing of speech signals*. Prentice Hall, New Jersey: Englewood Cliffs.
30. Levinson, N. (1946). The Wiener (root mean square) error criterion in filter design and prediction. *Journal of Mathematics and Physics*, *25*(1–4), 261–278.
31. Durbin, J. (1960). The fitting of time-series models. *Revue de l'Institut International de Statistique* 233–244.
32. El-Jaroudi, A., & Makhoul, J. (1991). Discrete all-pole modeling. *IEEE Transactions on Signal Processing*, *39*(2), 411–423.
33. Gray, R., Buzo, A., Gray, A., & Matsuyama, Y. (1980). Distortion measures for speech processing. *IEEE Transactions on Acoustics, Speech, and Signal Processing*, *28*(4), 367–376.
34. Liu, M., & Lacroix, A. (1996). Improved vocal tract model for the analysis of nasal speech sounds. In *Proceedings of the IEEE International Conference on Acoustics, Speech, and Signal Processing Conference Proceedings (ICASSP), Atlanta, Georgia, USA* (Vol. 2, pp. 801–804).
35. Alku, P. (1992). An automatic method to estimate the time-based parameters of the glottal pulseform. In *Proceedings of the IEEE International Conference on Acoustics, Speech, and Signal Processing (ICASSP), San Francisco, California, USA* (Vol. 2, pp. 29–32).
36. Vincent, D., Rosec, O., & Chonavel, T. (2005). Estimation of LF glottal source parameters based on an ARX model. In *Proceedings of the Ninth European Conference on Speech Communication and Technology (INTERSPEECH/EUROSPEECH), Lisboa, Portugal*.
37. Milenkovic, P. (1986). Glottal inverse filtering by joint estimation of an AR system with a linear input model. *IEEE Transactions on Acoustics, Speech, and Signal Processing*, *34*(1), 28–42.
38. Veeneman, D., & BeMent, S. (1985). Automatic glottal inverse filtering from speech and electroglottographic signals. *IEEE Transactions on Acoustics, Speech, and Signal Processing*, *33*(2), 369–377.
39. Childers, D. G., Hicks, D. M., Moore, G. P., & Alsaka, Y. A. (1986). A model for vocal fold vibratory motion, contact area, and the electroglottogram. *The Journal of the Acoustical Society of America*, *80*(5), 1309–1320.
40. Alku, P. (2011). Glottal inverse filtering analysis of human voice production—a review of estimation and parameterization methods of the glottal excitation and their applications. *Sadhana*, *36*(5), 623–650.
41. Rothenberg, M. (1977). Measurement of airflow in speech. *Journal of Speech and Hearing Research*, *20*(1), 155–176.
42. Alku, P. (1992). Glottal wave analysis with pitch synchronous iterative adaptive inverse filtering. *Speech Communication*, *11*(2–3), 109–118.
43. Drugman, T., Bozkurt, B., & Dutoit, T. (2011). Causal-anticausal decomposition of speech using complex cepstrum for glottal source estimation. *Speech Communication*, *53*(6), 855–866.
44. Childers, D. G., Skinner, D. P., & Kemerait, R. C. (1977). The cepstrum: A guide to processing. *Proceedings of the IEEE*, *65*(10), 1428–1443.
45. Tribolet, J. (1977). A new phase unwrapping algorithm. *IEEE Transactions on Acoustics, Speech, and Signal Processing*, *25*(2), 170–177.
46. Drugman, T., Thomas, M., Gudnason, J., Naylor, P., & Dutoit, T. (2012). Detection of glottal closure instants from speech signals: A quantitative review. *IEEE Transactions on Audio, Speech, and Language Processing*, *20*(3), 994–1006.

47. Drugman, T., & Dutoit, T. (2009). Glottal closure and opening instant detection from speech signals. In *Proceedings of the Tenth Annual Conference of the International Speech Communication Association (INTERSPEECH), Brighton, UK* (pp. 2891–2894).
48. Cheng, Y. M., & O'Shaughnessy, D. (1989). Automatic and reliable estimation of glottal closure instant and period. *IEEE Transactions on Acoustics, Speech, and Signal Processing, 37*(12), 1805–1815.
49. Wong, D., Markel, J., & Gray, A. (1979). Least squares glottal inverse filtering from the acoustic speech waveform. *IEEE Transactions on Acoustics, Speech, and Signal Processing, 27*(4), 350–355.
50. Brookes, M., Naylor, P. A., & Gudnason, J. (2006). A quantitative assessment of group delay methods for identifying glottal closures in voiced speech. *IEEE Transactions on Audio, Speech, and Language Processing, 14*(2), 456–466.
51. Naylor, P. A., Kounoudes, A., Gudnason, J., & Brookes, M. (2007). Estimation of glottal closure instants in voiced speech using the DYPSA algorithm. *IEEE Transactions on Audio, Speech, and Language Processing, 15*(1), 34–43.
52. Gerhard, D. (2003). Pitch extraction and fundamental frequency: History and current techniques. Technical Report TR-CS 2003-06. Department of Computer Science, University of Regina, Canada (pp. 0–22).
53. Seltzer, M. L., & Michael, D. (2000). Automatic detection of corrupt spectrographic features for robust speech recognition. Master of Science Thesis, Department of Electrical and Computer Engineering, Carnegie Mellon University, Pittsburgh, USA.
54. Scordilis, M. S., & Gowdy, J. N. (1989). Neural network based generation of fundamental frequency contours. In *Proceedings of the IEEE International Conference on Acoustics, Speech, and Signal Processing (ICASSP), Glasgow, Scotland* (pp. 219–222).
55. Han, K., & Wang, D. (2014). Neural network based pitch tracking in very noisy speech. *IEEE/ACM Transactions on Audio, Speech and Language Processing (TASLP), 22*(12), 2158–2168.
56. Su, H., Zhang, H., Zhang, X., & Gao, G. (2016). Convolutional neural network for robust pitch determination. In *Proceedings of the IEEE International Conference on Acoustics, Speech and Signal Processing (ICASSP), Shanghai, China* (pp. 579–583).
57. Ananthapadmanabha, T. V., & Fant, G. (1982). Calculation of true glottal flow and its components. *Speech Communication, 1*(3–4), 167–184.
58. Lucero, J. C., & Koenig, L. L. (2005). Phonation thresholds as a function of laryngeal size in a two-mass model of the vocal folds. *The Journal of the Acoustical Society of America, 118*(5), 2798–2801.
59. Titze, I. R. (1992). Phonation threshold pressure: A missing link in glottal aerodynamics. *The Journal of the Acoustical Society of America, 91*(5), 2926–2935.
60. Plant, R. L., Freed, G. L., & Plant, R. E. (2004). Direct measurement of onset and offset phonation threshold pressure in normal subjects. *The Journal of the Acoustical Society of America, 116*(6), 3640–3646.
61. Isshiki, N. (1981). Vocal efficiency index. In K. N. Steven & M. Hirano (Eds.), *Vocal fold physiology* (pp. 193–203). Press: University of Tokyo.
62. Klatt, D. H. (1987). Review of text-to-speech conversion for English. *The Journal of the Acoustical Society of America, 82*(3), 737–793.
63. Rosenberg, A. E. (1971). Effect of glottal pulse shape on the quality of natural vowels. *The Journal of the Acoustical Society of America, 49*(2B), 583–590.
64. Hedelin, P. (1984). A glottal LPC-vocoder. In *Proceedings of the IEEE International Conference on Acoustics, Speech, and Signal Processing (ICASSP), San Diego, California, USA* (Vol. 9, pp. 21–24).
65. Hedelin, P. (1986). High quality glottal LPC-vocoding. In *Proceedings of the IEEE International Conference on Acoustics, Speech, and Signal Processing (ICASSP), Tokyo, Japan* (Vol. 11, pp. 465–468).
66. Fujisaki, H., & Ljungqvist, M. (1986). Proposal and evaluation of models for the glottal source waveform. In *IEEE Proceedings of the International Conference on Acoustics, Speech, and Signal Processing (ICASSP), Tokyo, Japan* (Vol. 11, pp. 1605–1608).

67. Fant, G., Liljencrants, J., & Lin, Q. G. (1985). A four-parameter model of glottal flow. *Speech Transmission Laboratory Quarterly Progress and Status Report (STL-QPSR), 4*(1985), 1–13.
68. Gobl, C. (2003). The voice source in speech communication-production and perception experiments involving inverse filtering and synthesis. Doctoral dissertation, Institutionen för talöverföring och musikakustik, Royal Institute of Technology, Stockholm, Sweden.
69. Drioli, C. (2005). A flow waveform-matched low-dimensional glottal model based on physical knowledge. *The Journal of the Acoustical Society of America, 117*(5), 3184–3195.
70. Avanzini, F. (2008). Simulation of vocal fold oscillation with a pseudo-one-mass physical model. *Speech Communication, 50*(2), 95–108.
71. Frøkjaer-Jensen, B., & Prytz, S. (1976). Registration of voice quality. *Brüel and Kjaer Technical Review, 3*, 3–17.
72. Childers, D. G., & Lee, C. K. (1991). Vocal quality factors: analysis, synthesis, and perception. *The Journal of the Acoustical Society of America, 90*(5), 2394–2410.
73. Titze, I. R., & Sundberg, J. (1992). Vocal intensity in speakers and singers. *The Journal of the Acoustical Society of America, 91*(5), 2936–2946.
74. Alku, P., Strik, H., & Vilkman, E. (1997). Parabolic spectral parameter—a new method for quantification of the glottal flow. *Speech Communication, 22*(1), 67–79.
75. Murphy, P. J. (1999). Perturbation-free measurement of the harmonics-to-noise ratio in voice signals using pitch synchronous harmonic analysis. *The Journal of the Acoustical Society of America, 105*(5), 2866–2881.
76. Cummings, K. E., & Clements, M. A. (1995). Analysis of the glottal excitation of emotionally styled and stressed speech. *The Journal of the Acoustical Society of America, 98*(1), 88–98.
77. Laukkanen, A. M., Vilkman, E., Alku, P., & Oksanen, H. (1996). Physical variations related to stress and emotional state: A preliminary study. *Journal of Phonetics, 24*(3), 313–335.
78. Laukkanen, A. M., Vilkman, E., Alku, P., & Oksanen, H. (1997). On the perception of emotions in speech: The role of voice quality. *Logopedics Phoniatrics Vocology, 22*(4), 157–168.
79. Gobl, C., & Chasaide, A. N. (2003). The role of voice quality in communicating emotion, mood and attitude. *Speech Communication, 40*(1–2), 189–212.
80. Airas, M., & Alku, P. (2006). Emotions in vowel segments of continuous speech: Analysis of the glottal flow using the normalised amplitude quotient. *Phonetica, 63*(1), 26–46.
81. Waaramaa, T., Laukkanen, A. M., Airas, M., & Alku, P. (2010). Perception of emotional valences and activity levels from vowel segments of continuous speech. *Journal of Voice, 24*(1), 30–38.
82. Higgins, M. B., & Saxman, J. H. (1991). A comparison of selected phonatory behaviors of healthy aged and young adults. *Journal of Speech, Language, and Hearing Research, 34*(5), 1000–1010.
83. Sapienza, C. M., & Stathopoulos, E. T. (1994). Comparison of maximum flow declination rate: Children versus adults. *Journal of Voice, 8*(3), 240–247.
84. Sapienza, C. M., & Dutka, J. (1996). Glottal airflow characteristics of women's voice production along an aging continuum. *Journal of Speech, Language, and Hearing Research, 39*(2), 322–328.
85. Hodge, F. S., Colton, R. H., & Kelley, R. T. (2001). Vocal intensity characteristics innormal and elderly speakers. *Journal of Voice, 15*(4), 503–511.
86. Welham, N. V., & Maclagan, M. A. (2003). Vocal fatigue: Current knowledge and future directions. *Journal of Voice, 17*(1), 21–30.
87. Ozdas, A., Shiavi, R. G., Silverman, S. E., Silverman, M. K., & Wilkes, D. M. (2004). Investigation of vocal jitter and glottal flow spectrum as possible cues for depression and near-term suicidal risk. *IEEE Transactions on Biomedical Engineering, 51*(9), 1530–1540.
88. Stanek, M., & Sigmund, M. (2015). Psychological stress detection in speech using return-to-opening phase ratios in glottis. *Elektronika ir Elektrotechnika, 21*(5), 59–63.
89. Sigmund, M., Prokes, A., & Zelinka, P. (2010). Detection of alcohol in speech signal using LF model. In *Proceedings of the International Conference on Artificial Intelligence and Applications. Innsbruck, Austria* (pp. 193–196).

90. Koike, Y., & Markel, J. (1975). Application of inverse filtering for detecting laryngeal pathology. *Annals of Otology, Rhinology & Laryngology*, *84*(1), 117–124.
91. Deller, J. (1982). Evaluation of laryngeal dysfunction based on features of an accurate estimate of the glottal waveform. In *Proceedings of the IEEE International Conference on Acoustics, Speech, and Signal Processing (ICASSP), Paris, France* (Vol. 7, pp. 759–762).
92. Hillman, R. E., Holmberg, E. B., Perkell, J. S., Walsh, M., & Vaughan, C. (1990). Phonatory function associated with hyperfunctionally related vocal fold lesions. *Journal of Voice*, *4*(1), 52–63.
93. Howell, P., & Williams, M. (1988). The contribution of the excitatory source to the perception of neutral vowels in stuttered speech. *The Journal of the Acoustical Society of America*, *84*(1), 80–89.
94. Howell, P., & Williams, M. (1992). Acoustic analysis and perception of vowels in children's and teenagers' stuttered speech. *The Journal of the Acoustical Society of America*, *91*(3), 1697–1706.
95. Björkner, E., Sundberg, J., Cleveland, T., & Stone, E. (2006). Voice source differences between registers in female musical theater singers. *Journal of Voice*, *20*(2), 187–197.
96. Sundberg, J., Fahlstedt, E., & Morell, A. (2005). Effects on the glottal voice source of vocal loudness variation in untrained female and male voices. *The Journal of the Acoustical Society of America*, *117*(2), 879–885.
97. Schafer, R. W., & Rabiner, L. R. (1970). System for automatic formant analysis of voiced speech. *The Journal of the Acoustical Society of America*, *47*(2B), 634–648.
98. Kammoun, M. A., Gargouri, D., Frikha, M., & Hamida, A. B. (2004). Cepstral method evaluation in speech formant frequencies estimation. In *Proceedings of the IEEE International Conference on Industrial Technology (ICIT), Hammamet, Tunisia* (Vol. 3, pp. 1612–1616).
99. Kammoun, M. A., Gargouri, D., Frikha, M., & Hamida, A. B. (2006). Cepstrum vs. LPC: A comparative study for speech formant frequencies estimation. *GESTS International Transactions on Communication and Signal Processing*, *9*(1), 87–102.
100. Hunt, M. J. (1987). Delayed decisions in speech recognition—the case of formants. *Pattern Recognition Letters*, *6*(2), 121–137.
101. Lee, C. H. (1989). Applications of dynamic programming to speech and language processing. *AT & T Technical Journal*, *68*(3), 114–130.
102. Snell, R. C., & Milinazzo, F. (1993). Formant location from LPC analysis data. *IEEE transactions on Speech and Audio Processing*, *1*(2), 129–134.
103. Sandler, M. (1991). Algorithm for high precision root finding from high order LPC models. *IEE Proceedings I-Communications, Speech and Vision*, *138*(6), 596–602.
104. Fant, G. (1962). Descriptive analysis of the acoustic aspects of speech. *Logos*, *5*, 3–17.
105. Fitch, W. T. (1997). Vocal tract length and formant frequency dispersion correlate with body size in rhesus macaques. *The Journal of the Acoustical Society of America*, *102*(2), 1213–1222.
106. Laan, G. P. (1997). The contribution of intonation, segmental durations, and spectral features to the perception of a spontaneous and a read speaking style. *Speech Communication*, *22*(1), 43–65.
107. Zhou, G., Hansen, J. H., & Kaiser, J. F. (2001). Nonlinear feature based classification of speech under stress. *IEEE Transactions on Speech and Audio Processing*, *9*(3), 201–216.
108. El Ayadi, M., Kamel, M. S., & Karray, F. (2011). Survey on speech emotion recognition: Features, classification schemes, and databases. *Pattern Recognition*, *44*(3), 572–587.

Chapter 6
Qualitative Aspects of the Voice Signal

Of all the studies referenced in Chapters. 1 and 3, the majority have found positive correlations between various profile parameters and *voice quality*. The word "quality" is very loosely used in the context of audio processing. For example, one may refer to "perceptual quality," "speech quality," "audio quality," "recording quality" etc. These usages must not be confused with the subject at hand—voice quality. From both signal processing and information theoretic perspectives, voice quality is an elusive entity. There is no consensus in the scientific community about its precise definition—quantitatively or even in a descriptive sense. It is in fact a complex entity that comprises a set of many (mostly) subjectively described characteristics, or *sub-qualities* that collectively represent it. These characteristics, or attributes of voice, give it its particular auditory *flavor*, and can also be thought of as comprising the overall quality of someone's voice in a manner analogous to a sound mixer used in music production. Unfortunately, there is no consensus on even the *number* of sub-qualities that comprise voice quality. Regardless, some important ones are described in this chapter.

This chapter also touches upon the subject of voice disguise, which is almost entirely a game of emulating and/or masking voice quality. Voice disguise is also revisited in Chap. 7 from a computational profiling perspective.

6.1 Voice Quality in Broad Perspective

At the outset, it is important to differentiate between the quality of voice, the quality of the voice *signal*, and its *intelligibility*. The phrase "quality of the voice signal" refers to the fidelity of the signal capture, storage and transmission processes in preserving the information content of the original analog signal. Speech intelligibility (often loosely referred to as "quality of *speech*") relates to the *understandability* of its content and the *discriminability* of the sounds in it—criteria that are of importance in many speech applications, such as transmission and communication of voice through telephone and wireless channels that cut off much of its higher spectral components.

R. Singh, *Profiling Humans from their Voice*,
https://doi.org/10.1007/978-981-13-8403-5_6

Intelligibility is extensively considered in the design of codecs and other schemes for handling speech.

As alluded to above, voice quality is best characterized as the "flavor" of voice as perceived by the listener. For example, one may say "her voice was soft" or "his voice was tense" or "'what a melodious voice!"—in these descriptions, the words soft, tense and melodious refer to voice quality in general, but each, individually, is a sub-component of that general entity. Voice quality thus comprises a number of sub-components, or "sub-qualities." The sub-qualities are themselves rather subjectively differentiated (as in the examples above), each referring to a specific perceptual flavor of voice. There is no ontology or taxonomy of voice qualities. They have historically been named as such based on human observation. On the other hand, hundreds, if not thousands of scientific studies have related human bio-relevant parameters such as age, body size, pathologies etc. to voice quality. For example "it is observed that with increasing age, women's voices become heavier," or "...... nine out of ten subjects with x disease were observed to have unusually breathy and jittery voices," or "slurred speech is positively correlated with greater degrees of intoxication." From the sheer volume and compelling conclusions of such empirical studies, it is clear that defining and quantifying all of the sub-components of voice quality is highly desirable for profiling. If we could measure the level of breathiness in the voice, for example, we could build a predictor to identify illness x in a population that comprised people with and without illness x. In some cases, we may even be able to build such predictors without explicit computational learning procedures, based on human observations alone.

To quantify voice quality, we must look for causal processes and conditions in the human body and relate them to changes in the bio-mechanics voice production. We must then trace their effects on the voice signal, which must then be identified and measured using signal processing techniques. In part II of this book, we will discuss how all of this can be accomplished using machine learning and AI techniques in part (which have their own limitations), but for now, let us understand what is involved in taking the knowledge-based approach mentioned above.

There are extremely detailed accounts in the literature about the vital role played by the exact physics and mechanics of the vocal folds in determining many of these sub-qualities. In fact, voice quality as a whole varies due to the variation in the underlying bio-mechanics of voice production. Later in this chapter, when we address some selected voice quality sub-components, we will mention the corresponding causal mechanics briefly. However, this chapter does not focus on the bio-mechanics of voice production. Instead, we focus on measurable *signal characteristics* that relate to voice quality. This is motivated by the goal of laying quantitative groundwork for computational profiling, addressed in Part II of this book. Part of the reason for this choice is also the fact that we do not yet have perfect bio-mechanical models that we can readily use to *quantify* voice quality for computational profiling, and such models are difficult to formulate and verify due to the complexity of the bio-mechanisms involved.

From a signal processing perspective, it is important to note that when we speak of voice quality, we speak of voice quality *as it is perceived*. What we hear (and perceive) as a result of phonation is not just the signal produced by the vibration of the vocal folds, but also the interaction, or coupling, of this sound with the acoustics of the vocal tract. In fact, voice quality as we perceive it, *is a combination* of the characteristics of the speaker's articulatory, phonatory, resonatory, respiratory, and to some extent even prosodic, aspects of voice production.

It is also important to keep in mind that for computational profiling, our primary goal is to transfer the human perceptions of voice quality as accurately as possible to computational forms. This will allow us to continue to leverage human judgment (for which we do not yet have a replacement, despite recent advances in artificial intelligence) along with machine-discovered features, to perfect the processes and outcomes of profiling in the future.

6.2 Signal Characteristics Used in the Description of Voice Quality

Studies have related a variety of signal characteristics to perceptions of voice quality [1, 2]. Of these characteristics, some heavily implicated ones are listed and explained briefly below. Some of these properties relate to the time-domain waveform, but generally most relate to spectral-domain characterizations such as the fundamental frequency, noise in the spectrum, energy in the harmonics, spectral shape, spectral amplitudes, ratios of various spectral entities etc.

This section is sketchy, since the basic signal processing concepts have already been discussed in Chaps. 4 and 5. Only brief explanations are provided where necessary. In addition, the description switches between continuous and discrete signal domains for reasons of ease, clarity and brevity of presentation.

The following key notations are used frequently in this section:

- F0: Fundamental frequency, or pitch
- $\overline{F0}$: Average pitch
- T0: Time period of pitch, or the inverse of F0
- A0: Amplitude associated with pitch
- $\overline{A0}$: Average amplitude associated with pitch
- F1, F2, ..., Fn: first formant, second formant, ..., nth formant
- X_i: Any entity X measured for frame i of the signal
- t: Index for time
- N: Generally used to indicate the length of any sequence of entities, or number of objects in a set.

6.2.0.1 Pitch

Foremost, it is important to note that when we discuss pitch, we discuss it with reference to the glottal pulse waveform, and not with reference to the pitch estimated from the voice signal using other spectral analyses or algorithms such as [3–7]. Even if pitch is estimated using such algorithms, the glottal pulse waveform must be extracted in order to measure most of the entities mentioned below. This is because the voice signal as a whole, at the granularity at which the entities below are measured (e.g. cycle-to-cycle variations in the pitch period) may yield ambiguous values for such direct measurements.

Pitch, as mentioned in Sect. 6.4.1.1, is the fundamental frequency of phonation, denoted as F0. Pitch and its central tendencies have been correlated with the perception of many sub-components of voice quality.

The central tendencies of F0 are measured as follows:

1. **Mean (speaking fundamental frequency)**:

$$\overline{F0} = \frac{1}{N}\sum_{i=1}^{N} F0_i \tag{6.1}$$

2. **Geometric mean**:

$$\overline{F0}_g = \sqrt[N]{\prod_{i=1}^{N} F0_i} \tag{6.2}$$

3. **Root mean squared (RMS) value**:

$$F0_{rms} = \sqrt{\frac{1}{N}\sum_{i=1}^{N} F0_i^2} \tag{6.3}$$

5. **Dispersion (unbiased variance)**:

$$\sqrt{\frac{1}{N-1}\sum_{i=1}^{N}(F0_i - \overline{F0})^2} \tag{6.4}$$

6. **Coefficient of variation**:

$$\frac{\sqrt{\frac{1}{N-1}\sum_{i=1}^{N}(F0_i - \overline{F0})^2}}{\overline{F0}} \tag{6.5}$$

7. **Percentual variability**:

$$\frac{\frac{1}{N}\sum_{i=1}^{N}|F0_i - \overline{F0}|}{\overline{F0}} \tag{6.6}$$

8. **Coefficient of consistency**:

$$\frac{\overline{F0}}{\frac{1}{N}\sum_{i=1}^{N}|F0_i - \overline{F0}|} \tag{6.7}$$

Pitch *variability* is further gauged in two ways in the spectral domain: perturbations in frequency, and perturbations in the amplitude of F0. These perturbations can also be respectively regarded as variants of *jitter* and *shimmer*, descriptions of which follow next.

6.2.0.2 Frequency Perturbations of F0: Jitter

Perturbations in the pitch *frequency* can be gauged in many ways. A widely used measure is **jitter**. Perturbations in the pitch *amplitude*, on the other hand, are called **shimmer**. Jitter and shimmer are also often mentioned or used directly as sub-components of voice quality. However, they are measures of modulation, and as such are more appropriately categorized as signal characterizations. When speech has significant jitter or shimmer, *what it is perceived as* (such as *roughness* in the case of jitter, and *buzzing* in the case of shimmer) are rather the sub-components of voice quality as intentioned in this chapter, and described in the next section.

Jitter refers to the cycle-to-cycle variation in a speaker's pitch, for voiced speech.

$$\text{Jitter} = \frac{1}{N-1}\sum_{i=1}^{N-1}|T0_i - T0_{i+1}| \tag{6.8}$$

where $T0_i$ denotes the ith time period of pitch, and N is the number of periods considered.

Since *any* perturbation to F0 can be related back to a cycle-to-cycle variation in F0, pitch perturbation of any kind, if in frequency, can be considered to be a variant of jitter (and likewise for shimmer, in the context of the amplitude of F0). As such, several variants of jitter have been used in the literature—for example, a *local jitter* may be defined as the relative period-to-period variability (measured in percent), *absolute jitter* is defined as the absolute period-to-period variability (measured in microseconds). The RAPQ or the *relative average perturbation quotient* (measured in percent) is defined as the average jitter among a few consecutive periods. When measured based on absolute jitter, this is called the *pitch perturbation quotient*, or PPQ.

Some popular measures that gauge the perturbation of pitch are enumerated below. The list is non-exhaustive. A note on units: measures in time (such as pitch perturbation), are expressed in milliseconds (ms) or microseconds (μs, usually when based on T0); ratios of entities may be variously expressed as percent, dB etc.; measures in frequency are expressed in Hz (usually). Units are not mentioned below where obvious. In the following, different measures are computed either over successive periods or over successive windows. Wherever $T0$ is used, the computation is over successive periods, and wherever F0 is used, the computation is over successive windows. This is because a single period is not sufficient to robustly compute F0. Generally, both variants are given for each labeled entity.

1. **Normalized jitter**: This is measured in percent:

$$100 \times \frac{1}{N-1} \sum_{i=1}^{N-1} \frac{|T0_i - T0_{i+1}|}{T0_i} \tag{6.9}$$

or when measured over successive windows,

$$100 \times \frac{1}{N-1} \sum_{i=1}^{N-1} \frac{|F0_i - F0_{i+1}|}{F0_i} \tag{6.10}$$

2. **Percent jitter**: $100\times$ ratio of deviation in $T0$ to mean $T0$

$$100 \times \frac{\frac{1}{N-1} \sum_{i=1}^{N-1} |T0_i - T0_{i+1}|}{\frac{1}{N} \sum_{i=1}^{N} T0_i} \tag{6.11}$$

Jitter ratio (JR) is a commonly used variant of this, defined as $10 \times$ percent jitter, or $1000 \times$ ratio of deviation in $T0$ to mean $T0$.
When computed over successive windows, percent jitter is also called the **Jitter Factor (JF)**, computed as $100 \times$ ratio of deviation in F0 to mean F0

$$100 \times \frac{\frac{1}{N-1} \sum_{i=1}^{N-1} |F0_i - F0_{i+1}|}{\frac{1}{N} \sum_{i=1}^{N} F0_i} \tag{6.12}$$

3. **Mean absolute jitter**: Mean absolute period jitter is given by

$$\frac{1}{N-1} \sum_{i=1}^{N-1} |T0_i - T0_{i+1}| \tag{6.13}$$

Mean absolute frequency jitter is given by

$$\frac{1}{N-1} \sum_{i=1}^{N-1} |F0_i - F0_{i+1}| \tag{6.14}$$

4. **Normalized mean absolute jitter**: Normalized mean absolute period jitter is given by

$$\frac{1}{\max_j(T0_j)} \frac{1}{N} \sum_{i=1}^{N} |T0_{i+1} - T0_i| \tag{6.15}$$

Normalized mean absolute frequency jitter is given by

$$\frac{1}{\max_j(F0_j)} \frac{1}{N} \sum_{i=1}^{N} |F0_{i+1} - F0_i| \tag{6.16}$$

6.2.0.3 Related Perturbations in Pitch That Are Not Categorized as Jitter

The term jitter usually directly applies to changes that are measured over successive cycles or windows. when this is not the case, the entities measured are categorized as *perturbations* in pitch period, or in pitch frequency. In the following, we enumerate some measures that have been used in different studies in the literature.

1. **Pitch period perturbation**:

$$\Delta T0 = |T0_i - T0_{i-1}| \tag{6.17}$$

Based on this, the **Pitch perturbation factor (PPF or PF)** is defined as the ratio of the number of periods found perturbed (N_{Perturb}) to the number of periods considered for counting.

$$\text{PF} = \frac{N_{\text{Perturb}}}{N_{\text{Total}}} \tag{6.18}$$

$N_{Perturb}$ may be determined by setting a threshold for the perturbation (which may be subjective to the kind of signal analyzed). If for example, we choose to label any perturbation value greater than 0.3 ms as a valid perturbation, then the PPF is given by

$$\frac{\text{frequency of } \Delta T0 \geq 0.3 \text{ ms}}{\text{total number of } \Delta T0 \text{ considered}} \tag{6.19}$$

2. **Pitch period perturbation quotient (PPQ)**:

$$\frac{\frac{1}{N-4}\sum_{i=1}^{N-4}\frac{1}{5}\sum_{r=0}^{4}|T0_{i+r} - T0_{i+2}|}{\frac{1}{N}\sum_{i=1}^{N} T0_i} \tag{6.20}$$

A popular variant of this, measured as a percent and also called PPQ, is given by

$$\frac{100}{N-2}\sum_{i=2}^{N-1}\frac{|T0_i - \frac{T0_{i-1}+2T0_i+T0_{i+1}}{4}|}{\frac{T0_{i-1}+2T0_i+T0_{i+1}}{4}} \tag{6.21}$$

Yet another widely used variant of PPQ is:

$$\text{PPQ}_{\text{Kasuya}} = \frac{100}{N-k+1}\sum_{i=1}^{N-k+1}\left|\frac{1 - T0_{i+m}}{\frac{1}{k}\sum_{n=1}^{k} T0_{i+n-1}}\right| \tag{6.22}$$

A smoothed value of PPQ, the **smoothed pitch period perturbation quotient**, can be computed as

$$\frac{\frac{1}{N-k+1}\sum_{i=1}^{N-k+1}\frac{1}{k}\sum_{r=0}^{k-1}|TO_{i+r} - TO_{i+m}|}{\frac{1}{N}\sum_{i=1}^{N}TO_i} \tag{6.23}$$

where the smoothing factor k is an odd integer greater than 1, and $m = (k-1)/2$.

3. **Deviation from Linear Trend (DLT)**:

$$\text{DLT} = \frac{1}{N-5}\sum_{i=3}^{N-2}\left|\frac{TO_{i-2} + TO_{i+2}}{2} - TO_i\right| \tag{6.24}$$

4. **Perturbation factor without linear trend**: Measured in percent

$$\frac{\frac{1}{N-2}\sum_{i=2}^{N-1}\left|\frac{TO_{i-1}+TO_{i+1}}{2} - TO_i\right|}{\frac{1}{N}\sum_{i=1}^{N}TO_i} \tag{6.25}$$

5. **Pitch variability index (PVI)**: For this, $T0$ is extracted from a 10 Hz bandpass filter centered on the first harmonic.

$$\text{PVI} = 1000 \times \frac{\frac{1}{N}\sum_{i=1}^{N}(TO_i - \overline{TO})^2}{\overline{TO^2}} \tag{6.26}$$

6. **Relative average perturbation (RAP)**: Measured in percent

$$\text{RAP} = \frac{\frac{1}{N-2}\sum_{i=1}^{N-2}\left|\frac{TO_{i-1}+TO_i+TO_{i+1}}{3} - TO_i\right|}{\frac{1}{N}\sum_{i=1}^{N}TO_i} \tag{6.27}$$

When measured over windows, this is called the **Frequency perturbation quotient (FPQ)**, and is also measured in percent

$$\text{FPQ} = \frac{\frac{1}{N-2}\sum_{i=1}^{N-2}\left|\frac{FO_{i-1}+FO_i+FO_{i+1}}{3} - FO_i\right|}{\frac{1}{N}\sum_{i=1}^{N}FO_i} \tag{6.28}$$

7. **F0 modulation**: This is measured over a window of speech

$$\frac{max(TO) - min(TO)}{max(TO) + min(TO)} \tag{6.29}$$

8. **Perturbation magnitude (PM)**:

$$\text{PM} = \frac{1}{N_{\text{Perturb}}} \sum_{i=1}^{N_{\text{Perturb}}} \frac{|F0_{i+1} - F0_i|}{F0_i} \tag{6.30}$$

9. **Perturbation magnitude mean (PMM)**:

$$\text{PMM} = \text{PF} \times \text{PM} \tag{6.31}$$

6.2.0.4 Amplitude Perturbations of F0: Shimmer

Shimmer is the cycle-to-cycle variation in the amplitude of F0.

Shimmer in dB is defined as:

$$\text{Shimmer(dB)} = \frac{1}{N-1} \sum_{i=1}^{N-1} \left| 20 \log \left(\frac{A0_{i+1}}{A0_i} \right) \right| \tag{6.32}$$

where N is the number of extracted periods and $A0_i$ represents the peak-to-peak amplitude within the ith period. This is often also referred to a **mean shimmer** or **absolute shimmer**.

Like jitter, shimmer is also gauged in multiple ways. Some popular expressions used in the literature are given below:

1. **Normalized amplitude perturbation (normalized period shimmer)**:

$$100 \times \frac{1}{N-1} \sum_{i=1}^{N-1} \frac{A0_i - A0_{i+1}}{A0_i} \tag{6.33}$$

2. **Absolute amplitude perturbation (absolute period shimmer)**:

$$\frac{1}{N-1} \sum_{i=1}^{N-1} |A0_{i+1} - A0_i| \tag{6.34}$$

3. **Normalized mean absolute period shimmer**: Measured in dB SPL

$$\frac{1}{N \max_j (A0_j)} \sum_{i=1}^{N} |A0_{i+1} - A0_i| \tag{6.35}$$

4. **Percent shimmer (or shimmer factor)**:

$$100 \times \frac{\frac{1}{N-1}\sum_{i=1}^{N-1} |A0_i - A0_{i+1}|}{\frac{1}{N}\sum_{i=1}^{N} A0_i} \tag{6.36}$$

6.2.0.5 Related Perturbations in Amplitude of F0 That Are Not Categorized as Shimmer

Perturbations in amplitude of F0 that are measured over a window are generally not categorized as shimmer, since the latter explicitly characterizes variations in adjacent cycles. The following are some such measures.

1. **Peak amplitude modulation**:

$$\frac{max(A0) - min(A0)}{max(A0) + min(A0)} \tag{6.37}$$

2. **Peak-to-peak amplitude modulation**:

$$\frac{max(A0) - min(A0)}{max(A0)} \tag{6.38}$$

3. **Amplitude variability index (AVI)**:

$$\text{AVI} = \log_{10}\left(1000 \times \frac{\frac{1}{N}\sum_{i=1}^{N}(A0_i - \overline{A0})^2}{\overline{A0}^2}\right) \tag{6.39}$$

where

$$\overline{A0} = \frac{1}{N}\sum_{i=1}^{n} A0_i \tag{6.40}$$

is the mean amplitude.

4. **Amplitude perturbation quotient (APQ)**: This is the N-point average cycle-to-cycle amplitude difference from the mean amplitude across the N-point window (N is chosen between 5 and 10 in most studies)

$$\frac{\frac{1}{N-10}\sum_{i=1}^{N-10}\left|\frac{1}{11}\sum_{r=0}^{10} A0_{i+r} - A0_{i+5}\right|}{\frac{1}{N}\sum_{i=1}^{N} A0_i} \tag{6.41}$$

The **smoothed amplitude perturbation quotient (sAPQ)** is defined as

$$\mathrm{sAPQ} = \frac{\frac{1}{N-k+1}\sum_{i=1}^{N-k+1}\left|\frac{1}{k}\sum_{r=0}^{k-1} A0_{i+r} - A0_{i+m}\right|}{\frac{1}{N}\sum_{i=1}^{N} A0_i} \tag{6.42}$$

where k is a smoothing factor (odd number) and $m = (k-1)/2$. A popular variant of this is

$$\mathrm{APQ}_{\mathrm{Kasuya}} = \frac{100}{N-k+1}\sum_{i=1}^{N-k+1}\left|\frac{1 - A0_{i+m}}{\frac{1}{k}\sum_{r=1}^{k} A0_{i+r-1}}\right| \quad k = 3, \quad m = 1. \tag{6.43}$$

5. **Amplitude perturbation factor without linear trend**: Measured in percent.

$$100 \times \frac{\frac{1}{N-2}\sum_{i=2}^{N-1}\left|\frac{A0_{i-1}+A0_{i+1}}{2} - A0_i\right|}{\sum_{i=1}^{N} A0_i^2} \tag{6.44}$$

6.2.0.6 Perturbations of Formant Frequencies F1, F2, . . ., FN

Perturbations of formant frequencies are characterized by amplitude and frequency modulation (AM-FM) measurements. AM-FM measurements of formants are examples of *micro-modulation features*, which are measured using various micro-modulation speech models of voice [8]. These models are designed to capture rapid formant fluctuations.

The ith resonance can be modeled as an AM-FM signal in the continuous signal domain as:

$$r_i(t) = a_i(t) cos\Big(2\pi \int_0^T f_i(\tau)d\tau\Big) \tag{6.45}$$

where $i = 1, 2, \ldots, 6$ is the index of the resonance r, and T is the time window length. $f_i(t)$ is the instantaneous frequency, $|a_i(t)|$ is the instantaneous amplitude.

For a given neighborhood of a formant, a superposition of approximately six components given above is satisfactory to model the formant. Based on this observation, the AM-FM Modulation features for a formant may be defined as:

$$F_i = \frac{\int_0^T f_i(t)a_i^2(t)dt}{\int_0^T a_i^2(t)dt} \tag{6.46}$$

$$B_i = \frac{\int_0^T \Big[a_i^2(t) + \Big(f_i(t) - F_i\Big)^2 a_i^2(t)\Big]dt}{\int_0^T a_i^2(t)dt} \tag{6.47}$$

where B_i denotes the mean bandwidth and F_i represents the weighted mean frequency of the resonance r_i. The short-time weighted mean of the instantaneous frequency, $f_i(t)$ and the short-time mean of the instantaneous amplitude, $|a_i(t)|$, provide information about the instantaneous formant variations and the fine structure of the envelope of the signal amplitude, and can be used as features.

Another popular feature based on this is the **frequency modulation percentage**, which indicates how much the formant frequencies vary around a center frequency. It is defined as

$$\text{FMP}_i = \frac{B_i}{F_i} \tag{6.48}$$

6.2.0.7 Cepstral Peak Prominence (CPP)

Cepstra were discussed earlier in this book. There are many other variants of cepstra that have been used in the literature, one of which is the log power spectrum *of the log power spectrum* [9, 10]. In this domain, frequency is called "quefrency" and harmonics are called "rahmonics." CPP is the distance (or prominence) of the first rahmonic (peak) above a regression line that normalizes the energy of the cepstrum.

CPP in fact measures the intensity of the harmonics in the voice signal. It has been correlated with many voice sub-quality features observed in medical conditions such as dysphonia [11], vocal fold palsy [12] etc. Its quantification has been discussed in many articles, e.g. [13–15].

It is worth mentioning here that CPP forms a component of the widely used **Acoustic voice quality index (AVQI)** in clinical investigations. AVQI [15] is an index that draws upon cues in time, spectral and cepstral (quefrency) domains. In its most accepted form, it is based on a combination of six features: smoothed CPP, harmonics-to-Noise ratio (HNR), shimmer (two versions), slope of the long-term average spectrum, and tilt of the regression line through the long-term average spectrum. In clinical practice, AVQI is rated on a scale of 0-10, and is used in rating the severity of vocal disorders such as hoarseness, dysphonia etc.

6.2.0.8 Glottal to Noise Excitation (GNE) Ratio

GNE ratio is based on correlations between the Hilbert envelopes of different frequency bands in the voice spectrum. It is a measure of the degree of turbulent noise in a voice signal, and an estimate of the degree to which voice originates from vocal fold vibrations in the process of phonation as compared to turbulence associated with it.

The noise of turbulent airflow just above the glottis, caused due to volume displacement of air, is relatively small compared to the sound generated by the vocal fold vibrations. However, it significantly contributes to the perception of voice quality. The spectrum of this noise is not white (not flat), and its exact shape depends on glottal opening, flow rate and the shape of the vocal tract. Not surprisingly, GNE has been highly implicated in many voice disorders e.g. [16]. It is also more persistent in the voices of women than in men, and high for voices with a "breathy" quality.

The GNE ratio is calculated by first obtaining the excitation signal through inverse filtering of the voice signal typically, down-sampled to 10 kHz, and applying a filter bank to the linear spectrum of the excitation signal. The Hilbert envelopes of adjacent frequency bands are then cross-correlated. In this case, "adjacent" bands refer to each pair of bands that are separated by an amount greater than or equal to the difference in their center frequencies. The GNE ratio (or GNE factor) is the maximum of the set of peak values of all cross-correlations obtained in this manner.

A Hilbert envelope is an estimate of the contour trace of a signal waveform in the time domain. It is obtained using the Hilbert transform, which is a linear operator that operates on a signal (or function) $s(t)$, by convolving it with the function $\frac{1}{\pi t}$:

$$H(s(t)) = \hat{s}(t) = \frac{1}{\pi}\int_{-\infty}^{\infty} \frac{s(\tau)}{t-\tau} d\tau \tag{6.49}$$

where $H(s(t))$ is the Hilbert transform.

Note that the Hilbert transform in turn allows us to estimate the complex valued analytic signal, given a signal in the time domain. This allows us to calculate the phase, which is important for the intelligibility of the voice signal, and carries many cues for profiling. If $\hat{s}(t)$ is the Hilbert transform of a real-valued signal $s_r(t)$, then the complex valued analytic signal is given by

$$s(t) = s_r(t) + j\hat{s}(t) = a(t)e^{j\omega(t)} \tag{6.50}$$

where $a(t)$ is the envelope or the instantaneous amplitude of the signal $s(t)$, and $\omega(t)$ is its instantaneous phase. The instantaneous amplitude is used to find the envelope of the signal, which is then the Hilbert envelope, as shown in the example in Fig. 6.1.

The instantaneous frequency (traditionally useful in many applications that require monitoring of acoustic phenomena closely, such as sudden changes in a continuous sound) is given by:

$$f(t) = \frac{1}{2\pi}\frac{d\omega}{dt}(t) \tag{6.51}$$

Other methods of estimating signal envelopes, e.g. [17] can also be used for the computation of the GNE ratio. It is also worthwhile to note here that the GNE ratio is not affected by jitter and shimmer, contrary to expectation. The reasoning behind this is explained in [18].

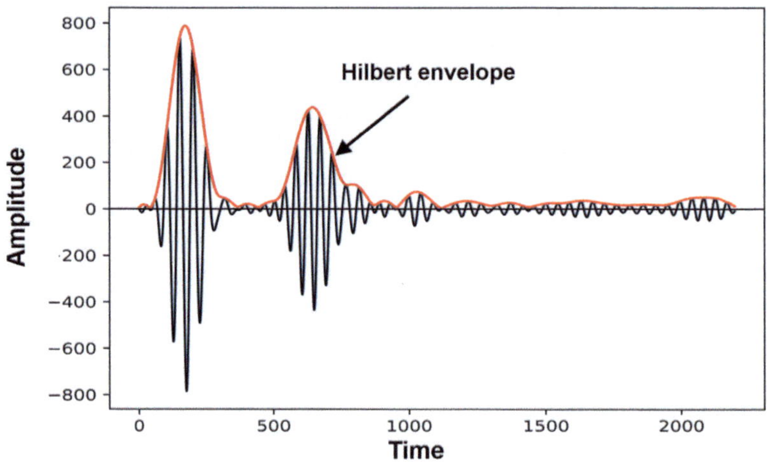

Fig. 6.1 Example of a Hilbert envelope

6.2.0.9 Pivot Frequency

Pivot frequency is the frequency at which the speech spectrum is partitioned into two bands, such that the lower band contains the formants F1 and F2 that play a significant role in the intelligibility of speech, and the upper band contains information that is thought to be relevant to the perception of speech quality. The Hammerberg Index, for example, is based on the pivot frequency of 2 kHz. Measurements and ratios of measurable entities within these two bands form the basis for many additional features (not discussed here, since many are non-standard and applicable only to specific studies in the context of which they have been mentioned).

6.2.0.10 Hammerberg Index (HI)

HI is defined to be the difference between the peak energy in the 0–2 and 2–5 kHz frequency bands. 2 kHz is the pivot frequency in this case. It is shown in Fig. 6.2. Since the peak energy in these bands is likely to coincide with the corresponding formant peaks, HI is essentially the difference between the peak energy associated with formants F3 and F2.

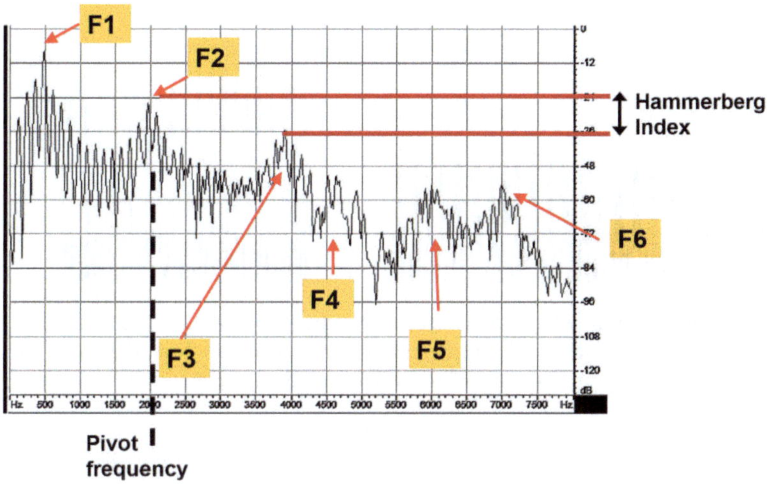

(a) The spectrum of the sound /EY/ spoken by a 29 year old healthy Caucasian male, 5'5" in height. The formant peaks are indicated by labels in the figure. The Pivot frequency (2 kHz) is very close to the F2 for this speaker, but this is only a coincidence. The Hammerberg index is the difference in peak values of F3 and F2 as shown. The horizontal axis is linear in frequency, and the vertical axis is intensity in dB below zero (i.e. the energy of the signal is normalized to 1).

Fig. 6.2 Illustration of Hammerberg index

6.2.0.11 Long-Term Average Spectrum (LTAS)

LTAS is a power spectrum oriented measure. The term itself refers to the spectrum of an entire utterance or a long window (comprising many phonemes) of the signal. Many different derivatives of it are used in the literature. An important set of derivatives comprises those that measure spectral energy changes across a pivot frequency: usually the frequency of 1000 Hz is chosen for this. The entities measured include drop-off of the spectral energy above 1000 Hz, ratio of energy below 1000 Hz (low-frequency region) to that above 1000 Hz (high-frequency region) etc.

6.2.0.12 Spectral Slope

Spectral slope measures have been widely implicated in the perception of voice quality [19]. It is the difference in the amplitude of specific harmonics in the speech spectrum. Variants of spectral slope measurement depend on the specific harmonics used, and the corrections/functions applied to them before computing the amplitude differences. Referring to the amplitudes of harmonics H1, H2, ... as A1, A2, ... respectively, spectral slope is usually defined as $A1 - A2$. Its variants include A2–

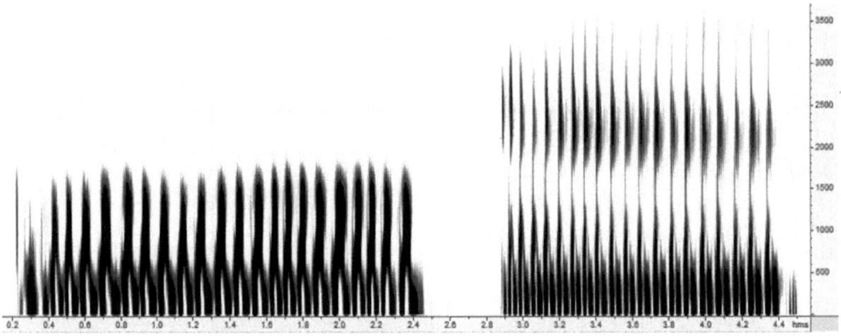

Pitch pulses for an adult female speaker for the sound /AA/ and the nasalized sound /AEN/. Darker shades represent higher energy. Up-close, the cycles are not entirely uniform and the cycle-to-cycle variations are measured as jitter.

Fig. 6.3 Pitch pulses for a nasalized and non-nasalized vowel

A4, A2 − A_{2kHz} (A_{2kHz} refers to the harmonic closest to the pivot frequency of 2 kHz), $A_{2kHz} - A_{5kHz}$ etc. Spectral slope is indicative of some key characteristics of the glottal flow dynamics. The energy difference between the first two harmonics is related to the time T_0 taken for the glottis to close (from a fully open position) within each cycle of phonation [20–22]. When $T_0 = 0$, i.e. the closure is abrupt, there is a strong discontinuity in the flow derivative. This causes spectral *roll-off* (described later in this section) at a rate of −6 dB per octave toward high frequencies. When T_0 is higher, the negative peak in the derivative gradually returns to 0. This acts as a low-pass filter with a frequency cutoff at $1/T_0$. Harmonics above this cutoff frequency are weak. The spectral slope also correlates with resonances in the vocal tract [23], and in practical use, must be corrected for the effects of the first formant F1 (at least).

For profiling, spectral slope measures are best obtained from the glottal flow waveform. Its estimate from the signal spectrum (where *pitch pulses* equate to the glottal flow waveform) may be affected by the resonances and anti-resonances caused within the vocal chambers. This is illustrated in Fig. 6.3 which shows the pitch pulses for two sounds uttered by the same individual. During phonation, the glottal flow amplitude is highest when the glottis is fully open. However, at this time the resonant energy in the voice signal is lowest, since the vocal tract is no longer the equivalent of a tube closed at one end. Once the glottis closes, these energies are highest. This leads to periodic energy highs and lows in the spectrum, at the rate of F0, and accounts for the energy patterns seen in the pitch pulses on the spectrogram. The *spread* of this energy over time relates to the time-derivative of the glottal flow waveform, while the spread of this energy over frequency relates to the glottal closure time (the greater this time, the higher the cumulative sound intensity, and therefore also the energy in the harmonics seen on the spectrogram). These can be seen clearly on wideband spectrograms, as shown in Fig. 6.3. The spectral holes (frequency bands of low energy) for the nasalized sound are due to the anti-resonances.

Of course, for the glottal flow waveform, the energy distribution does not *exactly* follow the pattern above. The energy (and its spread) for the first few harmonics in the spectrum of this waveform is determined by the open-quotient, $T_{\text{open}}/T0$, where T_{open} is the duration for which the glottis is open within each cycle, and T0 is pitch period, i.e. the duration of a complete open-close cycle [24–27]. The first harmonic H1 is dominant when this is high. Energy in the spectrum peaks at the frequency $1/T_0$. Accordingly, for speakers with very short T_0, H2 and H4 are often the most dominant harmonics. Generally, a relatively wider glottal opening leads to stronger low-frequency and weaker high-frequency components in the spectrum of the glottal flow waveform [21].

6.2.0.13 Spectral Roll-Off

Spectral roll-off is the negative or "downward" trend of the regression line that best fits the spectral energies in the speech spectrum. It is measured in dB (fall) per octave, and is based on the following definition, and the Riemann Lemma (stated below):

Definition: A function $G(\omega)$ is said to have a $\frac{1}{\omega^{n+1}}$ roll-off, or a negative net downward trend in successive values, if there exist ω_0 and some positive constant $K < \infty$ such that $|G(\omega)| < \frac{K}{\omega^{n+1}}$ for all $\omega > \omega_0$

(Riemann Lemma): If the derivatives up to order n of a function $w(t)$ exist and are of bounded variation, then its Fourier Transform $F(\omega)$ is asymptotically of order $\frac{1}{\omega^{n+1}}$, i.e.,

$$F(\omega) = \mathcal{O}\left(\frac{1}{\omega^{n+1}}\right), \quad (\text{as } \omega \to \infty) \tag{6.52}$$

Thus a discontinuity in the nth derivative corresponds to a roll-off rate of $1/\omega^{n+1}$ in the spectrum.

6.2.0.14 Spectral Tilt

Spectral tilt is a more accurate calculation of the approximate estimates of spectral slope defined above. Spectral tilt is the slope of the regression line modeling the natural negative roll-off of the speech spectrum, and can be found as follows:

The equation of a line through a data sequence $x[n]$, $y[n]$ is $y[n] = a + bx[n]$, where a is the y-intercept and b is the slope. If the data are spectral data, then $y[n]$ is an amplitude at frequency $x[n]$. The spectral tilt is given by:

$$a = \frac{\sum y \sum x^2 - \sum x \sum xy}{n \sum x^2 - (\sum x)^2} \tag{6.53}$$

$$b = \frac{n \sum xy - \sum x \sum y}{n \sum x^2 - (\sum x)^2} \tag{6.54}$$

Spectral slope and spectral tilt are related, but since the former is extremely locally calculated, the trends may sometimes be very different, or even opposite.

6.2.0.15 Spectral Difference (SD)

Given the STFT of a voice signal, the spectral difference is the geometrical distance between two spectral vectors with components x^f (where f is frequency) in the D-dimensional space of frequencies, where D is the number of frequency bins that the STFT resolves each frame into. The spectral distance can be based on any distance measure, and can be based on any band of frequencies in the spectrum or the entire range of frequencies in the spectrum (all D dimensions). For a frequency band limited by a lower frequency of f_{low} and a higher frequency of f_{high}, and using the Euclidean distance measure, SD at any frame (or time) t is given by:

$$\mathrm{SD}_t = \sqrt{\sum_{f=f_{low}}^{f_{high}} (x^{f_t} - x^{f_{(t-1)}})^2} \tag{6.55}$$

SD is used to detect sudden changes in the shape of the signal spectrum, e.g. when multiple sound events happen transiently in the signal, as in music and many natural soundscapes.

A variant of SD considers only the positive differences:

$$\mathrm{SD}_t^+ = \sqrt{\sum_{f=f_{low}}^{f_{high}} \frac{(x^{f_t} - x^{f_{(t-1)}}) + |x^{f_t} - x^{f_{(t-1)}}|}{2}^2} \tag{6.56}$$

6.2.0.16 Spectral Flux

The **normalized spectral flux** of the spectral vector **X** is the spectral distance of **X** normalized by its $L1$ or $L2$ norms at each time t.

$$\mathrm{SF}_t = \sum_{f=f_{low}}^{f_{high}} \left(\frac{x^{f_t}}{\mu_t} - \frac{x^{f_{(t-1)}}}{\mu_{t-1}} \right)^2 \tag{6.57}$$

where the normalization factors μ are:

$$\mu_t^{L1} = \sum_{f=f_{low}}^{f_{high}} (|x^{f_t}|) \tag{6.58}$$

$$\mu_t^{L2} = \sqrt{\sum_{f=f_{low}}^{f_{high}} (|x^{f_t}|)^2} \tag{6.59}$$

When $\mu = 1$, the spectral flux is said to be "unnormalized."

6.2.0.17 Harmonicity

This is the ratio of the number of peaks in the spectrum to the number of bins in the spectrum, or to the total energy in the spectrum. The peaks can be determined by any means, including simple peak-picking by rank-ordering the energies in each frequency bin and picking the top few.

6.2.0.18 Harmonics-to-Noise Ratio (HNR) and Noise-to-Harmonics Ratio (NHR)

HNR compares the energy in the harmonics in the speech spectrum to that in the non-harmonic (noisy) portions. It is measured in dB as:

$$\text{HNR} = 10\ \log_{10} E_{\text{harmonic}}/E_{\text{noise}} = 10\ \log_{10} \frac{R_{xx}[T0]}{R_{xx}[0] - R_{xx}[T0]} \tag{6.60}$$

where $R_{xx}[l]$ is the autocorrelation of the spectrum at lag l. At lag $l = 0$, $R_{xx}[l]$ represents the energy in the entire spectrum, while at lag $l = T0$, it represents the energy of its harmonic components.

A rough calculation of HNR can be done by finding the ratio and the energy in the harmonic portions (approximately between 70 and 1500 Hz for a pivot frequency of 1500 Hz) and noisy portions (approximately between 1500 and 4500 Hz) of speech.

If the energy in the harmonic potion of the signal is 100 times the energy in the noisy portions, then HNR is 20 dB, which is roughly the HNR for a sustained /AA/ or /IY/ vowel produced by a speaker with healthy vocal function. For /UW/, which has energy concentration in predominantly lower frequencies, HNR is around 40 dB. Vowels such as /AA/ and /IY/ with lower HNR—or relatively greater concentration of energy in the higher frequencies—are also more susceptible to more jitter.

HNR is related to the Signal-to-Noise Ratio (SNR): a signal with 0 dB HNR is also a signal with 0 dB SNR.

NHR is the inverse of HNR: it is the ratio of the energy in the noisy portions of the spectrum to that in the harmonic portions. Both HNR and NHR are widely used in the context of voice quality. HNR and NHR are both sensitive to additive noise, or turbulence.

6.2.0.19 Modulation Index

In signal processing, the terms *amplitude modulation index*, *frequency modulation index* and *phase modulation index* apply to the case where we have a signal frequency modulated by another (carrier) frequency. The amplitude modulation index in this context is defined as the ratio of the amplitudes of the modulated frequency with that of the carrier frequency. Similar definitions apply to frequency and phase modulations.

The term **modulation index** or **modulation depth** referred to here are likewise a measure of the degree of variation of a variable around its unmodulated level. In the spectral domain, the variables can be frequency, amplitude and phase. Modulation index values range from 0 to 1, since they are ratios. They can also be measured as percentages, where they range from 0 to 100.

6.2.0.20 Energy and Intensity

Denoting the amplitudes of the fundamental frequency F0 and the harmonics H1, H2, ... as A0, A1, A2, ..., the **Normalized harmonic energy (NHE)** computed on F0 is defined as

$$\mathrm{NHE}_{F0} = \frac{\sum_{i=3}^{N}(A0_i - 2A0_{i-1} - A0_{i-2})^2}{\sum_{i=1}^{N} A0_i^2} \tag{6.61}$$

It is similarly computed for the harmonics as (dropping the numerical index for the harmonics for generality)

$$\mathrm{NHE}_{H} = \frac{\sum_{i=3}^{N}(A_i - 2A_{i-1} - A_{i-2})^2}{\sum_{i=1}^{N} A_i^2} \tag{6.62}$$

More specifically, A_i is the **intensity**, defined as the integral of the sample values (or the RMS value of samples) over a single pitch period T0. Intensity in itself can be used as a feature, and can be used as such in many forms:

If x_t denotes sample values of the signal (or instantaneous amplitudes), the **mean intensity** (or central tendency of intensity) is defined as

$$\bar{I} = \sqrt[N]{\prod_{t=1}^{N} x_t} = \exp\left(\frac{1}{N}\sum_{t=1}^{N} \log_t x_t\right) \tag{6.63}$$

The **root mean squared (RMS)** value of intensity is:

$$\text{RMS} = \sqrt{\frac{1}{N}\sum_{t=1}^{N} x_t^2} \tag{6.64}$$

The **intensity envelope** is its trace over time.

The **Dispersion of intensity** can be measured as:

1. **Standard deviation**:

$$\sqrt{\frac{1}{N-1}\sum_{i=t}^{N}(A_i - \bar{A})^2} \tag{6.65}$$

2. **Coefficient of variation**:

$$\frac{\sqrt{\frac{1}{N-1}\sum_{i=t}^{N}(A_i - \bar{A})^2}}{A} \tag{6.66}$$

A measure based on NHE is the **Normalized noise energy (NNE)**, which is defined as the normalized energy in the noisy regions of speech.

6.2.0.21 Degree of Subharmonic Components

A subharmonic is a frequency for which F0 is a harmonic, i.e., subharmonics are the set $\{F0/N | N \in \mathbb{Z}\}$. DSH (in percent) is calculated as:

$$\frac{100 \times \text{time of intervals with subharmonics}}{\text{total phonation time}} \tag{6.67}$$

6.3 Sub-components of Voice Quality

There are many sub-components of voice quality. One doctoral dissertation [28] reports 142 terms used in this context. Table 6.1 lists these, with some necessary additions and omissions.

The specific sub-components described in this section are largely chosen from Titze [29], together with some terms from the GRBAS [30] scale used in the clinical rating of voice disorders. Due to the ambiguity in their descriptions in the literature,

Table 6.1 Terms used in relation to voice type or quality, based on the set reported in [28]

Quality-descriptors				
1 Airy	2 Animated	3 Aphonic	4 Babyish	5 Back
6 Balanced	7 Beautiful	8 Belted	9 Biphonic	10 Biting
11 Bleating	12 Brassy	13 Breathy	14 Bright	15 Brilliant
16 Broad	17 Buzzy	18 Chesty	19 Clear	20 Cloudy
21 Coarse	22 Cold	23 Constricted	24 Covered	25 Crackly
26 Creaky	27 Dark	28 Denasal	29 Diplophonic	30 Dulcet
31 Dull	32 Easy	33 Efficient	34 Falsetto	35 Firm
36 Flageolet	37 Flat	38 Flutelike	39 Focussed	40 Forced
41 Front	42 Full	43 Glottal	44 Grating	45 Gravelly
46 Gruff	47 Guttural	48 Harsh	49 Head	50 Heavy
51 High	52 Hoarse	53 Honky	54 Hushed	55 Husky
56 Icy	57 Intense	58 Labored	59 Lilting	60 Limpid
61 Loud	62 Low	63 Luxuriant	64 Mellow	65 Melodious
66 Metallic	67 Middle	68 Mixed	69 Modal	70 Modulated
71 Monotonous	72 Mouth	73 Muddy	74 Muffled	75 Musical
76 Naked	77 Narrow	78 Nasal	79 Noisy	80 Normal
81 Open	82 Opera	83 Pale	84 Pleasant	85 Pressed
86 Pulsed	87 Raspy	88 Relaxed	89 Resonant	90 Retracted
91 Rich	92 Ringing	93 Rough	94 Rounded	95 Scooping
96 Shaky	97 Shrill	98 Slack	99 Sliding	100 Smooth
101 Soft	102 Steady	103 Stiff	104 Strained	105 Strangled
106 Stressed	107 Strident	108 Strohbass	109 Strong	110 Sugary
111 Sultry	112 Supported	113 Sweet	114 Tense	115 Thick
116 Thin	117 Throaty	118 Tight	119 Tinny	120 Tremerous
121 Twangy	122 Ugly	123 Unforced	124 Unpleasant	125 Ventricular
126 Vibrant	127 Warm	128 Weak	129 Whining	130 Whistly
131 Wobbly	132 Wooden	133 Wooley	134 Yawny	
Quality-related terms				
1 Fry	2 Jitter	3 Lift	4 Shimmer	5 Speech
6 Whiteness				

the descriptions in this section are subjective and may differ slightly from descriptions elsewhere.

1. **Breathiness**
   Breathiness in voice is caused by issues with the involuntary glottal opening and closure mechanism during phonation [31]. It is characterized by increased open-quotient, which is the fraction of the total duration during which the glottis

is *not* closed (also discussed in Sect. 5.5.2.2). An increased open quotient causes increased turbulent noise during phonation. This effect is heard as a *breathy* sound. Incomplete glottal closure can happen due to structural, pathological or physiological issues that affect the muscles in the larynx (e.g. *vocal process separation*, surface bulging [31]) etc. which cause leakage of air through the glottis while speaking. This can also happen due to muscular exhaustion (e.g. during vigorous exercise). Increased breathiness is also characteristic of the progression of Parkinson's disease [32], Dysphonia [11] etc.
The quality of breathiness manifests differently over the course of an utterance. It increases for unstressed syllables, in utterance-final syllables (so that they take on a breathy-laryngealized phonation quality), and at the margins of voiceless consonants. On average, female voices are more breathy than males [33]. Breathiness is also characterized by increased bandwidths of lower formants, and extra poles and zeros in the transfer function of the vocal tract [33]. GNE is a good estimate of the level of breathiness in voiced sounds [34]. Some studies e.g. [35, 36], have found CPP to decrease with increasing vocal process separation (an inverse relationship): thus a relatively lower (than normal) CPP indicates a relatively high degree of breathiness. Other measures that have been correlated positively to breathiness are spectral slope, HNR ratio, AVQI etc. HNR, especially, has been found to be strongly correlated to breathiness (and roughness) [37]. The spectral slope of the glottal flow waveform (H1–H2, after correction for the effect of F1) is correlated with glottal closure [20–22], and thus with breathiness. However, while some articles in the literature support these observations e.g. [38, 39], others contest any relation between spectral slope and breathiness [40].

2. **Aesthenicity**
Aesthenicity refers to a general weakness perceived in voice. Bio-mechanically, like breathiness, aesthenicity is also related to the vocal fold contact during phonation. In healthy people with good vocal functionality, the vocal fold contact is sharp and abrupt. There is more energy in the higher-order harmonics, and the voice has a **sharp** or **bright** quality. In contrast, when the open-quotient is very high, the vocal fold contact is very gradual or absent, the time derivative is so smooth that the spectrum only exhibits some weak harmonics, if any. Voice is then perceived as being aesthenic in quality. The condition is called Asthenia. Studies have positively correlated the spectral slope with the perception of aesthenicity.

3. **Aphonicity**
Aphonicity is perceived in part as a **whisper**. In deliberate whispering, all sounds are aphonic—there is no phonation present in speech, and the resonance patterns needed to discriminate between sounds are produced by modulating the sound of airflow through an open glottis. There are also, however, involuntary causes of aphonicity. In some speakers, the ability to set the vocal folds into sustained vibration needed for phonation is diminished or absent due to issues such as muscular and respiratory diseases, injury or disease of the larynx etc. In these

cases, aphonicity may not be complete and may affect some sounds while not others, depending on contextual articulations. In these cases, voice may take on a "whisper quality," although it may not be fully perceived as a whisper. From an articulatory-phonetic perspective, aphonicity is most clearly evident in voiced phonemes.
In speech with aphonic quality, there is more than usual absence of a measurable fundamental frequency in continuous speech, relative absence of harmonics and increased entropy and turbulence in the upper frequency regions of the spectrum. These can all be quantified using spectral measures. When present in very mild degrees, the quality of aphonicity can sound like the quality of breathiness.
Interestingly, injury or disease of the larynx severe enough to lead to complete aphonicity affects our sense of smell and taste [41].

4. **Crackle/Shimmer**
A *crackly* voice has high shimmer and sounds **buzzy**. A detailed description of shimmer from a signal processing perspective is given in the previous section. Crackle, a.k.a shimmer is related to the cycle-to-cycle variations in the amplitude of the glottal flow waveform (i.e. in F0). It has often been associated with vocal pathologies, but even in normal subjects, it is observed to be higher for voices with lower F0 (natural or enacted). It also increases when the intensity of the sound produced is deliberately decreased.

5. **Hoarseness**
Hoarseness is also called **raspiness**. It normally co-occurs with the sub-quality of **roughness** to different degrees. Hoarseness is caused by a combination of irregular patterns of breathiness and aperiodicities in the pattern of vocal fold vibrations. It has been observed to be associated with several health conditions that directly or indirectly affect the vocal folds, such as unilateral or bilateral vocal fold paralysis [42], infections of the upper respiratory tract, vocal trauma and strain, allergies, neurological disorders etc. In voices with a hoarse quality, irregularities in the vibration of the vocal folds occur in all voiced sounds, but may manifest to different degrees in different phonemes. This pattern is likely to be highly speaker dependent.
Hoarseness has been found to be inversely related to CPP, and directly to jitter and HNR [43]. HNR and jitter, in fact, are often used together to measure it's degree, with HNR playing a significantly greater role in its measurement [44, 45].
Hoarseness is one of the three main sub-components of voice quality that comprise the RBH (Roughness, Breathiness, Hoarseness) clinical scales that are used for the assessment of dysphonia and other medical conditions of the vocal tract, e.g. [43, 46].

6. **Roughness/Jitter**
Voice with prominent jitter is often perceived as **rough** in quality. Roughness is perceived as an overall or long-term quality of unevenness and unsteadiness in the vocalization of sounds in speech. It is caused by irregular vocal fold

movements, and asynchronies (phase differences) in the manner of vibration of the vocal folds. These irregularities may not be consistent over any segment or unit of speech, and can randomly appear in different sounds uttered by a speaker. Additional causes for roughness may be the presence of mucus, or sporadic AM-FM modulation. The causal connection between roughness and irregular vocal fold movements has been established through videokymography [47], an imaging method used to visualize the vibrations of the vocal folds. Roughness is sometimes also referred to as **bumpiness**. It often occurs in conjunction with hoarseness, and with diplophonicity.

7. **Biphonicity**
   Biphonicity, or *biphonation*, refers to the production of two different pitches at the same time. It is due to bi-periodic or toroidal vibrations of the vocal folds [48, 49]. This happens when the frequencies of the vocal folds become desynchronized during phonation, which in turn may be caused by many factors such as laryngeal asymmetry, high sub-glottal pressure, incomplete glottal closure, attempting to speak in a very high pitch, etc. One of the leading causes of biphonicity in humans is partial vocal fold paralysis.

8. **Diplophonicity**
   Diplophonia (or *diplothongia*) is a special case of biphonicity, where an F0/2 subharmonic (apparent as a second pitch one octave lower in frequency) appears in the spectrum of some portions of specific voiced sounds. This is also sometimes referred to as a *period-doubling* in the frequency domain. In this case, the glottal pulses have two or more cycles of a different shape and/or amplitude. It is caused by the simultaneous vibration of the vocal folds and ventricular bands false vocal folds [50]. In clinical practice, diplophonia is recognized as a "double voice," an irregular pattern of phonation with the simultaneous production of two pitches. In severe cases, it is a condition that is sometimes treated with voice therapy, e.g. [51]. Diplophonic irregularities can be observed in the timing of glottal periods, and frequently occur at the end of an utterance in normal speech [33].
   In most cases the causes for diplophonicity are largely the same as those for biphonicity, where two different pitches are produced through quasi-periodic vibrations of the vocal folds [52].
   Diplophonicity is also measured as a perturbation in F0, much like jitter and shimmer [53]—In some studies, a **Diplophonia ratio (Mean DR)**: has been measured from the pitch period as:

$$\mathrm{DR} = \frac{\frac{1}{N-3}\sum_{i=2}^{N-1}\left|\frac{TO_{i-1}+TO_{i+1}}{2} - TO_i\right|}{\frac{1}{N-5}\sum_{i=3}^{N-2}\left|\frac{TO_{i-2}+TO_{i+2}}{2} - TO_i\right|} \tag{6.68}$$

Diplophonicity has been associated with pathologies of the vocal folds, e.g. unilateral vocal fold paralysis with incomplete glottal closure, mass lesions on the vocal fold, intracordal cyst, granuloma etc. [54, 55].

9. **Coveredness**
Coveredness, also referred to as **darkened** in the literature, is perceived as a **muffled** voice. In the production of voice with this quality, the lips are rounded in a pronounced manner leading to a greater protrusion of the lips. Accompanied with a lowering of the larynx, the net effect on all sounds that are produced in this configuration is that the formants are lowered in frequency. The pitch, or the fundamental frequency is also relatively stronger. This occurs frequently in speakers with smaller mouth opening who attempt to "project" their voice to make it sound "stronger."
This sounds similar to, but is differentiated from the voice quality perceived as **throaty**, which is characterized by an increase of $F1$, a decrease of $F4$ in all sounds, and a decrease of $F2$ in front vowels. These are thought to be resultant from a narrowing of the pharynx.

10. **Creakiness**
Creakiness can be likened to the sounds produced by bending stiff leather, or moving a door with ungreased hinges. Vocal creak is caused by the resistance of the vocal folds to movements. This can happen when the arytenoid cartilages in the larynx are drawn together—an action which compresses the vocal folds and causes them to become lax or slack. During phonation, this causes them to vibrate irregularly at a frequency much lower than the normal F0—in fact about two octaves below F0. Because of this, the pattern of vibration of the vocal folds also incorporates temporally separated frequency modulations. This creates a highly periodic and complex pattern of subharmonics. Vocal creak is also referred to as *laryngealization*, *glottal rattle*, *glottal fry* or *pulse phonation*.
Creakiness is characterized by a relatively greater (and often more irregular) separation between the periodic peaks in the glottal pulse waveform, as compared to non-creaky voices. The energy between individual glottal pulses dies down enough that formant energy subsides instantaneously between them. On spectrograms, this is also evident as zero-energy gaps in the spectrograms, or cleanly separated pitch pulses. The gaps occur at a frequency of 70 Hz or less.
Vocal creak is not a result of vocal pathologies—they are usually natural to speakers or a result of vocal training by learning to speak (or sing) in a "creaky" style. This is possible because to some extent the compression of the arytenoid cartilages is under voluntary control of the speaker—i.e. a speaker can often render a creaky voice at will. However, sustaining the creak requires training. In fact creaky voices have lately become a trend among young women in the United States [56].
Figure 6.4 shows examples of vocal creak produced by two different individuals. Variants of vocal creak may be as individual as the person's voice. Vocal creaks can be of different kinds, and are characterized by different patterns in the time and frequency domains [57].

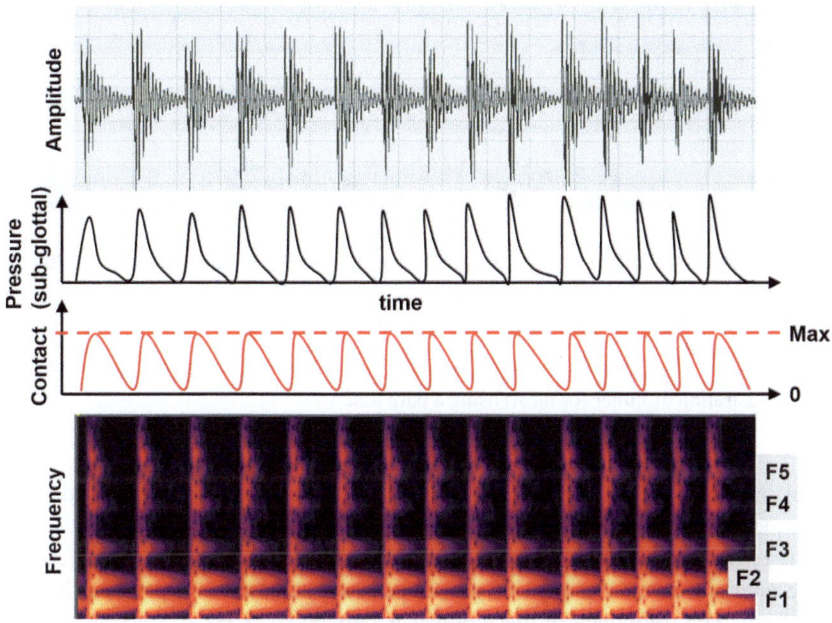

(a) **Top:** The speech waveform in time domain exhibiting vocal fry – this example represents one of many possible variants of vocal creak. **Middle:** Subglottal pressure vs. contact between vocal folds (0 $\Longrightarrow$ glottis completely open; max $\Longrightarrow$ glottis completely closed), **Bottom:** Characteristic appearance and disappearance of energy in all formants in vocal fry

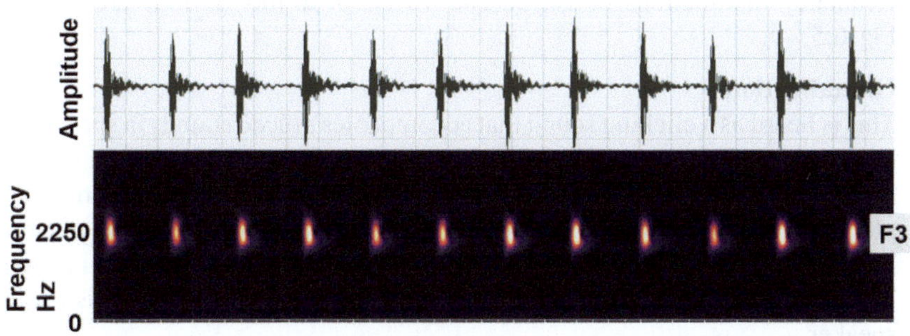

(b) Waveform and formant energy patterns of vocal fry produced on demand by a highly trained professional singer 12 of the 60 or so pitch cycles per second are shown.

Fig. 6.4 Spectral patterns corresponding to vocal creak

11. **Pulsed**
    Vocal pulsing is a term that is synonymous to the quality perceived in the singing techniques of vibrato or *vocal tremolo*. This is performed by singers by rapidly

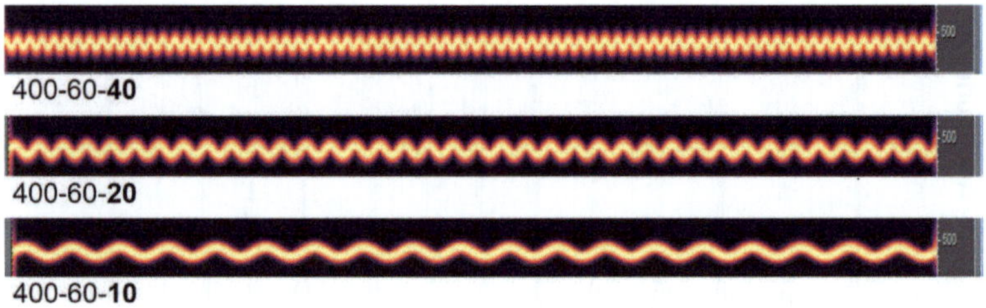

Modulated tones: 400 Hz tone, with 60 Hz amplitude of modulation and 40, 20 and 10 Hz frequency of modulation (top to bottom). Bleat is the equivalent of the 10 Hz case. The vertical axis is frequency, and the horizontal axis is time in each case.

Fig. 6.5 Illustration of flutter (or bleat) using a pure tone

pulsing the air expelled from the lungs while maintaining the same fundamental frequency.
The term "vocal pulses" is applied to the pulsed quality that is perceived when the energy between groups of glottal pulses fluctuates in a periodic high-low fashion. The frequency of these pulses is between 4–8 Hz.

12. **Flutter**
This quality is also often referred to as **bleat**, since it is often likened to the bleat of lambs. Flutter is characterized by frequency and/or amplitude modulations in the 8–12 Hz range, of the frequencies in the vocalized sounds. As an example, frequency modulations of 10, 20 and 40 Hz of a 400 Hz tone are shown in Fig. 6.5.

13. **Glottalization**
This is heard as a clicking sound that accompanies voiced sounds in speech. This happens during continuous speech when there is a momentarily greater degree of rigidness of the vocal folds. This results in an unusually tight obstruction in the glottis during the enunciation of the sound, in turn followed by an instantaneous but unusually forceful movement of the vocal folds. This is the clicking sound heard in glottalization. Glottalization patterns can be highly characteristic of the speaker.

14. **Nasality**
The nasal chamber of every speaker has a unique size and shape that depends on the speaker's facial bones and cartilages, which in turn are determined by genetic and age-related factors. Generally, during normal speech, the nasal passages are opened to a small degree for non-nasal sounds, and to a greater degree for nasal sounds, which are characterized by their pronounced nasal resonances. In some speakers, the nasal passages are open to a greater degree during the enunciation of even non-nasal sounds. This gives a nasal quality to the sounds throughout the speaker's utterances. Excessive nasality is often also referred to **honkiness**.

Sounds in nasal speech (also referred to as speech with high *nasalence*), exhibit a greater degree of anti-resonances than normal, non-nasal speech. This is easily explained through models of speech production, as in [58]. Later chapters in this book describe methods based on artificial intelligence that may be applied to measure it.

15. **Pressed**
Speech is perceived as **pressed** when it takes on a throttled quality that often also sounds **harsh** or **rough**. This happens due to an unusually greater degree of constriction of the glottis, that affects the volume of airflow through the vocal folds—it is lowered as in near-whispering. During the production of pressed speech, the arytenoid cartilages also cause a medial compression of the vocal folds. The resultant speech sounds strained and laborious, and has a forced-whisper quality to it. In studies of the glottal flow waveform, it has been observed that a voice with dominant H2, H4 and weak H1 exhibits a *pressed* voice quality [21]. This can be evident even in the spectrum of the voice signal, since the approximate glottal opening and closure patterns of the speaker's glottis also have a somewhat linear relationship to the relative energy in the harmonics of the spectrum of the voice signal. The difference in formant positions F1–F3 has also been found to be correlated with airflow declination through the glottis [20], and is thus can be indicative of pressed quality.

16. **Resonant**
This quality is often also referred to as **ringing** or **vibrancy**. Resonant or ringing/vibrant speech is caused when the resonances above the larynx—in the supralaryngeal region of the vocal tract—are enhanced. This manifests as higher-energy in the upper formants, easily seen on the spectrogram. Especially, the formants F3–F5 are seen to group together due to their increased bandwidths. Spectral peaks are strongest in the range 2500–3500 Hz. Resonant voices are known to carry well and are often associated with a strong "presence." Resonant voices can be felt by the speaker as higher vibratory sensations on the face/cheeks.
From a voice production perspective, resonant voices are thought to be due to lowered phonation threshold pressure. The vibratory sensations felt in such voices are a consequence of efficient conversion of aerodynamic energy to acoustic energy [59], and not only due to increased energy in formants. Lower volume of the oral cavity is correlated with reduced resonance.
Ringing can be measured as the ratio of formant bandwidths and energy of the higher formants to that of the lower formants. Higher values of this entity imply a higher resonant or ringing quality of speech.

17. **Strained**
Speech takes on a strained quality (often also referred to as "hyperfunctional use of voice") when excess energy is concentrated in the laryngeal region. This is observed to be a common response to pain and distress in humans. Sounds such

as groans made in pain are usually strained in quality. Usually vowels sound more strained than consonants.

18. **Strohbassness**
    Strohbassness is vocal fry that is often described as **popcorning**, **glottal rattle**, or **glottal scrape**, and is often indistinguishable from it. It is characterized by the same high-low energy patterns as in vocal fry, but may extend to frequencies well below (sometimes several octaves below) $F0$. It is more often encountered in singing voices.
19. **Tremor**
    Tremor, also described as **shaky** voice, is a quality that is perceived in speech when the fundamental frequency fluctuates by up to 15 Hz. This may also be accompanied by unusual patterns of amplitude perturbations. Note that different degrees of frequency and amplitude modulations are characteristic of even normal speech and singing, but are more controlled and regular than in a shaky voice. Tremor is generally a consequence of neurological problems, muscular problems, psychological stress such as lying, apprehension, fear, anger etc. It is often a natural consequence of aging.
20. **Twanginess**
    A *twangy* quality is perceived when there is a combination of a high degree of nasality and high degree of epilaryngeal resonance in the voice. This can be resultant from a narrowing of the pharynx [60], which is also characterized by a rise in F1. The overall shape of the vocal tract (and especially its effect on F1) plays a significant role in imparting this quality to voice [61]. The distance between F1 and F2 is increased (especially in vowels) in twangy voices.
21. **Ventricular**
    This is a quality that accompanies *Dysphonia Plicae Ventricularis* [62], which is a medical condition where phonation is produced by the vibration of ventricular bands of tissue in the vocal tract, rather than the vibration of the vocal folds. Ventricularity generally accompanies dysphonia as a compensatory mechanism. Dysphonia (where the vocal folds have trouble vibrating) in turn can result from speech problems, laryngeal surgery, damaged vocal folds and many other conditions. Perceptually, ventricular speech sounds harsh and abnormal in a non-specific manner.
22. **Wobble**
    Wobble refers to amplitude or frequency modulations in the 1–3 Hz range, often occurring irregularly. It can manifest in any frequency band that is relevant to voice.
23. **Yawniness**
    The overall shape of the vocal tract plays a significant role in imparting the quality of yawniness to voice. Yawny speech is produced in the configuration of the vocal tract that widens the oral cavity and increases the tract length [61, 63]. A voice that has a yawny quality to it sounds as if it is produced when the

speaker is yawning. The act of yawning causes the pharynx to be widened and also lowers the larynx. This brings F1 and F2 closer together.

24. **Instability/Unstableness**
This term is generic and rarely used in the literature, but refers to the fluctuation of overall voice quality over time. Quite cyclically, it qualifies to be a sub-component of voice quality itself.

Voice quality features are widely studied and correlated to medical conditions in multiple medical fields. From a voice-production point of view, it makes complete sense that medical conditions should cause changes in "normal" voice quality and manifest as one or many of its sub-components. Though many are not directly quantifiable, there are several subjective measurement scales that are used in diagnostic settings. One widely used example is that of the GRBAS scale, which grades voice quality on subjective ratings of roughness, breathiness, asthenia and strain. While variants are possible, each entity on this scale is usually given a rating of 0–3 (0: normal; 1: mild; 2: moderate; 3: high) [64]. Examples of other scales that use different voice quality sub-components are in [65, 66].

6.4 Manipulations of Voice Quality: Voice Disguise

Perhaps the greatest display of the importance of voice quality in human perception comes from the performing arts that involve qualitative manipulations of voice, i.e. changes that are brought about in the voice signal that cause it to be perceived differently from its true nature. These manipulations take different forms, such as voice disguise, voice illusions, ventriloquism, and even operatic singing.

As discussed in Chap. 1, in voice disguise, the perception of voice *source* is altered.

In **ventriloquism**, both the perception of voice source *and its location* is altered. Part of this is achieved through the McGurk effect (explained in Sect. 1.4.3.2), wherein the perception of the identity of a spoken sound is shown to be dependent on the *visual cues* provided to the listener. Ventriloquism is the art of producing intelligible speech *without lip or jaw movement*, essentially removing visual cues that point to the source and identity of sound. Phonemes such as labials, (/p/, /b/, /m/ etc.), some fricatives (/f/, /v/ etc.) that require the closure, opening or movement of lips and/or the jaw are "substituted" with sounds that have equivalent resonant characteristics [67] (but may not be phonemes on closer examination). This is achieved by creating alternate "equivalent" configurations within the vocal tract by adroit tongue and vocal muscle movements, altering the stiffness and constriction in different areas of the pharynx, and adjusting the laryngeal positioning during speech production. The removal of visual cues, accompanied by heightened pitch and an interplay of sounds based on psychoacoustic principles of spatial auditory perception, results in creating altered perceptions of the identity and location of the actual speaker.

In **voice illusions**, the perception of the source is the same, while interpretation of the *content* of the voice signal is altered. In the performance of various dramatic arts and in singing, voice illusions are used to achieve multiple goals such as personality projection and the creation of acoustic symphonies that have the desired effect of invoking (usually) pleasant sensations, or impressions of illusory orchestra such as in acapella singing.

In this section, we will focus on **voice disguise**, especially on impersonation, as an exemplary case. Voice disguise can be done in two ways—by altering the voice signal using external devices (such as mechanical or electronic voice transformers, or even muffling), or by altering voice characteristics to create a desired impression on listeners. A special case of the latter is impersonation (or mimicry), wherein voice characteristics are altered in a way that may cause the voice of one person to be perceived as that of another (real or fictitious person). We focus on the alteration of voice characteristics through impersonation.

6.4.1 Mechanisms of Control

We are able to control and alter many (but not all) aspects of voice production. This is largely achieved through muscular control exerted on the laryngeal musculature, and the muscles that move the articulators. Muscular control is used to change the configuration and dimensions of the vocal tract, and to cause more subtle changes, e.g. in the positioning of the larynx in the vocal tract while speaking. In this section, however, we discuss how some key aspects of voice are controlled at the *bio-mechanical* level. Many, but not all of these can be altered by a speaker on a voluntary basis.

6.4.1.1 The Fundamental Frequency: How Is It Controlled?

Pitch, or the fundamental frequency F0, corresponds to the frequency of the excitation signal produced by the vocal folds. As discussed in Chap. 2, F0 is correlated to the frequency of air pulses created by the opening and closing of the glottis—it is the frequency of the vocal fold eigenmodes that are synchronized and excited. This depends on many factors.

For vibrating strings, their length, tension and mass determines the fundamental frequency of vibration, which is directly proportional to tension (or stress—this refers to the mechanical state of vibrating entity), and inversely proportional to mass and length (this is discussed further in Chap. 2). The same holds for vocal folds, except that in vocal fold vibration, additional factors play significant roles. Their eigenfrequencies depend on their geometry (length, breadth, thickness etc.), and stress (which in turn includes stiffness and tension). Stiffness refers to the elastic restoring force in an entity, in response to deformation. All of these can vary between individuals, and can vary *for* the same individual, changing (voluntarily or involuntarily) under different conditions.

Across individuals, some of the variations are natural. For example, larynxes of adult men are larger than that of adult women, which are larger then in children. The lengths of the vocal cords vary proportionally. Short vocal folds correspond to higher eigenfrequencies and thus high F0 (as in children). More stress (extending to muscular stress in the larynx) also corresponds to higher F0.

For a given individual, stiffness contributes more to F0 than tension when the vocal folds are not fully stretched, and they both contribute equally when they are [68–70]. The stretching of vocal folds lowers F0, but the increase in stress counters that and often there is a net *increase* in F0 as a result.

As discussed in Chap. 2, *vocal registers* are characterized by their sustained F0 patterns. The process of registration (change of registers) is facilitated by the changes in vocal fold stiffness, closure properties and geometry [71]. F0 changes with changes in the thickness of the medial surface of the vocal folds, which is controlled by the CT and TA muscles, which also control stiffness and closure of the vocal folds. The IA/TA/LCA muscles are innervated by the same nerve, and get activated together [72]. TA muscle activation has a complex effect on F0. Depending on the level of involvement of the lamina propria in inducing vibrations and other vocal fold conditions, activation of TA muscles can either increase or decrease the eigenfrequencies [73], and F0 may increase or decrease. Joint activation of the IA/LCA muscles increases sub-glottal pressure and can increase F0. Increasing subglottal pressure alone can increase the F0 by as large as 20 Hz/kPa [74].

F0 also increases with the *collision force* during vocal fold contact. When this is high, the effective stiffness of the vocal folds increases. Because the closure and contact of vocal folds depends on many other factors, such as sub-glottal pressure, musculature of the vocal folds, air vortices etc., a change in any of these factors, and the balance between them, can result in different levels of changes in F0 [75].

6.4.1.2 The Vocal Intensity: How Is It Controlled?

Intensity (or loudness) of the voice is dependent on the sub-glottal pressure on one hand, and the overall resonant/lossy properties of the vocal tract, coupled with lip radiation, on the other hand.

Sub-glottal pressure is dependent on the alveolar pressure in the lungs, which in turn is dependent on the respiratory muscles and the lung volume. Increase in sub-glottal pressure increases the amplitude of vibration of the vocal folds, and also the negative peak of the time derivative of the glottal airflow. If the glottal closure is not complete (e.g. when a person is exhausted), the sub-glottal pressure drops, the glottal airflow is greater in volume and rate and there is a relatively faster decline in the volume of air in the lungs while speaking. The person breathes more frequently to maintain the sub-glottal pressure required for phonation, and appears to be panting. However, in spite of this, the vocal tract is the main *controller* of vocal intensity. What happens in the larynx is magnified by the vocal tract. This is not so much due to its shape than to the fact that it facilitates impedance matching between the glottis the external environment (via the lip opening), which increases the efficiency of sound radiation from the mouth [76, 77]. If the magnification of sound in the vocal tract is such that the harmonics close to a formant are amplified more than the first harmonic, the sound produced is most clearly distinguishable and loudly heard. This effect can be controlled by laryngeal (or vocal cord) and vocal tract shape adjustments [26, 78, 79].

Laryngeal adjustments: A low intensity breathy voice is produced by keeping the glottis relatively open. Higher intensities require increasing the subglottal pressure, which increases the amplitude of vocal fold vibration, or increasing both the surface-area and duration of the vocal fold contact [80]. The latter increases the energy in the harmonics that lie close to the first formant. The use of both these mechanisms makes it possible for humans to achieve desired intensities while controlling the pitch and articulation of sounds [81].

Vocal tract shape adjustments: The first formant can be adjusted by opening the mouth wider, and when the adjustment is such that the frequency of the first formant approaches or matches F0, this maximizes vocal output. This is often done by operatic singers [82]. Intensity can also be increased by the combination of a wide supra-glottal region with lower glottal resistance, or a narrow supra-glottal region with higher glottal resistance [83].

6.4.1.3 Voice Quality: How Is It Controlled?

Changes in voice quality can happen as a result of air-pressure changes in the intra-glottal region, changes in tissue composition of the vocal tract, muscular movements, changes in ligament elasticity and strength, changes in the vocal fold layer composition, especially of the lamina propria, and a slew of other factors. Voluntarily, however, we can only control some of these. For example, some aspects of the stiffness of the vocal folds can be controlled by lengthening or shortening them using the laryngeal muscles. However, the inherent strength and agility of those muscles is not under our voluntary control. Weak IA, TA and LCA muscles, for example, cause incomplete closure of the glottal opening, which causes noise of air turbulence during phonation [75, 84], imparting a breathy quality to the voice. Finer differences within this quality then come from other factors that influence glottal closure—e.g. factors that relate to the cartilaginous portion of the glottis, or to its membranous portion. In the former case, the breathy voice has strong high-frequency harmonics. In the latter case the higher frequency harmonics are feeble. When the vocal folds become too separated due to incomplete closure, each fold may vibrate at a different F0, leading to the sub-quality quality of biphonicity, which may in turn be perceived as beats, as discussed earlier in this chapter.

There are further nuances to this. For example, when the duration of glottal closure is longer, the voice sounds pressed, and the second harmonic is stronger than the first. The duration of glottal closure is largely determined by thickness of the vocal fold's medial surface [85]. Longer closure times are due to the action of TA and LCA muscles on the arytenoid cartilages. To achieve a pressed quality, high TA muscle activation is also thought to be necessary. With aging and other factors including diseases of the muscular system, the ability to achieve a pressed quality decreases in humans, since complete glottal closure itself becomes hard to achieve. Overtly strong activations of these muscles can occur in some diseases, such as in spasmodic dysphonia, and this may also render a rough quality to voice [84]. Physiologically, for constant values of vocal fold stiffness and area of glottal opening in a resting position, increasing sub-glottal pressure results in voice that has a rough quality to it [86]. This is again due to the airflow instabilities that occur in the supra-glottal region, and can result in erratic vibrations of the vocal folds [87].

Voice quality changes are often coupled with changes in F0. Its perception plays an extremely important role in vocal quality. As discussed above, F0 is dependent on many factors, including the thickness of the medial surface of the vocal folds and their stiffness. If the activation of the IA/TA/LCA muscles leads to greater medial surface thickness of the vocal folds, and consequently greater stiffness, F0 is likely to increase. The voice has a "chesty" quality to it. This configuration results in longer glottal closures, accompanied by complex and large phase differences in the vibration of vocal folds. Longer glottal closures also result in smaller flow rates, longer pressure build-up and strong harmonics in the spectrum. If the activations of the muscles are not uniform (as may be expected when a person's muscular system is affected by adverse factors), voice is further affected. For example, if TA activation is very strong, while the CT activations are weak, and the sub-glottal pressure is very low, the glottis can remain closed for longer than normal durations in the glottal cycles, leading to vocal fry, which is easily perceived by humans.

As we recall, falsetto is a register, but sometimes it is referred to as a quality. This quality (aslo called the *loft* register or *head voice*) is achieved through a voluntary attempt to increase F0 by using the CT and vocalis muscles to increase the stiffness of the vocal folds, by elongating them and making them thin along the edges. In this register, the glottal closure is incomplete, producing a smooth, nearly sinusoidal sound wave of high frequency (high F0), with very few harmonics, as is the case with a synthetically produced pure tone.

The discussion above involves only a couple of small examples. The mechanisms that vary all aspects of voice quality require a complete book (and a very thick one at that) to discuss.

6.4.1.4 Impersonation: What is Manipulated?

The first thing to understand about impersonation is that the voice rendered by the imitator (the *source*) is *not an exact reproduction* of the "target" speaker's voice. For two voices to be *exactly* the same, all characteristics, compared across all corresponding phonemes between the two voices, should be the same. The transitions between phonemes, the rate at which they occur, the intervocalic breath durations, the energy distribution across them, the manner in which pitch and energy change across them, the quality—in fact all such higher level characteristics of the signal must also be identical. This is close to impossible. As we have discussed earlier, myriad and diverse factors cause two people's voices to differ—differences in learning, exposure to language, cognitive capabilities, use of articulators, dimensions of articulators, muscle agility, tissue elasticity, moisture levels, length and thickness of the focal folds, lung capacity, tracheal diameter etc represent a small collection of these. All such factors cumulatively cause voice to be unique to each speaker, and make exact mimicry almost impossible.

For two voices to be *perceived* as identical, however, the requirements are different. Since human auditory perception is limited in many ways, only those characteristics that *can* be differentiated audibly by the human ear need be used for voice disguise. Of these, only the characteristics that are under the speaker's voluntary control can, in reality, be varied during voice disguise by impersonation.

Studies have found that impersonators frequently mimic or caricature speech style, dialect, prosody, speaking rate, pitch, pitch register, vowel formant frequencies, loudness, breathing patterns etc.—many of which are entities that also strongly reflect in voice quality [88, 89]. Pitch is often used to create the impression of gender and age during voice disguise. It can be easily changed by simple mechanisms, such as exerting control over the length of the vocal folds and the size of the glottal opening, which (as discussed above) are significant factors in producing F0 [90, 91]. Interestingly, even within the set of properties that can be voluntarily manipulated by mimics, they fail to replicate the finer details of these. For example, mimics successfully imitate fundamental frequency patterns and the global speech rate of the target, but often fail to replicate segmental timing patterns [92]. Mimics are also reportedly able to imitate key formant patterns [93, 94] for several phonemes; however successful imitation of formant *bandwidths*, which are shaped by the vocal tract's tissue composition, has not been reported.

The distinction between the success in imitating high-level patterns, and the relative intransigence of fine-level detail in being replicated may be attributable to three reasons.

1. **Intent**: From the qualitative standpoint of a human listener, a successful impersonation is that which creates an impression of the target that is sufficient for the listener to believe that the voice is indeed that of the target speaker. Once this objective is achieved, any additional effort in imitating fine detail is unnecessary. Thus, most successful impressions are auditory illusions generated by the impersonator, and are created by taking advantage of the limitations and shortcomings of human hearing and cognition [88, 95].
2. **Biological**: As discussed above, it is physically impossible for the mimic, with an entirely different vocal apparatus, to replicate every detail of the speech produced by another speaker. This makes the creation of aural illusions using voluntarily changeable factors all the more necessary for impersonation.
3. **Cognitive**: Cognitively, impersonation is not a natural act and, regardless of training and skill, will not duplicate natural speech. This is supported by studies such as [96], which is based on brain images and shows that mimicry is a deliberate activity invoking regions of the brain that are not activated similarly during natural speech. Thus the brain may not be able to orchestrate all the finer details of the target speech.

At the bio-mechanical level, as discussed above, voluntary changes in voice quality (for impersonation) can be made in many ways. Examples include controlling the glottal opening and airflow, inducing qualities of breathiness, asthenia, nasalization (by controlling the nasal passage opening during speech production) etc. However, when such changes are forced, they can cause the vocal folds to become asynchronous or falter in multiple ways, inducing qualities of roughness, jitter etc.

which may or may not be present in the target voice. Other voluntary changes to voice quality can be made by changing the pitch and formant positions voluntarily or by shifting the entire natural spectral range. When this is done by changing the position of the larynx in the vocal tract, it results in *raised-larynx* and *lowered-larynx* voices, which are characterized by full-spectrum raising and lowering of frequency patterns respectively. Raised larynx voice qualities include pronounced voiceless fricatives, voiced approximants, voiced and voiceless trills and voiceless plosives [97].

Other mechanisms involve *pharyngealization* [98], *faucalization* etc. Pharyngealization is done by narrowing the upper pharynx—by retracting the tongue and constricting the muscles of the pharynx walls in the region of or close to the epiglottis [99, 100]. This causes an increase in F1 and a lowering of F2. Faucalization is the narrowing of the upper regions of the pharynx, and is done by an approximation of the palatoglossal and palatopharyngeal muscles (faucal pillars). This again results in an increase in F1 and a lowering of F2. Together, pharyngealization and faucalization allow the speaker to exert a very fine level of voluntary control over their fundamental frequency. These voluntary changes in pitch are different from those that happen due to molecular, tissue-level and other changes—those may not result in linear shifts in the spectrum. The formants stay positioned, while the energy distribution in the spectrum and over the harmonics may change.

The manipulation of formants during impersonation is voluntary to some extent, governed by the relationship of formant positions to the vocal tract configuration. For instance, a constriction in the velum will result in a short section between the vocal folds and the velum, and a longer section from the velum to the lips, giving rise to one higher frequency formant from the short section, and one lower formant from the longer one. The *exact* frequency of the two formants would depend on the actual length of the section from glottis to velum and velum to lips. While this would depend on the structure of the speaker's face, neck etc., there is some pliability possible in positioning the constriction, augmented by lowering the larynx, extending the lips etc.

In general, formant manipulations for changing voice quality are extremely effective in portraying different people. This is because the actual position and width of spectral peaks of formants, the rate with which they change, etc. depend on the details of the vocal production mechanism of a speaker, and thus carry very specific information about the speaker. This ability of formant features to capture speaker-specific details, and their usefulness in capturing the differences between speakers, is well known. An example is given by the time-course of changes in vowel formant frequencies in the transition between phonemes. These are referred to as formant transitions. An example illustrating their speaker specificity is given in Sect. 7.2.

6.4.2 *The Range of Alterations of Voice*

6.4.2.1 The Range of Voluntary Variations in Voice

In this section we extend the discussion on voluntary variations in voice to include those that potentially address the question of gauging the range of human voice variations. The range of voluntary variations in voice while *singing* have been extensively studied in the context of the dramatic arts and operatic singing. In the performing arts, voluntary variations comprise learned vocal styles, which are categorized as different types of *vocal techniques*. These are extensively described in the literature on vocal pedagogy. These include techniques for voice projection, vocal resonation, articulation, registration, sostenuto, legato, vibrato, coloratura, vocal range extension, control of tone quality etc. Other examples include ingressive speech (speech produced while inhaling), falsetto (singing in a high vocal register), glottalization (producing glottal fry while speaking), yodeling, ululation, vocal trill, Inuit throat singing, rekuhkara (a style of singing where the glottis is not used as a sound source, rather other sound sources are projected into and 'shaped' by the singer's vocal tract), buccal speech, overtone and undertone (period doubling or tripling) production, multiphonic sound production (by straining or asymmetrically contracting the laryngeal muscles to produce multiple fundamental frequencies), screaming etc.

From the perspective of the vocal fold movements during phonation, the range of voluntary variations involved in voice manipulations through training is probably best categorized by the different vocal registers, from vocal fry to sharp whistling sounds that are reached at the extreme of screaming. From an articulatory-phonetic perspective, the range of voluntary variations is best categorized by the ensemble of sounds that are produced in the world's languages. From an acoustic perspective, the range can be based on energy or intensity of the sounds produced, degrees of amplitude and frequency modulation etc. Such categorizations would include whispers, breathing, turbulence, sharp reports such as clicks and smacks, frication etc. From a geometrical perspective, the range of voluntary variations would include those produced by changing the shape and size of the different parts and chambers of the vocal tract, variations induced by shifting the position of the entire larynx, variations that result from moving abdominal muscles and diaphragm to change the respiratory dynamics during phonation etc. It is easy to see that when all of these are considered together, the space of voluntary variations of human voice (which can be achieved by voice training) is combinatorial.

Finally, we also note that both ventriloquism and the creation of voice illusions are forms of vocal art that involve voice training. In both of these, alterations of voice and voice quality are done using the same underlying mechanisms that are mentioned above in the context of voice disguise. Only their *effects* and goals are different.

6.4.2.2 The Range of Involuntary Variations in Voice

Involuntary variations in voice span the space of finer-level variations induced dynamically by myriad (often transient) profile parameter changes, such as disease, intoxication, exhaustion, emotions etc. These transient factors can influence different parts of the vocal tract, changing, enhancing or hampering their function. Involuntary changes in voice quality occur with changes in the responses of the nervous system, respiratory system, endocrine system etc., and significantly so with changes in the condition and responsiveness of the brain. Currently, we do not have any studies that establish the limits and ranges (or even causes, sometimes) of such involuntary variations.

Some involuntary variations are induced in specific contexts, such as voice breaks that occur during singing or voice disguise. Voice breaks are often observed during vocal manipulations that shift the pitch (especially as it is raised), or lower the F1 (to project gravity, dominance or leadership qualities). As the frequency of F0 approaches that of F1, the vibration of the vocal folds is destabilized. It is maximally destabilized when F0 crosses the frequency of F1 [101]. When this happens, the synchronization between coupled eigenmodes breaks down. This *temporal desynchronization* leads to chaotic vocal fold vibration [102–104] and reflects in the perceived voice quality.

6.4.3 On Voice Quality Synthesis

Do synthetic voices have the "natural" range of voice qualities? Is the synthetic production of voice qualities possible? Currently, many physical models of the vocal tract exist. The best of these are parametric models [25, 27] in the time domain, which are capable of producing voice source waveforms of varying F0, amplitude, open quotient, and degree of abruptness of the glottal flow shutoff, and thus are capable of synthesizing different voice qualities. Mechanical models of the vocal folds allow us to synthesize excitation signals. Varying their parameters can produce different patterns of glottal airflow, and different qualities of excitation signals. However, speech synthesized through such mechanical models does not sound natural, due to the relative simplicity of these models.

Speech synthesized using neural networks (or AI systems) sounds much more natural. For the synthesis of voice *qualities*, AI systems have to trained in a data-driven fashion for the most part. They can be supplemented by parametric systems for the synthesis of excitation signals, or for the variation of respiratory and other vocal control patterns that occur while speaking in specific contexts and emotional states. They can also be guided by other external knowledge about voice quality as it relates to the synthesized speech signal, resulting in speech that can invoke specific perceptions of voice quality. All of this is possible today.

However, the micro-variations in human voice that result from fine variations in the underlying bio-mechanical processes that produce it, are hard to emulate. On closer examination, speech signals that are synthesized using current AI systems have much more regularity at finer levels than speech produced by humans. For the same speaker within the same settings, the latter is also characterized by a high degree of inconsistency across repeated sounds across time.

Nevertheless, synthesized voices today can evoke the perception of different voice qualities, if the generative AI systems are trained well. At the time of writing this book, however, there are many aspects of speech production that are not fully understood (and therefore cannot be incorporated into generative models). These include many aspects of the physiology and bio-mechanics the vocal tract, especially of the vocal folds, under both normal and pathological conditions. Their variation over the human populace remains to be charted and understood. Without a good understanding, or a wide representation (in the case of neural networks) of these, the nuances of human-like speech at finer levels might be impossible to produce with high fidelity, due to the limitations of parametric models, and the overt generalizations that neural network and other data-driven models intrinsically make while training from large amounts of data.

6.4.3.1 Voice Manipulations and Speaker-Verification Systems

Lately, voice (or voice property) cloning efforts have led to their adversarial use—to fool automatic voice processing systems that perform speech recognition, speaker matching, authentication etc. We discuss some aspects of this below. We will however not discuss voice *cloning* in the context of human profiling, since entirely synthetically generated voices can be easily identified by computational means. They lack the finer level nuances of true human speech, as discussed earlier in this section.

Can a human break into a voice authentication (or voice biometric) system by performing expert mimicry? The answer is no. Although expert mimics may deceive human listeners, they cannot similarly deceive state-of-art speaker-verification systems, even though they (currently) consider the aggregate distribution of the features in the speaker's speech signal. It has been shown that even a simple GMM-based verification system cannot be deceived by expert voice mimics [105]. The detection-error tradeoff curves of a state-of-art speaker-verification system stay essentially unchanged when the impostor is an expert voice mimic [106]. As a matter of fact, even *human* listeners who may be deceived when they listen to impersonations by the mimic in isolation will not be deceived when they can also hear the actual speech by the target for comparison [107]. Hence, the risks posed to voice-based authentication systems by voice mimicry are not great.

However, it must be noted that *algorithmic* deception of speaker verification systems is very much possible. This is embodied by all of the current research on *adversarial systems* that can breach or alter the performance of speech processing systems. The fact that speaker matching systems (e.g. those that perform speaker verification or speaker identification) use aggregate distributions of features, or may use algorithms that have other shortcomings, can be exploited. For example, if a voice authentication system uses features that are based on statistical averages, then any sound signal that has the same statistical averages may be able to break into the system, *and it need not even contain voice*. This is the general principle that adversarial efforts employ. Fortunately, while machines can exploit these, humans cannot. This is because mimicry is based on human perception, and not dynamic computing that matches the derivatives from voice, as is done by machines.

As expected from the discussion above, the most successful attempts at breaking into voice-authentication systems have been with synthetic voices, obtained either by direct synthesis or by modifying an actual voice [108]. It may be very interesting to note that, paradoxically, these voices would not fool a human ear.

6.4.3.2 Voice Manipulations and Intelligibility

Voice quality does not directly relate to the intelligibility of its content in all cases—some voice qualities may mask intelligibility, while others may enhance it or not affect it at all. Intelligibility can be measured to some extent by its (subjectively observed) associations to signal properties. For example, it is thought that increased energy in the 0–1000 Hz range of long-term average LTAS and greater depth of the low-frequency modulations of the intensity envelope contribute strongly to intelligibility. Specifically, for clarity of speech, a modulation depth of frequencies less than 3 Hz in the intensity envelopes of the 250, 500, 1000, and 2000 Hz octave bands (an octave is merely a doubling of base frequency) is especially important [109]. Measures such as PESQ (Perceptual Evaluation of Speech Quality) [110–112] have been defined based on similar observations, but are not discussed here since they are not very relevant to profiling.

To summarize, the human voice, in all its fine details and nuances, cannot currently be imitated at all levels by completely artificial means. The algorithms that profile human voices can profile artificial voices too. If set up correctly, profiling algorithms should return inconclusive or no results from such signals. The subject of automatic detection of voice disguise is discussed in the second part of this book, which deals with computational profiling.

References

1. Kreiman, J., Shue, Y. L., Chen, G., Iseli, M., Gerratt, B. R., Neubauer, J., et al. (2012). Variability in the relationships among voice quality, harmonic amplitudes, open quotient, and glottal area waveform shape in sustained phonation. *The Journal of the Acoustical Society of America*, *132*(4), 2625–2632.
2. Kreiman, J., & Gerratt, B. R. (2012). Perceptual interaction of the harmonic source and noise in voice. *The Journal of the Acoustical Society of America*, *131*(1), 492–500.
3. Gonzalez, S., & Brookes, M. (2014). PEFAC-a pitch estimation algorithm robust to high levels of noise. *IEEE/ACM Transactions on Audio, Speech and Language Processing (TASLP)*, *22*(2), 518–530.
4. Vincent, E., Bertin, N., & Badeau, R. (2010). Adaptive harmonic spectral decomposition for multiple pitch estimation. *IEEE Transactions on Audio, Speech, and Language Processing*, *18*(3), 528–537.
5. Duan, Z., Pardo, B., & Zhang, C. (2010). Multiple fundamental frequency estimation by modeling spectral peaks and non-peak regions. *IEEE Transactions on Audio, Speech, and Language Processing*, *18*(8), 2121–2133.

6. Bach, F. R., & Jordan, M. I. (2005). Discriminative training of hidden markov models for multiple pitch tracking [speech processing examples]. In *Proceedings of the IEEE International Conference on Acoustics, Speech, and Signal Processing (ICASSP), IEEE, Philadelphia, PA, USA* (Vol. 5, pp. v–489).
7. Hermes, D. J. (1988). Measurement of pitch by subharmonic summation. *The Journal of the Acoustical Society of America, 83*(1), 257–264.
8. Dimitriadis, D., Maragos, P., & Potamianos, A. (2005). Robust AM-FM features for speech recognition. *IEEE Signal Processing Letters, 12*(9), 621–624.
9. Bogart, B. P., Healy, M. J., & Turkey, J. W. (1963). The quefrequency analysis of time series for echoes; cepstrum, pseudo-autocovariance cross cepstrum and saphe cracking. In M. Rosenbaltt (Ed.) *Proceedings of the Symposium on Time Series Analysis* (pp. 209–243) New York: Wiley.
10. Oppenheim, A. V., & Schafer, R. W. (2004). From frequency to quefrency: A history of the cepstrum. *IEEE Signal Processing Magazine, 21*(5), 95–106.
11. Heman-Ackah, Y. D., Michael, D. D., & Goding, G. S, Jr. (2002). The relationship between cepstral peak prominence and selected parameters of dysphonia. *Journal of Voice, 16*(1), 20–27.
12. Balasubramanium, R. K., Bhat, J. S., Fahim, S., & Raju, R. (2011). Cepstral analysis of voice in unilateral adductor vocal fold palsy. *Journal of Voice, 25*(3), 326–329.
13. Heman-Ackah, Y. D. (2004). Reliability of calculating the cepstral peak without linear regression analysis. *Journal of Voice, 18*(2), 203–208.
14. Heman-Ackah, Y. D., Sataloff, R. T., Laureyns, G., Lurie, D., Michael, D. D., Heuer, R., et al. (2014). Quantifying the cepstral peak prominence, a measure of dysphonia. *Journal of Voice, 28*(6), 783–788.
15. Maryn, Y., & Weenink, D. (2015). Objective dysphonia measures in the program Praat: Smoothed cepstral peak prominence and acoustic voice quality index. *Journal of Voice, 29*(1), 35–43.
16. Godino-Llorente, J. I., Osma-Ruiz, V., Sáenz-Lechón, N., Gómez-Vilda, P., Blanco-Velasco, M., & Cruz-Roldán, F. (2010). The effectiveness of the glottal to noise excitation ratio for the screening of voice disorders. *Journal of Voice, 24*(1), 47–56.
17. Venkitaraman, A., & Seelamantula, C. S. (2013). Temporal envelope fit of transient audio signals. *IEEE Signal Processing Letters, 20*(12), 1191–1194.
18. Michaelis, D., Gramss, T., & Strube, H. W. (1997). Glottal-to-noise excitation ratio - a new measure for describing pathological voices. *Acta Acustica United with Acustica, 83*(4), 700–706.
19. Garellek, M., Samlan, R., Gerratt, B. R., & Kreiman, J. (2016). Modeling the voice source in terms of spectral slopes. *The Journal of the Acoustical Society of America, 139*(3), 1404–1410.
20. Holmberg, E. B., Hillman, R. E., Perkell, J. S., Guiod, P. C., & Goldman, S. L. (1995). Comparisons among aerodynamic, electroglottographic, and acoustic spectral measures of female voice. *Journal of Speech, Language, and Hearing Research, 38*(6), 1212–1223.
21. Hanson, H. M. (1997). Glottal characteristics of female speakers: Acoustic correlates. *The Journal of the Acoustical Society of America, 101*(1), 466–481.
22. Hanson, H. M., & Chuang, E. S. (1999). Glottal characteristics of male speakers: Acoustic correlates and comparison with female data. *The Journal of the Acoustical Society of America, 106*(2), 1064–1077.
23. Iseli, M., Shue, Y. L., & Alwan, A. (2007). Age, sex, and vowel dependencies of acoustic measures related to the voice source. *The Journal of the Acoustical Society of America, 121*(4), 2283–2295.
24. Gobl, C., & Chasaide, A. N. (2010). Voice source variation and its communicative functions. *The handbook of phonetic sciences, 1*, 378–423.
25. Fant, G. (1979). Glottal source and excitation analysis. *Speech Transmission Laboratory—Quarterly Progress and Status Report (STL-QPSR), 20*(1), 85–107.
26. Fant, G. (1982.) Preliminaries to analysis of the human voice source. *Speech Transmission Laboratory—Quarterly Progress and Status Report (STL-QPSR), 4*(1982), 1–28.

27. Fant, G., Liljencrants, J., & Lin, Q. G. (1985). A four-parameter model of glottal flow. *Speech Transmission Laboratory—Quarterly Progress and Status Report (STL-QPSR)*, *4*(1985), 1–13.
28. Barone, N. A. (2011). Acoustic, Aerodynamic, and Electroglottographic Properties of True Vocal Fold Body-Cover Conditions. Doctoral dissertation, Misericordia University, Pennsylvania, USA.
29. Titze, I. R. (1994). *Principles of voice production*. Englewood Cliffs, New Jersey, USA: Prentice-Hall Inc.
30. Dejonckere, P. H., Remacle, M., Fresnel-Elbaz, E., Woisard, V., Crevier-Buchman, L., & Millet, B. (1996). Differentiated perceptual evaluation of pathological voice quality: Reliability and correlations with acoustic measurements. *Revue de Laryngologie-otologie-rhinologie*, *117*(3), 219–224.
31. Samlan, R. A., Story, B. H., & Bunton, K. (2013). Relation of perceived breathiness to laryngeal kinematics and acoustic measures based on computational modeling. *Journal of Speech, Language, and Hearing Research*, *56*(4), 1209–1223.
32. Logemann, J., Boshes, B., Fisher, H., & Siegfried, J. (1973). The steps in the degeneration of speech and voice control in Parkinson's disease. In Siegfried, J. (Ed.), *Parkinson's Diseases: Rigidity, Akinesia, Behavior*; *Proceedings of the 4th International Symposium on Parkinson's Disease, Zurich, Switzerland* (pp. 101–112).
33. Klatt, D. H., & Klatt, L. C. (1990). Analysis, synthesis, and perception of voice quality variations among female and male talkers. *The Journal of the Acoustical Society of America*, *87*(2), 820–857.
34. Frohlich, M., Michaelis, D., & Strube, H. W. (1998). Acoustic "breathiness measures" in the description of pathologic voices. In *Proceedings of the International Conference on Acoustics, Speech and Signal Processing (ICASSP), IEEE. Seattle, Washington, USA* (Vol. 2, pp. 937–940).
35. Hillenbrand, J., Cleveland, R. A., & Erickson, R. L. (1994). Acoustic correlates of breathy vocal quality. *Journal of Speech, Language, and Hearing Research*, *37*(4), 769–778.
36. Kumar, B. R., Bhat, J. S., & Prasad, N. (2010). Cepstral analysis of voice in persons with vocal nodules. *Journal of Voice*, *24*(6), 651–653.
37. Krom, G. D. (1995). Some spectral correlates of pathological breathy and rough voice quality for different types of vowel fragments. *Journal of Speech, Language, and Hearing Research*, *38*(4), 794–811.
38. Wayland, R., & Jongman, A. (2003). Acoustic correlates of breathy and clear vowels: The case of Khmer. *Journal of Phonetics*, *31*(2), 181–201.
39. Huffman, M. K. (1987). Measures of phonation type in Hmong. *The Journal of the Acoustical Society of America*, *81*(2), 495–504.
40. Simpson, A. P. (2012). The first and second harmonics should not be used to measure breathiness in male and female voices. *Journal of Phonetics*, *40*(3), 477–490.
41. Strome, M., Stein, J., Esclamado, R., Hicks, D., Lorenz, R. R., Braun, W., et al. (2001). Laryngeal transplantation and 40-month follow-up. *New England Journal of Medicine*, *344*(22), 1676–1679.
42. Lu, F. L., Casiano, R. R., Lundy, D. S., & Xue, J. W. (1996). Longitudinal evaluation of vocal function after thyroplasty type I in the treatment of unilateral vocal paralysis. *The Laryngoscope*, *106*(5), 573–577.
43. Moers, C., Möbius, B., Rosanowski, F., Nöth, E., Eysholdt, U., & Haderlein, T. (2012). Vowel- and text-based cepstral analysis of chronic hoarseness. *Journal of Voice*, *26*(4), 416–424.
44. Yumoto, E., Sasaki, Y., & Okamura, H. (1984). Harmonics-to-noise ratio and psychophysical measurement of the degree of hoarseness. *Journal of Speech, Language, and Hearing Research*, *27*(1), 2–6.
45. Yumoto, E., Gould, W. J., & Baer, T. (1982). Harmonics-to-noise ratio as an index of the degree of hoarseness. *The Journal of the Acoustical Society of America*, *71*(6), 1544–1550.
46. Ptok, M., Schwemmle, C., Iven, C., Jessen, M., & Nawka, T. (2006). On the auditory evaluation of voice quality. *HNO*, *54*(10), 793–802.

47. Verdonck-de Leeuw, I. M., Festen, J. M., & Mahieu, H. F. (2001). Deviant vocal fold vibration as observed during videokymography: The effect on voice quality. *Journal of Voice*, *15*(3), 313–322.
48. Tigges, M., Mergell, P., Herzel, H., Wittenberg, T., & Eysholdt, U. (1997). Observation and modelling of glottal biphonation. *Acta Acustica*, *83*(4), 707–714.
49. Mergell, P., & Herzel, H. (1997). Modelling biphonation—the role of the vocal tract. *Speech Communication*, *22*(2–3), 141–154.
50. Ward, P. H., Sanders, J. W., Goldman, R., & Moore, G. P. (1969). Diplophonia. *Annals of Otology, Rhinology & Laryngology*, *78*(4), 771–777.
51. Rozen, A. L., & Lucas, E. V. (1977). A therapy approach for alleviating diplophonic voice quality. *International Journal of Language & Communication Disorders*, *12*(1), 69–74.
52. Aichinger, P., Schneider-Stickler, B., Bigenzahn, W., Fuchs, A.K., Geiger, B., Hagmüller, M., et al. (2013). Double pitch marks in diplophonic voice. In *Proceedings of the International Conference on Acoustics, Speech and Signal Processing (ICASSP), IEEE. Vancouver, Canada* (pp. 7437–7441).
53. Pinto, N. B., & Titze, I. R. (1990). Unification of perturbation measures in speech signals. *The Journal of the Acoustical Society of America*, *87*(3), 1278–1289.
54. Schreibweiss-Merin, D., & Terrio, L. M. (1986). Acoustic analysis of diplophonia: A case study. *Perceptual and Motor Skills*, *63*(2), 755–765.
55. Hong, K. H., & Kim, H. K. (1999). Diplophonia in unilateral vocal fold paralysis and intracordal cyst. *Otolaryngology - Head and Neck Surgery*, *121*(6), 815–819.
56. Yuasa, I. P. (2010). Creaky voice: A new feminine voice quality for young urban-oriented upwardly mobile American women? *American Speech*, *85*(3), 315–337.
57. Keating, P., Garellek, M. , & Kreiman, J. (2015, August). Acoustic properties of different kinds of creaky voice. In *Proceedings of the 18th International Congress of Phonetic Sciences, Glasgow, UK* (pp. 0821-1).
58. Lim, I. T., & Lee, B. G. (1993). Lossless pole-zero modeling of speech signals. *IEEE Transactions on Speech and Audio Processing*, *1*(3), 269–276.
59. Titze, I. R. (2001). Acoustic interpretation of resonant voice. *Journal of Voice, Elsevier*, *15*(4), 519–528.
60. Titze, I. R., & Story, B. H. (1997). Acoustic interactions of the voice source with the lower vocal tract. *The Journal of the Acoustical Society of America*, *101*(4), 2234–2243.
61. Story, B. H., Titze, I. R., & Hoffman, E. A. (2001). The relationship of vocal tract shape to three voice qualities. *The Journal of the Acoustical Society of America*, *109*(4), 1651–1667.
62. Feinstein, I., Hilger, P., Szachowicz, E., & Stimson, B. (1987). Laser therapy of dysphonia plica ventricularis. *Annals of Otology, Rhinology & Laryngology*, *96*(1), 56–57.
63. Xu, J. H., Ikeda, Y., & Komiyama, S. (1991). Bio-feedback and the yawning breath pattern in voice therapy: A clinical trial. *Auris Nasus Larynx*, *18*(1), 67–77.
64. Wood, J. M., Athanasiadis, T., & Allen, J. (2014). Laryngitis. *British Medical Journal Publishing. Group*, *349*, g5827.
65. American Speech-Language-Hearing Association. (2002). *Consensus auditory-perceptual evaluation of voice (CAPE-V)* (p. 3). Rockville: ASHA Special Interest Division.
66. Kreiman, J., & Gerratt, B. R. (2010). Perceptual assessment of voice quality: Past, present, and future. *Perspectives on Voice and Voice Disorders*, *20*(2), 62–67.
67. Vox, V. (1981). *I can see your lips moving: The history and art of ventriloquism*. London, UK: Kaye and Ward Publishers.
68. Titze, I. R., & Hunter, E. J. (2004). Normal vibration frequencies of the vocal ligament. *The Journal of the Acoustical Society of America*, *115*(5), 2264–2269.
69. Yin, J., & Zhang, Z. (2013). The influence of thyroarytenoid and cricothyroid muscle activation on vocal fold stiffness and eigenfrequencies. *The Journal of the Acoustical Society of America*, *133*(5), 2972–2983.
70. Yin, J., & Zhang, Z. (2014). Interaction between the thyroarytenoid and lateral cricoarytenoid muscles in the control of vocal fold adduction and eigenfrequencies. *Journal of Biomechanical Engineering*, *136*(11), 111006.

71. Titze, I. R. (1980). Comments on the myoelastic-aerodynamic theory of phonation. *Journal of Speech, Language, and Hearing Research, 23*(3), 495–510.
72. Hirano, M., Kakita, Y., & Daniloff, R. G. (1985). Cover-body theory of vocal fold vibration. In R. G. Daniloff (Ed.), *Speech science* (pp. 1–46). San Diego, California: College-Hill Press.
73. Hirano, M., Ohala, J., & Vennard, W. (1969). The function of laryngeal muscles in regulating fundamental frequency and intensity of phonation. *Journal of Speech, Language, and Hearing Research, 12*(3), 616–628.
74. Zhang, Z. (2016). Cause-effect relationship between vocal fold physiology and voice production in a three-dimensional phonation model. *The Journal of the Acoustical Society of America, 139*(4), 1493–1507.
75. Van den Berg, J., & Tan, T. S. (1959). Results of experiments with human larynxes. *Journal for Oto-Rhino-Laryngology, Head and Neck Surgery (ORL), 21*(6), 425–450.
76. Rabiner, L. R., & Schafer, R. W. (1978). *Digital processing of speech signals*. Prentice-hall, New Jersey: Englewood Cliffs.
77. Titze, I. R., & Sundberg, J. (1992). Vocal intensity in speakers and singers. *The Journal of the Acoustical Society of America, 91*(5), 2936–2946.
78. Sundberg, J., & Rossing, T. D. (1990). *The science of singing voice*. Monroe, Michigan: ASA Publishing Corporation.
79. Sundberg, J., & Högset, C. (2001). Voice source differences between falsetto and modal registers in counter tenors, tenors and baritones. *Logopedics Phoniatrics Vocology, 26*(1), 26–36.
80. Henrich, N., d'Alessandro, C., Doval, B., & Castellengo, M. (2005). Glottal open quotient in singing: Measurements and correlation with laryngeal mechanisms, vocal intensity, and fundamental frequency. *The Journal of the Acoustical Society of America, 117*(3), 1417–1430.
81. Stathopoulos, E. T., & Sapienza, C. (1993). Respiratory and laryngeal function of women and men during vocal intensity variation. *Journal of Speech, Language, and Hearing Research, 36*(1), 64–75.
82. Joliveau, E., Smith, J., & Wolfe, J. (2004). Acoustics: Tuning of vocal tract resonance by sopranos. *Nature, 427*(6970), 116.
83. Titze, I. R. (2002). Regulating glottal airflow in phonation: Application of the maximum power transfer theorem to a low dimensional phonation model. *The Journal of the Acoustical Society of America, 111*(1), 367–376.
84. Isshiki, N. (2013). *Phonosurgery: Theory and practice*. Springer Science & Business Media.
85. Van den Berg, J. W. (1968). Register problems. *Annals of the New York Academy of Sciences, 155*(1), 129–134.
86. Isshiki, N. (1998). Mechanical and dynamic aspects of voice production as related to voice therapy and phonosurgery. *Journal of Voice, 12*(2), 125–137.
87. Berry, D. A., Herzel, H., Titze, I. R., & Krischer, K. (1994). Interpretation of biomechanical simulations of normal and chaotic vocal fold oscillations with empirical eigenfunctions. *The Journal of the Acoustical Society of America, 95*(6), 3595–3604.
88. Zetterholm, E. (2003). *Voice imitation: A phonetic study of perceptual illusions and acoustic success. Computational linguistics* (Vol. 44). Lund University Press.
89. Amin, T. B., Marziliano, P., & German, J. S. (2014). Glottal and vocal tract characteristics of voice impersonators. *IEEE Transactions on Multimedia, 16*(3), 668–678.
90. Harries, M., Hawkins, S., Hacking, J., & Hughes, I. (1998). Changes in the male voice at puberty: Vocal fold length and its relationship to the fundamental frequency of the voice. *The Journal of Laryngology & Otology, 112*(5), 451–454.
91. Hollien, H., Brown, W. S, Jr., & Hollien, K. (1971). Vocal fold length associated with modal, falsetto and varying intensity phonations. *Folia Phoniatrica et Logopaedica, 23*(1), 66–78.
92. Eriksson, A., & Wretling, P. (1997). How flexible is the human voice? A case study of mimicry. In *Proceedings of the Fifth European Conference on Speech Communication and Technology (EUROSPEECH), Rhodes, Greece* (pp. 1043–1046).

93. Eriksson, A., Llamas, C., & Watt, D. (2010). The disguised voice: Imitating accents or speech styles and impersonating individuals. *Language and Identities*, *8*, 86–96.
94. Kitamura, T. (2008). Acoustic analysis of imitated voice produced by a professional impersonator. In *Proceedings of the Ninth Annual Conference of the International Speech Communication Association (INTERSPEECH), Brisbane, Australia* (pp. 813–816).
95. Deutsch, D. (1983). Auditory illusions, handedness, and the spatial environment. *Journal of the Audio Engineering Society*, *31*(9), 606–620.
96. McGettigan, C., Eisner, F., Agnew, Z. K., Manly, T., Wisbey, D., & Scott, S. K. (2013). T'ain't what you say, it's the way that you say it—left insula and inferior frontal cortex work in interaction with superior temporal regions to control the performance of vocal impersonations. *Journal of Cognitive Neuroscience*, *25*(11), 1875–1886.
97. Esling, J. H., & Clayards, J. A. (1998). Laryngoscopic analysis of pharyngeal articulations and larynx-height voice quality settings. In *Proceedings of the Fifth International Conference on Spoken Language Processing (ICSLP), Sydney, Australia.*
98. Esling, J. H. (1999). Voice quality settings of the pharynx. In *Proceedings of the 14th International Congress of Phonetic Sciences, San Francisco, USA* (Vol. 3, pp. 2449–2452).
99. Laver, J. (1980). The phonetic description of voice quality. *Cambridge Studies in Linguistics London*, *31*, 1–186.
100. Laufer, A., & Condax, I. D. (1981). The function of the epiglottis in speech. *Language and Speech*, *24*(1), 39–62.
101. Titze, I. R. (2008). Nonlinear source-filter coupling in phonation: Theory. *The Journal of the Acoustical Society of America*, *123*(4), 1902–1915.
102. Herzel, H., Steinecke, I., Mende, W., & Wermke, K. (1991). Chaos and bifurcations during voiced speech. In E. Mosekilde & L. Mosekilde (Eds.), *Complexity, chaos, and biological evolution* (pp. 41–50). Boston, MA: Springer.
103. Herzel, H. (1993). Bifurcations and chaos in voice signals. *Applied Mechanics Reviews*, *46*(7), 399–413.
104. Berry, D. A., Zhang, Z., & Neubauer, J. (2006). Mechanisms of irregular vibration in a physical model of the vocal folds. *The Journal of the Acoustical Society of America*, *120*(3), EL36–EL42.
105. Mariéthoz, J., & Bengio, S. (2005). *Can a professional imitator fool a GMM-based speaker verification system? EPFL-REPORT-83202*. Martigny, Switzerland: IDIAP Research Institute.
106. Hautamäki, R. G., Kinnunen, T., Hautamäki, V., Leino, T., & Laukkanen, A. M. (2013). I-vectors meet imitators: On vulnerability of speaker verification systems against voice mimicry. In *Proceedings of the Annual Conference of the International Speech Communication Association (INTERSPEECH), Lyon, France* (pp. 930–934).
107. Schlichting, F., & Sullivan, K. P. H. (1997). The imitated voice - a problem for voice line-ups? *Forensic Linguistics*, *4*, 148–165.
108. Kinnunen, T., Wu, Z. Z., Lee, K. A., Sedlak, F., Chng, E. S., & Li, H. (2012). Vulnerability of speaker verification systems against voice conversion spoofing attacks: The case of telephone speech. In *Proceedings of the International Conference on Acoustics, Speech and Signal Processing (ICASSP), IEEE. Kyoto, Japan* (pp. 4401–4404).
109. Krause, J. C., & Braida, L. D. (2004). Acoustic properties of naturally produced clear speech at normal speaking rates. *The Journal of the Acoustical Society of America*, *115*(1), 362–378.
110. Rix, A. W., Beerends, J. G., Hollier, M. P., & Hekstra, A. P. (2001). Perceptual evaluation of speech quality (PESQ)—a new method for speech quality assessment of telephone networks and codecs. In *Proceedings of the International Conference on Acoustics, Speech, and Signal Processing (ICASSP), IEEE. Salt Lake City, Utah, USA* (Vol. 2, pp. 749–752).
111. Beerends, J. G., Hekstra, A. P., Rix, A. W., & Hollier, M. P. (2002). Perceptual evaluation of speech quality (PESQ) the new ITU standard for end-to-end speech quality assessment part II: Psychoacoustic model. *Journal of the Audio Engineering Society*, *50*(10), 765–778.
112. Rix, A. W. (2003). Comparison between subjective listening quality and P.862 PESQ score. *Proceedings of Meetings on Acoustics*, *4*(1), 17–25.

Part II
Computational Profiling

Part II of this book deals with computational profiling. It comprises four chapters, numbered 7–10. Chapter 7 discusses micro-features and their discovery. Chapter 8 deals with mechanisms for the deduction of bio-relevant parameters: techniques for mapping the measured micro-features to human and environmental parameters. It briefly outlines some machine learning and AI-based techniques that are especially relevant for building models and predictors for profiling. Chapter 9 is about the reconstruction of the human persona in 2D and 3D, and its reverse: the reconstruction of voice from facial, structural, and other bio-relevant information about the speaker. Chapter 10, the final chapter of this book, deals with practical, real-life applications of voice-based profiling. A significant portion of this chapter is dedicated to the subject of establishing the reliability and confidence bounds of profile parameter deductions, since these issues are *extremely* important in real-life contexts. The chapter, and this book, closes with a brief discussion of the privacy and ethical issues associated with profiling from voice, and the future outlook for this science.

Chapter 7
Feature Engineering for Profiling

Humans are able to perceive many profile parameters from voice. Such perceptual relationships between voice and parameters show that there *is* information relevant to profiling in the voice signal, but may not tell us *what* that information is. If a study finds perceptual relations between some parameter and specific features derived from voice, then that indicates the existence of a *statistical* relationship between those features and voice, which may be directly causal, or relate to a common underlying cause. Such features will usually be directly useful in profiling.

Ideally, for effective computational profiling, causal relationships between parameters and voice must be sought. Perceptual relationships must be reasoned out and transformed to causal relationships. Once the cause is found or reasonably guessed, information about it—in the form of features—can be derived from voice. For example if we know that the cause of age-related changes in voice is the decline in vocal fold elasticity, then the features that reflect elasticity most can be derived to predict age. If this fact were not known, the hypothesis that there must be a feature that carries the (observed) signatures of age in voice is still valid, but the feature itself must be searched out by other means. Thus, in the absence of reasoning about causal relationships, various *feature discovery* methods must be used. This is the focus of this chapter.

In feature discovery for profiling, a key assumption we make is that if there exists a pathway whereby any part or system of the human body can exert an influence on the voice production system, then the subtlest influence on that part or system *must* lead to *some* changes in the voice produced. The changes may be intangible—too fine to be seen (in a spectro-temporal sense) or heard (in a perceptual sense) by humans. For example, a change in fundamental frequency from 201 Hz to 205 Hz from morning to evening in an adult female may go unnoticed by the human ear. Nevertheless, those changes exist. If they exist, they must be discoverable. They may exist in hypothetical spaces: their signatures may manifest in high-dimensional mathematical spaces that may not be open—for reasons of dimension and scale—to manual inspection or interpretation. The mere hypothesis that such changes are present in some space can drive the design of powerful techniques that can find them, *even in the absence of any knowledge of the underlying causal relationships*. This is

R. Singh, *Profiling Humans from their Voice*,
https://doi.org/10.1007/978-981-13-8403-5_7

not to say that such knowledge is now redundant—it can in fact make a tremendous positive difference to the outcome of profiling mechanisms. Rather, this is to say that more can be discovered from voice than was thought possible before.

In this chapter we will first discuss micro-articulometry, a process that allows us to transition from a causal relationship-based world to a discovery-ready one, by helping identify the fragments of voice that are most likely, in the first place, to contain the signatures we seek to discover. We will then discuss two important approaches to feature discovery, giving one example under each: one is that of *feature selection* based on micro-articulometry, where we explore all existing features at our disposal to automatically find out which ones are most relevant for the parameter under consideration. This is done when there is no knowledge of the causatives to guide the choice of features. The second is that of *feature engineering*, where we actually put together mechanisms based on concepts in automated learning and artificial intelligence that *discover or create* the most relevant features for use in the profiling task at hand. If any causatives are known, we factor those into the discovery process to yield even better results. It is not necessary for this process to be strictly sequential, though. With advanced AI implementations of feature engineering (and large amounts of data), the most relevant micro-articulometric fragments and the signatures contained therein, can be concurrently discovered.

7.1 Micro-articulometry

Let us begin with the example of Alzheimer's disease, a progressive and severely debilitating neurological condition. This is not a muscular disease, and on the surface of it, a very unlikely candidate for profiling from voice. Does it affect voice? Alzheimer's disease is characterized by progressive loss of memory and other cognitive function. The production of voice is as much a cognitive function as a biomechanical one, as we have discussed earlier. The effects of Alzheimer's must therefore extend to voice—they may manifest as impairment to the conscious coordination required within the vocal tract structures, from the vocal fold muscles to the articulatory movements, to produce a cogent sequence of sounds in continuous speech [1–3]. This is indeed the case, and multiple studies have correlated voice changes with Alzheimer's, e.g. [4–8]—there are many more equally good ones that are not referenced here.

For argument's sake (since at the time of writing this chapter we have not actually performed the experiments needed to support this), let us assume that because the disease causes impaired judgment of the sequence of muscle actions needed for proper phonation, the muscles do not control the vocal folds in a normal manner. This would change the level of control the speaker has over the vocal cord geometry, on the modes of vibrations, on their synchronization, on the movements between registers, on glottal geometry and open-close cycles, positioning of the larynx, and many other finer factors. The resultant changes would appear in the spectrographic domain in many ways—in the energy distribution among harmonics, in turbulence patterns, in frequency patterns produced during phonation etc. They may also be expected to be uncharacteristically distributed within the different voiced sounds produced in continuous speech (each sound requires a different configuration of the articulators and vocal tract, and a somewhat different pattern of vocal fold movements). It is also reasonable to expect that the signature patterns of Alzheimer's disease would not just be confined to spectrographic representations of speech—that other (even abstract) representations may capture them in their own characteristic ways. It is further reasonable to assume that even if these patterns are expected to manifest in the spectrographic domain, they may be too faint or obscured (by other influences) to be discernable by visual or auditory examination.

The question that naturally follows is—where and how (in the speech signal) must we to find these changes in *any* representation? Continuing with the spectrographic example, due to the nature of the dynamics explained above, the majority of these (if not all) would be expected to be very fine, appearing as *micro* variations in the frequency composition of different sounds. They may also be expected to be slightly deviant from those represented as a "healthy" canonical set of patterns in normal speech. They are also expected to occur *within* the time a particular pattern of phonation lasts before moving to the next sound, due to the explicit involvement of the bio-mechanical structures in the vocal tract in phonation. In fact, they may occur in a *part of the duration* of that phonation (and furthermore, probably in the first few milliseconds than later, since the practice of a lifetime may weigh in, overcome the deficiencies due to the disease, and restore some normalcy to the pattern in a short time). Such conjectures and expectations lead us to hypothesize that the specific patterns that might correlate (in this case) to Alzheimer's could lie in fractions of each sound produced, and if so, must by logical argument, manifest differently in each sound produced.

This is a set of conjectures based on *articulometric* considerations, clearly. This set of conjectures also points to the space of *micro-features* as the one likely to hold the most relevant information for parameters that clearly have a physiological pathway to affecting voice. These parameters may actually have patterns in voice in response to them, but their effects may be highly likely to *not* be perceivable by the human ear.

They nevertheless (arguably) exist, and if found, could be used for profiling.

This set of conjectures forms the basis for what we call ***Micro-articulometry***—the science of discovery and measurement (hence the suffix "-metry") of micro-properties or micro-variations in the voice signal, based on and guided by the principles of articulatory phonetics.

The resolution and analysis of the human speech production process in terms of articulatory-phonetic units is in fact based on strong cognitive and developmental foundations. As humans learn to manipulate and use their vocal structures to produce sounds, they do so in the context of communicating meaning through language or other vocalizations. Every sound is thus practiced and learned so that it can be best discriminated *individually*, so that words and sentences are collectively more communicable and understandable by listeners [9]. In other words, phonemes comprise the basic units of speech *communication*, and speech production adheres to their auditory structure over a speaker's lifetime.

Micro-articulometry is also supported by the theories of categorical perception [10], which has been shown to relate to speech, and to color perception by the human brain. Categorical perception can be differentiated from continuous perception by the continuity of how changes in the underlying entity are perceived. In the latter, the entity perceived can take a continuum of values, and changes can be smooth and gradual. In the former, changes are abrupt and quantized (as in categories). Speech falls under the set of entities that are categorically perceived by the human brain. These categories are individual sounds. Thus, even from a perception viewpoint, an articulometric approach makes complete sense.

7.1.1 Articulometric Sectioning of the Voice Signal

Although not a necessity, for micro-articulometry it is useful to identify the sections of different sounds that might be affected by the parameter under consideration for the purpose of profiling. We may want to study the interior stable "locus" regions of the phonemes in some cases, study the onset regions, or the offset ones for others, transition regions for yet others and so on. We may want to extract features from them for automated analysis, or manipulate them in different ways. Regardless of the purpose, it is useful to have automated, scalable and robust techniques to identify these regions accurately. This is a difficult task, since phonemes are short in duration and the margins of (allowed) error are small. This is further complicated by the fact that in continuous speech, phonemes seamlessly "flow" or merge into each other within words, and do not have actual "boundaries."

Accordingly, the first important step in micro-articulometry is to isolate the sections of articulated sounds that may reflect the influence of the parameter under consideration. In this context, it is useful to be guided by the locus theory of phonemes.

7.1.1.1 The Locus Theory, and its Significance in Micro-articulometry

The locus theory of phonemes states that every phoneme has a "locus," corresponding to a canonical arrangement of the vocal tract for that phoneme, and that the articulators move towards it in the production of the phoneme [11, 12]. In continuous speech, the loci of phonemes may not be fully reached as the articulators move continuously from one set of configurations to another. Figure 7.1a shows an example.

While the locus theory does not explain all the variations observed in different instantiations of a phoneme, a key, valid insight that may be derived from it is that the interior regions of the phoneme are representative of its locus, and are much more invariant to contextual and shorter-term effects than the boundaries. These contribute the most to the individual discrimination of the particular phoneme, and may thus also reflect the relatively stable profile parameters of a speaker more consistently.

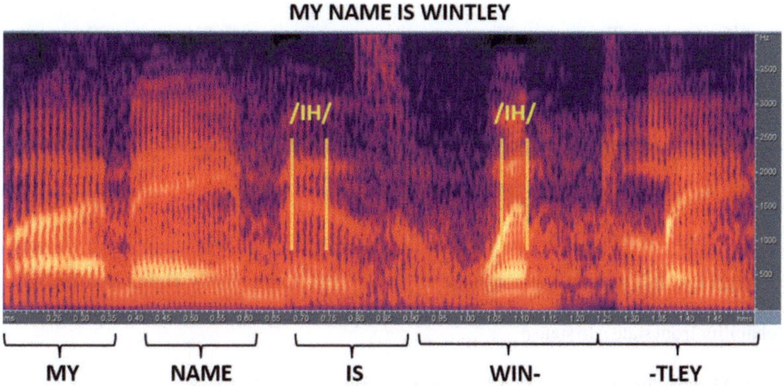

(a) The sentence “My name is Wintley” spoken by an adult male. Note the long-term spectral patterns that flow continuously across the entire utterance while moving towards different phoneme loci in the process. The phoneme /IH/ is marked . It occurs in two different phonemic contexts, and the influences can be seen at the beginning and end of the phoneme in each instance.

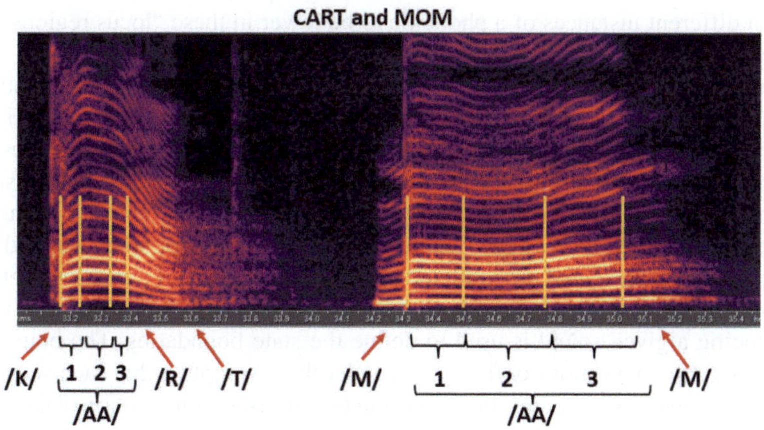

(b) State level segmentations for the phoneme/AA/in the American English pronunciation of the word CART (K AA R T) and MOM (M AA M)

Fig. 7.1 Illustration of phoneme loci

On the other hand, for some profile parameters (e.g. of Alzheimer’s, following the reasoning given above), the extreme portions of the phoneme may carry more information. We refer to the interior stable region of a phoneme as the “locus” of the phoneme (see Fig. 7.1b). This does not imply that the locus of the phoneme is actually *achieved* in this segment; rather that it is the region that is most typical of the phoneme and least influenced by context and flow.

Let us look at one mechanism (of many possible) of how different regions of a phoneme may be automatically identified. For this, let us only consider the problem of identifying the loci of phonemes. The exact interior region that represents the

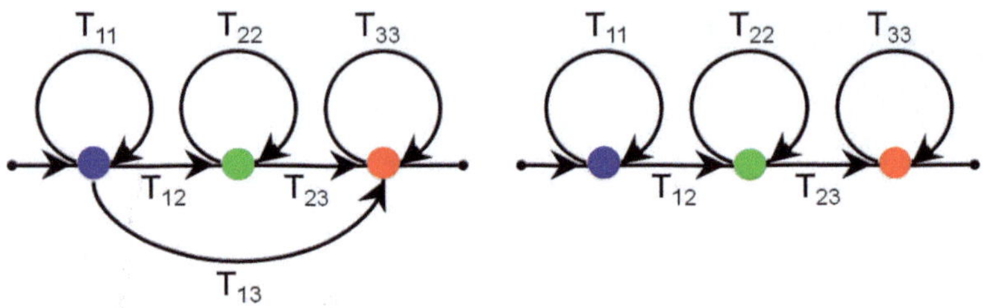

(a) A generic left-to-right topology for a hidden Markov model. T_{ab} is the transition probability from state a to state b

(b) A strict left-to-right topology

Fig. 7.2 HMM with three states and a left-to-right (Bakis) topology

locus may vary with different instances of the phoneme. It need not be at the center of the phoneme, and its position must be carefully estimated. In order to identify this section accurately, we build on the fact that we may expect the inter-instance variation between different instances of a phoneme to be lower in these "locus regions" than in other regions.

To automatically segment phonemes, algorithms that are used in large vocabulary continuous speech recognition (LVCSR) systems may be effectively used. Hidden Markov Models (HMMs) are especially well-suited for this purpose. These are generative models of various sound units that, in the LVCSR system, are assumed to comprise continuous speech. In a typical LVCSR system, each sound unit is modeled by an HMM with a specified number of states. The sound is assumed to be the sequentially aggregated probabilistic output of a sequence of transitions between these states. In practice, the number of states transitioned through by the model in producing a given sound is used to define the state boundaries. The boundaries demarcate different sections of the sound, under the assumption that each state has a different distribution of features that represent the corresponding section of the sound. An HMM-based LVCSR algorithm can thus be designed to identify the different segments of phonemes.

In order to identify the *steady-state* of a phoneme, however, without relying entirely on the assumption that the central segment contains all of the steady state (which may be true in most instances, but not all), the standard LVCSR algorithm must be modified. This is described in greater detail below.

In an algorithm for segmentation based on HMMs, each phoneme is modeled (or assumed to be generated) by a Hidden Markov Model (HMM), with Gaussian-mixture state output densities. To identify the loci, it is sufficient to model each phoneme by a three-state HMM with a strict left-to-right topology for transitions. Such an HMM is illustrated in Fig. 7.2.

This HMM topology ensures that any sequence of states drawn from the HMM must include a segment from the central state. Conventional maximum-likelihood HMM training utilizes the Baum-Welch algorithm to estimate the parameters $\hat{\Lambda}$ of the HMM for a stochastic process [13] (in this case, one that generates a phoneme) to maximize the log likelihood of the training instances **X**:

$$\hat{\Lambda} = \arg\max_{\Lambda} \log P(\mathbf{X}; \Lambda) \tag{7.1}$$

This is modified to include an *entropic prior* on the distribution of the central state:

$$\hat{\Lambda} = \arg\max_{\Lambda} \log P(\mathbf{X}; \Lambda) - \alpha H(\Lambda_c) \tag{7.2}$$

where Λ_c represents the parameters of the Gaussian-mixture density of the *central* (i.e. the second) state, and $H(\Lambda_c)$ represents the entropy of a Gaussian-mixture density with parameters Λ_c. α is a regularization parameter.

The optimization above simultaneously attempts to both maximize the likelihood of the observations, and minimize the entropy of the distribution of the central state. This has the effect that the observations from the central state, as modeled by the HMM, have least variation and are hence maximally consistent across different observation sequences. Combined with the fact that every state sequence through the HMM must visit the central state, for phonemes modeled by such entropically trained HMMs we may expect the central state to represent the consistent locus region of the phoneme.

The parameter estimation of Eq. 7.2 can be performed using a modified version of the Baum-Welch algorithm described in [14]. Given a trained HMM for any phoneme, the locus region of a new instance of the phoneme can be segmented out by aligning the instance against the states of the HMM. The segment of the instance that aligns against the central state represents the locus region of the phoneme. An example is shown in Fig. 7.1b.

It must be noted that the entropic HMM conservatively focuses on matching the steady-state regions of the phonemes; however this comes at the cost of accurate identification of the boundaries of these regions, and hence the durations of the steady state. This is useful in many profiling contexts, especially when decisions are to be made based on relatively fewer available instances of phonemes.

7.1.1.2 What Micro Variations Must We Look For?

In the micro-segments that we identify (using the above or any other equivalent technique for phoneme fragmentation), we could look for any of the large set of traditional features used for voice analysis that remain valid and relevant within micro-durations of speech. It makes sense, though, to first apply human knowledge to hypothesize the causal relationships between the parameter to be profiled, and the voice production process. This would guide the choice of features very effectively. When this is not possible, all available features can be used and one of many *feature selection* techniques may be applied to eliminate the ones that do not contribute to the profiling for the parameter under consideration. Alternatively, one can use what in non-scientific terms may be called "designer" features—which are feature representations discovered by carefully engineered processes designed to create features that optimally allow for the profiling of the given parameter. These processes comprise *feature engineering* techniques.

In this section we will see some examples of traditional features that remain relevant within micro-durations of speech. In the subsequent two sections we will discuss feature selection and feature engineering respectively.

7.1.2 Micro and Macro Features

Independently of all of the discussion above, **micro-features** can be defined as *measurements that are made within the average duration of each unit of sound*, typically within time intervals of the order of 1–20 ms. Measurements at this scale capture the fine-grained properties of the voice signal. These properties could be distributed over the entire signal, may be specific to a sound unit within it, or may be transient and related only to specific parts of specific sounds. Due to their fine-grained nature, these features are also usually difficult to measure with precision, and customized mechanisms must be devised to accurately measure each of them.

In contrast, **macro-features** (e.g. i-vectors [15], or voice quality, explained in Chap. 6) capture the statistical properties of entire utterances (or collections of sounds) as a whole. These are immensely useful in many applications, but may be of limited use in profiling for parameters that exert very subtle, often imperceptible influences on the speech signal. For other kinds of parameters, e.g. gender, age etc. that have gross influences on voice production, macro-features could be very useful.

Micro-features are not a new concept (although their extensions through feature discovery are new, as described later in this chapter). Many macro-features are discussed in Chaps. 5 and 6. Of the others, many qualify to be micro-features. In the section below, four such traditional features are listed as illustrative examples of micro-features: formants, harmonic bandwidths, voicing onset time and frequency modulation.

7.1.2.1 Formants

Formants contain discriminatory information for profiling. For example, a study of voices of identical twins in [16] shows that analysis of F4 in most *vowels*, and of particular aspects of *consonant* production is able to discriminate among closely related similar-sounding speakers. This is just a randomly selected example, but we can see clearly how it supports micro-articulometric contentions. Many other examples exist in the literature.

Because the nature of formants has been explained earlier in Chaps. 2 and 5, we only discuss some key computational settings for their computation here. To reiterate, formants represent the resonances of the vocal tract. We note that resonances occur for *all* sounds that are produced in the vocal tract (and heard by listeners). However, their positions (on the frequency axis) can be computationally calculated more reliably for the voiced phonemes—and of those they are best estimated for vowels and phonemes of longer duration. This is more a shortcoming of the computational techniques used for measurement than of the sounds themselves. For these phonemes in turn, estimates of resonant frequencies are likely to be more reliable within their locus regions. Once formant peaks are calculated from these regions using the techniques discussed in Chap. 5, other formant-based features, such as formant bandwidths, formant dispersion etc. can be estimated. Formant transition slopes can be calculated from adjacent formant peaks between any two regions that yield formant measurements.

To give an idea of typical settings that may be used in practical systems: if we use the LPC method (see Chap. 5) for estimating formant peaks [17], for broadband speech we would typically fit a 12th order LPC spectrum to 50 ms slices of the signal, with a 40 ms overlap between adjacent frames. The segments must be pre-emphasized, and windowed with a Hamming window. The twelve LPC pole frequencies represent candidate formant positions. To derive the actual formants, we would have to run a simple dynamic programming algorithm to track the formant values through time, subject to constraints on the actual values of the formants. The first formant F1 can be typically assumed to lie in the range 200–900 Hz, F2 to lie within 900–2800 Hz, F3 to lie between 1500–3500 Hz, and F4 and F5 can be assumed to lie above 3500 and 4500 Hz respectively. These guesses can be made more accurate a-priori by visual or any other means of inspection of the spectral composition of the actual data at hand. As mentioned before, for more reliability, all formants are best tracked only within speech regions. These can be identified through voice activity detection (VAD) procedures, or other indirect means, such as ASR systems.

The dynamic program may not always find all formants in every frame, although F1 and F2 may be generally found for most phonemes with this method (enough for dialect identification using vowel formant planes, for example). Estimates of higher formants (F4, F5 ...) are likely to be more unreliable. Once formant positions are determined, the bandwidth of each formant (see Chap. 5) can be determined by finding the frequencies on either side of the spectral peak at the formant, at which the spectral energy falls by 3 dB with respect to the peak frequency. Other formant-based features can be similarly calculated.

7.1.2.2 Harmonic Bandwidths and Frequency Modulation

Figure 7.3 shows an example of two micro-features in the spectrographic domain. The features appear in *frequency* and *time-frequency*. The top panel, Fig. 7.3a, shows a narrow-band spectrogram of a rendition of Popeye the Sailor Man [18] by Jack Mercer. This was recorded in 1935, at a time when there were few audio processing techniques available to artificially render the unusual spectral characteristics seen in the spectrogram at the time of audio capture. The voice actor had to actually produce

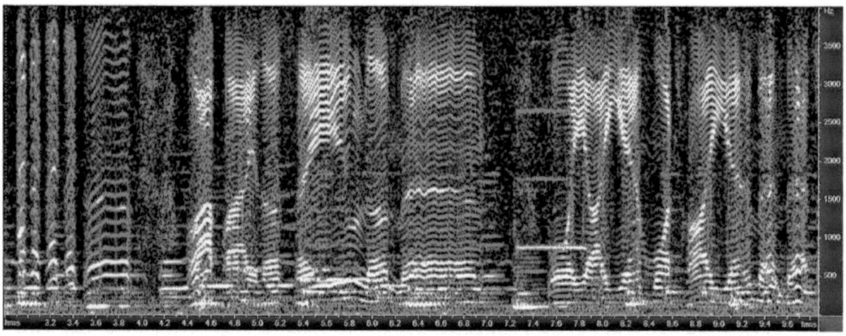

(a) Spectrogram of a voice signal showing sound-specific frequency modulation patterns

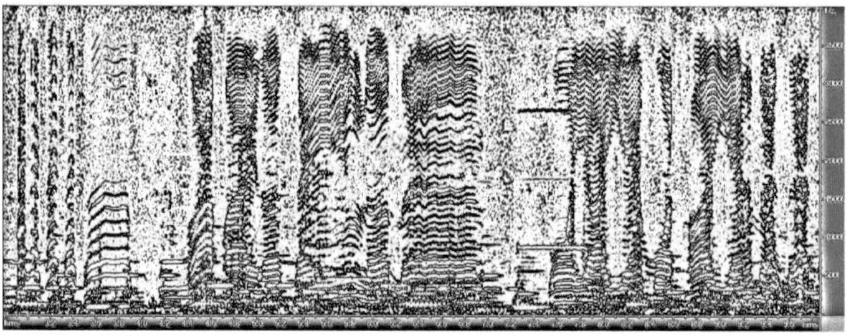

(b) Spectrogram showing the variations of instantaneous harmonic bandwidths across sounds. Each harmonic has been outlined along a contour that represents a 3dB fall relative to the peak energy in that harmonic. Contours have been traced automatically using image processing techniques.

Fig. 7.3 Spectrograms showing two visualizable micro-features, and their variations across different sounds

the sounds as seen in this figure. The spectrogram shows frequency modulation (FM) (explained in Chap. 6) of each harmonic clearly. FM, when measured over short windows of the signal, is an example of a micro-feature. For a normal person, different vocal registers can show different degrees of modulation for the same section of a sound. The lower panel of Fig. 7.3b shows a micro-feature in the time-frequency domain. This is the *bandwidth* of each harmonic, and is outlined by fine black contours which mark a boundary around the peak frequency of each harmonic, where the energy is attenuated to 3 dB below the peak value. We see that the bandwidths are different for each harmonic. As we have discussed before, the energy distribution across harmonics (which decides these harmonic bandwidths) is very characteristic of each speaker and is correlated with the shape of the glottal pulse waveform, as explained in Part I of this book.

7.1.2.3 Voicing Onset Time

In continuous speech, phonemes follow one another in rapid succession. When a voiced phoneme follows an unvoiced one, the vocal folds must rapidly transition from a non-vibrating state (or a state of rest) to a state of phonation (self-sustained vibrations). Due to the inertia of the vocal structures, a short time elapses between the two states. This is called the *voicing onset time* (VOT). It is considered to be a potential indicator of the ability of the vocal tract to move from one configuration to another [19], i.e. of the *agility* of the vocal tract [20, 21].

A note on inertia: the vibration of the vocal cords is concurrent with the change of geometry of the vocal folds themselves. Depending on the phoneme that is intended to be produced, they could be compressed or stretched in different ways by the laryngeal muscles that control their motion. Achieving each configuration takes a finite amount of time that is likely to be related to the biological factors that control and influence muscle movement. This characterizes the inertia of the vocal tract. It is difficult to measure directly, and VOT is only used as its indicator.

Specifically, in practice and by definition, VOT is the time taken for the self-sustained vibrations that are characteristic of phonation to begin, following the release of a plosive sound. A plosive sound (such as /T/, /P/, /B/ etc.) consists of two parts—an initial *stop*, in which the vocal tract is constricted to near-total closure at some location, stopping all airflow from the vocal tract to the exterior causing a pressure build-up within the vocal tract, and a subsequent release where the air is released in a *burst*. The stop and burst together form a plosive sound. An example is shown in Fig. 7.4.

Similarly, the stop of phonation (especially when succeeded by a silence or stop consonant) is characterized by continued resonance in the vocal tract due to reverberation. The time taken for the sound intensity to fall below a given threshold can be defined as the *voicing offset time* (VOFT) . Alternate definitions may include entities such as the time taken for the peak energy in a vowel to reduce to the energy just before the burst of a plosive, or the duration between the end of a phonated sound (if the voicing has clearly stopped) to the beginning of an unvoiced phoneme etc. While VOT is indicative of vocal tract *inertia*, when defined as above, VOFT can be indicative of the "lossiness" or *loss-momentum* of the vocal tract. VOFT is illustrated in Fig. 7.5.

VOT is of three kinds: (a) Zero VOT: the duration between the burst and the subsequent voicing pattern is zero, (b) Positive VOT: there is a measurable duration between the two sounds and (c) Negative VOT: in rare cases, voicing begins *before* the onset of the stop. Often, for voiced stop consonants in particular, the VOT may not be present at all, as the voicing may simply proceed through the closure and beyond. In these cases the convention is to consider the VOT to be negative. The measurement of VOT is discussed in Chap. 8.

VOT is a good example of the potential of a micro-feature to carry profiling relevant information. It is closely related to the specific physical and physiological characteristics of the speaker that contribute towards vocal inertia, such as vocal tract geometry and muscle control. Some studies report such correlations, e.g. [22].

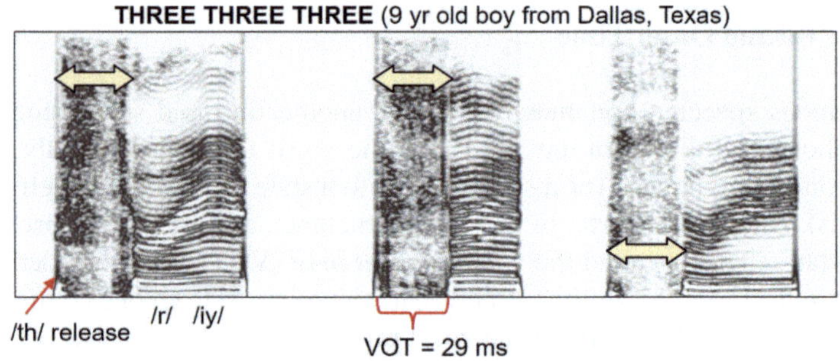

(a) Three instances of the word "three" spoken by a 9 year old boy from Texas (source of data: [22]). Although the durations of the words are different, the voicing onset time α remains the same. This is because VOT is dependent on factors that are not under the voluntary control of the speaker. In practice, for a specific combination for phonemes, the VOT changes within a very small range around the mean for each speaker.

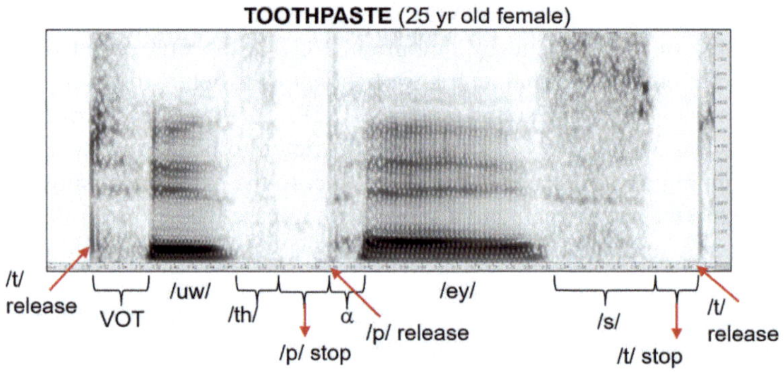

(b) Voicing onset time within the word "Toothpaste" spoken by a 25-year old female. The duration of VOT is different for the two combinations of plosive and vowel in this word: /t/-/uw/ and /p/-/ey/.

Fig. 7.4 Illustration of voicing onset time

Some studies have found joint correlations of VOT with age and other parameters, e.g. gender [23], hearing loss [24], age of learning (a second language) [25], age of learning *and* speaking rate [26]. Along another dimension, joint correlations of VOT and other measures such as formants and their bandwidths have been found with age [27]. A large number of studies have shown VOT to be predictive of Parkinson's disease [28, 29], depression [30] etc.

These were examples of just a few of the many available micro-features. We note that even within micro-features, the signatures of various profile parameters may be embedded at different (micro)levels, or (micro)resolutions of the voice signal. These could be revealed by appropriate transformation of the features *a-posteriori*. Though not discussed further, all of these fall within the purview of micro-articulometry.

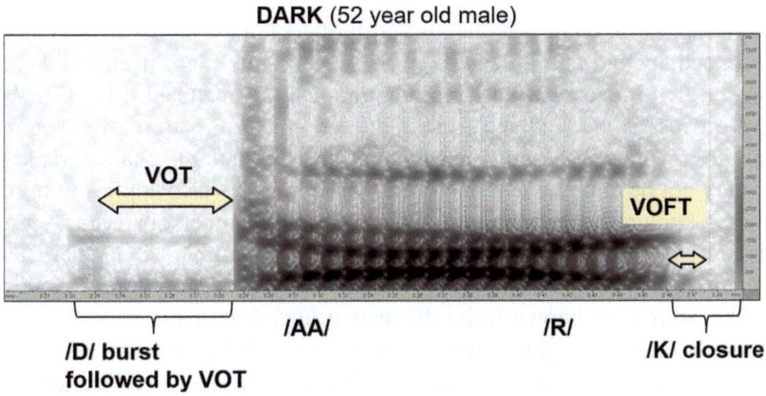

VOT and VOFT on the spectrogram of an instance of the word DARK, as obtained by the structured-prediction algorithm. In the speech of an adult speaking American English, VOT is of the order of 25 ms following voiced plosives and 95 ms following unvoiced plosives.

Fig. 7.5 Illustration of voicing onset and voicing offset times

7.2 Micro-articulometry in Voice Disguise

One of the most difficult challenges in profiling is that of profiling accurately in the presence of voice disguise. The goal of this section is twofold: (a) to discuss how micro-articulometry is a logical practical approach to solving this problem, and (b) to discuss the usefulness of some of the micro-features discussed above.

There is a strong reason why this topic precedes further discussion of feature engineering techniques in this chapter. Micro-features are viewed as "high-resolution" or "fine-detail" features derived from the articulometric units of speech. A question that naturally arises, and *must* be addressed before we discuss their further development and discovery, is whether their fine nature makes them vulnerable to voluntary obfuscation or destruction by the speaker. If that were the case, further work on their discovery would not be justified. However, as we see below, micro-features are in fact persistent under voluntary variations of voice as exemplified by voice disguise.

7.2.1 The Problem of Breaking Voice Disguise

In Sect. 6.4.2 we discussed the subject of voluntary and involuntary variations of voice. For profiling accurately in the presence of voice disguise, we must obviously focus on the characteristics of voice that *cannot* be varied. Interestingly, the majority, if not all of these characteristics relate to phonemic productions at articulometric levels.

Let us again focus on impersonation (or mimicry) in the realm of voice disguise. Human speech, including mimicked speech, is the result of bio-mechanical and cognitive *generative* processes that adapt continuously to feedback [31–33]. When a cognitive process is induced and forced into an unnatural pattern, the demands on the brain are greater [34], and it is expected that the patterns characteristic to the process may not be consistently sustained over time. This also applies to mimicked speech. It is known through studies that in successful impersonation, even among the characteristics that can be voluntarily varied by the mimic, some characteristics of voice are successfully mimicked while others are not. Most of the successfully mimicked features are known to be distinctive *high-level* patterns such as accent, style, rhythm, rate, prosody etc. It is generally through the imitation of these characteristics that mimics create their auditory illusions of their target's speech. More information on this subject is given in Sect. 6.4.1.4 of Chap. 6.

We refer to the process of detecting voice disguise and profiling successfully in the presence of it as *breaking through voice disguise* or *breaking voice disguise*. The process of detecting whether the given voice signal represents disguise or not, in the absence of prior voiceprints for reference, is called *in-vacuo detection of voice disguise*. Both, the detection of voice disguise, and profiling accurately in the presence of voice disguise, are tied in concept. Detection involves the identification of inconsistencies in sound production during impersonation. Profiling involves the identification of consistent characteristics—those that cannot be changed by the impersonator and persist throughout the impersonation—that can be used for analysis. Since they are obviously complements of each other, the discussion below is unilateral.

Detection of impersonation is based on two hypotheses:

1. **Temporal inconsistencies**: Even for speech characteristics that the mimic *does* successfully imitate from the target, they will revert, at least partially, to their "natural" mode when the mimic is not alert, or when doing so does not appear to affect the perceived quality of the mimicry. Thus from this perspective, the variation in patterns of even well-mimicked features are expected to be greater than in the natural voices of either the target or the mimic.
2. **Articulometric inconsistencies**: At articulometric levels, some phonemes are not changeable, while others are changeable to various degrees. For example, the sound of frication as we say /ch/ as in the word "catch," or "/jh/" as we say "just," or of turbulence during intervocalic respiration etc. are generally not changeable. These do have resonance patterns, though, and can be used to deduce the true profile parameters of the speaker [35]. The sounds that are changeable to various degrees may be inconsistently changed, leading to a motley collection of articulometric units in the mimic's speech with relatively disparate properties.

While temporal and articulometric inconsistencies can be used to automate the detection of voice disguise, the invariance of many micro-characteristics (and micro-features) under voice disguise can also be exploited. Let us discuss the specific example of breath sounds first, and then phonemes in general as exemplary cases.

7.2.1.1 Breath Sounds

The spectral details in the low and mid frequencies of different phonemes in speech are important in the judgment of naturalness of voice [36, 37]. Changing these is important for successful mimicry. However, while this is possible in the case of vowels and some consonants, it is not for some consonants, and for some sounds that are categorized as neither but still play an important role in speech production. One example is the intervocalic breath sound (explained in Sect. 2.6).

A breath sound is treated as a sound unit of continuous speech, just like a phoneme, in LVCSR systems. This distinction is also based on articulometric considerations. For every speaker, during inhalation, the articulators assume a specific lax position that is habitual to the speaker in which the speaker's vocal tract is in its most natural shape. The lips are not protruded, nor do the articulators obstruct the vocal tract in any way.

The vibration of vocal folds is not involved in the production of breath sounds. Instead, the source of breath sounds is turbulence related to aspiration. Like all other sound waves that originate in the vocal tract, breath sounds also resonate in the vocal chambers of the speaker, and consequently also exhibit spectral peak patterns that are derivable from the physics of wave motion in such cavities (see Sect. 2.3.2). The *intensity* and duration of the intervocalic breath sounds, as explained in Chap. 2, depends on the respiratory dynamics of the speaker, rather than the speaker's muscular control. Any attempt to mimic someone else's breath in isolation is likely to sound *more* unnatural and inconsistent as compared to other mimicked sounds.

Fortunately for mimics, breath sounds are lower in intensity than other phonemic sounds, and do not hamper the listener's perception. However, many of their characteristics remain invariant under disguise, and their analysis can reveal the true identity of the speaker.

Figure 7.6 shows the spectrogram of the breath sounds of four adult speakers, three of who were attempting to impersonate the fourth speaker. This example is in fact extracted from a database of impersonated speech of US presidential candidates. The impersonation was done by expert mimics and was convincingly representative of the target. We see that despite the impersonation, the qualitative differences in the turbulence and resonance patterns of their breaths remain clear.

For breath sounds of two speakers to resonate at the same frequencies, they must have similar facial skeletal proportions, and similar volumes and relative sizes of the different vocal cavities (in the least). In addition, they must have similar respiratory dynamics. If they have different lung capacities and are in different states of health, the *number of words* they speak after which they need to inhale air could be very different. In addition, the *volume of air* that they inhale could be very different. Given these considerations, it is reasonable to expect that the breath sound of one person cannot be easily impersonated by another (except in an flagrant manner), unless both speakers are in identical state of health, have identical lung capacities and volumes, identical trachea diameters and identical vocal fold elasticity and control, among many other factors.

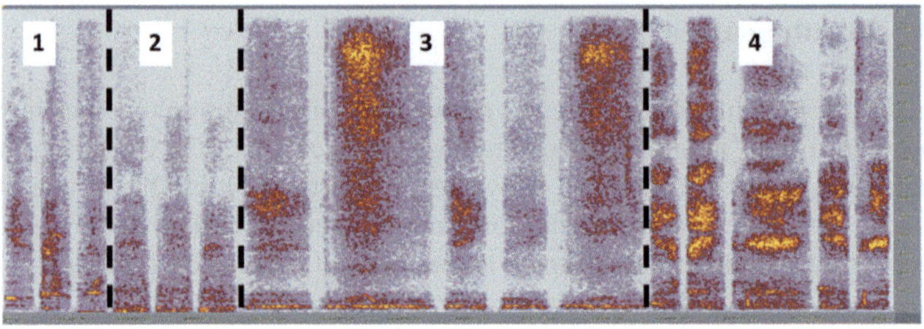

Breath sounds of a target speaker (a presidential candidate from our database), and of expert public performers while impersonating his voice. Label 3 shows the target's natural breath sounds, Labels 1, 2 and 4 are the breath sounds of his impersonators. All signals were energy-normalized and are displayed on exactly the same scale.

Fig. 7.6 Spectral patterns of breath sounds in the case of impersonation

In impersonation, the *constancy* of breath durations and intervals between breath sounds is also disturbed. The impersonator may actually strain to speak more words than they are used to speaking in their natural state before they inhale (in an attempt to emulate a larger person), or breathe unnaturally frequently and erratically if they attempt to impersonate a smaller person (such as a child). Impersonating someone else's words and styles also imposes a cognitive load and affects breath patterns [38].

7.2.1.2 Phonemes in Voice Disguise

The articulometric considerations applied to breath can be extended to other sounds. Vowels are most easily varied during impersonation, since these depend on tongue height and backness. The tongue is a highly pliable articulator, and it can be used to imitate the vowel sounds made by a different speaker effectively with practice. Of this category, the front vowels are easier to manipulate then the back vowels.

Of the consonants, the glottal (involving only the larynx), and velar (involving the back of the tongue and the velum) consonants are often the specific consonant categories that most mimics find hardest to change.

In general, consonants that involve at least one rigid or two relatively fixed structures in the vocal tract, and positions of articulators that have relatively fewer degrees of freedom (such as the back of the tongue) are harder to change than others. This category includes the alveolar (produced by touching the tongue against the alveolar ridge), and palatal (involving the middle of the tongue and the hard palate) consonants. Within the alveolars, the ones that involve the tongue blade (such as /ch/, /sh/, /jh/) are less easily modified than those that involve the tongue tip (such as /r/).

The dental consonants (articulated with the tongue against the back of the upper incisors) can be changed by touching the tongue further backwards onto the hard palate or by touching the portions of the tongue slightly further down from the tip, to the upper incisors. Interdental consonants (produced by placing the tip of the tongue between the upper and lower front teeth, also called non-sibilant fricatives), can be modified to some extent by using the tongue differently. Bilabial consonants involve both lips, but the lips are pliable in shape, and and are easily manipulated.

Within the generalizations made above [35], mimic-specific differences exist. Also, based on place of articulation, finer distinctions can be made based on the manner of articulation. For example, the liquids are most easily modified, since they involve significant movements of the tongue. Fricatives and nasals are less easily modified, affricates and glides are the least easy to modify, and stop sounds are only modified through a change in the duration of the stop of airflow before their characteristic burst. All of these observations are agnostic of language.

In addition to understanding which phonemes are likely to be manipulated in voice disguise, it is important to know what characteristics of those may be modified. Some commonly modified characteristics are the voicing source, relative durations of phonemes, their relative loudness, formant frequencies, and formant transitions between phonemes.

Formant frequency values (or positions) are important for the intelligibility of speech sounds. Being the primary characteristic of the sounds produced by any given configuration of the vocal tract, they are also key to disambiguating phonemes (typically, the first three formants F1–F3 are sufficient to disambiguate vowel sounds). During impersonation, the mimic must maintain them in a manner that the intelligibility of speech is not lost. Formant positions must also be maintained because specific consonants derive their identities from them (see Sect. 2.3.4.2).

The timing and duration of transitions between formant positions of phonemes in continuous speech are related to the articulatory agility of the speaker's vocal tract in changing configuration from one phoneme to the next. As a result, the formant energy patterns during these transitions capture many of these speaker characteristics. Figure 7.7 shows an example. Figure 7.7 is the spectrogram of multiple instances of the word "great" spoken by a target and his impersonators. The transitions between the phonemes /r/and /ey/ in this word are visible as the high-energy sloping contours in the figure. We see that the slopes of these formant transitions are different across different speakers. They are distinctive for each speaker, and remarkably similar for the same speaker regardless of the actual duration of the transitioning phonemes. Formant transition patterns have in fact been observed to persist for specific combinations of phonemes for a speaker in spite of the within-utterance variations in fundamental frequency that occur during normal conversational speech.

Once voice disguise is detected using these features, profiling must be done using only those parts of the voice that are likely to reflect the true parameters of the speaker. This relies on a data-driven analysis of phonemes to answer the question: "which are the sounds that a speaker *cannot* vary during impersonation?" This analysis is easily done by comparing similar sounds at micro-levels from data that represent impersonated and normal speech from the same speaker. The comparison could be done across multiple micro-features, e.g. formant-based measurements as in [35]. When such analysis is not feasible, a judicious choice of phonemes and features can still yield accurate profile information, regardless of the fact that the voice is disguised. For example, the voicing onset time is very characteristic of a speaker [39], and hard to change voluntarily unless done in an instance-specific egregious

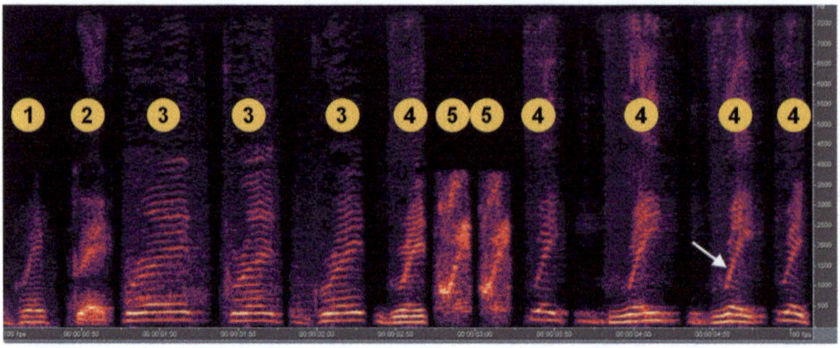

Spectrogram of the word "great" spoken by multiple people. 1,2 and 3 are impersonators of the target 4, and 5 is the target's voice over a noisy, nonlinear, lossy channel. Formant transitions between phonetic units R and EY are seen as the high-energy slopes within each instance of the word. One of these is indicated by the white arrow.

Fig. 7.7 Illustration of the uniqueness of inter-phoneme formant transition patterns across speakers

manner (it does change under the influence of intoxicants or diseases, but those changes are not voluntary). Efficient algorithms to find and measure it exist [40], making this feature useful for profiling in the presence of disguise.

For in-vacuo detection, articulometric cadence is also a useful feature. Since impersonators must capture a multitude of obvious patterns like accent, cadence, prosodic and linguistic style during impersonation, maintaining a specific cadence or speech pattern in a sustained fashion over all instances of each sound over time can be extremely difficult.

7.3 Feature Selection

The example above was given to illustrate the applicability and importance of micro-articulometric considerations in profiling. The features (such as formants and their transition patterns) were chosen based on prior knowledge in this specific example.

How do we decide which features are best for a profiling task? Or for any task, for that matter? If we have a number of possible candidate features, we must select those that are effective for the task. If we don't have features at all, we must design or discover them.

The exercise of choosing the right features for a task is called *feature selection*. In most data-driven learning schemes, a large number of feature types are concatenated into a high-dimensional feature and thereafter, the dimensionality of this composite feature is reduced to yield more effective yet lower-dimensional features. However, this is *not* feature selection. This is *dimensionality reduction*.

Technically, dimensionality reduction techniques such as Linear discriminant Analysis (LDA), singular Value Decomposition (SVD), Principal Components analysis (PCA) etc seek to *compact* the information represented by the original high-dimensional features. The original features are transformed in this process. Feature selection techniques, on the other hand, seek to *downselect* a set of features from a larger set of features. Usually the downselected set is equally, if not more, effective for the task at hand. The goal of downselection is to improve computational efficiency while at the same time eliminating confusing or meaningless features.

Feature selection methods are of three kinds: filter, wrapper and embedded methods. **Filter methods** rank features (often independently of other features, and thus are also called univariate methods) based on statistical scores computed using tests such as chi-squared tests, correlation analysis etc., and eliminate low-scoring features. **Wrapper methods** seek to search for the most optimal combination of features for a model or task. Model accuracy and task completion metrics may be used to assess all possible feature combinations, or a judiciously chosen subset of those, for optimality. The search methods involved may be of different kinds—rule-based or heuristic, stochastic, hill-climbing etc. **Embedded Methods** select features based on model accuracy while the model parameters are being learned. Examples of such methods are regularization methods that impose additional constraints (or regularizers) on regression algorithms to guide the learning process to reach better optima. Specific examples of such regularized feature selection algorithms are those used in sparse optimization, and include LASSO, elastic net, ridge regression etc.

In feature selection for profiling, all three types of algorithms can be highly relevant. The choice depends on the application context. Traditional features may capture the effects of some parameters, but may still need to be downselected for maximum impact and efficiency. However, due to the nature of micro-signatures, and the difficulty in measuring them directly using basic signal processing or other considerations, most feature selection strategies in profiling are likely to be those that are based on task outcomes: if we have a set of features to downselect from, they can be individually tested and ranked based on task completion metrics, and selected for further use based on the rankings.

In other cases features may need to be discovered, or devised, or both. From a feature selection perspective, the methods used may still be thought of as wrapper or embedded methods for feature selection, but they digress from these in complex ways, enough that we treat these cases separately as that of *feature discovery* or *feature engineering*: the discovery or creation (respectively) of features that are optimally suited for a specific task.

For profiling, perhaps the greatest promise lies in engineered features – those that are designed to discover the effects of different profile parameters on speech, starting only with the hypothesis that such effects must be present in the voice signal, regardless of how nebulous we think they are, whether or not we are able to hear them as humans, or visualize/interpret them on standard/traditional representations of the voice signal (such as the spectrogram and others discussed in earlier chapters). We discuss this important topic below.

7.4 Feature Engineering

As a reminder, we reiterate that there are two facets to micro-articulometry. In one, well-studied effects of articulatory-phonetic variations on different sounds are found and measured. This is a necessary step in all bottom-up approaches to profiling (see Chap. 1). In the second, which relates to the top-down approach to profiling, micro-features must be automatically discovered or engineered. In this section we illustrate the latter in two different contexts: one where we know of effective features, but cannot measure them; and another where no feature exists *a-priori* for the profiling task at hand.

7.4.1 Proxy Features

As we have noted from the outset, the human voice has been a subject of curiosity, and of scientific study, for literally centuries. Often, in existing scientific literature, we come across carefully curated studies that report useful results and provide valuable insights into the relation of voice quality to various speaker parameters. For example, age has often been correlated with increased raspiness and creak in voice. Breathiness has been related to various physical conditions, including exhaustion and general physical health, intoxication has been related to wobble in voice etc.

If the voice qualities referred to in these studies (breathiness, raspiness, creak etc.) could all be quantified, it would be possible to leverage the results of these studies to build much more accurate profiling mechanisms than are possible in a purely data driven setting—a fact that will become more evident in Chap. 8, which deals with mechanisms for profiling.

The problem, however, is that many such features (especially voice-qualitative features) cannot be quantified, either for lack of a precise definition, or a precise methodology for their measurement from the signal.

To find a solution to this problem, we turn to the concept of transfer learning. Transfer learning is often used to deal with situations wherein data needed to learn the components of the algorithm(s) required for the completion of a primary task are scarce (or missing). In transfer learning, we learn components of the algorithm(s) from a different but related task (secondary task) for which we have sufficient data and then "transfer" the knowledge gained from the secondary task to the primary task. For an appropriately chosen secondary task, we expect optimal features to be similar to those needed for the primary task. This paradigm can be rearchitected to measure features that are not quantifiable. One way to do this is through *proxy* features.

Proxy features may be defined as quantifiable features that can be reasonably used in the place of features that may not be quantifiable for some reason (often the reason is simply the absence of a method for quantification). For example, if we observe that a specific voice quality component could be a base feature for profiling for some parameter, but is difficult to quantify or measure, then we try to find an alternative feature that *can* be measured, and thus can proxy for the unquantifiable entity. However, for such a feature to qualify as a proxy feature, it must satisfy some minimum statistical criteria:

1. It must have high mutual information with the base feature
2. It must have a high correlation with the base feature

In addition, a proxy feature must be *learnable* from data, and must be *verifiable* using tagged (labeled) data.

Once proxy features are found, machine learning algorithms can then learn to quantify the base features they proxy for. The quantified base features can then be used for profiling.

In order to find a proxy feature for an unquantifiable voice quality, we begin with some proxy speech characteristics that are measurable and are believed, to be directly relatable to the target voice quality, and use these to construct a predictor for those proxy characteristics from appropriately labeled data. Let c be a proxy characteristic for a quality feature Q. Let $\mathscr{C}$ be the set of values that c may take. For numeric proxies $\mathscr{C}$ is the set of numeric values that c can take. For categorical c, $\mathscr{C}$ is the set of categories. We train a predictor $P : S \longrightarrow \mathscr{C}$, which operates on a speech signal S to attempt to predict the value of c. Let f_S be an *intermediate* feature value computed by P from S to make the prediction. For instance, if P is a neural network, f_S might be the output of the penultimate (or any other) layer of the network (Fig. 7.8a).

The output $P(S)$ of the predictor, or the intermediate feature f_S may now be used as proxy features for the voice quality. The effectiveness of this proxy for the target voice quality can now be evaluated using appropriate statistical tests, such as the point-biserial correlation between the proxy feature and the actual voice quality Q.

When we have more than one such proxy characteristic that may be suitable for building the proxy feature, we must select the subset of proxy characteristics that best predicts Q. Let $c_1, \cdots, c_k$ be the set of candidate proxy characteristics, and let $\mathscr{C}_j$ be the set of values that may be taken by c_j. Let $C = \{c_i, \ldots, c_l\}$ be any subset of the k proxies. Let $\mathbf{C}$ be all possible subsets of $c_1, \cdots, c_k$. Let $\hat{\mathscr{C}}_C = \prod_{j:c_j \in C} \mathscr{C}_j$ be the set of all combinations of values that the proxies in C can take.

We can now compose a predictor $P_C : S \longrightarrow \hat{\mathscr{C}}_C$ to predict the values of all $c \in C$ from S (Fig. 7.8b). Let F_C (short for $F_C(S)$) be a proxy feature derived from P_C. We define a *proxy quality measure* $X(F_C, Q)$, which quantifies the statistical relation between F_C and Q, e.g. the point-biserial correlation between the two.

We can now choose our optimal proxy feature as

$$F = \arg\max_{C \in \mathbf{C}} X(F_C, Q) \tag{7.3}$$

In practice searching over all the subsets in $\mathbf{C}$ will often not be possible, either because it is computationally infeasible, or because the requisite training data to train the predictors are unavailable. In such cases, the choice of subsets to evaluate $\mathbf{C}$ must be restricted through heuristics that take into account factors such as the

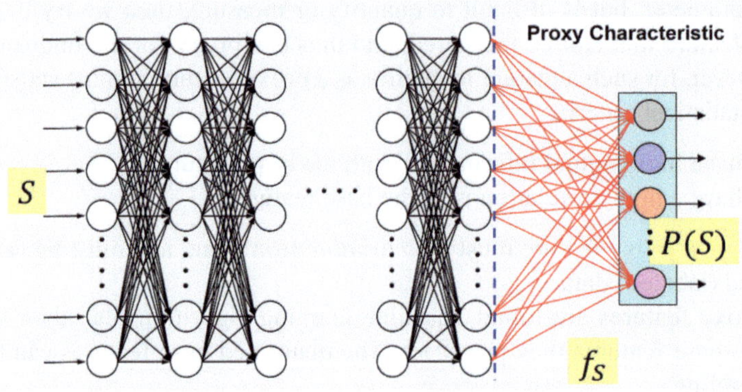

(a) Composing proxy features for a voice quality feature Q based on a single proxy characteristic. A neural-net classifier operates on speech input S to compute the probabilities $P(S)$ for the various values of the characteristic. Either $P(S)$ or f_S, the output of the penultimate layer of the network, may be used as a proxy feature for Q.

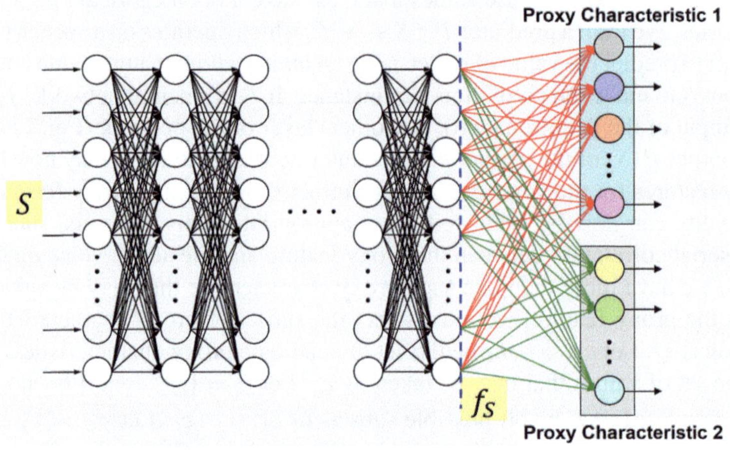

(b) Composing proxy features from a set of proxy characteristics. The network now simultaneously predicts the values of all the characteristics. The output f_S of the penultimate layer of the network acts as the proxy feature for voice quality Q.

Fig. 7.8 Generation of proxy features

availability of training data, and *a priori* knowledge such as phonetic or acoustic groupings derived from linguistic principles or the theory of voice production.

As specific examples that have been implemented in practice, we take the voice qualities of breathiness and nasality. It has been found that among phonemes in a language, proxy features generated from nasals and nasal-vowel combinations are predictive of nasality. The verification of this comes from the observations that (a) nasal-sound classifier scores have high point-biserial correlation with nasality, and (b) Pearson's correlation when one of the variables is binary, is greater than 0.7. For breathiness, features from groupings that separate unvoiced fricatives from voiced fricatives (or in general voiced sounds) have high proxy value, and show high point-biserial correlation with breathiness. They are also highly correlated with high ratio of frication to harmonic energy in speech. Similar proxies can be found for several other voice quality features, using voice characteristics that lead to judgments that are observed to be made from those voice quality features, e.g. masculinity and femininity.

The predictors trained with such groupings, as mentioned before, can be used to yield high proxy-quality features. These can then be used within a transfer learning framework to deduce profile parameters. This entire pipeline is shown in Fig. 7.8. The figure shows a specific instance where we train a non-linear logistic classifier for proxy classes, which is a deep neural network with softmax output, from more readily available training data. The proxy classes for nasality, for example, could be the linguistic groupings of nasal phonemes and non-nasal phonemes. A classifier that is able to discriminate between them is likely to yield features that correlate highly with the quality of nasality. The network output scores are thus the ideal proxy features for the target quality feature. Once trained (ideally with a lot of data), the network parameters are frozen, and a new prediction network is attached to it that is trained to predict the profile parameter under consideration. Since the larger proxy-generating network is frozen, the new attached network can be smaller, with fewer parameters, and would thus require much less data to train. This entire network is thus a transfer-learned network for profiling with proxy features.

As we can imagine, many different embodiments of the concept of proxy are possible. The discussion here was only an illustrative example.

7.4.2 Feature Discovery Using Artificial Intelligence Techniques

Often, all that is known in the scientific literature about how a particular parameter affects the human body seems to also indicate beyond doubt that those effects must extend to the human voice. A simple example is that of caffeine. Caffeine is known to be a mental and physical stimulant. It isn't hard to reason that the effects of this stimulation must also extend to the muscles, nervous activations and other bio-mechanisms in the vocal tract and thus affect the entire voice production process. This leads to the simple but well-founded hypothesis that the effects of caffeine must be embedded in some form (e.g. characteristic patterns) in the voice signal. However, an examination of standard representations of the speech signal may not reveal these effects, and neither may subjective human evaluation. Human hearing may be oblivious to these effects.

Yet, if the effects are reasonably present, where are they? Can the effects be isolated? Can they be captured in the form of some feature that can be extracted from the voice signal? If so, can it (or they) be used to identify the presence or absence of caffeine in the bloodstream? How can we identify and extract such a feature? Can it be "pure" enough that it captures primarily the effects of caffeine, and not other influences? Can it help us cleanly disambiguate the effect of caffeine from the myriad other factors that influence the voice signal at the same time?

In most cases, we may not already know of such an ideal feature. When this is the case, AI techniques are often the most appropriate and effective mechanisms to use. These can help us *discover and design* such custom features. *Feature discovery* involves designing mechanisms to create custom or engineered features that are specifically designed for a particular task (in this example case, deducing whether or not a person has ingested caffeine).

There are multiple frameworks and paradigms under which one can approach this problem. For instance, traditional analyses such as principal component analysis (PCA) and independent component analysis (ICA) can derive low-dimensional representations of the signal which capture its salient features in the sense of the objective optimized by the analysis technique. The optimized objective may have different goals, e.g correlation in the case of PCA, statistical independence in the case of ICA, etc. However, the derived features may bear no direct relation to any profiling-relevant signal semantics. For our purpose, we require an approach that learns features that are forensically interpretable. This requires appropriately constrained mechanisms to derive such variables. For the purpose of illustration, we describe *one* of many possible ways to do so in the paragraphs below.

This example of feature engineering is a very simple yet powerful one, based on a neural-network framework. Within this framework, an encoder transforms any standard representation of speech (such as spectrograms or others described in Chaps. 3 and 4) into a "custom" or "engineered" feature representation in a structured latent space via adversarial matching. A decoder within the framework ensures that the custom feature faithfully incorporates all the information present in the original representation of speech. The adversarial matching provides a semantic link to the custom feature, enabling us to ensure that the feature does indeed capture the effects of the desired parameter that is hypothesized to influence the voice signal. The framework also enables us to impose any additional constraints, e.g. on the nature of the distribution of the parameter, which may be derived from a priori knowledge of the characteristics of the feature. It also provides mechanisms that are meant to disambiguate the effects of this parameter from all others that may simultaneously also influence the voice signal. A more detailed explanation follows.

7.4.2.1 Feature Discovery: A Specific Example

The specific example explained here is that of an *adversarial autoencoding learning framework*. It establishes a continuous bi-directional mapping between the original (e.g. spectral) space in which the speech signal is represented, and a latent space.

It also imposes a parametric structure onto the representation in the latent space, providing a tool to semantically manipulate the mapping between the latent space and the original space.

A specific implementation of such a network is shown in Fig. 7.9. This feature discovery framework is predicated on the hypothesis that underlying any voice signal X is a latent characterization Z that holds the "essence" of the signal, and from which the signal can be reconstructed. This is also the desired feature that fulfills the conditions of having specific statistical properties, and optimally and unambiguously representing the effects of the profile parameter in question. Once discovered, the feature representation Z is expected to lead to effective performance for at least one classifier (used for profiling) that seeks to detect the presence, absence or degree of the parameter affecting the voice signal.

The model in Fig. 7.9 (showing only inputs and not the update pathways) comprises four primary components, as described below. In the expressions below, the subscript to any symbol for a function represents the parameters of the function.

- An *encoder* $E_\theta(X)$ with parameters θ that operates on clean speech X to extract latent representations $Z = E_\theta(X)$ from it.
- A *decoder* $G_\phi(Z)$ with parameters ϕ that operates on the latent representation to reconstruct the original speech from Z.
- An optional *constraint enforcer* $H_\psi(Y)$ with parameters ψ that takes in random signals Y and outputs a vector $\tilde{Z}$ that conforms to statistical or topological constraints that Z must ideally follow.
- A *discriminator* $D_\delta()$ with parameter δ that attempts to distinguish between latent representations $\tilde{Z}$ derived from Y, and the latent representation Z derived from speech.

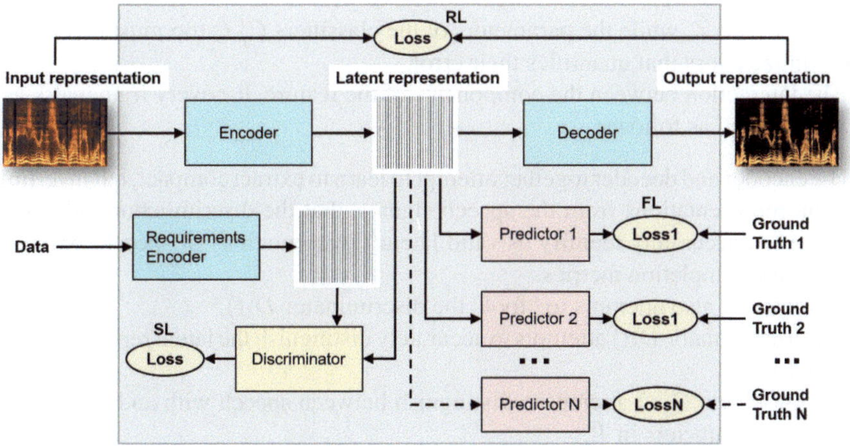

Fig. 7.9 A neural model for feature discovery

- A bank of *classifiers* $C_\rho^P(Z)$ with parameters (collectively represented as) ρ that operates of subspaces of the latent space to detect a set P of targeted voice parameters that are encapsulated in Z. These voice parameters may be bio-relevant parameters that are expected to have signatures in the voice.

For the discovery or creation of effective features for profiling according to the criteria mentioned earlier, the components have the following objectives.

1. Ideally, any latent encoding Z produced from the speech signal X must be perfectly recomposed to the original signal by the decoder, i.e. $G_\phi(E_\theta(X)) = X$.
2. The latent representations $\tilde{Z}$ derived from Y by $H_\psi()$ must be statistically indistinguishable from the representation Z that would be derived by $E_\theta()$ from the speech signal X.
3. The discriminator $D_\delta()$ must be able to accurately distinguish between latent representations Z and $\tilde{Z}$.
4. $C_\rho^P(Z)$ must be able to operate on subspaces of Z and perfectly distinguish between the signals having the effects of the parameters in P, and those not having them in $\bar{P}$.

The model (or rather the parameters of the model) must be learned from a large labeled corpus, i.e. (X, L) pairs—where L is a *set* of labels relating to the features P. The encoder $E_\theta()$ and decoder $G_\phi()$ are trained such that a divergence $Div(X, G_\phi(E_\theta(X)))$ between the speech X and the *autoencoded* version $G_\phi(E_\theta(X))$ is minimized, thereby ensuring that the decoder $G_\phi()$ is able to accurately reconstruct speech signals from the latent representations Z derived from speech by $E_\theta()$. Concurrently, the encoder $E_\theta()$ must also learn to "fool" the discriminator $D_\delta()$, to ensure that Z is statistically indistinguishable from $\tilde{Z}$, and thus has the statistical characteristics imposed by $H_\psi()$. $E_\theta()$ is also learned to minimize the losses computed by the classifiers $C_\rho^P(Z)$. $D_\delta()$ must in turn be adversarially trained to distinguish between Z and $\tilde{Z}$, while the parameters of the classifiers $C_\rho^P()$ too must be updated to minimize a loss that quantifies their error.

The interaction between the components of the feature discovery framework can be summarized as follows:

- The encoder and decoder together attempt to learn to extract compact, but invertible latent representations from the speech signals, that the discriminator and classifier(s) too accurately identify as valid latent representations of speech, based on their task completion metrics.
- The encoder also attempts to "fool" the discriminator $D_\delta()$.
- The discriminator $D_\delta()$ attempts to accurately distinguish the latent representations Z and $\tilde{Z}$.
- The classifiers $C_\rho^P()$ attempt to distinguish between speech with and without the targeted parameters in P.

This leads to the following learning process. Let P_X represent the probability distribution of the speech recordings. Let P_Y represent the distribution of $\tilde{Z}$. We define the following losses:

- *Reconstruction loss (RL)*:

$$L_{reconstruction} = E_{X \sim P_X} Div_r(X, G_\phi(E_\theta(X)))$$

The expectation $E_{X \sim P_X}$ is over the probability distribution P_X. The divergence function $Div_r()$ is any typical divergence function, e.g. ℓ_2, that may be used to compare two signals.

- *Discriminator loss (DL)*:

$$L_{discriminator} = E_{X \sim P_X} D_\delta(E_\theta(X)) - E_{Y \sim P_Y} D_\delta(H_\psi(Y))$$

- *Scatter loss (SL)*: If we want to train the parameters of $H_\psi()$ as well, we define the scatter loss

$$L_{scatter} = E_{Y \sim P_Y} D_\delta(H_\psi(Y))$$

- *Classifier loss (Forensic loss or FL)*: We define the classifier loss as the sum of the cross-entropy losses for the individual classifiers

$$L_{\text{classifier}} = \sum_P w_P \text{Xent}(C_\rho^P(X), P(Z))$$

where w_P is a weight that assigns an importance to parameter P and $P(Z)$ is the specific value of the parameter for the signal from which Z has been extracted.

The signal encoder, classifiers, and decoder are optimized as

$$\phi^*, \theta^*, \rho^* = \arg\min_{\phi,\theta,\rho} L_{reconstruction} + L_{discriminator} + L_{classifier} \tag{7.4}$$

If $H_\psi()$ must also be optimized, the above objective includes the scatter loss and is additionally also optimized with respect to ψ.

$$\phi^*, \theta^*, \rho^*, \psi^* = \arg\min_{\phi,\theta,\psi,\rho} L_{reconstruction} + L_{discriminator} + L_{classifier} + L_{scatter} \tag{7.5}$$

The discriminator is optimized as

$$\delta^* = \arg\min_\delta L_{discriminator} \tag{7.6}$$

The above framework in fact represents a typical instance of a generative adversarial formalism [41], of which variants exist that attempt to improve its generalizability and stability [42, 43]. In practice the losses will not be actual expectations, but rather be empirical estimates obtained from training-data collections $\mathcal{X} = \{X_1, X_2, \ldots, X_N\}$ of speech recordings and $\mathcal{Y} = \{Y_1, Y_2, \ldots, Y_M\}$ of Y. The minimization of Eq. 7.4 (or Eq. 7.5) and Eq. 7.6 is performed through gradient descent, where the gradients of the losses with respect to the various network parameters are computed using the backpropagation algorithm. As is customary in such models, parameter updates for Eq. 7.4 (or Eq. 7.5) and Eq. 7.6 are alternately performed.

Figure 7.9 and the solution outlined above are only intended to be illustrative of a more generic framework. Various extensions, to impose more direct feedback from $D_\delta()$ to $E_\theta()$ and $G_\phi()$, to ensure that $G_\phi(\tilde{Z})$ represents a valid speech signal, etc. may be imposed by simple extension.

The classifiers ensure that Z leads to optimal classification performance for the parameter whose effects are expected to be captured by Z. This ensures that Z is meaningful for profiling for the parameter. Further, the projections into subspaces for the individual P may now be explored (or inverted) to identify the specific speech characteristics that relate to parameter P. In order to ensure that this feature disambiguates between competing effects, the optimization may be augmented by losses that ensure that while a projection into a subspace relevant to a feature P maximizes the classification of that parameter, the same subspace is ineffective (i.e. results in random classification) for other parameters.

The bank of classifiers can ideally be constructed to resolve all ambiguities in the discovery of the correct Z. In practice, however, the size of this bank may be limited by the availability of training data.

As we see, feature engineering in this manner represents a paradigm shift in the derivation of meaningful features from the voice signal. Traditionally, feature representations were designed from basic signal considerations, as explained in Chap. 4.

In the next chapter, we will discuss mechanisms that use these features for profiling. These may be different from the classifiers used in the feature discovery framework. Since the framework itself learns to generate optimal features from data, the coupling of discovered features with other downstream mechanisms for profiling is essentially *transfer learning* [44, 45] for profiling in the broadest sense.

References

1. Fowler, C. A., & Saltzman, E. (1993). Coordination and coarticulation in speech production. *Language and Speech*, *36*(2–3), 171–195.
2. Maeda, S. (1990). Compensatory articulation during speech: Evidence from the analysis and synthesis of vocal-tract shapes using an articulatory model. In W. J. Hardcastle & A. Marchal (Eds.), *Speech Production and Speech Modelling* (pp. 131–149). Dordrecht: Springer.
3. Ziegler, W., & Von Cramon, D. (1985). Anticipatory coarticulation in a patient with apraxia of speech. *Brain and Language*, *26*(1), 117–130.
4. Nicholas, M., Obler, L. K., Albert, M. L., & Helm-Estabrooks, N. (1985). Empty speech in Alzheimer's disease and fluent aphasia. *Journal of Speech, Language, and Hearing Research*, *28*(3), 405–410.
5. Cummings, J. L., Darkins, A., Mendez, M., Hill, M. A., & Benson, D. F. (1988). Alzheimer's disease and Parkinson's disease: Comparison of speech and language alterations. *Neurology*, *38*(5), 680–680.
6. Mendez, M. F., Clark, D. G., Shapira, J. S., & Cummings, J. L. (2003). Speech and language in progressive nonfluent aphasia compared with early Alzheimer's disease. *Neurology*, *61*(8), 1108–1113.

7. Ahmed, S., Haigh, A. M. F., de Jager, C. A., & Garrard, P. (2013). Connected speech as a marker of disease progression in autopsy-proven Alzheimer's disease. *Brain*, *136*(12), 3727–3737.
8. Forbes, K. E., Venneri, A., & Shanks, M. F. (2002). Distinct patterns of spontaneous speech deterioration: An early predictor of Alzheimer's disease. *Brain and Cognition*, *48*(2–3), 356–361.
9. Menn, L. (1983). Development of articulatory, phonetic, and phonological capabilities. *Language Production*, *2*, 3–50.
10. Harnad, S. (2003). *Categorical perception. encyclopedia of cognitive science*. Macmillan: Nature Publishing Group.
11. Delattre, P. (1969). Coarticulation and the locus theory. *Studia Linguistica*, *23*(1), 1–26.
12. Ferrand, C. T. (2006). *Speech science: An integrated approach to theory and clinical practice*. Boston: Allyn & Bacon.
13. Singh, R., Raj, B., & Vitanen, T. (2012). The basics of automatic speech recognition. *Techniques for Noise Robustness in Automatic Speech Recognition (T. Virtanen, R. Singh & B. Raj (Eds.)*. New York: Wiley.
14. Brand, M. (1999). Structure learning in conditional probability models via an entropic prior and parameter extinction. *Neural Computation*, *11*(5), 1155–1182.
15. Dehak, N., Kenny, P. J., Dehak, R., Dumouchel, P., & Ouellet, P. (2011). Front-end factor analysis for speaker verification. *IEEE Transactions on Audio, Speech, and Language Processing*, *19*(4), 788–798.
16. Loakes, D. (2006). A forensic phonetic investigation into the speech patterns of identical and non-identical twins, Doctoral dissertation, School of Languages. Australia: University of Melbourne.
17. Snell, R. C., & Milinazzo, F. (1993). Formant location from LPC analysis data. *IEEE Transactions on Speech and Audio Processing*, *1*(2), 129–134.
18. Mercer, J. (1935). "Popeye the Sailor Man." Popeye—Original Motion Picture Soundtrack Album (Vinyl, LP, Album: Voice Art Rendition). USA: The Boardwalk Entertainment Co.
19. Auzou, P., Ozsancak, C., Morris, R. J., Jan, M., Eustache, F., & Hannequin, D. (2000). Voice onset time in aphasia, apraxia of speech and dysarthria: A review. *Clinical Linguistics and Phonetics*, *14*(2), 131–150.
20. Klatt, D. H. (1975). Voice onset time, frication, and aspiration in word-initial consonant clusters. *Journal of Speech, Language, and Hearing Research*, *18*(4), 686–706.
21. Kent, R. D., & Rosenbek, J. C. (1983). Acoustic patterns of apraxia of speech. *Journal of Speech, Language, and Hearing Research*, *26*(2), 231–249.
22. Sweeting, P. M., & Baken, R. J. (1982). Voice onset time in a normal-aged population. *Journal of Speech, Language, and Hearing Research*, *25*(1), 129–134.
23. Morris, R. J., & Brown, W. S, Jr. (1994). Age-related differences in speech variability among women. *Journal of Communication Disorders*, *27*(1), 49–64.
24. Tremblay, K. L., Piskosz, M., & Souza, P. (2003). Effects of age and age-related hearing loss on the neural representation of speech cues. *Clinical Neurophysiology*, *114*(7), 1332–1343.
25. Flege, J. E. (1991). Age of learning affects the authenticity of voice-onset time (VOT) in stop consonants produced in a second language. *The Journal of the Acoustical Society of America*, *89*(1), 395–411.
26. Stölten, K., Abrahamsson, N., & Hyltenstam, K. (2015). Effects of age and speaking rate on voice onset time: The production of voiceless stops by near-native L2 speakers. *Studies in Second Language Acquisition*, *37*(1), 71–100.
27. Decoster, W., & Debruyne, F. (1997). The ageing voice: Changes in fundamental frequency, waveform stability and spectrum. *Acta Oto-Rhino-Laryngologica Belgica*, *51*(2), 105–112.
28. Fischer, E., & Goberman, A. M. (2010). Voice onset time in Parkinson disease. *Journal of Communication Disorders*, *43*(1), 21–34.
29. Ackermann, H., & Hertrich, I. (1997). Voice onset time in ataxic dysarthria. *Brain and Language*, *56*(3), 321–333.

30. Flint, A. J., Black, S. E., Campbell-Taylor, I., Gailey, G. F., & Levinton, C. (1992). Acoustic analysis in the differentiation of Parkinson's disease and major depression. *Journal of Psycholinguistic Research*, *21*(5), 383–399.
31. Heald, S., & Nusbaum, H. C. (2014). Speech perception as an active cognitive process. *Frontiers in Systems Neuroscience*, *8*, 35.
32. Hashimoto, Y., & Sakai, K. L. (2003). Brain activations during conscious self-monitoring of speech production with delayed auditory feedback: An fMRI study. *Human Brain Mapping*, *20*(1), 22–28.
33. McGettigan, C. (2015). The social life of voices: Studying the neural bases for the expression and perception of the self and others during spoken communication. *Frontiers in Human Neuroscience*, *9*, 129.
34. Raichle, M. E., MacLeod, A. M., Snyder, A. Z., Powers, W. J., Gusnard, D. A., & Shulman, G. L. (2001). A default mode of brain function. *Proceedings of the National Academy of Sciences*, *98*(2), 676–682.
35. Singh, R., Gencaga, D., & Raj, B. (2016). Formant manipulations in voice disguise by mimicry. In *Proceedings of the Fourth International Workshop on Biometrics and Forensics (IWBF)* (pp. 1–6). Limassol, Cyprus: IEEE.
36. Klatt, D. H., & Klatt, L. C. (1990). Analysis, synthesis, and perception of voice quality variations among female and male talkers. *The Journal of the Acoustical Society of America*, *87*(2), 820–857.
37. Kreiman, J., Garellek, M., Chen, G., Alwan, A., & Gerratt, B. R. (2015). Perceptual evaluation of voice source models. *The Journal of the Acoustical Society of America*, *138*(1), 1–10.
38. Henderson, A., Goldman-Eisler, F., & Skarbek, A. (1965). Temporal patterns of cognitive activity and breath control in speech. *Language and Speech*, *8*(4), 236–242.
39. Allen, J. S., Miller, J. L., & DeSteno, D. (2003). Individual talker differences in voice-onset-time. *The Journal of the Acoustical Society of America*, *113*(1), 544–552.
40. Sonderegger, M., & Keshet, J. (2012). Automatic measurement of voice onset time using discriminative structured prediction. *The Journal of the Acoustical Society of America*, *132*(6), 3965–3979.
41. Goodfellow, I., Pouget-Abadie, J., Mirza, M., Xu, B., Warde-Farley, D., Ozair, S., et al. (2014). Generative adversarial nets. In *Advances in Neural Information Processing Systems 27 (NIPS 2014)* (pp. 2672–2680).
42. Arjovsky, M., Chintala, S., & Bottou, L. (2017). Wasserstein generative adversarial networks. In *Proceedings of the Thirty-Fourth International Conference on Machine Learning (ICML)* (pp. 214–223). Australia: Sydney.
43. Neyshabur, B., Bhojanapalli, S. & Chakrabarti, A. (2017). Stabilizing GAN training with multiple random projections. arXiv:1705.07831.
44. Pan, S. J., & Yang, Q. (2010). A survey on transfer learning. *IEEE Transactions on Knowledge and Data Engineering*, *22*(10), 1345–1359.
45. Dai, W., Jin, O., Xue, G. R., Yang, Q., & Yu, Y. (2009). Eigentransfer: A unified framework for transfer learning. In *Proceedings of the Twenty-sixth Annual International Conference on Machine Learning (ICML)* (pp. 193–200). Montreal, Canada: ACM.

Chapter 8
Mechanisms for Profiling

So how is profiling actually done? Most of this book has been dedicated to developing the basic understanding needed for it. We have seen that the knowledge of how a parameter affects the vocal production mechanism can help us identify the most relevant representations from which we may extract the information needed for profiling. We have also seen how such knowledge can help us reason out why certain parameters may exert confusable influences on the voice signal. All of this knowledge can then help us design more targeted methods to discover features that are highly effective for profiling.

The next step, and the subject of this chapter, is to find ways to computationally relate these features to the parameters to be deduced. The greatest challenge of computational profiling is to do this correctly *in the presence of* confounding influences of myriad other parameters that may affect voice at the same time. The learning of relationships, and prediction from those, are aspects of this technology that must draw from the fields of machine learning and artificial intelligence.

That brings us to the obvious question: what *exactly* is artificial intelligence? For our purposes, any computational mechanism for performing "intelligent" tasks such as learning and decision-making, whether or not completely intentioned by a human at every stage of its functionality, falls under the aegis of artificial intelligence. The difference between AI and machine learning is simply that while machine learning produces AI, not all AI requires machine learning. AI in fact could be strictly rule-based, based on a compendium of hand-crafted expert rules. Machine learning algorithms, on the other hand, *actively* learn from experience and data.

Techniques designed for profiling must combine both: rule-based approaches that build upon knowledge of the processes and causal influences that underlie voice production, and data-driven machine learning approaches that learn relationships between various entities. Moreover, they must be designed to address and solve for the inherent challenges (such as the disambiguation of concurrent influences) *through and within the process of learning*. For these reasons, profiling techniques fall under the broader umbrella of AI.

In this chapter we will discuss one exemplary machine-learning algorithm for the measurement of micro-features that must be derived from standard representations,

R. Singh, *Profiling Humans from their Voice*,
https://doi.org/10.1007/978-981-13-8403-5_8

two exemplary AI-based techniques that are especially useful for profiling for bio-relevant parameters, and point the reader to the literature for others that could be used. AI is a rapidly evolving field, so what is presented here may be superseded in detail very soon. Not in concept, though—those presented in this chapter will remain relevant.

We will also discuss the subject of environmental profiling. Profiling for environmental parameters is bifurcated in approach: for tracking the parameters through their influence on the *voice* produced by the speaker, the mechanisms are exactly the same as those used for profiling for bio-relevant parameters. For tracking environmental parameters through the modifications they make on the *audio* signal, the influence of which may extend to the voice contained in it, the mechanisms are completely different. They range from techniques to isolate the influence of the environment from the signal, followed by analysis of the "cleaned" or "de-voiced" signal, to analysis of the voice component of the signal to gauge environmental parameters from the changes that happened to it *after* it was generated—between the processes of generation and storage. Between those end-points, a number of influences may have played a part—from the devices used for audio capture, to the coding schemes and channels used for transmission, to the mechanisms used for formatting the received signal for storage. This pipeline is so varied and complex in itself that this chapter can only provide a to-do list, rather than explanations. The section on profiling for environmental parameters will therefore just skim through topics at a superficial level, giving some level of detail to the *very* key approaches.

8.1 Mechanisms for the Deduction of Bio-Relevant Parameters

To deduce profile parameters, computational techniques must learn to map or infer the parameters from given features. Essentially, these are then *prediction* mechanisms. They can follow two paradigms. In one, (components of) the underlying physical processes that produce speech are directly modeled in order to relate features to parameters. For instance, one may analyze the effect of parameters such as age or illnesses on the periodicity of the speech signal by solving or simulating the coupled differential equations that describe the movements of the vocal folds, and replicating the same effects through appropriate settings of the variables and constants of these equations, which in turn comprise the features for this task. Such models can then be used to predict age and illness parameters in many ways.

In the second, the determination of the relation between features and parameters is treated purely as a problem of learning a mathematical mapping between the two. The learned models may bear no relation to the underlying physical processes that produced the data, e.g. a neural network model that attempts to predict age or illness from the given features. The boundary between the two approaches is sometimes fuzzy. For instance auto-regressive models of the speech signal are derived purely from data, but the model coefficients obtained can nevertheless be related to acoustic-tube approximations of the vocal tract, e.g. [1, 2].

In the literature, mathematical models for machine learning that relate input and output variables are often also categorized in other ways. The problem of learning a relationship may categorized as **classification** or **regression**, depending or whether the output variable is *categorical*, e.g. gender, or *numeric*, e.g. age (although some researchers consider categorical variables to be a special case of numeric variables, and classification to be an instance of regression). The learned models may be **generative**, where they attempt to accurately learn the joint probability distributions of features and parameters—or they may be **discriminative**, where they may attempt to find a relation from the one to the other without considering their actual distributions. The models may also be **probabilistic** or **deterministic**, depending on whether they invoke (models for) the statistical distributions of the variables, or simply learn a relationship without modeling the uncertainty in the values.

The models may use a variety of constraints and learning strategies. We will not describe these further in this chapter. Many detailed expositions of machine learning models exist in the literature, such as [3–7], among dozens of others. Many (generative) models of the vocal tract, and models of other aspects of vocal dynamics have been proposed in the literature as well. Some representative examples are described in [8–14].

8.1.1 Machine Learning Algorithms for the Estimation of Micro-Features

Micro-features that are based on traditional representations of the speech signal, such as spectrograms, correlograms and others described in Chap. 5, are by definition fine-grained signatures found within those representations. They must be measured with precision for profiling. Often, signal processing techniques are not sufficient for this, due to inherent ambiguities in their precise placement.

A good example is that of voicing onset time (VOT), described in Chap. 7. VOT is a micro-feature in the time domain that corresponds to vocal fold inertia. It is best observed as the time elapsed between the end of a voiceless consonant, and the beginning of a voiced phoneme. Due to the variable nature and transitional continuities of speech sounds, the exact location of the boundaries of VOT in time are not defined. They must be probabilistically located, learned and determined. Machine learning algorithms are best suited for problems such as these. For illustrative purposes, the measurement of VOT is described below. This serves the dual purpose of describing an exemplary machine learning process, and of providing information specifically about the measurement of VOT, which is an important micro-feature for profiling. The algorithm is fairly generic, and can be applied for the measurement of a wide variety of micro-features with minimal modification.

8.1.1.1 Estimation of VOT: A Structured-Prediction Algorithm

To measure a micro-feature, its precise "location" in the mathematical space that it exists in must be determined. In many cases, the "indicators" that can pinpoint its location may be weak. A real life far-left-field (intentionally so) analogy may be that of trying to locate a small break in an underground water pipe. A patch of green grass that is greener than the grass around may be an indicator of the location of the break, but anomalous shades of green are not always indicators of underground pipe-leaks. A wet patch of soil may be an indicator, but again, all wet patches do not indicate a broken pipe. If homes on either side of the patch have slightly unequal water pressure in their water supply, then the evidence combined with the anomalously green grass, and the wet patch in its vicinity may make it more likely that the location in question is the location of the fault, although the causes for low water pressure don't always have to do with breaks in water pipes. In this example, joint evidence from three weak indicators is used to solve a difficult fault-location problem. When a sufficiently large number of weak indicators are used, their joint evidence may help locate the fault almost exactly.

Structured-prediction is a powerful machine-learning approach that uses the same principle to locate micro-features in mathematical spaces. It is extremely useful for the measurement of micro-features when multiple indicators of their presence are available, although they may be weak indicators. We illustrate this algorithm using the measurement of VOT as an example.

VOT is measured in terms of its duration. Automatic measurement of VOT is a challenging problem—the onset of voicing is typically a faint cue that is easily missed, as is the initial burst that signals a plosive. The typical duration of a VOT is often between 10–20 ms. Of the many techniques available for its measurement, e.g. [15–17] the structured-prediction method has been consistently shown to result in VOT estimates with errors less than 5 ms [18], provided the algorithm is cued about the approximate location of the VOT. These approximate location cues that roughly "choose" the plosive-vowel combinations where a VOT is most likely to be clearly exhibited, are derived using a speech recognition system. Once found, the structured-prediction algorithm is used on those segments as described below.

For the measurement of VOT, we make the assumption that while several features derived from the speech signal can act as indicators for the presence of VOT, no single feature is sufficiently informative to be relied upon entirely. The structured prediction (SP) model combines the evidence from a number of such features to determine the boundaries of VOT. These features are derived from the voice signal, and their exact nature for the measurement of VOT is described in [18].

In the first step, from a given speech segment $X = \{x_1, x_2, \ldots, x_n\}$ that includes adjacent phonemes with a VOT (e.g. a plosive-vowel combination), we compute a number of numeric features $\phi_i(X, T_p, T_v), \ i = 1 \ldots K$ from the signal.

Consider a *training* signal segment X that includes a plosive-vowel conjunction with a VOT. Let $t_{X,p}$ be the time of the release of the plosive, i.e. the true *beginning* of the VOT, and $t_{X,v}$ the instant of voicing, i.e. the true *end* of the VOT period. Let

$\phi_i(X, T_p, T_v),\ i = 1 \ldots K$ be K *feature maps* computed on the signal, assuming the (estimated) beginning and end times of the VOT to be T_p and T_v.

The SP model computes a score

$$S(X, T_p, T_v) = \sum_i w_i \phi_i(X, T_p, T_v) \tag{8.1}$$

from the time segment (T_p, T_v) of the signal. In vector notation,

$$\mathbf{w} := [w_1, w_2, \ldots, w_K]^\top \tag{8.2}$$
$$\Phi(X, T_p, T_v) := [\phi_1(X, T_p, T_v), \phi_2(X, T_p, T_v), \ldots, \phi_K(X, T_p, T_v)] \tag{8.3}$$
$$S(X, T_p, T_v) := \mathbf{w}^\top \Phi(X, T_p, T_v) \tag{8.4}$$

The SP model estimates the start and end of the VOT as

$$\hat{t}_{X,p}, \hat{t}_{X,v} = \arg\max_{T_p, T_v} S(X, T_p, T_v) \tag{8.5}$$

To learn the optimal VOT estimator, the weight vector $\mathbf{w}$ must be estimated such that the error of the above estimate is minimized.

To do so we define the following loss:

$$L_X = \max\{|(t_{X,p} - t_{X,v}) - (\hat{t}_{X,p} - \hat{t}_{X,v})| - \varepsilon, 0\} \tag{8.6}$$

where ε is a margin. The weight $\mathbf{w}$ is estimated to minimize the empirical average of the above loss over a large number of training instances, leading to an iterative estimate with the following update rule:

$$\mathbf{w} \longleftarrow \mathbf{w} + \sum_X \Gamma(\Phi(X, t_{X,p}, t_{X,v}) - \Phi(X, \hat{t}_{X,p}, \hat{t}_{X,v})) \tag{8.7}$$

where Γ is a diagonal matrix whose ith diagonal entry represents the learning rate corresponding to the ith feature map of $\Phi(X, t_p, t_v)$. This update rule has been proven to converge to a local optimum in [19]. The learned weights may be used in conjunction with Eq. 8.5 to estimate the VOT on test instances.

To summarize, the structured prediction algorithm involves computing a feature map ϕ_i from all the features derived from the signal, the score of which may be expected to peak strongly when computed on the segment between the start and end instants of the estimated VOT (with boundaries T_p and T_v), and have lower values over other segments of the signal. The VOT may then be estimated simply by searching for the segment boundaries at which this joint score peaks—a task which may be performed through dynamic programming. The challenge of training this predictor is simply that of optimally determining the weights with which the feature maps must be combined. To summarize, the SP model computes a weighted combination $\sum_i w_i \phi_i(X, T_p, T_v)$ of the evidence from all of the feature maps. The boundaries of the true VOT are estimated as the instants $\hat{t}_{X,p}$ and $\hat{t}_{X,v}$ at which this score peaks. The learning phase for the detector estimates the weights $\mathbf{w} = [w_1, w_2, \ldots, w_K]^\top$ such that the expected error between the estimates given by the above predictor and the *true* boundaries of the VOT is minimized.

The structured prediction algorithm outlined above may be easily modified to estimate other micro-features as well. What is required is a set of features derived from the voice signal that are expected to assume optimal values in the presence of the micro-feature, and below-threshold values when the micro-feature is not present. In other words, the only difference would in fact be in the feature maps used, which must then be responsive to the specific micro-feature being measured, rather than to VOT.

8.1.2 AI Algorithms for the Estimation of Parameters

In this section we outline two AI-based mechanisms that use AI in its broadest sense. They are particularly useful for profiling because they can include (and be intrinsically guided by) any available profiling-relevant knowledge, while using powerful machine learning techniques such as neural networks. One is *corrective learning*: essentially a learning process that corrects its learning strategy based on current outcomes and prior knowledge. The second is *regression by classification* that uses knowledge-based covariates to guide the learning process.

8.1.2.1 Corrective Learning

Classification and regression algorithms must handle the task of making a prediction of a vector Y, from an input X, where Y may either be a vector of scores associated with a number of classes, or a prediction of a continuous valued variable. For instance, X might be an image of an object, and Y may be a vector of probabilities of object classes for the image. Or X might be a speech recording, and Y may be a prediction of the body parameters of the speaker. The classifier or predictor may generally be represented as a function $f()$, with parameters W, such that the prediction Y in response to an input X may be written as $Y = f(X; W)$. The parameters W are generally learned from training data—(X, Y) pairs that are known to be associated with one another—through various optimization algorithms. The function $f()$ may take any form, such as linear or non-linear regressions, autoregressive models, or, more recent neural network structures of various kinds, such as multi-layer perceptrons, convolutional neural networks, or recurrent neural network structures that take an input X and make a prediction Y. We will generally refer to the function $f()$ as a predictor, without making a distinction about whether it attempts to predict a class probability or a continuous-valued regression, as this distinction is immaterial to the rest of this discussion. In many problems, the predictor may be presented with multiple independent inputs, for which the same output is expected. For instance, an image classifier may be presented with multiple pictures $X_1, X_2, \ldots, X_N$ of the same object, where it is known *a priori* that $X_1, X_2, \ldots, X_N$ represent the same object. A speech-based height predictor may be provided with multiple spoken recordings from the same speaker. Note that $X_1, X_2, \ldots, X_N$ do not represent a single time-series of inputs. Rather, they are unrelated independent inputs which could individually be used to make a prediction of Y. In these cases, the multiple separate inputs provide separate evidence to the predictor variable. It naturally makes sense that the evidence from all of the inputs must be combined to make the prediction. A common theme in these situations has been to determine how best to combine the evidence from the multiple inputs in predicting the output.

Several solutions to this problem have the following format: The individual predictions $Y_i = f(X_i; W)$ are assumed to be independent of one another. A final overall prediction is made as a weighted average of the individual predictions:

$$Y = \sum_i w_i Y_i \tag{8.8}$$

where the weight w_i (with a value between 0 and 1) is assumed to represent a confidence in the output of the prediction Y_i and may be derived in various ways. The drawback in such approaches is that here, each input X_i effectively ignores the information derived from the other inputs X_j, $j \neq i$, and effectively starts "afresh."

A more efficient method takes a "corrective" approach to prediction. It stipulates that while an initial prediction may be obtained from one or more inputs in the traditional manner, subsequent inputs must be used to refine the prediction and improve it directly, rather than to make entirely novel predictions, ignoring all that has been learned thus far from current inputs.

To this end, a *corrective estimator* may be used, such as the following bank of predictors:

$$\begin{aligned} Y &= f_0(X; W_0) \\ \Delta Y &= f_1(X, Y; W_1); \quad Y \longleftarrow Y_{init} + \Delta Y \\ \Delta Y &= f_2(X, Y; W_2); \quad Y \longleftarrow Y_{init} + \Delta Y \\ &\cdots \end{aligned}$$

Here, the function $f_0(X; W_0)$ predicts the value Y as before, but the subsequent functions $f_1(.)$, $f_2(.)$ etc. are corrective functions. Given an initial estimate Y_{init} of the value Y, they output a *correction* ΔY that must be added to Y_{init} to give a better value of Y.

The operation of the corrective predictor is as follows: Given a series of independent inputs $X_1, X_2, \ldots, X_N$, all corresponding to the same dependent variable Y, an initial set of inputs $X_1, X_2, \ldots, X_{K_0}$ are used to make an initial prediction of Y as follows:

$$Y = \frac{1}{K_0} \sum_{i=1}^{K_0} f_0(X_i; W_0) \tag{8.9}$$

This initial prediction may be based on a small number of initial instances, or even on a single instance (i.e. $K_0 = 1$).

Subsequently, from each of the next K_1 instances, corrections are made using function $f_1()$.

for $i = K_0 + 1, \ldots, K_0 + K_1$ **do**
  $\Delta Y = f_1(X_i, Y; W_1)$
  $Y = Y + \Delta Y$
end for

$f_2()$ is a finer correction function than $f_1()$, and is applied when the corrections recommended by $f_1()$ either converge or oscillate in a chaotic-like manner. Thus for each of the subsequent K_2 inputs, corrections may be made using $f_2()$.

for $i = K_0 + K_1 + 1 \ldots K_1 + K_2$ **do**
  $\Delta Y = f_2(X_i, Y; W_2)$
  $Y = Y + \Delta Y$
end for

Further refinements may be made by additional functions when these corrections converge, and so on. Figure 8.1 below explains the overall structure of the computation.

The discussion above is agnostic to the nature of the functions $f_0()$, $f_1()$, $f_2()$, etc., and a variety of different functions and models may be used.

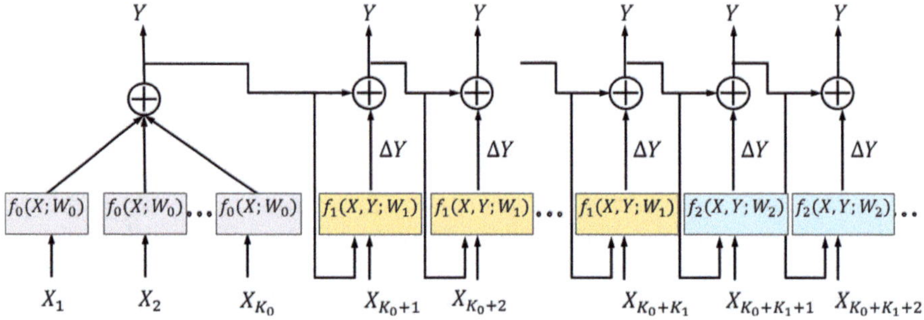

Fig. 8.1 Overall structure of corrective computation. Each "block" is indicated by a different central color

8.1.2.2 Training the Functions

The functions are trained through a modification of standard training procedure. Training requires a number of sets of training examples $\{T_1\}, \{T_2\}, \ldots, \{T_K\}$ of the kind:

$T_1 = \{(X_{11}, Y_1), (X_{12}, Y_1), \ldots, (X_{1N_1}, Y_1)\}$
$T_2 = \{(X_{21}, Y_2), (X_{22}, Y_2), \ldots, (X_{2N_2}, Y_2)\}$
...
$T_K = \{(X_{K1}, Y_K), (X_{K2}, Y_K), \ldots, (X_{KN_K}, Y_K)\}$

where all the training instances in each set have the same value for the dependent variable Y, and K is the number of such sets.

We define a schedule $K_0, K_1, \ldots$, for training, where K_0, K_1, etc. comprise a sequence of non-decreasing numbers and may be experimentally or heuristically set.

We define a loss function for $f_0()$ computed on a training instance (X, Y) as

$$L_0(X; W_0) = Div(f_0(X; W_0), Y) \tag{8.10}$$

where $Div()$ is a non-negative divergence function which has the property that $Div(f_0(X; W_0), Y) = 0$ if and only if $f_0(X; W_0) = Y$.

As seen in Fig. 8.1, at each corrective stage a function $f_i()$ takes in a current input X_t, and a current estimate of the output (which we will generically represent as Y_{t-1}, to distinguish it from the *desired* target output Y), to compute a correction $f_i(X_t, Y_{t-1}; W_i)$, and a corrected output $Y_t = Y_{t-1} + f_i(X_t, Y_{t-1}; W_i)$. We define a loss function quantifying the discrepancy between the estimate Y_t and the target training output Y as

$$L_i(X; W_i) = Div(Y_t, Y) \tag{8.11}$$

We can now define a set loss L_k^{set} as the loss on the kth training set:

$$L_k^{set} = \sum_{i=1}^{K_0} L_0(X_i; W_0) + \sum_{j=1}^{J_{max}} \sum_{i=K_{j-1}^{cum}+1}^{K_j^{cum}} w_i L_j(X_i; W_j) \tag{8.12}$$

where $K_j^{cum} = \sum_{l=0}^{j} K_l$. The second sum terminates when the total number of terms that are summed equals N_k, the number of instances in the set, and w_i is an error weight term that is typically set to be increasing with i to reflect the fact that ideally, the error in prediction should decrease rapidly as the number of inputs used to make the (corrected) predictions goes up. In other words, with more evidence, the function must be more certain of its predictions. The overall loss function on the training input is given by

$$L(W_0, W_1, \ldots) = \sum_k L_k^{set} + R \tag{8.13}$$

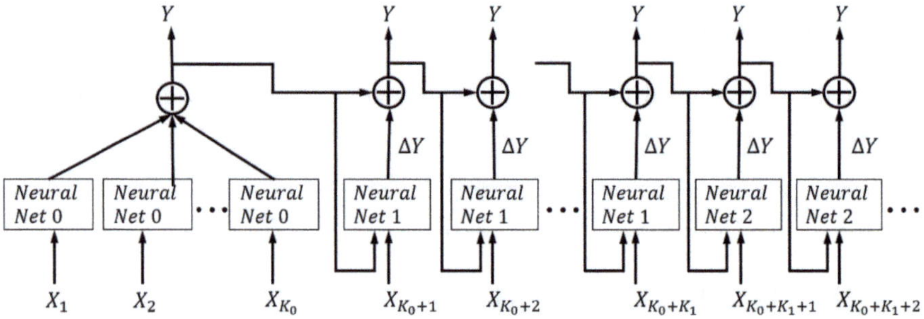

(a) Each predictor network makes predictions of the correction term using only the current input and the latest estimate of Y.

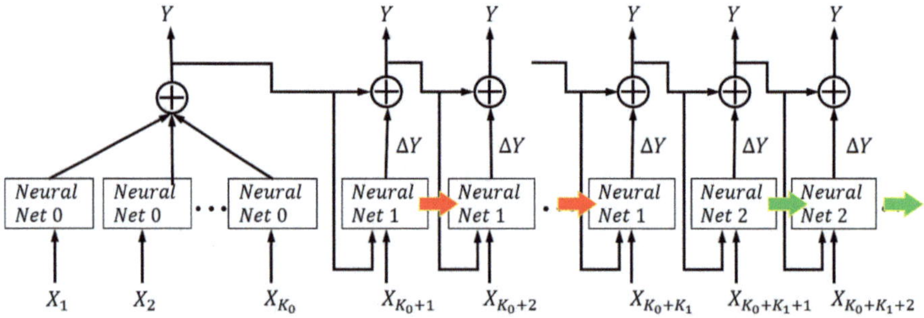

(b) Each predictor also takes, the hidden state values computed from the previous input, as input to its hidden layers. Other mechanisms of carrying information across predictions are also possible.

Fig. 8.2 Two specific implementations of corrective training of a predictor

where R is a regularization term that includes any regularization imposed on the parameters of the functions. The parameters of the functions are learned to minimize the regularized loss function.

$$W_0, W_1, \ldots = \text{argmin}_{\hat{W}_0, \hat{W}_1, \ldots} L(\hat{W}_0, \hat{W}_1, \ldots) \tag{8.14}$$

The minimization may be performed using any appropriate optimization algorithm, such as gradient descent, Newton's method, expectation maximization etc.

The framework above is a generic corrective learning framework. Let us look at a specific example of an implementation based on it using neural networks. In such an implementation, $f_0()$, $f_1()$, etc. are neural networks within this framework. Figure 8.2a, b show two specific implementations of these networks.

In Fig. 8.2a, each network operates independently, only accepting as input the current estimate Y and the actual input X to make a prediction ΔY. In Fig. 8.2b, the network accepts as input not only the current estimate, but also its own internal state from the previous update. The internal state here would be the state of the hidden neurons in the network.

The networks used may be neural networks, convolutional neural networks, or any combination of these. Network parameters may be learned through the well-known back-propagation algorithm. In learning the network parameters, in an example implementation, we could use the following values for the error weights w_i:

- **Uniform**: $w_i = 1$. All errors contribute equally to the total loss regardless of how many inputs have been factored into making the prediction that the error resulted from.
- **Linear**: $w_i = \alpha i$. Here, errors are considered increasingly more important as more and more inputs are obtained. In effect, this weighting scheme claims that making an error of 1 after observing 100 inputs is 100 times as bad as making that same error after observing only one input.
- **Quadratic**: $w_i = \alpha i^2$
- **Exponential**: $w_i = e^{\alpha i}$

Other such regimes may be specified.

8.1.2.3 Regression via Classification

Regression refers to the statistical modeling of the relationship between different variables (explained in more detail in Chap. 10). For example, the variation of height with age, or the variation of height with a feature extracted from a voice signal, the variation of fashion with time and culture—these are all examples of regression problems. In these, the variable with respect to which the relationship is modeled is called the *independent variable*[1] or the *predictor*, and the variable that varies in response to it is called the *dependent* variable. Regression analysis is the term given to any procedure used to estimate such relationships.

Classification involves categorical prediction. It may be viewed as a special instance of regression to a discrete-valued variable. For example, predicting the gender of a person given features from the voice signal is a classification problem. The underlying relationship that is learned between the gender (dependent variable) and the feature (independent variable) is also a regression. All classification problems that are solved using learning-based procedures are regressions (direct rule-based predictions may not involve learning, but may nevertheless be inferred based on statistical methods). In other words, in standard practice, *classification is done via regression.*

This is not always straightforward. For example take the case where both the dependent and independent variable take a continuum of values, as in the case of most profile parameters such as height, weight, degree of an illness etc. Learning a regression for these in a data-driven fashion requires ample data to learn from. Ideally, the data must adequately span, or represent, the entire continuum. Often, this is not possible. For example, to learn a good regressor that can estimate a person's height

[1]The variables are more appropriately called *explanatory* variables, since they may not be independent of one another.

from relevant features derived from voice, the data used must include the voices of people of all heights. Even when we do manage to obtain such a dataset, there may be underlying trends that correlate with some other factor, and are so important that the data must also span the heights within those co-factors adequately. For example, the manner in which heights are distributed across humanity is predicated on age, as well as gender (and possibly other variables). We refer to these variables as **co-factors**. In this example, a good regressor must learn from samples that span all possible heights within the entire range of each co-factor. When we think about such requirements, it is easy to see that a straightforward learning of a regressor from small amounts of data (or data that are not properly distributed across target cases) would yield suboptimal decisions for many profile parameters.

For profiling, a different approach, called **Regression-via-Classification (RvC)** may be more effective, for the reasons explained below.

Traditional regression models such as linear regression, or even neural networks, learn a single set of model parameters for all of the data. Alternately stated, they assume that a single function $y = f(x; \theta)$ with a single global parameter θ is valid over the entire (x, y) data space.

The problem here is that the model parameters that are appropriate for one subset of the data may not be appropriate for another subset. Statistical fits of the model to the data will minimize a measure of the overall prediction error, but may not be truly appropriate for any specific subset of the data. Non-parametric regression models such as kernel regressions and regression trees attempt to deal with this by partitioning the feature space (the space of x), and computing separate regressors within each partition. For high-dimensional data, however, any computed partition runs the risk of overfitting to the training data, necessitating simplifying strategies such as axis-aligned partitions [20, 21].

An alternate strategy is to *partition* the space based on the dependent, or *response* variable, i.e. given a response variable y that takes values in some range $(y_{\min}, y_{\max})$, we find a set of threshold values $t_0, \ldots, t_N$, and map the response variable into bins as $y \mapsto C_n$ if $t_{n-1} < y \leq t_n$ for $n = 1, \ldots, N$. This process, which effectively converts the continuous-valued response variable y into a categorical one C, is often referred to as *discretization*. The new response variable C is, in fact, not merely categorical, but *ordinal*, since its values can be strictly ordered. In order to determine the value y for any x we must now only find out which bin C_n the feature x belongs to; once the appropriate bin has been identified, the actual estimated y can be computed in a variety of ways, e.g. the mean or median of the bin. Consequently, the problem of regression is transformed into one of classification. This process of converting a regression problem to a classification is called *Regression-via-Classification (RvC)* [22].

The concept of RvC is in itself very generic, and a variety of implementations of RvC are possible. In the discussion below, a specific AI-based algorithm for RvC is described for illustrative purposes. Other implementations of RvC can be devised based on the underlying principles that guide such an algorithm, and the task at hand.

An Example of RvC: Neural Regression Trees (NRT)

The Neural Regression Tree (NRT) algorithm [23] builds a hierarchical tree-structured model for RvC—termed a *regression tree* [20].

At each node n of this tree, a partition for the data (i.e. a categorical separation of the data, or classification) is done based on a learned optimal decision boundary for the data corresponding to that node of the tree. The separation is based on a threshold t_n (of a score, applied to the predicted variable), and is a binary partitioning of data—i.e., the data are divided into two groups depending on whether or not the score crosses a threshold. The tree is grown by recursively (hierarchically) partitioning its nodes in a top-down fashion. The strategy of hierarchical binary partitioning of the dependent variable is a greedy one, where the partitioning is locally optimized. This is done by a simple margin-based linear classifier. The strategy has the advantage that at each node, the optimal features (from among a set of available ones) can also be independently chosen for optimality. This is done using a neural network (hence the name *Neural Regression Tree* (NRT)).

The algorithm finds an optimal criterion based on which the data at each node can be partitioned by evaluating all possible partitions of the data based on the dependent variable. The optimal criterion is obtained using a binary *margin-based linear classifier* $h_n : \mathbb{X} \rightarrow \{-1, 1\}$. The optimal criterion itself is the threshold t_n^* that minimizes the classification error

$$t_n^* = \operatorname*{argmin}_{t_n} \mathscr{E}_{h_n}(\mathbb{D}_n, t_n) \tag{8.15}$$

where $\mathscr{E}_{h_n}$ is the classification error of h_n, and $\mathbb{D}_n$ is the data on node n. The number of times this procedure is done within a tree determines the number of nodes N in the full tree. The depth of the tree is determined by this. It stops increasing when the classification performance saturates.

Once learned in this manner, the NRT can be used for prediction. To determine the value of the dependent variable y given some value of the independent variable (such as a feature derived from the voice signal) x, the NRT model first determines which leaf node the feature x belongs to, by following the partition tree. It then computes a value $\hat{y}$ for y based on some preferred statistic (such as mean or median of the data at the leaf node). The generic structure of NRT is shown in Fig. 8.3. Some algorithmic details of NRT that are generic to a variety of profiling data are further explained in the following paragraphs.

Algorithm Details

The elements used in the NRT algorithm are defined below:

- A **partition** Π on a set $\mathbb{Y} \subset \mathbb{R} := \Pi(\mathbb{Y}) = \{C_1, \ldots, C_N\}$. This set is the set of partitions over an entire tree (defined below).

- The **disjoint set** $\{C_1, \ldots, C_N\}$ represented by **C**. C_n represents an element of this set: $\bigcup_{n=1}^{N} C_n = \mathbb{Y}$
- The **dependent variable** $y \in \mathbb{Y}$ that is assigned to a set in C_n through a classification rule $\Pi(y) := C_n \; ; t_{n-1} < y \leq t_n$ for $n = 1, \ldots, N$
- A **classification function** h_θ with parameters θ that estimates the probability of a data instance x being classified into bin C_n, $h_\theta^n(x) = P(C_n \mid x)$, i.e. $h_\theta : x \mapsto \mathbb{P}_N$, where $\mathbb{P}_N$ represents the probability simplex in the N-dimensional space. Thus, $h_\theta^n(x)$ is a vector of probabilities in $\mathbb{P}_N$, and its components sum to 1
- A **regression function** r that maps $r : [x, C_n] \mapsto (t_{n-1}, t_n]$
- An **RvC function** $E_{C_n}[r(x, C_n)]$
- An **error** $\mathscr{E}(y, \hat{y}(x))$ between the true y and the estimated $\hat{y}(x)$, where $\hat{y}(x)$ denotes that $\hat{y}$ is an estimate derived from x
- The **thresholds** $\{t_0, \ldots, t_N\}$ and the parameters $\{\theta_0, \ldots, \theta_N\}$ such that the expected error is minimized
- A **tree** ending in **leaves**. For any leaf l, $l_0 \rightarrow \cdots \rightarrow l_p$ represent the chain of nodes from root to leaf, where l_0 is the **root** and $l_p \equiv l$ is the leaf itself

The RvC rule is given by

$$\hat{y}(x) = E_{C_n}[r(x, C_n)] = \sum_{n=1}^{N} h_\theta^n(x) r(x, C_n) \tag{8.16}$$

Where $E[.]$ is the expectation.

The thresholds are learned such that the expected error over the entire dataset is minimized:

$$\{t_n^*\}, \{\theta_n^*\} \longleftarrow \underset{t,\theta}{\operatorname{argmin}} \, \mathbb{E}_x \left[\mathscr{E}(y, \hat{y}(x)) \right]. \tag{8.17}$$

Note that instead of the above, the *empirical* average error $\text{avg}(\mathscr{E}(y_i, \hat{y}(x_i)))$ can also be minimized.

The joint optimization of $\{t_n\}$ and $\{\theta_n\}$ is a hard problem that scales exponentially with n. A binary tree (shown in Fig. 8.3) makes this tractable, since it can be optimized in a greedy fashion.

The threshold t_n partitions each node n into child nodes n' and n''. To achieve this, a classifier assigns any instance with features x to one of n' or n'': $h_{\theta_n} : x \mapsto \{n', n''\}$. In the NRT model, the classifier h_{θ_n} is a neural classifier that not only classifies the features but also adapts and refines the features to each node.

Once the entire tree is built, the *a posteriori* probability of any leaf is given by $P(l \mid x) = \prod_{r=1}^{p} P(l_r \mid l_{r-1}, x)$, where each $P(l_r \mid l_{r-1}, x)$ is given by the neural classifier on node l_{r-1}. Substitution into (8.16) gives the regressed (estimated) value of y as

$$\hat{y}(x) = \sum_{l \in \text{leaves}} P(l \mid x) r_l(x) \tag{8.18}$$

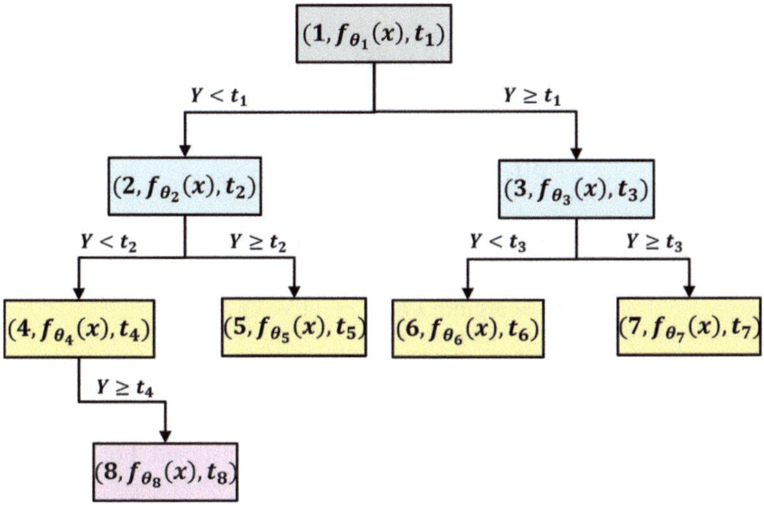

A neural regression tree: the neural classifier h_θ at each node is used to determine the splitting threshold t, which is predicated on the dependent variable y and is locally optimized

Fig. 8.3 Illustration of neural regression tree

where $r_l(x) := r(x, l)$, may be the mean value of the leaf bin or some other statistic (or function) that can be computed from it.

To grow the tree, the average regression error $\mathscr{E}(\mathbb{D}_n; t_n, \theta_n) = \text{avg}\left(\mathscr{E}(y, \hat{y}_n(x))\right)$ between y and its estimated value $\hat{y}_n(x)$ is minimized. In practice, this average regression error is not computable, since the subtree at n is as yet unknown. It is therefore approximated as the *classification* accuracy of the classifier at n in a way that the resultant classification is not trivial and permits useful regression.

A **classification loss** at node n is defined as

$$E_{\theta_n, t_n} = \frac{1}{|\mathbb{D}_n|} \sum_{(x,y) \in \mathbb{D}_n} \mathscr{E}(y(t_n), h_{\theta_n}(x)). \tag{8.19}$$

where $y(t_n) = \text{sign}(y - t_n)$ is a function that indicates whether an instance (x, y) must be assigned to child n' or n'', and $\mathscr{E}(y(t_n), h_{\theta_n}(x))$ is a quantifier of the classification error (which can be binary cross-entropy loss or hinge loss) for any instance (x, y). The classification loss E_{θ_n, t_n} cannot be directly minimized w.r.t t_n, since this can lead to trivial solutions, e.g. setting t_n to an extreme value such that all data are assigned to a single class. While such a setting would result in perfect classification, it would not be ideal for inference. To prevent such solutions, a *triviality* penalty $\mathscr{T}$ is used. It is defined as the entropy of the distribution of instances (membership of each of the partitions at the node) over the partition induced by t_n (other triviality penalties such as the Gini index [21] may also be chosen in a different implementation)

$$\mathscr{T}(t_n) = -p(t_n)\log p(t_n) - (1 - p(t_n))\log(1 - p(t_n)), \tag{8.20}$$

where

$$p(t_n) = \frac{\sum_{(x,y)\in\mathbb{D}_n}(1 + y(t_n))}{2|\mathbb{D}_n|}.$$

The overall optimization of node n is performed as

$$\theta_n^*, t_n^* = \underset{\theta_n, t_n}{\operatorname{argmin}}\, \lambda E_{\theta_n, t_n} + (1 - \lambda)\mathscr{T}(t_n), \tag{8.21}$$

where $\lambda \in (0, 1)$ is used to assign a relative importance to the two components of the loss.

The details of such optimization must be carefully designed. For the NRT algorithm, further details are given in [23].

8.2 Mechanisms for the Deduction of Environmental Parameters

Some environmental parameters influence humans, and thereby their voice. These have been discussed in Chap. 3. Examples of these are physical motion of platforms on which the speaker may be located while talking (e.g a moving car or boat), pollutants in the air breathed in, etc. These **intrinsic influences** can be profiled using the same techniques that apply to profiling for other bio-relevant parameters.

However, some environmental parameters exert other **extrinsic influences** on the voice *signal*. They modify voice *after* it is generated. In this section we specifically focus on some methods that can be used to deduce such environmental parameters from their signatures in the audio signal.

Since the subject of this book is *human* profiling, and not environmental profiling from audio, this is not a very detailed section. The goal here is to give an idea of the kind of influences one can profile for, and the techniques that may be useful in the process.

Voice signals are also audio signals. However, not all audio signals contain voice, and even if they do, voice (in the form of speech) is not continuous. Interspersed with voice are periods of silence, where the influence of the environment on the audio signal may be prominent. Other sounds-emitting objects may leave signatures that may help us deduce their presence, type and nature. This also holds for sound-modifying objects or factors, such as those that reflect sound, cause echo and reverberation, and cause other changes in the audio signal. Environmental profiling includes the deduction of all of these influences.

Making such inferences from general audio signals is a vast field in itself. It comes within the purview of *computational auditory scene analysis* (CASA), and the areas of audio signal processing, audio engineering, machine learning and other fields that CASA draws upon. It is well beyond the scope of this book to discuss those technologies.

Listed below, however, are some key environmental factors that exert an influence on the audio (and voice) signals. Tracking these can help us deduce some facts about the environment in which the captured signal was generated.

1. **Reverberation**: Reverberation affects the voice signal by smearing its spectral content.
2. **Recording devices**: Microphones and associated devices used to capture voice, modify its spectral characteristics.
3. **Transmission channels**: The characteristics of the channel over which the voice signal has been transmitted after recording and prior to storage can modify the voice signal in complex ways.
4. **Storage devices**: The devices used to store voice can modify voice in the process of storage.
5. **Power sources**: At any stage, if an AC power supply is used, it modifies the signal.
6. **Sound emitting objects**: The sounds emitted by objects in the environment are superimposed on the signal. The objects can range from fans and air conditioners in the vicinity of the speaker, to airplanes and other noise sources at considerable distance.

In each of the cases above, the source of the influence can be deduced by learning appropriate detectors (the same mechanisms as the predictors discussed earlier in this chapter) from data that are representative of these influences. For example, to detect the source of a specific sound in an audio signal, a classifier that can identify different types of sound-emitting objects from their acoustic signatures may be used. Such a classifier would need to be trained from (labeled) examples of sound signals from different types of sound sources. To identify sound capture/transmission devices—which can be of many kinds e.g. push-to-talk, open-channel, voice-activated etc.—a database of the most relevant device signals, and their signatures must be used. Their usage signatures can help gauge the skill of the user in using these devices. Examples include usage patterns, sounds introduced by wrong usage, signatures introduced by specific usage settings such as feedback, change in usage-induced sounds (e.g. electronic echo) over time etc.

Similar approaches may be used to identify power sources, storage devices, transmission channels, transmission equipment etc. Transmission of sound over long distances is a complex process. The signal is coded in multiple ways for efficient transmission over channels that may cause distortion, attenuation, dispersions, drop-outs, errors etc. They modify the signal bandwidth and spectral composition of sounds, which the codecs used try to preserve at least in the range of relevance for the sound being transmitted. But the codecs are usually lossy in other ways. The patterns of such effects for different channels can help identify them. They can also help identify the specific codecs used in the transmission process.

The physical dimensions and the structural/material composition of enclosures can be deduced from an analysis of reverberation signatures. For this, the response function of the enclosure must be first deduced using appropriately chosen dereverberation techniques. This function comprises an estimate of how the different frequency components of the sound signal are attenuated by reverberation, and together with information about how different materials reflect and absorb sound (from fields like architectural acoustics, physics and materials science) the signatures can be reverse-engineered to yield an estimate of the type, composition and dimensions of the physical obstructions that result in the specific pattern of reverberation observed in the sound signal under analysis.

While many more methods are being used for profiling and geo-locating sound sources, we will not dwell on those here. We move on to a topic that is very important in the analysis of sound for profiling—that of removing unwanted influences, or noise from the signal. We discuss this topic in more detail here since the techniques are sometimes highly desirable even in human profiling contexts.

8.2.1 Enhancing Audio Signals for Profiling

Enhancing an audio signal is synonymous with removing unwanted influences from it so that the key content of interest is highlighted or made more prominent.

Often the question arises whether it may make sense to use noise removal techniques to enhance voice to highlight the effects of parameters on it, before profiling. Most noise removal techniques modify the signal and leave artifacts in it, and are likely to modify the micro-traces in the signal that are necessary for direct profiling. While they are not recommended for direct human profiling, they can nevertheless be highly desirable in the analysis of the contextual influences of voice during the process. For the same reason, *data enhancement* techniques must not be used in the machine learning stages of profiling. Human profiling is best done *in the presence of noise*, without algorithmic data enhancement.

There are many techniques that may be used in profiling for sound emitting objects in the environment. Since sounds in the environment are additively superimposed on the voiceprint, **signal separation** methods may be used to separate the voice from the background audio signal, prior to analysis. These do not reduce noise, they only attempt to separate it out from the voice. From the perspective of the voice content, they reduce noise by simply discarding the separated background signal while recomposing a new one from the voice components alone. However, this again leaves artifacts in the recomposed signal that may be detrimental to profiling.

In the discussion below, signal separation is explained in some detail, since this is an important step in many object identification tasks within environmental profiling. Currently, there is ongoing research on identifying objects without explicit signal separation or enhancement, most of those use neural networks. Once successful, it would be preferable to not use signal separation, since (as we will see from the discussion below), it not completely lossless and leaves artifacts in the signal. It has many other drawbacks which are mentioned inline.

8.2.1.1 Signal Separation: Spectral Factorization through Compositional Models

The STFT of noisy speech can be shown to be the sum of the STFT of clean speech and that of the noise:

$$Y[f,t] = S[f,t] + N[f,t] \tag{8.22}$$

where $Y[f,t]$, $S[f,t]$ and $N[f,t]$ are the value at frequency f in the tth analysis frame of the STFT of the noisy speech, clean speech and noise respectively. We further assume that their magnitudes add, i.e., $|Y[f,t]| = |S[f,t]| + |N[f,t]|$ (in theory the powers add, but it has been empirically shown that addition of magnitudes is as effective for analysis of short segments of audio, sometimes more). The magnitude spectrum of clean speech can be estimated from this in many ways. Once estimated, to re-synthesize clean speech, the phase spectrum of the noisy speech is "borrowed" and used along with the magnitude.

In following, we will work with magnitudes only, but drop the $|.|$ notation for simplicity. We represent all the spectral components $S[f,t]$ at a single time as a vector S for brevity. Assuming D frequency components in the STFT, S is a D-dimensional vector.

The **compositional model** represents the magnitude spectral vector S as a linear *non-negative* combination of K D-dimensional "basis vectors" $B_k,\ k = 1 \ldots K$:

$$S = \sum_k w_k B_k, \tag{8.23}$$

where w_k is the weight assigned to the kth basis. The bases are intended to be meaningful compositional components of sound; hence they must reside in the same domain as the data S and have the same interpretation as S. Because S is a spectral magnitude (and magnitudes are strictly non-negative), the bases B_k also represent the spectral magnitudes and are also non-negative.

The weights w_k are also strictly non-negative. This is because most sound is compositional in nature—sound waves add constructively, e.g. music may be composed by additive composition of the notes that comprise it. *Cancellation*, which is a major component of any decomposition in terms of additive bases, rarely factors into the composition of a sound, except by careful design. Thus all terms in Eq. 8.23 are strictly non-negative.

The bases B_k can be learned, or simply chosen from representative data. A variety of methods have been proposed in the literature to learn the bases, of which the two most commonly used methods are variants of latent variable decomposition [24] and non-negative matrix factorization (NMF) [25]. The two methods have been shown to be equivalent [26].

In the following we use the NMF approach. Let $\mathbf{S}$ represent a $D \times N$ matrix whose N columns are D-dimensional spectral vectors from some training examples of a signal. We can write each vector S in $\mathbf{S}$ as a linear combination of bases as expressed in Eq. 8.23. In matrix terms, Eq. 8.23 can be rewritten as $S = \mathbf{Bw}$, where $\mathbf{B}$ is a $D \times K$ matrix whose columns are the K basis vectors B_k, and $\mathbf{w}$ is a K-dimensional vector comprised of the weights w_k. The complete matrix $\mathbf{S}$ can now be expressed in matrix form as

$$\mathbf{S} = \mathbf{BW} \tag{8.24}$$

where $\mathbf{W}$ is a $K \times N$ matrix whose columns are the weight vectors corresponding to the individual spectral vectors in $\mathbf{S}$. The bases $\mathbf{B}$ can now be found by minimizing a divergence measure between $\mathbf{S}$ and $\mathbf{BW}$. A variety of divergence measures can be used for this [27, 28]. A generalized Kullback-Leibler divergence is used below as an illustrative example. In this case, the bases (and weight matrix) can be estimated through iterations of [25]:

$$\mathbf{B} \longleftarrow \mathbf{B} \otimes \frac{\left(\frac{\mathbf{S}}{\mathbf{BW}}\right)\mathbf{W}^T}{\mathbf{1}\mathbf{W}^T} \tag{8.25}$$

$$\mathbf{W} \longleftarrow \mathbf{W} \otimes \frac{\mathbf{B}^T\left(\frac{\mathbf{S}}{\mathbf{BW}}\right)}{\mathbf{B}^T\mathbf{1}} \tag{8.26}$$

where $\mathbf{1}$ is a $D \times N$ matrix of ones. The multiplication $\otimes$ is an element-wise multiplication, and the divisions are all also element-wise. The NMF decomposition produced by the iterations above is typically a dimensionality-reducing decomposition—the number of bases K must be lesser than D, otherwise the decomposition becomes indeterminate and solutions such as (for $K = D$) $\mathbf{B} = \mathbf{I}$ (the identity matrix), and $\mathbf{W} = \mathbf{S}$ may be obtained. When the number of bases K exceeds the number of dimensions D, we get an *overcomplete* decomposition. To obtain non-degenerate solutions in this case, we must now impose an additional constraint. We do so by specifying that the columns of $\mathbf{W}$ must be sparse, i.e. in each column the number of non-zero elements must be small. Intuitively, this stipulates that while the total number of compositional elements (bases) that can combine to compose the signal can be arbitrarily high, the number of bases that compose a specific segment must be small. Music is a good simile—we have a very large number of notes in music, but those that are struck together at any particular instant are small. To model this constraint, the update rule for W given in Eq. 8.26 must be modified:

$$\mathbf{W} \longleftarrow \mathbf{W} \otimes \frac{\mathbf{B}^T\left(\frac{\mathbf{S}}{\mathbf{BW}}\right)}{\mathbf{B}^T\mathbf{1} + c} \tag{8.27}$$

Where c is a non-negative "sparsity coefficient" [29, 30]. Larger values of c result in more sparse solutions.

8.2.1.2 Separating Speech From Noise

In speech enhancement and robustness applications, noise must be eliminated from speech. In profiling for environmental parameters, the opposite must be done. In both cases, the discussion below is generically applicable.

Let's assume we have learned (from training data), a matrix of J bases $\mathbf{B}_n$ for the noise and another of K bases $\mathbf{B}_s$ for speech. The two sets of bases can be concatenated into a single matrix of bases: $\mathbf{B}_a = [\mathbf{B}_s \mathbf{B}_n]$. Representing (the magnitudes of) all the spectral components in the spectrogram of the noisy signal as the matrix $\mathbf{Y}$, the latter can be modeled as $\mathbf{Y} = \mathbf{S} + \mathbf{N}$, where $\mathbf{S}$ is the spectrogram for the underlying clean speech and $\mathbf{N}$ is the spectrogram of the corrupting noise.

We can now learn a decomposition

$$\mathbf{Y} = \mathbf{B}_a \mathbf{W}_a \tag{8.28}$$

by repeated iteration of Eq. 8.26 (or Eq. 8.27, if $\mathbf{B}_a$ is overcomplete, i.e. if $D < K + J$). Note that to only estimate $\mathbf{W}_a$ it is not necessary to iterate Equation 8.24. The estimated $\mathbf{W}_a$ is a $(K + J) \times N$ matrix (where N is the number of spectral vectors in $\mathbf{Y}$) and can be written as

$$\mathbf{W}_a = [\mathbf{W}_s^T \mathbf{W}_n^T]^T \tag{8.29}$$

where $\mathbf{W}_s$ is a $K \times N$ matrix that comprises the upper K rows of $\mathbf{W}_a$ which correspond to $\mathbf{B}_s$ and $\mathbf{W}_n$ is a $J \times N$ matrix that comprises the bottom J rows of $\mathbf{W}_a$ which correspond to $\mathbf{B}_n$. The contribution of speech to $\mathbf{Y}$ can now be computed as the estimated contribution $\widehat{\mathbf{S}}$ of the speech bases as:

$$\widehat{\mathbf{S}} = \mathbf{Y} \otimes \frac{\mathbf{B}_s \mathbf{W}_s}{\mathbf{B}_s \mathbf{W}_s + \mathbf{B}_n \mathbf{W}_n} \tag{8.30}$$

This simple solution is effective when used carefully, but has many drawbacks that quickly hamper it if not used under the ideal conditions. It is highly specific—it can be employed to separate signals for which we have bases. The bases for the speech are speaker-specific and noise-specific. It cannot be blindly applied in a speaker-independent or noise-independent manner. However, this does provide a perfect framework for a generic noise-compensation system that will work on a large range of noises and speakers without requiring highly specific knowledge about the set of noises. The basis-composition based technique provides a framework within which estimation of noise reduces to the estimation of weights for the bases.

The algorithm outlined above is a simple one. Many advanced versions of it have been proposed, including those that use neural networks, e.g. [31]. The underlying principle of **composition** behind them, however, remains the same. An alternate set of techniques use *masking*, where a "mask" comprises information that identifies the regions in a spectrogram that are corrupted by noise. Masking algorithms begin by first estimating a mask for the given signal representation (such as a spectrogram).

Once a mask is estimated, subsequent processing removes or fills in those regions within the representation used. For profiling, however, most such advanced techniques must be carefully vetted and used, since their artifacts may distort profiling-relevant information or destroy important micro-characteristics of the signal.

8.2.2 *Profiling for Environmental Sound Emitting Objects*

As discussed earlier in this section, the one set of profile parameters that *can* be reliably determined in spite of voice or noise removal artifacts are the environmental sound emitting objects. To do this, we may or may not choose to first separate out the background signal from the voice for analysis. Once the signal is separated and available for analysis, sound object detection and matching techniques must be used to identify the source of the sound. These comprise algorithms that learn to identify different sound emitting objects by learning from training data of examples of their signatures. The areas of acoustic monitoring for fault detection in mechanical devices, and CASA are both rich sources of information on this topic.

They are many categories of algorithms that are applied to object detection. They range from simple classification and regression algorithms to complex object discovery algorithms based on deep neural networks. Aside from these, to identify sound emitting objects, generally databases of their signatures must be maintained for comparison. The illustrative comparison procedure that is outlined briefly below can also be used for voiceprint matching (speaker matching) for speaker identification and verification.

A generic process for identifying the sources of background noises is depicted in Fig. 8.4. There are two stages involved: training and profiling (or detection). Figure 8.4a shows the training process. Sample sounds from the objects to be detected, and of the background sound environment (sound textures) are collected and provided with labels in a database. Features are extracted from the signal samples, and along with the appropriate labels (e.g. texture labels for training texture models, type labels for training type models etc.), are used to train either a classifier or a regression. A classifier is appropriate when there are distinct categories to be identified, a regression is appropriate when values along a continuum must be deduced. The two can be combined in many ways within more complex architectures. The parallel dashed arrows in Fig. 8.4a indicate that there could be multiplicity of classifiers and regressors trained in this manner. the outcome of this process is a trained model that can map a voiceprint to the corresponding label.

Figure 8.4b shows one possible implementation of the object identification process that uses different models trained in this manner sequentially (other architectures can be easily created using these). The voiceprint presented for analysis is first processed to separate out the signal of interest (the background, in this case) from it. Features are extracted from the separated signal, and the texture (a broad classification of the granularity of the sonic background) is identified. This provides information about

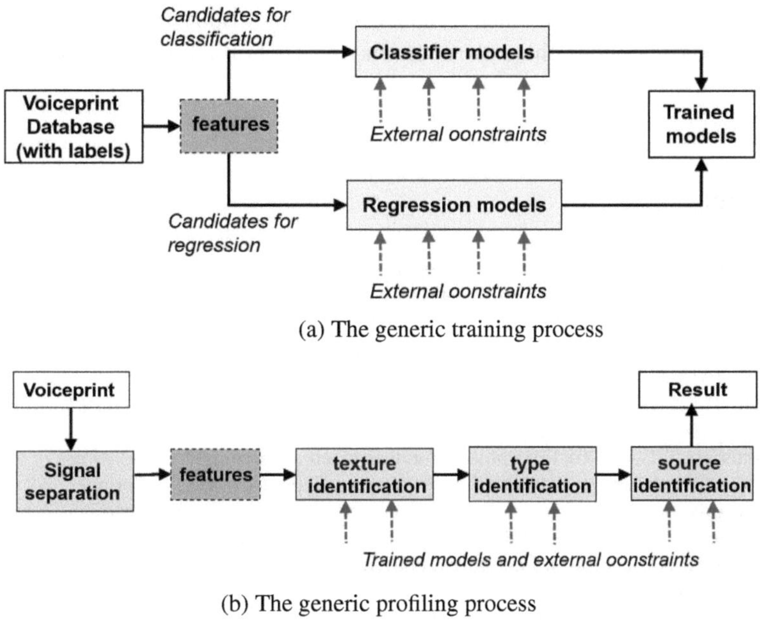

Fig. 8.4 Key steps in a simple generic object detection process for environmental profiling

the general nature of the background, so that the search for objects can be more narrow and directed.

This is followed by the identification of the *type* of sound—e.g. an automobile versus construction equipment—and then the actual object(s) that may have possibly generated the sound(s) in the voiceprint are identified. The actual process is not quite this simple, since multiple objects may be present in the signal, and many of those may have overlapping signatures. Further pre-processing of the signal may be needed to handle these problems completely automatically.

There are many standard classifiers and regressors that could be used in this context [32]. Some key algorithms used for sound object classification are Naïve Bayes [33–35], Multi-kernel support vector machines [36], random forest tees [37], neural networks of different architectures etc. For regression, some commonly used ones are simple linear regression [38], model trees [39, 40], ridge regression [41], lasso-lars and lasso algorithms [42, 43] etc. Neural nets and deep nets can also be used for this purpose.

It is important to note that acoustic signatures of objects that emit sound *could be out of the human hearing range*. One must look for these only if the signal, at *every* stage of its processing, from capture and digitization to storage has been maintained at sampling rates that allow for these sounds (which may be 20 kHz or below 20 Hz) to persist unchanged in the signal. This is unfortunately often *not* the case. For accurate profiling for subsonic [44] or ultrasonic signatures, the correct channels must also be maintained. Frequencies below 100 Hz and above 4 kHz are cut off by most telephone channels, some channels cut off all frequencies below 300 Hz.

References

1. Gath, I., & Yair, E. (1988). Analysis of vocal tract parameters in Parkinsonian speech. *The Journal of the Acoustical Society of America*, *84*(5), 1628–1634.
2. Grenier, Y., & Omnes-Chevalier, M. C. (1988). Autoregressive models with time-dependent log area ratios. *IEEE Transactions on Acoustics, Speech, and Signal Processing*, *36*(10), 1602–1612.
3. Adeli, H., & Hung, S. L. (1994). *Machine learning: Neural networks, genetic algorithms, and fuzzy systems*. New Jersey: Wiley.
4. Dietterich, T. G. (2000). Ensemble methods in machine learning. In *Proceedings of the International Workshop on Multiple Classifier Systems* (pp. 1–15). Berlin: Springer.
5. Kotsiantis, S. B., Zaharakis, I. D., & Pintelas, P. E. (2006). Machine learning: A review of classification and combining techniques. *Artificial Intelligence Review*, *26*(3), 159–190.
6. Nasrabadi, N. M. (2007). Pattern recognition and machine learning. *Journal of Electronic Imaging*, *16*(4), 049901.
7. Strobl, C., Malley, J., & Tutz, G. (2009). An introduction to recursive partitioning: Rationale, application, and characteristics of classification and regression trees, bagging, and random forests. *Psychological Methods*, *14*(4), 323.
8. Childers, D. G., Hicks, D. M., Moore, G. P., & Alsaka, Y. A. (1986). A model for vocal fold vibratory motion, contact area, and the electroglottogram. *The Journal of the Acoustical Society of America*, *80*(5), 1309–1320.
9. Wilhelms-Tricarico, R. (1995). Physiological modeling of speech production: Methods for modeling soft-tissue articulators. *The Journal of the Acoustical Society of America*, *97*(5), 3085–3098.
10. Steinecke, I., & Herzel, H. (1995). Bifurcations in an asymmetric vocal-fold model. *The Journal of the Acoustical Society of America*, *97*(3), 1874–1884.
11. Deng, L. (1999). Computational models for speech production. In *Computational Models of Speech Pattern Processing (K. Ponting Ed.)* (pp. 199–213). Berlin: Springer.
12. Alipour, F., Berry, D. A., & Titze, I. R. (2000). A finite-element model of vocal-fold vibration. *The Journal of the Acoustical Society of America*, *108*(6), 3003–3012.
13. Drechsel, J. S., & Thomson, S. L. (2008). Influence of supraglottal structures on the glottal jet exiting a two-layer synthetic, self-oscillating vocal fold model. *The Journal of the Acoustical Society of America*, *123*(6), 4434–4445.
14. Sagisaka, Y., Campbell, N., & Higuchi, N. (Eds.). (2012). *Computing prosody: Computational models for processing spontaneous speech*. Berlin: Springer Science & Business Media.
15. Stouten, V. (2009). Automatic voice onset time estimation from reassignment spectra. *Speech Communication*, *51*(12), 1194–1205.
16. Lin, C. Y., & Wang, H. C. (2011). Automatic estimation of voice onset time for word-initial stops by applying random forest to onset detection. *The Journal of the Acoustical Society of America*, *130*(1), 514–525.
17. Hansen, J. H., Gray, S. S., & Kim, W. (2010). Automatic voice onset time detection for unvoiced stops (/p/,/t/,/k/) with application to accent classification. *Speech Communication*, *52*(10), 777–789.
18. Sonderegger, M., & Keshet, J. (2012). Automatic measurement of voice onset time using discriminative structured prediction. *The Journal of the Acoustical Society of America*, *132*(6), 3965–3979.

19. Keshet, J., Shalev-Shwartz, S., Singer, Y., & Chazan, D. (2007). A large margin algorithm for speech-to-phoneme and music-to-score alignment. *IEEE Transactions on Audio, Speech, and Language Processing*, *15*(8), 2373–2382.
20. Quinlan, J. R. (1986). Induction of decision trees. *Machine Learning*, *1*(1), 81–106.
21. Breiman, L. (2017). *Classification and regression trees*. Routledge Press, Taylor & Francis Group.
22. Torgo, L., & Gama, J. (1997). Regression using classification algorithms. *Intelligent Data Analysis*, *1*(4), 275–292.
23. Memon, S. A., Zhao, W., Raj, B., & Singh, R. (2018). *Neural regression trees*. arXiv:1810.00974.
24. Shashanka, M., Raj, B., & Smaragdis, P. (2008). Probabilistic latent variable models as non-negative factorizations. *Computational Intelligence and Neuroscience*. Article ID 947438.
25. Lee, D. D., & Seung, H. S. (2001). Algorithms for non-negative matrix factorization. In *Advances in Neural Information Processing Systems 13 (T.K. Leen, T.G. Dietterich & V. Tresp (Eds.)), Proceedings of the Neural Information Processing Systems (NIPS)* (pp. 556–562).
26. Gaussier, E., & Goutte, C. (2005). Relation between PLSA and NMF and implications. In *Proceedings of the Twenty-Eighth Annual International Conference on Research and Development in Information Retrieval (SIGIR)* (pp. 601–602). Salvador, Brazil: ACM.
27. Cichocki, A., Zdunek, R., & Amari, S. I. (2006). Csiszar's divergences for non-negative matrix factorization: Family of new algorithms. In *International Conference on Independent Component Analysis and Blind Signal Separation (ICA)* (pp. 32–39). Charleston, SC, USA.
28. Cichocki, A., Lee, H., Kim, Y. D., & Choi, S. (2008). Non-negative matrix factorization with α-divergence. *Pattern Recognition Letters*, *29*(9), 1433–1440.
29. Heiler, M. & Schnörr, C. (2006). Controlling sparseness in non-negative tensor factorization. In *Proceedings of the European Conference on Computer Vision (ECCV)* (56–67). Graz, Austria.
30. Virtanen, T. (2007). Monaural sound source separation by nonnegative matrix factorization with temporal continuity and sparseness criteria. *IEEE Transactions on Audio, Speech, and Language Processing*, *15*(3), 1066–1074.
31. Huang, P. S., Kim, M., Hasegawa-Johnson, M., & Smaragdis, P. (2014). Deep learning for monaural speech separation. In *Proceedings of the International Conference on Acoustics, Speech and Signal Processing (ICASSP)* (pp. 1562–1566). lorence, Italy: IEEE.
32. Kumar, A. (2018). Acoustic Intelligence in Machines, Doctoral dissertation. School of Computer Science: Carnegie Mellon University.
33. Lewis, D. D. (1998). Naive (Bayes) at forty: The independence assumption in information retrieval. In *Proceedings of the European Conference on Machine Learning (V. Barr, & Z. Markov (Eds.))* (pp. 4–15). Heidelberg: Springer.
34. Zhang, H. (2004). The optimality of naive Bayes. In *Proceedings of the Seventeenth International Florida Artificial Intelligence Research Society Conference (FLAIRS)*. Florida, USA: AAAI.
35. Ng, A. Y., & Jordan, M. I. (2002). On discriminative versus generative classifiers: A comparison of logistic regression and naive Bayes. In *Advances in Neural Information Processing Systems* (pp. 841–848).
36. Scholkopf, B., & Smola, A. J. (2001). *Learning with kernels: Support vector machines, regularization, optimization, and beyond*. Massachusetts, USA: MIT press.
37. Ho, T.K. (1995). Random decision forests. In *Proceedings of the Third International Conference on Document Analysis and Recognition* (Vol. 1, pp. 278–282). Montreal, Canada: IEEE.
38. Myers, R. H., & Myers, R. H. (1990). *Classical and Modern Regression with Applications* (Vol. 2). Belmont, California: Duxbury Press.Classical and Modern Regression with Applications.
39. Frank, E., Wang, Y., Inglis, S., Holmes, G., & Witten, I. H. (1998). Using model trees for classification. *Machine Learning*, *32*(1), 63–76.
40. Landwehr, N., Hall, M., & Frank, E. (2003). Logistic model trees. In *Proceedings of the European Conference on Machine Learning (ECML)* (pp. 241–252). Cavtat-Dubrovnik, Coratia.

41. Hoerl, A. E., & Kennard, R. W. (1970). Ridge regression: Biased estimation for nonorthogonal problems. *Technometrics*, *12*(1), 55–67.
42. Efron, B., Hastie, T., Johnstone, I., & Tibshirani, R. (2004). Least angle regression. *The Annals of Statistics*, *32*(2), 407–499.
43. Park, T., & Casella, G. (2008). The Bayesian lasso. *Journal of the American Statistical Association*, *103*(482), 681–686.
44. Bedard, A., & Georges, T. (2000). Atmospheric infrasound. *Acoustics Australia*, *28*(2), 47–52.

Chapter 9
Reconstruction of the Human Persona in 3D from Voice, and its Reverse

It seems magical that we are at a point in time where it is possible to discuss the subject of accurate, in-vacuo generation of a three dimensional image of the human form from the voice signal alone. From the discussion in this book so far, it should be evident that both direct and indirect relationships exist between voice and the human form. For example, voice can be related to bone structure. It can also at the same time be related to the person's height, weight, age, gender and many other factors. These relationships can be transformed into predictive mechanisms. From predictions of the body dimensions and the weight, the person's body mass index may be deduced; from predictions of the skull type, and the length of the vocal tract, the person's likely skeletal proportions can be deduced. These alone are enough for the approximate reconstruction of the human form from voice. *Then* we address the problem of "filling in" the finer details. In this chapter, we discuss the challenge of reconstructing the human persona from voice from different perspectives, and also its reverse—the estimation of the human voice from an image of their persona.

9.1 Links Between the Human Form and Voice

The subject matter in this chapter involves navigating through an intricate network of (sometimes tenuous or complex) relationships between bio-physical entities.

Many parameters have both a direct and indirect affect on the voice signal. For example, as discussed in Chaps. 2 and 3, age affects voice *directly* by affecting the tissue composition of the vocal cords. It also affects bone and tissue structure, which affect the properties and dimensions of the vocal chambers, which affect the harmonic structure of the signal. Age affects our hearing, which in turn affects the voice production process. Thus age also exerts an *indirect* influence on voice.

The link between voice and facial structure has likewise been strongly established from many perspectives. As with the rest of the human body, the structure of vocal

R. Singh, *Profiling Humans from their Voice*,
https://doi.org/10.1007/978-981-13-8403-5_9

tract (size and shape of larynx, neck, throat etc.) is determined by both genetic and epigenetic factors. While genetic factors directly influence voice, epigenetic factors can exert both direct and indirect influence on it. The latter category includes factors such as vocation, atmospheric pollution, respiratory diseases etc. Other factors may directly affect the facial structure, and through those, may in turn affect the vocal structures. For example, populations whose diets comprise tougher foods requiring more powerful mastication have larger jaws. Similarly other (epigenetic) factors influence the breadth of the face, the size of the cheek bones and many other facial parts. All these are known to affect the dimensions of the vocal chambers, and through them, the voice. Even soft-tissue cartilaginous structures are not exempted from such a web of relations. The structure of the nose is genetically determined, but epigenetic factors such as climate [1]—and thus distance from the equator—also affect its shape. The structure of the nose, in turn, affects the quality of the voice by affecting nasalence and nasal resonances.

The web of relations deepens, and reinforces those between voice and the human face when myriad other links are considered. For example facial structure is related to a person's facial appearance [2]. The relations of facial appearance to aggression [3, 4], and to race, and of aggression and race independently to voice, thereby connect face to voice. However we will not traverse such deep relations for now. Statistics, when used for prediction, must only be stretched so far!

Similar inter-relationships link voice to the human body form. Facial structure is directly linked to the overall skeletal structure, mediated by many factors such as age, genetics, ethnicity etc. Indirect links between voice and the human form emerge from the strong relation of voice to these mediating factors. Even though the strength of relationships formed between voice and the entire human form through each mediating factor may be relatively weak, the sheer number of these relationships is large, and they reinforce to yield a strong relationship between voice and the human body form in its entirety. The hint for the existence of such a strong relationship was indirectly provided as far back as 1916 by Walter Swift in an article [5] that explicitly stated: *"I have shown that voice quality depends absolutely upon bone structure, that bone structure is inherited, and that therefore vocal quality is inherited."*

We have mentioned only a few mediating factors in the paragraphs above. In reality, a number of external environmental, social, demographic, cultural and other influences are known to affect both physical structure and voice, thus establishing many direct and indirect routes of dependency between the two. While it may be impossible to enumerate all the routes through which voice and physical structure are related, what *can* be established is that the two are linked, with possibly unique, injective, and possibly bijective relations between the two.

9.1.1 *Cognitive and Perceptual Evidence of Links Between Form and Voice*

It has been well documented in the literature that humans judge and perceive multiple aspects of a speaker from their voice alone. Of the entities that we perceive from voice, physical appearance is a prominently studied one. Historically, there have been many studies across different scientific disciplines that have reported positive correlations between a person's appearance and their voice. In many of these, human subjects were given photographs of speakers, and were asked to listen to voice recordings, based on which they were then asked to select the corresponding speaker. In all these studies, human subjects were able to correlate voices with faces with surprising accuracy.

What aspects of appearance do we specifically focus on? Although we do not consciously notice doing so, we form strong associations between people's voices and their facial characteristics. This was demonstrated in a series of surprising eyewitness experiments that demonstrated that showing the wrong face with a voice causes people to misrecognize the *content* of the voice to a greater degree [6–8].

These associations are formed because they exist. And the fact that they exist must then have a scientific basis which, when found, could be used to estimate one entity from the other—face from voice, and voice from face. That facial appearance is not isolated from the appearance of the rest of the body leads us to believe that inferences about the body can be made from inferences about the facial structure. In fact, through the discussion in earlier chapters of this book, we have seen that many other facts about the human form can be deduced from voice, such as height, weight, age, lung capacity etc. Together with all of these, and the broader anthropometric inferences that can be made from an estimate of the facial structure, it is reasonable to expect that the entire human form can be reconstructed from voice.

9.2 Reconstructing the Human Form from Voice: General Considerations

The task of accurately reconstructing the human **face** from voice poses many challenges. A face that is reconstructed from voice must look like the face of the speaker in real life. Of course, not all details might be possible to estimate or reconstruct, now or perhaps ever in the future. One example is hairstyle, or choice of facial grooming—these entities have no correlation with voice. For the reverse—reconstruction of voice from face—the challenges are deeper. Facial structure cannot predict the language spoken by a person, cannot predict the personality, learning, style, experiences, and thought processes of the speaker—all of which influence and modulate the speech produced. In the reverse task, perhaps only the *quality* of the speaker's voice can be estimated. Or, in other words, only the *voice*, in its strictest definition as a non-modulated signal, can be reconstructed accurately.

When it comes to the reconstruction of the **full body** from voice, there are deeper challenges. It may be virtually impossible to reconstruct even a person's face, let alone their body structure in *perfect* detail from voice alone. Nevertheless, there is arguably enough information in voice to make *reasonably good physically and statistically supported guesses*.

What information do we exactly need to reconstruct the human body geometrically in 3-dimensional (3D) space—from voice? What might be the best approach? A hint to a viable approach may come from human intuition—exemplified by the art of *perspective drawing* of the human form. When asked to do this on paper, typically an artist draws the outline of the face, followed by the shoulders, some way down the arms, down the torso, legs, and then fills in the arms. This is not a random choice. In fact, this sequence is definitive enough in its effectiveness that it is taught in schools of art. There have been multiple case studies of how visual artists create artwork e.g. [9], which have used different research methods, such as eye movement and sketch recordings, X-ray and brain scan analysis. These studies have sought to understand the cognitive reasons behind the creation processes and underlying patterns in how artists create conceptual themes, how they select pictorial elements, and *why*. Software tools e.g. [10] for surface drawing of the human form exploit these understandings. It is reasonable to assume that this specific sequence in perspective drawing is cognitively chosen because it is effective. It is also reasonable to assume that it is effective because there is probably an injective relationship between the proportions and appearance of the preceding and succeeding parts of the body that are drawn.

There is a subtle clue in this—for us, intuitively, the face *defines* the body. We have all experienced this. If photographs are lined up and the faces are hidden, it would be difficult for us to imagine the corresponding faces. If on the other hand, the bodies were obscured, and only the face shown, it would be easier to imagine what the corresponding bodies may look like. This is because the face provides relatively more clues—of the person's age, ethnicity, bone structure, state of health etc. and it is easier to guess the body structure with the aid of all this supplemental information.

Like an artist, then, we must begin the reconstruction of the human persona with first solving the problem of facial reconstruction from voice. There is another (algorithmic) reason for this: it is prudent to begin with the parts of the body that have more evidence of them embedded in voice. This reasoning drives algorithmic design in other areas of computer vision/graphics/modeling that tackle equivalent tasks, such as reconstructing 3D bodies from 2D images. Most algorithms used for this locate and use joint positions on 2D images, and use those to estimate body configuration and shape in 3D [11]. This is because those key positions carry more information that is relevant to the task.

Following this reasoning, when we think of reconstructing the entire human form from voice, we must begin with the face and not the joints, since voice carries relatively less implicit evidence of body geometry, or joint positions.

When it comes to actual methodologies, there are multiple ways to approach the problem of reconstruction. One approach is based on leveraging models that emulate the physics and bio-mechanics of voice production (discussed in Chap. 2). Physical models of the vocal tract can explain the resonant characteristics of the voice signal. Bio-mechanical models of the vocal tract can explain the energy distribution and other spectral characteristics of voice, including many voice qualities.

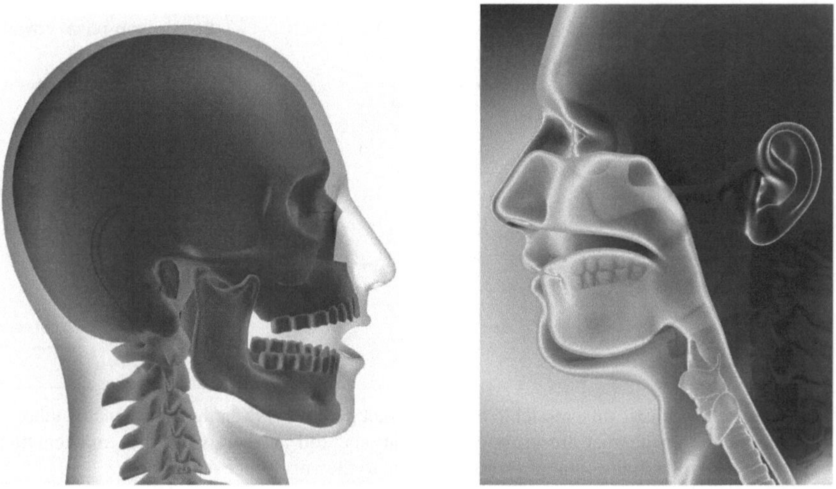

(a) The skull defines the dimensions of the face

(b) The face defines the dimensions and boundaries of the vocal tract

Fig. 9.1 Geometrical relationships between the skull, face and vocal tract

Let us take the example of physical models, of which the acoustic tube model is the most widely studied one. Such a model can approximate the resonant characteristics of voice, given the dimensions of the vocal tract of the speaker. Conversely, given the former, it can be used to estimate the latter. The three-dimensional geometry of the vocal tract is highly correlated to the speaker's skull, and thereby to the speaker's facial structure. The relations between these is illustrated in Fig. 9.1.

If we can estimate one of these entities, it is possible to estimate the other two through these geometrical relationships. It follows that if one of these can be related to voice, then the other two can also be related to voice by corollary.

To reconstruct facial structure from voice, the vocal tract geometry can be invoked as an intermediary. Vocal tract geometry can be estimated by matching the spectral patterns computed from the voice signal to the dimensions and properties of the model that best approximates them. Once the dimensions are known, the skull and facial dimensions can be estimated.

To reconstruct a voice from facial structure, the reverse process can be followed: estimate the dimensions of the vocal tract that best fits the facial structure (in 3D), then estimate the spectral characteristics of the voice such that it best matches those of the model that best approximates the estimated vocal tract. The model can incorporate both the dimensions and properties of the vocal tract, resulting in estimates of a number of voice qualities as well, which can then be used to synthesize a voice signal that has the estimated characteristics.

Figure 9.2 illustrates the fitting of a concatenated tube acoustic model (see Chap. 2) to the three dimensional rendering of a speaker's face.

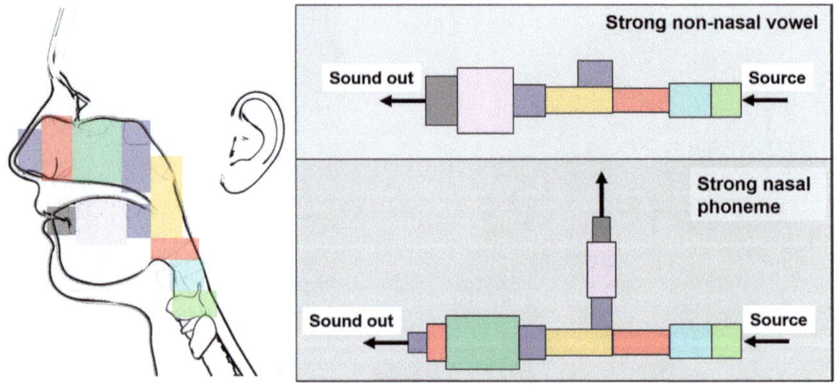

(a) Fitting the acoustic tube model to the vocal tract geometry: even the concatenated tube models of the vocal tract are gross approximations, and are best viewed as mathematical abstractions rather than actual physical moldings to the vocal tract shape.

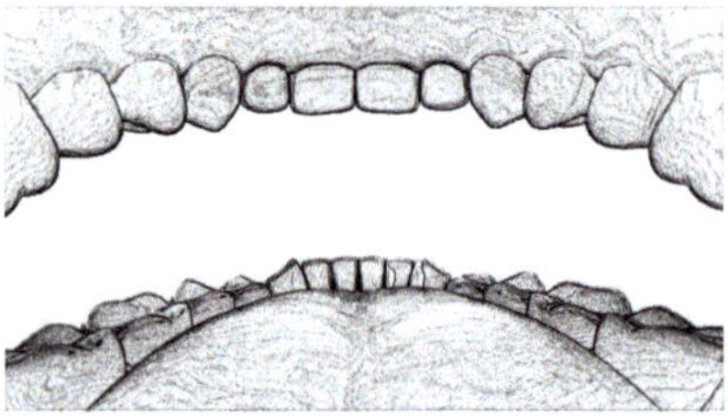

(b) View from the inside of the mouth: even when the mouth is wide open the opening is not really circular or oval. The opening follows the contours of the teeth rather than of the lips.

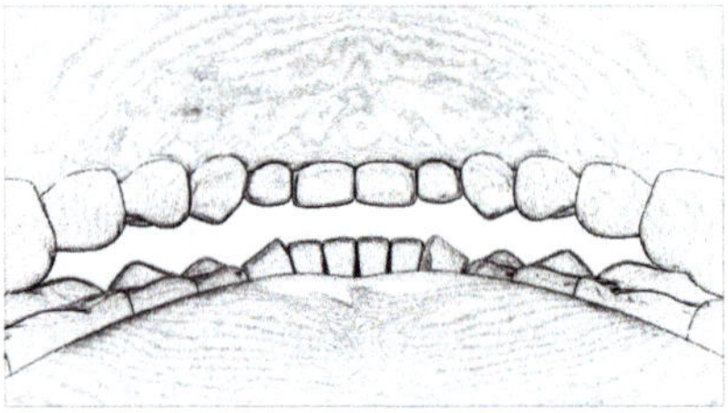

(c) Mouth opening for many consonants. The area of the opening especially affects the estimation position of the third formant.

Fig. 9.2 Fitting the concatenated tube model to the vocal tract to compute spectral patterns

The concatenated tube model is in fact one of the primary sources of information about the relationship of a speaker's vocal tract to their face. The tube model must fit in the face as illustrated by the figure. Given a speech signal and an initial estimate of the number of concatenated tubes in the model, the reflection coefficients at the various junctions can be estimated and thereby the ratios of the areas of the adjacent tubes can be estimated. Conversely, given a tube model with specification of tube lengths and area ratios, the transfer function of the vocal tract and thereby the characteristics of the speech signal can be estimated. More accurate relationships are given by lossy tube models which also consider tissue characteristics and other known characteristics, augmented by model extensions to account for the anti-resonances of the nasal passage and its area ratios.

There are other approaches to reconstructing facial structure from voice and vice-versa. A set of exemplary approaches based on AI techniques are described in the sections below.

9.3 Reconstruction of Face from Voice

At the outset, it is important to note that **mapping** is different from **reconstruction**. Given a voiceprint, a face (depicted in 2D or 3D) can be chosen from an ensemble of faces, such that the voice of the chosen person matches the given voiceprint. This process is *true mapping*, and requires that an ensemble of face-voice pairs be available to choose from. When such an ensemble is not available, the face of the speaker must be estimated in-vacuo from the voice sample. This is *true reconstruction*.

When mapping is an intermediate stage in estimating a face, or vice-versa, we call it **transformative reconstruction** and **transformative mapping**, respectively. In a transformative reconstruction framework, a face is chosen from an ensemble based on voice-matching criteria, and thereafter modified according to voice-based corrections to its features. The reverse is done in transformative mapping.

At the time of writing this chapter, techniques for mapping, transformative mapping, and transformative reconstruction have all been effectively devised. Techniques for true reconstruction, however, are in nascent stages of development.

In this section, we will discuss some of these categories of techniques. True reconstruction is (also) further discussed in the next section, dealing with the generation of the voice of a person from their facial images.

There are two broad approaches to the problem of estimating a face from voice:

1. A *knowledge-driven approach* where inferences are made about the size and relative proportions of the articulators and vocal chambers, and other relevant parts of the vocal tract, from their known correlations with properties derived from the voice signal. The majority of such correlations that are usable currently relate to various sub-components of voice quality. Facial features are then inferred from the estimated geometry of the vocal tract, and other co-factors that can be estimated from voice, such as gender, age, ethnicity, state of health, height, weight etc.
2. A *data-driven approach* that can learn (directly or indirectly) to map or reconstruct the image of a face from voice, given examples of both.

While these can be combined, we will discuss them separately for now. For reconstruction in a knowledge-based setting, facial features must be estimated from voice-derived ones through their known relationships. Estimates of facial features must then be algorithmically coalesced to form a complete image of a face.

For reconstruction in a data-driven setting, a *predictor* must be trained to generate either the image of the face directly, given voice, or to generate facial features that can be used by a *reconstructor* to reconstruct the face. The reconstructor must also either be knowledge-based or data-driven. In the former case, algorithms to generate heat maps (contour maps) of faces can be used, in conjunction with those that can geometrically fill-in the facial characteristics. In the latter case, the reconstructor must be trained using pairs of facial features and corresponding full (complete) faces.

9.3.1 Mapping: An Example Algorithm

A mapping requires a binary yes/no decision to be generated, given face-voice pairs. In data-driven settings, decisions can be arrived at by directly matching the derived features, or learning common embeddings from voice and face that can be matched e.g. [12–14].

The important point to note in the context of mapping is that if we directly try to infer one from the other, we make the implicit assumption that the two modalities directly inform about one another. This is not always the case. For example, facial structure is dependent on gender, age etc. The same is true, independently of the face, for voice. Males of a certain age range have distinctly male voices and faces, females of a certain age range have distinctly female voices and faces. A large part of the relationship between the two, thus, is predicated on gender, age etc. Many such factors that link voice and face exist [15].

A model or framework for mapping must properly account for these underlying linking factors (or *covariates*) between voice and face. One way to do this is to use *embeddings* that permit voices and faces to be compared to one another in common mathematical spaces. One such type of embedding could be one that can be used to predict covariates—which would ideally be the same covariates—from either voice or face data from the same person. As an example, a model that can estimate such embeddings [16] is explained below:

The model assumes that a voice sample $\mathscr{V}$ is related to the image of a face $\mathscr{F}$ through a (discrete-valued, for simplicity) covariate $\mathscr{C}$. $C(v)$ and $C(f)$ refer to the covariates related to voice v and face f. This applies to every covariate $C \in \mathscr{C}$. The following are then defined

- A voice embedding function $F_v(v; \theta_v) : v \mapsto \mathbb{R}^d$ with parameters θ_v, that maps v into a d-dimensional vector
- A face embedding function $F_f(f; \theta_f) : f \mapsto \mathbb{R}^d$, that maps f into a d-dimensional vector

The goal is to learn the parameters θ_v and θ_f in a manner that the corresponding embeddings are comparable.

For each covariate, a classifier $H_C(x; \phi_C)$ with parameters ϕ_C, assigns any input $x \in \mathbb{R}^d$ to a value for C. The classifier is agnostic to which modality x was derived from: x could be the d-dimensional vector output (feature) of either the voice embedding function or the face embedding function.

For each v (or f) and each covariate C, a loss $L(H_C(F_v(v; \theta_v); \phi_C), C(v))$ is computed between the covariate predicted by $H_C(.)$ and the true value of the covariate for v, $C(v)$. The *total loss* $\mathscr{L}$ over the set of all voices $\mathscr{V}$ and the set of all faces $\mathscr{F}$, over all covariates, is given by

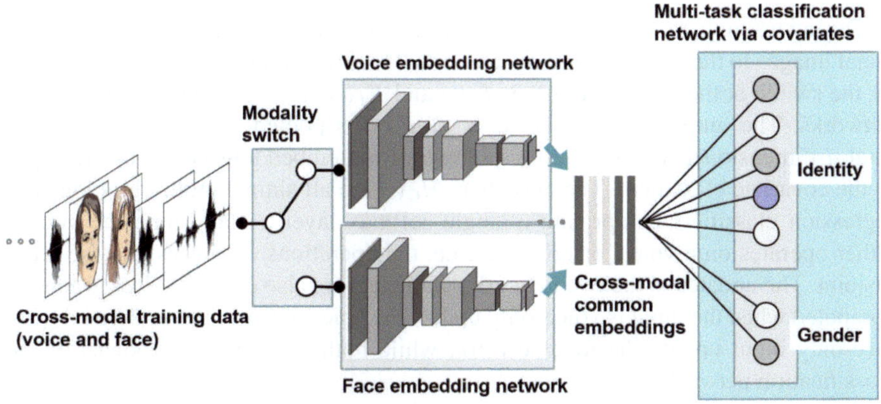

The DIMNet framework: the input training data can be voice or face, pairing is not needed. A modality-switch channels the data to the embedding network for the given modality. A supervised multi-class classification network is concurrently learned with the embeddings.

Fig. 9.3 Illustration of a DIMNet framework

$$\mathscr{L}(\theta_v, \theta_f, \{\phi_C\}) = \sum_{C \in \mathscr{C}} \lambda_C \left(\sum_{v \in \mathscr{V}} L(H_C(F_v(v; \theta_v); \phi_C), C(v)) + \sum_{f \in \mathscr{F}} L(H_C(F_f(f; \theta_f); \phi_C), C(f)) \right) \tag{9.1}$$

The parameters of the embedding functions, θ_f and θ_v, are learned by performing the optimization:

$$\theta_v^*, \theta_f^* = \arg\min_{\theta_v, \theta_f} \min_{\{\phi_C\}} \mathscr{L}(\theta_v, \theta_f, \{\phi_C\}) \tag{9.2}$$

The voice and face embedding functions could potentially take any form. In a specific formulation called a *Disjoint Mapping Network* (DIMNet), $F_v(v; \theta_v)$, $F_f(f; \theta_f)$ and $H_C(x; \phi_C)$ are instantiated as neural networks. The network's architecture is shown in Fig. 9.3. It comprises three components:

1. A *Voice Network* that represents $F_v(v; \theta_v)$, which extracts d-dimensional embeddings of the voice recordings
2. A *Face Network* that represents $F_f(f; \theta_f)$, which extracts d-dimensional embeddings of face images
3. A bank of *Classification Networks*, one network per covariate considered. Each classification network serves to classify one covariate (e.g. gender), and operates on the d-dimensional features from the embedding networks.

During training, voice recordings are input to the voice-embedding network, and facial images to the face-embedding network. This switching operation is illustrated by the switch at the input in Fig. 9.3. $F_v(\cdot)$ and $F_f(\cdot)$ are both convolutional neural networks. The output of the final layer of $F_v(\cdot)$ is pooled over time to obtain a final d-dimensional representation. Outputs of the embedding networks are input to the covariate classifiers. The classifiers $H_C(\cdot)$ are all simple multi-class logistic-regression classifiers comprising a single softmax layer. At any time the system either operates on a voice, or on a face, i.e. the operations on voices and faces are disjoint. The updates of the two network parameters are also disjoint—loss gradients computed when the input is voice-only update the voice network, those derived from face-only input update the face network, while both contribute to updates of the classification networks.

An appropriate loss (e.g. a cross-entropy loss) may be used to optimize the networks. In practice, instead of the optimization in Eq. 9.2, the actual optimization is performed as below. The difference is inconsequential.

$$\theta_v^*, \theta_f^*, \{\phi_C^*\} = \underset{\theta_v, \theta_f, \{\phi_C\}}{\arg\min}\, \mathscr{L}(\theta_v, \theta_f, \{\phi_C\}) \tag{9.3}$$

The parameters of such a network can be trained through backpropagation, using the stochastic gradient descent method. For voice input, voice embeddings are computed using $F_v(v;\theta_v)$. Losses are computed using classifiers $H_C(\cdot)$ for all the covariates, and the loss gradient is backpropagated to update the voice network as well as the covariate classifiers. The same is done for the face data—the backpropagated loss gradients are used to update the face network and the covariate classifiers. Thus, the embedding functions are learned using the data from their modalities individually, while the classifiers are learned using data from all modalities. For better generalization, more covariates can be used by this model.

Once trained, the embedding networks $F_v(v;\theta_v)$ and $F_f(f;\theta_f)$ can be used to extract embeddings from any voice recording or face image.

This is directly useful for mapping: given a voice recording v and a face image f, one can now compute a similarity between the two using an appropriate metric, e.g. through the cosine similarity $S(v, f) = \frac{F_v^\top F_f}{|F_v||F_f|}$. This is used to evaluate the match of any face image to any voice recording. This also enables one to rank a collection of faces $f_1, \ldots, f_K$ in the order of their estimated match $S(v, f_i)$ to a given voice recording v, or conversely, to rank a collection of voices $v_1, \ldots, v_K$ according to their match to a face f, in the order of decreasing $S(v_i, f)$.

9.3.2 Transformative Reconstruction: An Example

The embeddings generated though a model such as the one described above, serve as features that can be used within a transfer learning framework (described in Chap. 5) to generate faces. By our definition, this process would fall under the category of transformative reconstruction, where a matched face is modified using additional procedures.

In a transfer learning framework, the model above is used to generate embeddings, which form the input for another model that learns the relationships of these to a faces. The model can either learn to use the embeddings to directly reconstruct a face, or to transform the mappings to specific facial features that can then be geometrically extrapolated to yield the image of a face. Details of this extension will become clearer in the next section, as we discuss the notion of meaningful facial features, and through them, address the problem of generating voices from faces. The explanation better serves both sections when provided in that context.

9.3.3 Reconstruction: An Example

While transformative mapping serves the goal of true reconstruction: once the embedding network is trained, given a voice signal the algorithm can generate the speaker's face, from a scientific perspective it does not solve the problem of true reconstruction. For *true* reconstruction, ideally no intermediate mappings must be needed. By definition, such a mechanism must be able to accurately generate a completely new type of face (face types are described later in this chapter) without having learned from explicit examples of such a type. With some thought, it should become clear that true reconstruction is not possible in a purely data driven fashion, using machine learning or AI techniques. Let us see why this is so.

The simplest data driven mechanism that might learn to generate a face from a voice is one that takes in the voice signal as input, and yields the facial image as output. Such a mechanism still implicitly learns mappings from voice to face, and to be able to function, must be trained using matched voice-face pairs. In all other extensions and embodiments of a data driven mechanism, such pairs are needed in the training phase in one form or another.

A system for true reconstruction must necessarily be a knowledge-based one. One such embodiment might be where voice quality features from the voice signal are used to deterministically infer the measurements of parts of the vocal tract, and those to measurements of the face (such as facial proportions from landmarks), and those in turn are used to recreate the geometry and features of the face. However, such deterministic methods are not yet fully studied or recorded in the literature. Only the methodologies for geometrical reconstruction of faces *from* estimates of landmarks or other measurements and transformations, in 2D and 3D, are well developed in the field of computer vision, e.g. [17–20].

A necessary step in such reconstructions is the computation of appropriate features from the face and voice. Feature computation from voice has been extensively dealt with in earlier chapters. Feature extraction from images of the face is discussed briefly in later sections, but is limited to transfer learning frameworks, since in most data driven systems currently, the image of the face is directly used after minimal normalizations of size (preserving aspect ratio), color, texture and other variations that may be spurious to the estimation process.

For most practical purposes, transformative mapping is of course sufficient, since it achieves the goals of being able to create new faces in response to new voices, once trained.

9.4 Reconstruction of Voice from Face

Let us now discuss the reverse problem: that of generating voices from facial images. A person's face is predictive of their voice. Biologically, this is to be expected: the same genetic, physical and environmental influences that affect the face also affect the voice. More directly, the vocal tract that generates voice also partially defines the geometry of the face and *vice-versa.*

Human perception apparently recognizes this association. Voices and faces alike are characterized by physical features that are related to the structure of the vocal apparatus, and integrated models for voice and face perception have been proposed on this basis [21]. Humans have been shown to be able to associate voices of unknown individuals to pictures of their faces [22]. They also show improved ability to memorize and recall voices when previously exposed to pictures of the speaker's face, but not imposter faces [6, 23], also confirmed by event-related brain potential (ERP) readings [8]. Studies indicate that neuro-cognitive pathways for voices and faces share common structure [24], possibly following parallel pathways within a common recognition framework [21, 25]. In fact, the ability to match faces to voices is thought to be a key stage of child development [7]. Studies such as these lend credence to the hypothesis that it may be possible to estimate voices *from faces* algorithmically as well.

In concept, this should be easier than the reconstruction of face from voice (it is quite the opposite in reality, as we shall see later). This is simply because the *given* entity is now the face, and given the images of a specific face, the dimensions and ratios of its parts can be precisely measured, and possibly be made accurate to scale in a real 3D space.

The computation of features is now vastly simplified: the specific measurements of facial parts (in 2D or 3D) now comprise the features. The facial measurements may be manually derived or automatically obtained. In the latter case, facial landmark detection could be used to obtain the measurements. These must be mapped to voice quality measurements (described in Chap. 6). This is so because those are the only aspects of voice that are expected to be predictable from the image of a face. As mentioned earlier in this chapter, the content and style of speech cannot be predicted in any manner from facial structure.

The problem of voice generation from face then becomes that of *prediction of voice quality features*, and of generating a voice signal with those specific qualities. In other words, measurements of the voice qualities of the generated voice must match the algorithmic estimates (or predictions) of the measurements. Note that the term "quality" includes all voice signal characteristics such as vocal resonances and anti-resonances, pitch, estimated glottal flow waveform etc.

The following paragraphs explain the methodology for face-to-voice generation in generic term.

9.4.1 General Methodologies

A fully automated pipeline for the generation of voice from facial structure (and body where available in its advanced implementations) requires two key components, applied serially, or implicitly: a **content generator** and a **quality generator**.

A **content generator** is a mechanism that, given a 2D or 3D images of a face (or the full body in advanced versions), generates a speech signal in plausible *language* that is also *period-appropriate*, *origin-appropriate*, *personality-appropriate*, *context-appropriate*, *style-appropriate* etc. Mechanisms for the synthesis of human speech and its linguistic content have been a subject of research for decades. Current methods that are based on artificial intelligence techniques are excellent content generators in a syntactic sense—they can generate word sequences that are appropriate in all the respects above—provided they have ample data to train with. Automatic generation of long-term consistency and semantic/pragmatic language content remains an unsolved problem though, and we will ignore that for now.

Some examples of such voice synthesis systems that are in vogue at the time of writing this book are WaveNet [26], VoiceLoop [27], ClariNet [28], Tacotron [29] etc. **WaveNet** directly generates voice signals from given text. Given the voice samples of a speaker (from a closed set of speakers), it synthesizes voice that sounds like the given speaker's voice speaking the text provided. It can switch between speakers when given speaker identities, and thus potentially generate conversations with non-overlapping speech. **VoiceLoop** transforms, or "speaks out" given text in a voice that perceptually matches the quality and style of a given sample of voice form any unknown speaker. Currently this system accepts text in only a limited number of languages. **ClariNet** is a text-to-speech system with capabilities along the same lines as WaveNet. **Tacotron** [29] is yet another text-to-speech system that operates on text alone to transform it to natural-sounding voice.

Note that in all of these systems, text must be provided. Automatic generation of meaningful text has also been a subject of extensive research and methodologies exist [30–33] to generate compositions in text that make sense locally, in a completely automated fashion. As of the time of writing the chapter, the state-of-art is represented by a system that is based on [34].

Today, the potential generation of speech that is appropriate in all of the aspects mentioned above (other than pragmatic) is possible, *provided we have large amounts of data to train the corresponding neural-net based (AI) systems with.* If such data are not available, then content generation must be done manually: the appropriate content should be researched and rendered into voice form with as much stylistic and person-fidelity as possible. This signal (or the aggregate of a collection of such signals), comprises the input for the second key component of an automated face-to-voice generating system: the quality generator.

The function of the **quality generator** is to deduce the specific perceptual voice qualities of a speaker, given the image of a face. This could be an automated system that maps some specific features derived from facial images to estimates of different sub-components of voice quality. Once the voice quality estimates are available, a subsequent mechanism can learn to impart the given voice qualities to the signal generated from the content generator (which could be automated, or be human-generated, as we discussed). Transforming voice to have a desired set of qualities falls under the aegis of voice *transformation*. Note that the content generator falls under the aegis of voice *synthesis*.

Voice transformation can be entirely data-driven, or knowledge-based, or a combination of both. In a data-driven approach, a system must learn to transform signals with specified sets of voice quality values to voice signals with a different set of voice quality values. Essentially, they must learn to transform the voice of one speaker to that of another. The content of the speech must either be the same for both speakers, or, if different, more complex training mechanisms must be used (requiring even more data). The latter can be done within existing AI frameworks, however, if desired. The availability of large amounts of training data is however a strong and stringent requirement in these approaches. If training data are scarce or unavailable, then data-driven mechanisms cannot be used.

A knowledge-based approach to voice transformation involves using traditional signal processing methods to transform the quality of a voice signal. The estimate for each specific voice quality is obtained from the facial image of the speaker. The estimate itself may be obtained using any data-driven regression mechanism, which can be trained using image-voice-quality pairs. In each pair, the voice quality features used are derived from the voice of the speaker corresponding to the face. When such a regressor is used, the input would be face, while the output would be the estimated voice quality values.

Alternatively, a system that takes a data-driven approach to voice transformation could be trained with {face,voice-quality} pairs as input, and appropriate audio samples (or an invertible representation of audio such as a spectrogram) as both input and output. The input voice-quality features are estimated from the face using a mapping that can be learned during training. The system is constrained such that these estimated qualities are transferred to the input audio signal, yielding the output audio signal.

This completes the generic end-to-end process for generation of voice from an image of a speaker's face. The complete pipeline discussed above is shown in Fig. 9.4.

In the rest of this section, we will focus on methodologies used within the quality generation component of the pipeline in Fig. 9.4, namely the face to voice-quality estimation and voice transformation. Let us first discuss the estimation of voice quality from facial images.

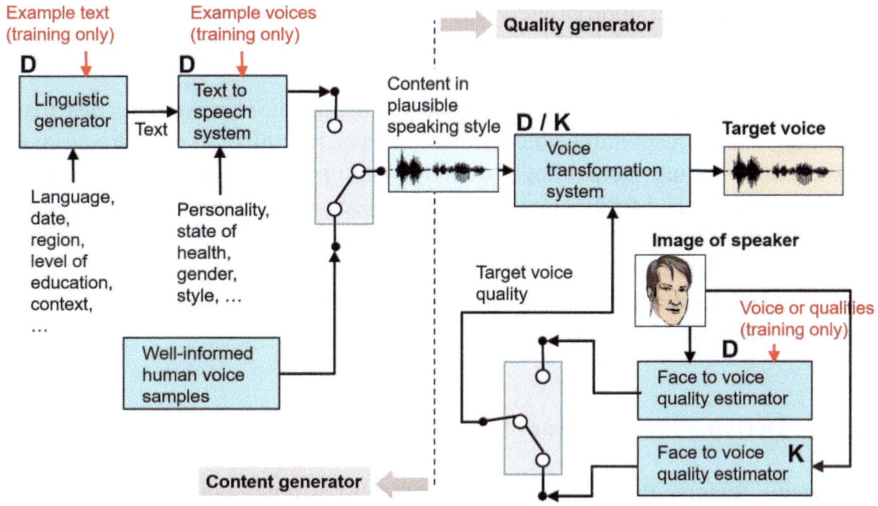

A generic face to voice framework: the components marked **D** involve data intensive training. The components marked **K** are based on traditional signal processing and/may be rule based. Different embodiments of this pipeline may consolidate different parts of it into common modules.

Fig. 9.4 A generic end-to-end framework for the generation of voice from face

9.4.2 Estimation of Voice Quality from Facial Features

Why voice quality? Why do we not choose to work directly with more basic signal parameters that can be measured and altered, such as the formant positions, HNR etc? The reason is pragmatic, looking into the future of profiling. Voice quality is a perceptual entity. If we predicate our technologies on this, we can continue to leverage human judgment and perception to generate data. The value of this resource to the future of this technology cannot be overstated. Besides, owing to the vast number of prior studies, the relations between profile parameters with voice quality are much more tangible than those with basic signal parameters todate.

As shown in Fig. 9.4, there are two ways to approach the problem of voice quality estimation from facial features: a data-intensive one (labeled **D**), and a knowledge-based one (labeled **K**) that exploits known relations between facial features and voice qualities.

The data driven estimator is a regressor (which could be neural network) that is trained using a large number of voice-quality (features) and facial image pairs, where each pair belongs to the same speaker. During training, the system takes in facial images as input, and learns to correctly estimate the corresponding voice qualities as output. Obviously, to derive the voice qualities for training, each speaker's voice must be tagged with the corresponding voice qualities, for which a knowledge-based approach is necessary and unavoidable. If tagged data are not available, a face-to-voice mapper may be appropriately engineered and trained to yield features that proxy for the voice qualities. The measurement of voice qualities from voice signals has been discussed to some extent in Chap. 4.

Alternatively, one can use a knowledge-driven estimator. This estimator also maps facial images to corresponding voice quality features, but unlike its data-driven counterpart, it does not use facial images directly. Appropriate features must be derived from these images and formulaic mappings between these features and voice qualities, devised offline from separately conducted studies, must be used to map the derived facial features to voice quality estimates. The topics of derivation of facial features and that of mapping these to voice quality estimates from a knowledge-based perspective are discussed below.

9.4.2.1 Deriving Features from Facial Images

The specific features that we derive from a face to estimate voice quality must be meaningful or informative in this respect. As we have discussed before, the shape and dimensions of the vocal tract are estimated by those of the skull, but not all parts of the skull are relevant for this. Those that play a part in deciding the vocal tract anatomy are the ones whose measurements are likely to relate to voice quality.

Figure 9.5a shows the anatomy of a human skull, focusing on the major bones only. The bones on the upper part of the skull have little to do in shaping the vocal tract, but they do give an indication of the shape and dimensions of the rest of the skull, including those that comprise the articulators and define the shape and dimensions of the vocal chambers (oral, nasal and nasopharyngeal) of the skull.

The most meaningful facial features for voice generation are the measurements of different relevant parts of the face, as shown in Fig. 9.5b, which shows some meaningful direct measurements in the horizontal and vertical directions, and of area over some relevant bones, respectively. These can be obtained using one of many available automated facial landmark detection algorithms, e.g. [35, 36], or even measured manually if only a few faces must be processed in a given situation. *When 3D scans are available, similarly motivated measurements can be defined and used.* 3D scans can yield measurements such as the cephalic index (a scaled ratio of the width of the head to its length, explained later in this chapter), which is correlated to voice through its correlations to the lengths of the vocal chambers.

The features used for profiling in the knowledge-based framework are a combination of the basic facial measurements, and their ratios. Note that such measurements can also be useful in reconstructing the face or skull—the mapping from these to voice quality can be reversed to obtain facial/skull structure from voice quality. The voice quality used in this process can be estimated from the voice signal using data driven or knowledge driven techniques (see Fig. 9.6c).

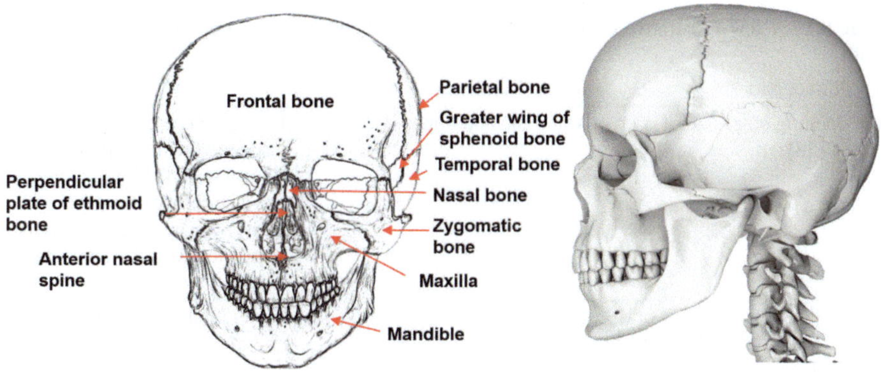

(a) Facial bones

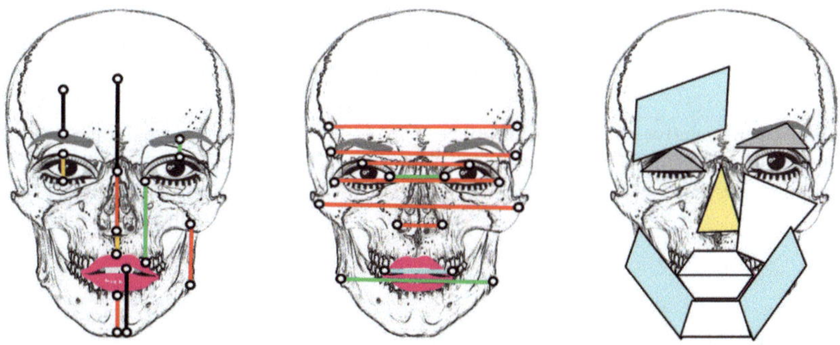

(b) Skull-related vertical, horizontal and area measurements of the face. Facial proportions can be defined by ratios of such attributes and used as features. The ones below the level of the eyes are most useful for the reconstruction of voice from face.

Fig. 9.5 Structure of the human skull, and facial measurements related to it

9.4.2.2 Relating Facial Features to Voice Qualities

The relationship between facial features and voice quality can no doubt also be learned in a data driven fashion—however, in some situations we may not have enough data to do so. In such situations, functional relationships must be established between these features and voice quality. Relationships between various parts of the vocal tract, and properties of the voice signal (which in turn relate to voice quality, as explained in Chap. 6), have been discussed earlier in this book. These relationships are directly applicable to those facial measurements that are linearly related to the corresponding parts of the vocal tract. They can thus be used to estimate the corresponding voice quality values. Table 9.1 lists a few examples, with references. Many others have been reported in prior studies. For example, tongue volume is correlated to the size of the oral cavity [37], which is related to skull size and facial dimensions. Tongue volume is also a factor in some voice qualities.

Table 9.1 A few illustrative examples of relationships between vocal tract geometry and voice quality

Vocal tract part	Related skull part	Voice quality
Lip opening size, size of oral cavity [38]	Mandible and upper jaw	Twangy voice quality is associated with shortened tract length, widened lip opening, and a slightly constricted oral cavity; yawny voice quality is associated with the opposite of these
Lower jaw, lower half of oral cavity, tongue size, lower teeth [38, 39]	Mandible	Vowel qualities, resonance
Size of pharynx [40, 41]	Distance between Nasion and Gnathion (see Fig. 9.7) i.e., top of the nasal bone to the lower tip of the mandible; and length of neck	Pitch; voice qualities dependent on formant positions, such as resonant, ringing etc.
Position of larynx relative to the pharynx [42–44]	Same as above	Raised larynx voice qualities (Sect. 6.4.1.4), or lowered larynx voice qualities (coveredness, yawniness etc.)

Such studies, however, do not exist in the context of all voice qualities. Those that do not exist can be estimated (at least roughly) from a few examples of face-voice pairs. The simplest way to estimate these is to chart the variation of each voice quality feature with each kind of facial measurement, and fit a function to the graph. Given a new face, ensembles of such functions can be used to estimate each voice quality.

9.4.3 *Imposing Voice Quality Features on a Given Voice Signal: Voice Transformation*

For voice transformation, the mechanism used could either be data-driven, or knowledge based. In a knowledge-based mechanism, signal processing methods are used to impose the desired characteristics of the source voice signal on the target voice signal. We will not discuss these methods here.

In a data-driven mechanism, pairs of voice signals—from the source and target speaker—are used to train the system. The goal is to transform the source signal to have the style and quality of the target signal, while retaining the content of the source signal. One method to achieve this goal is shown in Fig. 9.6a.

The process shown in Fig. 9.6a **transfers** the voice quality of the target signal to the source signal, resulting in a generated signal that has the content of the source signal and the voice qualities of the target signal. To achieve this, the training of the

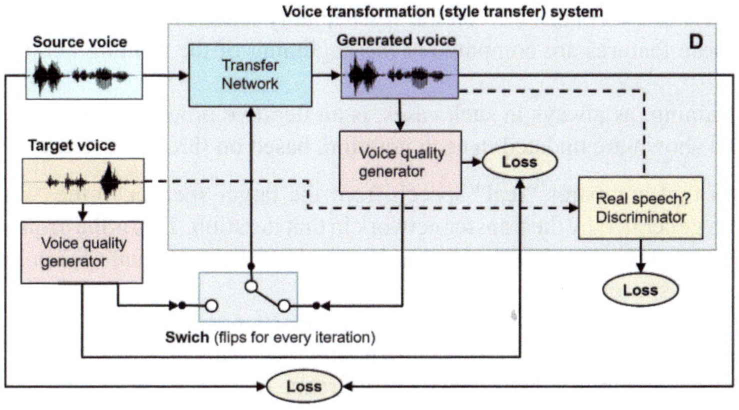

(a) A design for voice transformation.

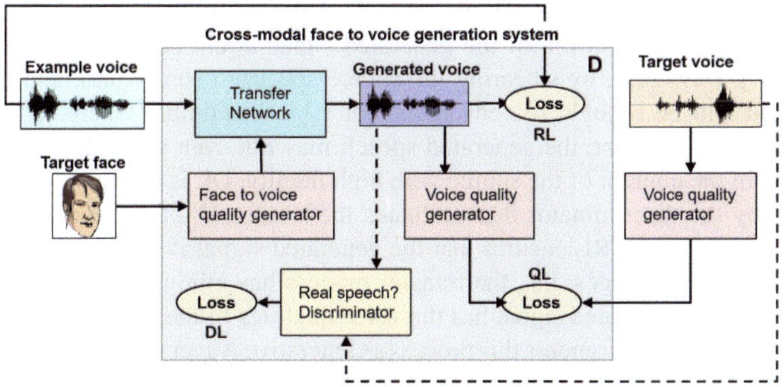

(b) Face to voice transfer training: a cross-modal transfer system for generating voice from face. It needs an example voice to transform, to which it can transfer the information derived from a given face. This voice may be generated by a different system.

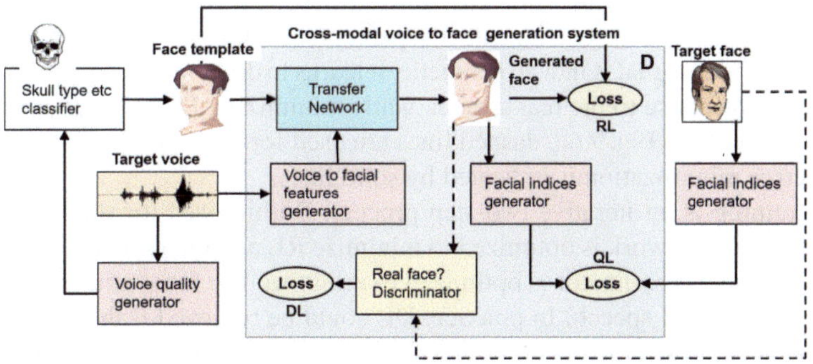

(c) Voice to face transfer training: a cross-modal transfer system for generating face from voice. It needs an example face, which may be an average facial template based on other profile detections.

Fig. 9.6 System architectures for cross-modal reconstruction

transfer network is conditioned on the voice quality features derived from the target signal (these features are computed at the beginning of the training iterations, and fixed).

The training, as always in such cases, is an iterative process. The parameters of the model shown are updated in each iteration, based on three objectives:

1. A loss that compares "real" speech from the target speaker to the "synthetic" speech generated by the transfer network in that iteration. This is the *discriminator loss* (DL). The output of the discriminator is ideally 1 for real speech and 0 for generated speech.
2. A loss that compares the "distance" (e.g. an L^2 norm) between the input and generated speech. This is the *reconstruction loss* (RL).
3. A loss that computes the distance or difference (e.g. an L^1 norm) between voice quality vectors of the target and generated voices. This is the *quality loss* (QL).

We will refer to these terms as DL, RL and QL respectively. DL is necessary because in its absence, the spectrum of the generated signal could potentially be changed in arbitrary ways (e.g. by smearing the source spectrum somewhat), and the process could still be brought to convergence at a local optimum where RL and QL are minimum. However, the generated speech may not even sound like speech, let alone retain the content of the source with high fidelity. DL is a binary (yes/no; 0/1) decision by the discriminator that evaluates the "realness" of the generated signal in a perceptual sense. RL ensures that the generated signal is the same as the input signal if nothing changes, i.e., the transfer process has minimum content loss. QL ensures that the generated signal has the voice qualities of the target signal.

The presence of DL renders this process a Generative Adversarial Network (GAN) based one. The transfer network generates a signal that sounds like the "real" speech. The goal is to "fool" the discriminator into deciding that the synthetic speech is in fact "real." It does so by adjusting the network parameters in each iteration so that RL and QL are minimized, while DL is maximized. The discriminator's goal is to discriminate between the two signals in a perceptual sense, i.e. to recognize the fact that the generated signal is indeed synthetic. It learns to do so by learning to maximize its output in response to the real speech, while minimizing DL (to 0) in response to synthetic speech. In Fig. 9.6a, dashed lines are used for the GAN component, while regular error minimization is indicated by solid lines.

The training is an iterative two-step process. Within each iteration, in the first step the transfer network is optimized to minimize RL and QL and maximize DL. In the second, the discriminator is optimized to minimize DL for synthetic speech and maximize it for real speech. In practice, DL could be removed if the training starts from very accurate estimators for both, voice quality and closeness of reconstruction. GAN setups are more complex and slower to train. The discriminator can also be made generic in practice, designed only to decide whether the generated signal is human speech or not.

Note that in this setup, there is no attempt to directly minimize the differences between the generated and *target* signal. They are disconnected. This renders the framework independent of content of the source and target signals, ensuring that only the desired voice qualities are transferred from the target signal to the source signal.

9.4.4 Face to Voice Generation: The Process

It is very easy to now see that the same setup can be used to transfer the **style** of speaking (and other mannerisms) of the target to the source signal as well. Such frameworks are in fact called *style-transfer* frameworks e.g. as in [45, 46], and in principle can be used to transfer the style of any pairs within the same modality, e.g. style of photographs to paintings, an image of a face to an image of another face etc. Style transfer frameworks can in fact be used in a wide variety of applications.

This voice transformation network can be extended to create a complete face to voice generation system, as shown in Fig. 9.6b. Since this system effectively transforms one modality into another (image to audio), this is an example of a cross-modal transfer system. Figure 9.6b in fact shows how such a system is trained. A database of face-voice pairs (called the "target face" and "target voice" in the figure) is used to train the system. The goal is to learn, given the target face, to generate a signal that has the qualities of the target voice. An example voice (with the desired linguistic and other content) is used as input, and the system is trained to transfer the target voice qualities to it, while retaining its content. Much of this framework is similar in principle to that of the voice transformation (or style transfer) system discussed above, except for that fact that the voice qualities imparted to the example voice *must be those that are deduced from the target face*. They must, ideally, also be the same as the voice qualities of the target voice. To this end, a face to voice-quality generator is used in the manner shown in the figure. The generated signal is transformed to have the voice qualities output by this generator. QL in this case ensures that the imparted qualities are the same (or close to) those that can be directly extracted from the target voice.

Note that since generally all components of these systems are neural networks, the training process involves updating the system parameters through standard back-propagation of the losses (not explained here, as there are many excellent tutorials in the literature on how to train neural networks, e.g. [47–49]).

For face to voice generation, the use of voice quality features is motivated by the fact that they are meaningful in a profiling context, and through associations that can be formed from prior studies, they are also directly relatable to facial measurements. If this were not the case, *any* encoding into a latent representation would suffice instead of the voice quality estimator, provided it were meaningful in some manner.

9.4.5 Voice to Face Generation: The Process Revisited

As explained above, a system that transforms face to voice is a cross-modal transfer system. The architecture of this system can be effectively used for voice to face generation, with minimal changes. The underlying principles of course remain the same as for any generic cross modal information transfer.

Figure 9.6c shows how a system that reconstructs face from voice may be set up and trained. The architecture is the same as in the face to voice generation system, except now the modalities are switched. Instead of an example voice, now an *example face* must be provided as input, on which the characteristics of the face estimated from a given voice signal must be algorithmically imposed.

The system in Fig. 9.6c is trained using target voice-face pairs. A template face is first estimated from voice in a procedure that first estimates the skull structure from voice. A pre-trained skull-type classifier may be used for this. A template face is then estimated on the basis of skull type and other profile parameters, such as gender, ethnicity etc. This comprises the example face. Facial features (such as landmarks) estimated from the target voice influence the transfer network to generate a face which is essentially a transformed version of the template face. DL, QL and RL evaluate the realness of this generated face, its closeness to the target face in terms of facial features, and it closeness to the template face.

When trained, the parameters of the system are fixed. During operational use, this system takes in a voice signal as input, and generates a face as output. Note that the generic architecture for cross modal transfer outlined above is necessary for face to voice generation. However, for voice to face generation, while simple data-in data-out systems may provide realistic-looking reconstructions, they may not accurately match the actual speakers, unless the intermediate representations implicitly capture the relationships above.

9.5 Reconstruction of the Full Body from Voice

The human body form is defined by many factors, of which the skeleton, and the body's external shape, posture and appearance are the most important. The skeleton largely determines the stature and relative proportions of the torso and limbs. The body shape and external appearance are strongly influenced by many factors such as gender, age, weight etc. There are of course many other factors that play significant but incidental roles, such as state of health, presence of disabilities, habits, environment, exercise levels, profession, nutrition etc. All of these factors are collectively decisive of body structure. As a result, estimates of body structure must be jointly conditioned on them, with further refinements conditioned on the available knowledge of incidental factors (which may also be derived through profiling).

The relation between voice and body structure is implicitly acknowledged by humans. Humans are often able to formulate a non-random guess about a person's physical structure from their voice—for instance, various studies have shown that humans can guess a person's height with better than random accuracy from their voice. It is well known that we associate personalities to voices based on the physical characteristics they imply or relate to. The relations need not be direct; indirect inferences are also often made. While one set of studies (e.g. [22, 50]) have shown that people are able to put faces to voices, others show that people perceive body size in relation to the face (e.g. [51–53]). Many other studies collectively lead to the conclusion that voice, to different degrees, correlates with enough physical/structural parameters of the speaker to potentially enable a full reconstruction of the human form factor.

But how do we go about this computationally? To reconstruct the body from voice, we exploit the more direct relations between the skull structure and voice. We first estimate the skull type and dimensions, and facial morphology from voice [54, 55]. Known relationships between the skull and the body morphology, augmented by relevant profile parameters that relate to body structure, such as age, height, weight etc. are then used to estimate the full body structure.

In this section we discuss how these two steps can be cast into computational mechanisms for the automatic reconstruction of the human persona from voice.

9.5.1 *Estimating Skull Dimensions and Facial Morphology from Voice*

Facial morphology can be estimated once the skull type and dimensions are specified, as shown in Fig. 9.6c. For computational profiling purposes, estimation of broad categories of skull types, and relating those to face types is sufficient in most cases. The finer details of the face can be further learned through appropriate data-driven AI algorithms. Some broad-level categorizations of the skull and face are described below.

9.5.1.1 Anthropometric Categorizations: Skull and Face

In classical anthropometry, human skulls are categorized into three types: *brachycephalic*, *mesocephalic* and *dolichocephalic*. This categorization is based on the *cranial index* [56]. Similarly, facial morphology is divided into three types: *euryprosopic*, *mesoprosopic* and *leptoprosopic*, based on the *facial index* [57, 58].

However, variations among these exist. Other categorizations that relate skull type to facial type classify faces as *brachyfacial*, *mesofacial* and *dolichofacial*, following the skull-type classification. Implicitly, these use the cranial index rather than the facial index. Some studies use four categories of facial types, e.g. [59], two of which are defined in terms of anterior-posterior measurements, and two in terms of vertical measurements of the skull/face. Many others exist in different contexts such

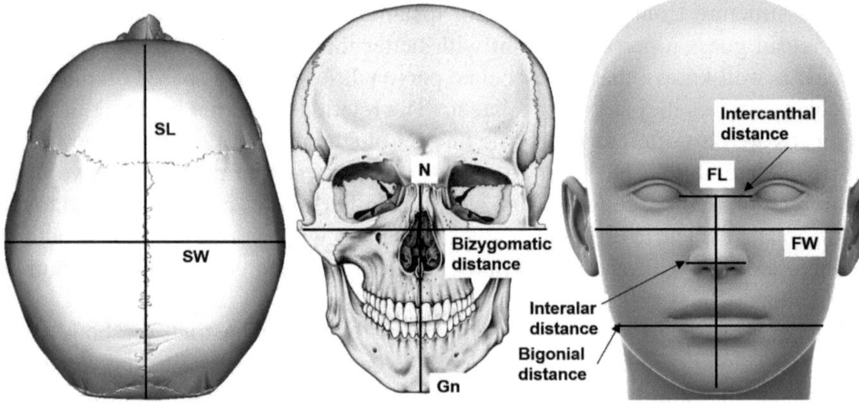

Fig. 9.7 Measurements used for the calculation of cranial and facial indices

as evolutionary biology, forensic anthropology, orthodontics etc. (e.g. [60–62]). For profiling purposes, [63] may be especially useful.

Skull morphology is reported to be related to facial morphology, as expected [64, 65]. Some studies seem to refute such an association, but these are relatively much smaller in number. We will therefore disregard the latter based solely on prevalence statistics in the literature.

For computational purposes, the information that there does exist a positive correlation between the two is enough to support the automatic inference of mappings between facial and skull types, for which appropriate classifiers can be trained.

Cranial and Facial Indices

Figure 9.7 shows the measurements used for defining the cranial and facial indices. The *cranial index* (CI) is defined in terms of the maximum skull width (SW) and the maximum skull length (SL) as $CI = 100(SW/SL)$. In living individuals, when measured from the top of the head, the same ratio is called *cephalic Index*. The *facial index* (FI) *of the skull* is defined in terms of the facial height (FH), measured from the Nasion (N) (midpoint of nasofrontal suture on the root of the nose) to Gnathion (Gn) (midpoint on the lower border of the mandible) anatomical landmarks, and the bizygomatic width (connecting the most lateral points of the zygomatic bone) (FW), measured from the right to the left Zygion as $FI = 100(FH/FW)$. On a face, the corresponding measurements are the face height and width. Other distances shown in Fig. 9.7 may be used to define more specific indices.

Tables 9.2 and 9.3 show the categorizations of skulls and faces into different types based on these indices, following [57, 58, 66].

Table 9.2 Head classification based on cephalic index

Type	Cephalic index range
Ultradolichocephalic	–64.9
Hyperdolichocephalic	65.0–69.9
Dolichocephalic	70.0–74.9
Mesocephalic	75.0–79.9
Brachycephalic	80.0–84.9
Hyperbrachycephalic	85.0–89.9
Ultrabrachycephalic	90.0–

Table 9.3 Face classification based on facial index

Type	Cephalic index range
Hypereuryprosopic	–79.9
Euryprosopic	80.0–84.9
Mesoprosopic	85.0–89.9
Leptoprosopic	90.0–94.9
Hyperleptoprosopic	95.0–

As mentioned earlier in this book, based on the overall proportions of the skull *and physical structure*, human population is divided into three categories: *Caucasoid*, *Mongoloid* and *Negroid* [67].

Some researchers believe that a fourth category—*Aboriginal* [68]—must be added to this list. Each skull type is on average distinguishable from the others through the proportions, angles and prominence of various bones.

Within each of these basic skull-types across the human population, there are epigenetic variations, often caused by variations of genetically determined but environmentally shaped growth patterns of tissues, muscles, nerves etc., and of bone growth. An extensive account and atlas of such epigenetic variations across humanity is given in [69]. This work could literally form the road map for skull reconstruction in the profiling context, provided the relations of these variations to voice are known.

According to [69], the specific characteristics that vary include facial cavities, sutures (shape, persistence), sutural and fontanelle ossicles, spines, tubercles, tori, notches, depressions, grooves, osseous bridges, foramina, canals and their finer structures. Figure 9.8 shows the key differences between the basic skull types.

Many of these differences affect the size and shape of vocal chambers. However, only some of these characteristics have been studied for their relation to voice. The studies have only focused on the standalone relationships, and have not taken all covariates of the skull into account due to various study constraints. Some examples of such standalone relationship studies include paranasal sinuses [72], oropharyngeal cavities [73] and generic facial structures [74]. Many others exist.

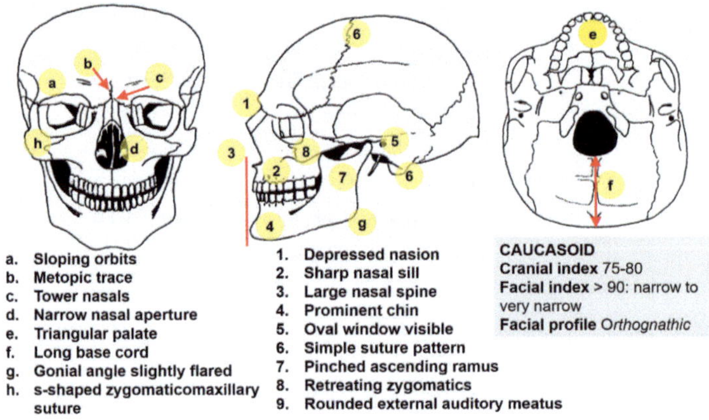

(a) Caucasoid

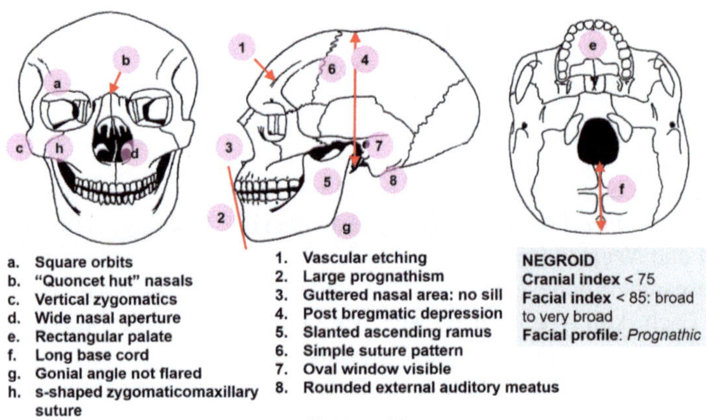

(b) Negroid

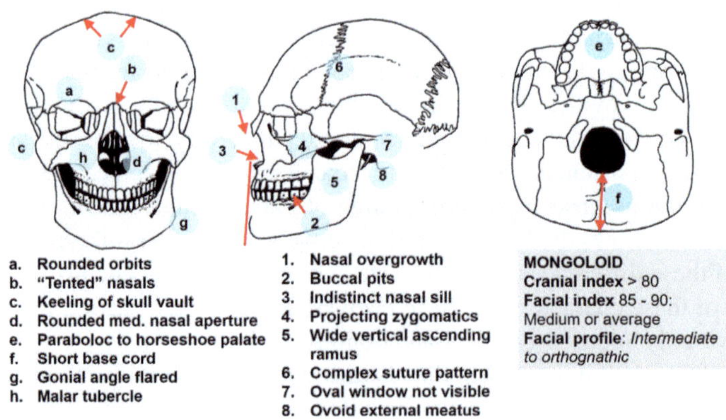

(c) Mongoloid (drawings modified from [70] and [71].

Fig. 9.8 Features of the three types of human skulls

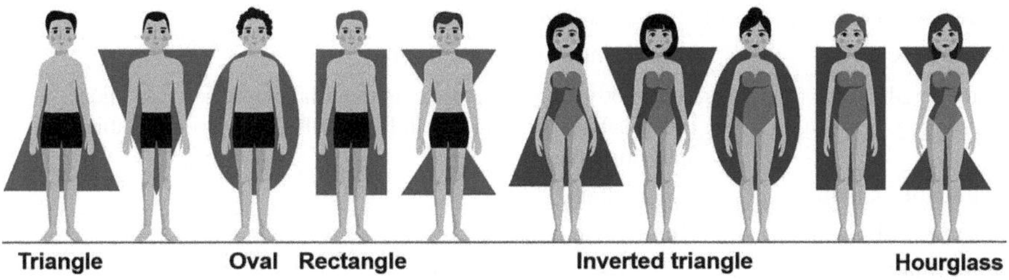

Fig. 9.9 Human body shapes

9.5.2 Estimating the Body Structure from the Skull/Face

The goal of full body reconstruction is to estimate the complete form factor (or *morphology*) of the body accurately.

To do this, the first step is to estimate the skeletal proportions from the estimated skull/facial morphology, aided by other profile parameter estimates. These estimates are based on known relationships between specific measurements of the body/skeleton to measurements of the skull/face. The most appropriate measurements for this purpose are those based on anthropometric guidelines. Anthropometry is the science of measuring the human body.

The skull *size* is generally highly correlated with skeletal size. The skull *proportions* have a weaker but positive correlation with skeletal proportions. Skeletal proportions however do not alone account for body shape, which depends on many other factors, including gender, age, body mass index (BMI), fat and weight distribution, musculature, health conditions etc.

Standard categorizations of body *type* and *shape* are not very useful in the profiling context. Categorizations of body shape may be used for rough estimates in the process, but body type categorizations are unlikely to be useful. The latter categorizations are controversial in the literature and not stable enough to form a clear basis for use in reconstruction. We will not discuss them here.

Body shapes are of five kinds—by most accounts—as shown in Fig. 9.9: pear (or triangle), inverted triangle, apple (or round), rectangle, and hourglass. These are genetically determined to a large extent—ethnicity plays a great role in determining them. However, the environment also exerts considerable long-term influence on a person's body shape over a lifetime [75].

9.5.2.1 Body Measurements

As in the case of the skull and face, anthropologists use specific measurements of the body to describe the full body morphology of a human. Anthropometric traits of the body are gauged in terms of *body indices*. As in the case of cephalic and facial indices described earlier in this chapter, these are usually defined in terms of

Table 9.4 List of body measurements: a study example

Measurement type	What is measured (Description)
Body length	Total body length, stature, suprasternal, acromial, dactylion, iliospinal anterior (shown in Fig. 9.10b), symphysion, tibial, and sphyrion tibial
Extremities length	Thigh, tibial, foot
Extremities breadth	Knee, ankle, foot
Limb lengths	Upper arm , forearm, hand
Limb breadths	Hand, wrist
Body trunk breadths	Biacromial, biiliac, chest
Body trunk depth	Chest
Body trunk length	Trunk

subsets of body measurements. Body measurements for such studies are based on *anthropometric points*, which are specified as in Fig. 9.11. As an example of how these are selected for different studies, the list in Table 9.4 shows a subset used in a specific study about the genetics of human body size and shape [76] (based on [77]). Body indices are usually specifically designed for different studies, but for profiling, they may be selected from widely used standard sets.

Body indices may be defined to measure both 2D and 3D geometrical form and proportions of the human body. As mentioned above, these tend to be subjectively defined for different studies. One example, from [76], is the **index of torso volume.** The index of torso volume (I_{TV}) is based on a measure of the volume of a cylinder that encloses the torso. The specific body measurements involved in its calculation are as labeled in Fig. 9.10a. I_{TV} is calculated by dividing it into subcomponents—thoracic volume (V_1) and abdominal pelvic volume (V_2). Both V_1 and V_2 are calculated using a triple integral to describe the specific torso contour involved (details are given in [76]). I_{TV} is then given by:

$$V_1 = \frac{\pi}{6} L_1(2(B_2 - B_1)(D_2 - D_1)) + 3(B_2 - B_1)D_1 + 3(D_2 - D_1)B_1 + 6B_1D_1) \tag{9.4}$$

$$V_2 = \frac{\pi}{6}(L_2 - L_1)(2(B_3 - B_2)(D_3 - D_2)) + 3(B_3 - B_2)D_2 + 3(D_3 - D_2)B_2 + 6B_2D_2) \tag{9.5}$$

$$I_{TV} = V_1 + V_2 \tag{9.6}$$

A vector of subjectively defined indices can be used as a feature vector in the process of full body reconstruction. If 3D reconstruction and motion is desired, 3D and other kinds of measurements must be used. For rendering motion, for example, the study above defines an equilibrium index [76] based on angles measured with respect to the line that defines the Iliospinal anterior height (shown in Fig. 9.10b).

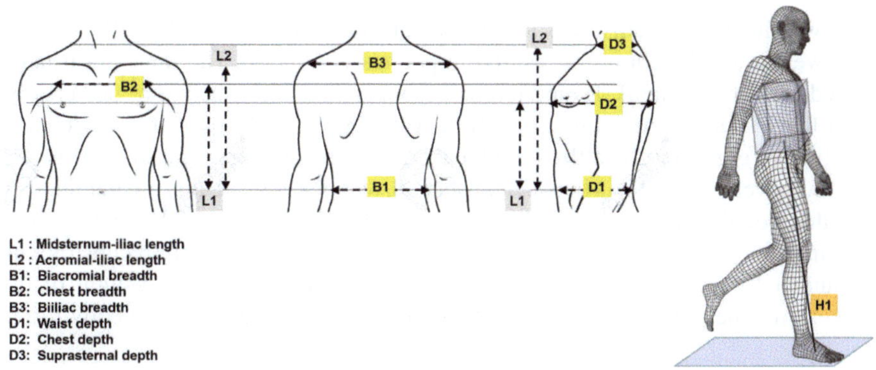

(a) Torso measurements

(b) H1: Iliospinal anterior height

Fig. 9.10 Measurements involved in the calculation of some example body indices

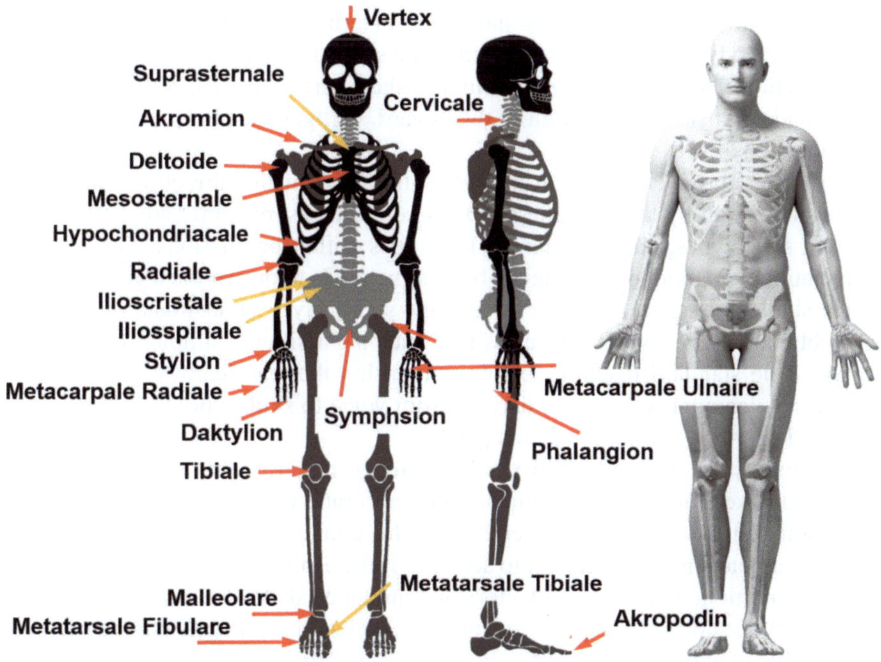

Fig. 9.11 Anthropometric points

Such indices can be very useful in determining static body pose (given a form factor for the body) as well.

Another important variable that plays a significant role in defining the body form is the **body mass index**, or BMI, measured in kg/m^2 as BMI $= w/h^2$, where w

is the weight in kilograms, and h is the height in meters; or BMI $= 703 \times w/h^2$ if the weight and height are measured in pounds and inches.The units are usually omitted while reporting BMI measurements, but understood to be kg/m^2 universally. According to current guidelines, a BMI between 18.5 and 25 kg/m^2 is considered to be optimal weight for a healthy individual. A person is underweight if BMI is less than 18.5, overweight if it is between 25 and 30, and obese when it is greater than 30. Note that people with high muscle:fat ratio tend to have higher BMIs which may sometimes be misleading (as in the case of professional athletes). In reconstructing the human form, an estimate of BMI (from estimated height and weight of the speaker) may help in determining the general leanness or fatness of the body form.

9.5.2.2 Relating Skull Measurements to Body Measurements

Cues for the estimation of body measurements from the skull/face can be drawn from many fields such as anatomy, biological anthropology (e.g. [78]), art etc. The field of forensic anthropology is perhaps the richest in terms of the information needed to estimate the body skeletal structure from facial structure, and for its reverse—facial reconstruction from skeletal parts, including skull parts, e.g. [79, 80]. Other fields also provide valuable information in this regard, e.g. genetics [81], or the study of human neoteny [82].

Skull-type classification is an important first step in this direction. Skull proportions (as given by the cephalic index and other measurements) are different for different skull types, and variants of skull types within populations. The simplest computational method to estimate the body structure is to determine a general shape for the skull (a template skull, or an average skull) given an estimate of the skull type, obtained using a skull-type classifier. Such a classifier is simple to train using any available standard machine learning algorithm that can learn to map features (typically voice quality features) derived from voice to skull type. For this, it must be trained on voice-skulltype labeled pairs.

The proportions of the template skull can then be used to estimate the proportions of the body skeleton. For this estimate to be accurate, the co-dependence of these proportions on the age, gender, height, weight, ethnicity and such other relevant profile parameters must be taken into account. If the estimates are probabilistic, they must be conditioned on those co-factors that influence skeletal proportions.

Figure 9.12 shows the dependence of body/skeletal proportions on one such co-factor, called **Neoteny**. Figure 9.1b shows how the proportions of the face (and by extension the skull) relate to the dimensions of the vocal chambers. Figure 9.11 shows anthropometric points that define the measurements that matter for the estimation of the relation between the skeletal structure and the skull (and thereby to the vocal chambers). More detailed accounts of these dependencies are available in the literature. In practical implementations, these must be used. This section only provides a couple of examples to illustrate the general methodology.

Once the full body proportions are determined, they can be refined. Some factors such as the weight distribution across the body can be carefully reconsidered, in

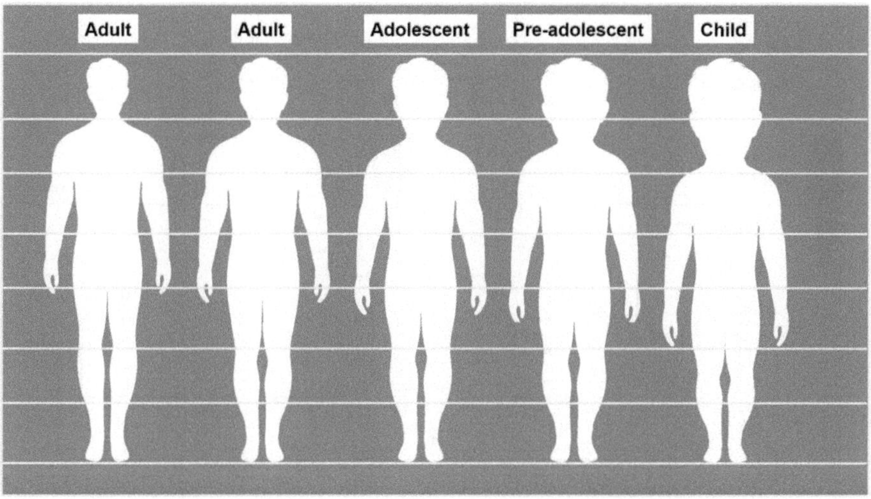

Variation of body proportions with age: Neoteny in adults refers to the degree to which their body proportions resemble those of younger individuals.

Fig. 9.12 Neoteny

the light of other profile information. Again, prior studies are likely to be helpful in developing guidelines for this. The proportions of the body are determined in part by the skeleton, but its shape and appearance (the persona) is also determined by muscle and tissue structure, which varies across different skeletal types [83].

9.5.3 Algorithmic Process for Body Reconstruction from Voice

Estimation of the full body structure from voice can now be achieved through a chaining mechanism, where the skull/face is first reconstructed from voice as explained in Sect. 9.4.5, and its characteristics are used to form an initial estimate, or template, of the body of the speaker. As before, other profile information, such as age, gender etc. of the speaker (concurrently derived from voice), may be used to guide this process.

The architecture of a system for full body reconstruction is shown in Fig. 9.13a. Once the template body is generated, it forms the input to this system. The system is trained with a database of body images, from which target face-body image pairs are used. During training, the goal of the system is to transfer the body characteristics of the target body to the template.

The principles underlying this architecture are the same as those used in the system for voice to face generation. DL, QL and RL indicate the humanness of the generated body morphology, closeness of the generated and target bodies in terms of body characteristics (such as type, shape, measurements, ratios, indices etc. extracted by the body features extractor), and the fidelity of reconstruction respectively. A face

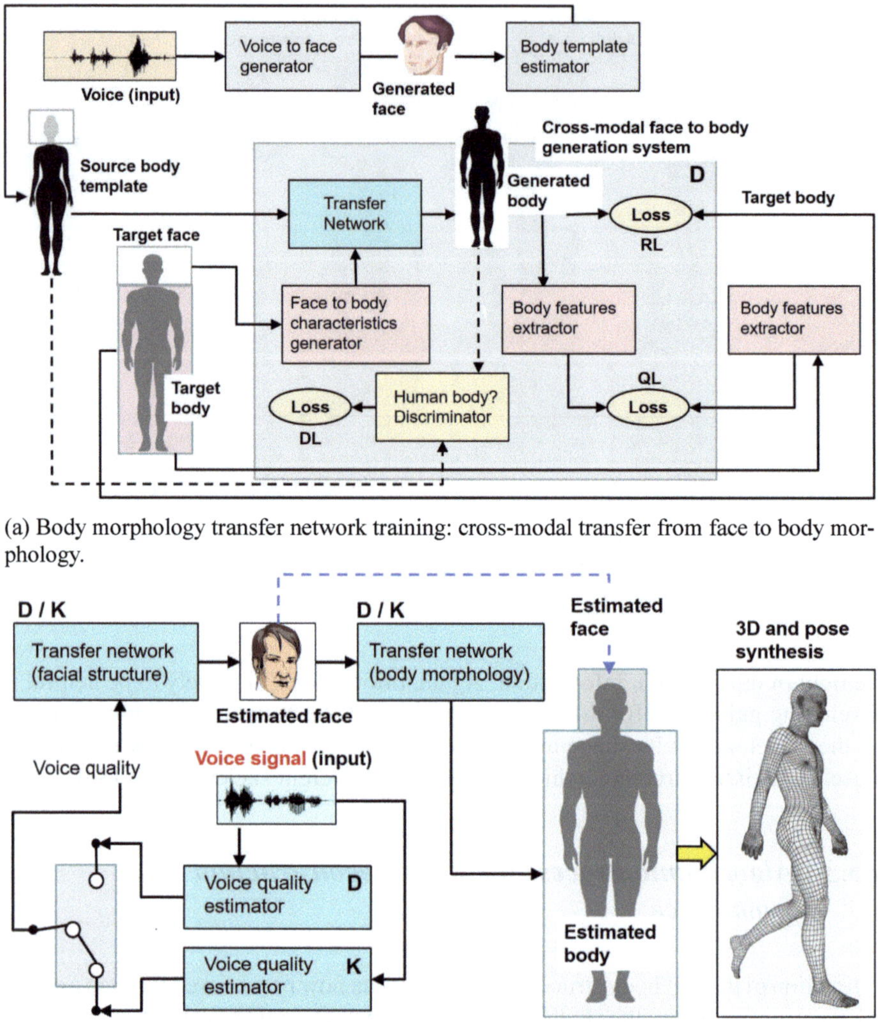

(a) Body morphology transfer network training: cross-modal transfer from face to body morphology.

(b) Full body reconstruction

Fig. 9.13 Key components in voice to full body reconstruction: the complete process

to body characteristics generator is used to estimate the body dimensions from the facial structure, as explained in Sect. 9.5.2.2.

In an operational setting, the general process for estimating the full body structure from voice is shown in Fig. 9.13b. Given a voice signal, a voice to face reconstruction system is used to generate the estimated face of the speaker. Based on this (and concurrent profile information derived form the voice signal), a template for the body is estimated, and the system then generates a body with the characteristics inferred from voice.

9.5.3.1 Subsequent Processing of the Generated Body

Subsequent to full body generation, texture filling and other details may be filled in using existing knowledge-based techniques. Pose estimation can also be done, imparting the body different plausible poses based on equilibrium index and other considerations. 2D forms can be extrapolated to render 3D forms, and motion vectors may be imposed on it to create videographic renderings. These technologies are well developed today, but we will not discuss them further in this book.

9.5.4 Alternate Algorithmic Approaches

Two alternative approaches to learn to predict physical parameters from skull proportions are:

Posterior prediction: In this model, the physical structure of a speaker is represented as a composite of "basic" patterns, akin to principal component analysis. For example, the face can be represented through a set of hand-marked keypoints, or measurements shown as in Fig. 9.5, or proportions of such measurements etc. The measurements can be extended to the full body.

If the training data comprises standardized, well-registered 3-D images of the body, these keypoints, in fact, represent an invertible measurement from which a reasonable reconstruction of the physical structure may be obtained. The posterior prediction model learns a direct regression from features derived from the corresponding voice signals to these features.

The actual regression models must consider the fact that information is embedded both in the instantaneous and temporal progression of the voice signal. When multiple independent inputs are presented to the regressor, it is inefficient for it to be obtaining independent estimates. Instead, it is more effective for additional inputs to be used to *refine* the estimates obtained from previous inputs. It must effectively utilize auxiliary information, such as gender, race, etc. to improve predictions, when presented; however, the predictions need not be contingent on the availability of these inputs.

This approach treats the prediction of different body parameters as separate, but coupled predictions within a multi-task framework. The framework allows a great deal of latitude in the specific mechanisms used, the constraints applied, the learning paradigms, etc.

Generative models: In this framework, the model learns the conditional probability distribution of the body parameter variables from training data. While it is conditioned on voice, the auxiliary measurements are treated as random variables whose distribution must also be learned. During inference, if their value is known, the distribution can be further conditioned on them; otherwise they can be marginalized out of the estimate.

The probability distributions to be learned are complex and not easily modeled through conventional means. Standard maximum-likelihood estimators are unable to capture the complexity of the model. This requires us to use variants of one of two types of frameworks that enable the learning of learning complex distributions through transformations of a random variable drawn from a simpler distributions.

The first variant is that of a GAN framework, wherein the parameters of the transform are optimized by an adversarial model that distinguishes between random draws from the model and actual samples of data. For training, the former could be transformed draws of the random variable, while the latter could be instances of actual physical characterizations of speakers. The transforms are optimized to obtain accurate approximations to the true conditional distribution of physical measurements for the target speaker given his/her voice signals, such that draws from it are indistinguishable from the actual physical measurements of the speaker. Once learned, the model can be used to derive maximum *a-posteriori* (MAP) estimates of the physical measurements, given the voice signal. The actual nature of the transformation, which is modeled by an appropriately designed neural network, the learning algorithms and the optimal inference algorithms must be carefully designed keeping data availability and computational constraints in mind.

The second variant is that of *normalizing flows*, wherein standard parametric models are converted to more complex ones through a series of invertible transformations. As in the generative adversarial framework, the transforms themselves are modeled by a neural network, however here the actual learning is performed through variational approximation. This framework is excellent for capturing conditional probability distributions of the voice signal, given conditioning measurements such as the physical measurements of the speaker, and the content of their speech. The model is easily inverted to obtain the distribution of physical measurements conditioned on the voice and additional auxiliary input. As before, the actual structure of the transformations, the learning algorithms, and the inference algorithms must again be designed based on practical considerations.

In Summary

Finally, as a roundup of the information that leads into this final chapter on core profiling methods, one can state that any end-to-end technology for mapping voice to a physically accurate avatar (embodiment or visage) of the speaker, and possibly elements of the environment around them as they speak, requires the development and application of many techniques in parallel. These include techniques for (a) micro-feature discovery guided by prior knowledge (b) disambiguating the mixed influences of multiple factors on voice (c) modeling and mapping the discovered features to judiciously chosen physical parameters in 3D space that can lead to further inferences of fuller structures, and d) the actual creation or rendering of images (holographic or otherwise) based on deduced information.

Figure 9.14 shows the current state-of-art in voice to 3D-face reconstruction, at the time of writing this chapter (October 2018). The first panel shows a live system demonstrated and used by the attendees of the World Economic Forum in Tianjin, China, in September 2018. This system was built at Carnegie Mellon University

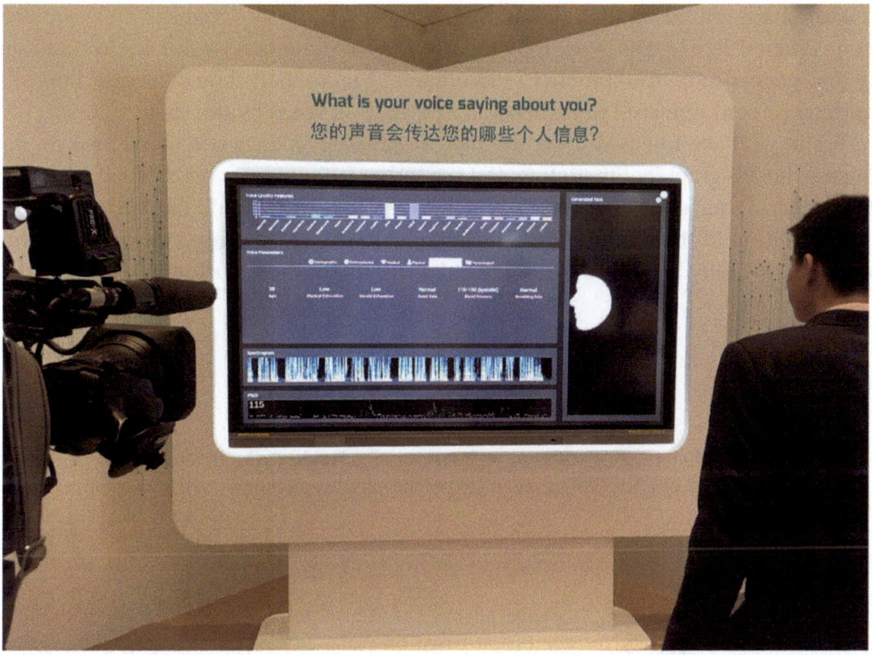

Profiling and 3D facial reconstruction shown on a screen at the World Economic Forum in Tianjin, China from 18-20 September 2018. The 3D reconstructed face of a speaker (from his voice) is seen in profile in this snapshot, and is shown being filmed by a television camera. Color, texture, eye color, hairstyle etc. were not rendered since they do not have direct relations to voice.

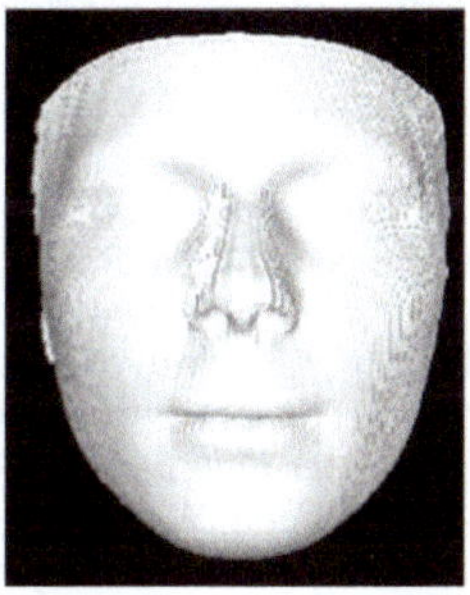

A close-up of a face on the screen during a live 3D reconstruction at the WEF. The facial features are now clearly visible.

Fig. 9.14 State of art in 3D voice to face reconstruction, as of September 2018

in early 2018, based on the methodologies described in this chapter. The face was reconstructed in 3D, and was open to 720°(or 4π steradians) solid angle rotations by the speaker in a virtual reality environment. Figure 9.14 shows the close-up of a face (on screen) rendered in response to a speaker's voice by the same system.

While we have touched upon the first three issues mentioned above, we do not discuss the 3D graphical modeling and rendering technologies here. Note, however, that rendering techniques are heavily tied to the hardware used, and both the speed and resolution of the renderings is constrained by these. Gaming systems use some of the most highly hardware-optimized algorithms for these currently.

References

1. Zaidi, A. A., Mattern, B. C., Claes, P., McEcoy, B., Hughes, C., & Shriver, M. D. (2017). Investigating the case of human nose shape and climate adaptation. *PLoS Genetics*, *13*(3), e1006616.
2. Subtelny, J. D. (1959). A longitudinal study of soft tissue facial structures and their profile characteristics, defined in relation to underlying skeletal structures. *American Journal of Orthodontics*, *45*(7), 481–507.
3. Short, L. A., Mondloch, C. J., McCormick, C. M., Carré, J. M., Ma, R., Fu, G., et al. (2012). Detection of propensity for aggression based on facial structure irrespective of face race. *Evolution and Human Behavior*, *33*(2), 121–129.
4. Carré, J. M., McCormick, C. M., & Mondloch, C. J. (2009). Facial structure is a reliable cue of aggressive behavior. *Psychological Science*, *20*(10), 1194–1198.
5. Swift, W. B. (1916). The possibility of voice inheritance. *Review of Neurology and Psychiatry*, *14*, 103.
6. McAllister, H. A., Dale, R. H., Bregman, N. J., McCabe, A., & Cotton, C. R. (1993). When eyewitnesses are also earwitnesses: Effects on visual and voice identifications. *Basic and Applied Social Psychology*, *14*(2), 161–170.
7. Campanella, S., & Belin, P. (2007). Integrating face and voice in person perception. *Trends in Cognitive Sciences*, *11*(12), 535–543.
8. Schweinberger, S. R., Kloth, N., & Robertson, D. M. (2011). Hearing facial identities: Brain correlates of face-voice integration in person identification. *Cortex*, *47*(9), 1026–1037.
9. Locher, P. (2010). *How does a visual artist create an artwork* (pp. 131–144). The Cambridge handbook of creativity. Cambridge, UK: Cambridge University Press.
10. Schkolne, S., Pruett, M., & Schröder, P. (2001). Surface drawing: Creating organic 3D shapes with the hand and tangible tools. In *Proceedings of the SIGCHI Conference on Human Factors in Computing Systems* (pp. 261–268). Seattle, WA, USA: ACM.
11. Mori, G., & Malik, J. (2002). Estimating human body configurations using shape context matching. In *Proceedings of the European Conference on Computer Vision (ECCV)* (pp. 666–680). Denmark: Copenhagen.
12. Kim, C., Shin, H. V., Oh, T. H., Kaspar, A., Elgharib, M., & Matusik, W. (2018). On learning associations of faces and voices. arXiv:1805.05553.
13. Nagrani, A., Albanie, S., & Zisserman, A. (2018). Learnable PINs: Cross-modal embeddings for person identity. arXiv:1805.00833.
14. Nagrani, A., Albanie, S., & Zisserman, A. (2018). Seeing voices and hearing faces: Cross-modal biometric matching. In *Proceedings of the IEEE Conference on Computer Vision and Pattern Recognition (CVPR), Salt Lake City, Utah, USA* (pp. 8427–8436).
15. Lippert, C., Sabatini, R., Maher, M. C., Kang, E. Y., Lee, S., Arikan, O., et al. (2017). Identification of individuals by trait prediction using whole-genome sequencing data. *Proceedings of the National Academy of Sciences*, *114*(38), 10166–10171.

16. Wen, Y., Ismail, M. A., Liu, W., Raj, B., & Singh, R. (2018). Disjoint mapping network for cross-modal matching of voices and faces. arXiv:1807.04836.
17. Blanz, V., & Vetter, T. (1999). A morphable model for the synthesis of 3D faces. In *Proceedings of Siggraph* (Vol. 99, pp. 187–194).
18. Cootes, T. F., Edwards, G. J., & Taylor, C. J. (2001). Active appearance models. *IEEE Transactions on Pattern Analysis and Machine Intelligence*, *6*, 681–685.
19. Sela, M., Richardson, E., & Kimmel, R. (2017). Unrestricted facial geometry reconstruction using image-to-image translation. In *Proceedings of the International Conference on Computer Vision (ICCV)* (pp. 1576–1585). Venice, Italy: IEEE.
20. Feng, Y., Wu, F., Shao, X., Wang, Y., & Zhou, X. (2018). Joint 3d face reconstruction and dense alignment with position map regression network. In *Proceedings of the European Conference on Computer Vision (ECCV), Munich, Germany* (pp. 534–551).
21. Belin, P., Fecteau, S., & Bedard, C. (2004). Thinking the voice: Neural correlates of voice perception. *Trends in Cognitive Sciences*, *8*(3), 129–135.
22. Kamachi, M., Hill, H., Lander, K., & Vatikiotis-Bateson, E. (2003). Putting the face to the voice: Matching identity across modality. *Current Biology*, *13*(19), 1709–1714.
23. Schweinberger, S. R., Robertson, D., & Kaufmann, J. M. (2007). Hearing facial identities. *The Quarterly Journal of Experimental Psychology*, *60*(10), 1446–1456.
24. Ellis, A. W. (1989). Neuro-cognitive processing of faces and voices. In A. W. Young & H. D. Ellis (Eds.), *Handbook of research on face processing* (pp. 207–215).
25. Belin, P., Bestelmeyer, P. E., Latinus, M., & Watson, R. (2011). Understanding voice perception. *British Journal of Psychology*, *102*(4), 711–725.
26. Van Den Oord, A., Dieleman, S., Zen, H., Simonyan, K., Vinyals, O., Graves, A., et al. (2016). WaveNet: A generative model for raw audio. arXiv:1609.03499.
27. Taigman, Y., Wolf, L., Polyak, A., & Nachmani, E. (2018). VoiceLoop: Voice fitting and synthesis via a phonological loop. arXiv:1707.06588.
28. Ping, W., Peng, K., & Chen, J. (2018). ClariNet: Parallel wave generation in end-to-end text-to-speech. arXiv:1807.07281.
29. Wang, Y., Skerry-Ryan, R. J., Stanton, D., Wu, Y., Weiss, R. J., Jaitly, N., et al. (2017). Tacotron: A fully end-to-end text-to-speech synthesis model. arXiv:1703.10135.
30. Danlos, L. (1987). *The linguistic basis of text generation*. Cambridge, UK: Cambridge University Press.
31. Oberlander, J., & Brew, C. (2000). Stochastic text generation. *Philosophical Transactions of the Royal Society of London A: Mathematical, Physical and Engineering Sciences*, *358*(1769), 1373–1387.
32. Zhang, Y., Gan, Z., Fan, K., Chen, Z., Henao, R., Shen, D., et al. (2017). Adversarial feature matching for text generation. arXiv:1706.03850.
33. Semeniuta, S., Severyn, A., & Barth, E. (2017). A hybrid convolutional variational autoencoder for text generation. arXiv:1702.02390.
34. Dehghani, M., Gouws, S., Vinyals, O., Uszkoreit, J., & Kaiser, K. (2018). Universal transformers. arXiv:1807.03819.
35. Zhang, Z., Luo, P., Loy, C. C., & Tang, X. (2014). Facial landmark detection by deep multi-task learning. In *Proceedings of the European Conference on Computer Vision (ECCV), Zurich, Switzerland* (pp. 94–108).
36. Perakis, P., Passalis, G., Theoharis, T., & Kakadiaris, I. A. (2013). 3D facial landmark detection under large yaw and expression variations. *IEEE Transactions on Pattern Analysis and Machine Intelligence*, *35*(7), 1552–1564.
37. Oliver, R. G., & Evans, S. P. (1986). Tongue size, oral cavity size and speech. *The Angle Orthodontist*, *56*(3), 234–243.
38. Story, B. H., Titze, I. R., & Hoffman, E. A. (2001). The relationship of vocal tract shape to three voice qualities. *The Journal of the Acoustical Society of America*, *109*(4), 1651–1667.
39. Thornbury, S. (1993). Having a good jaw: Voice-setting phonology. *ELT Journal*, *47*(2), 126–131.

40. Hynes, W. (1953). The results of pharyngoplasty by muscle transplantation in "failed cleft palate" cases, with special reference to the influence of the pharynx on voice production: Hunterian lecture delivered at the Royal College of Surgeons of England on 12th February 1953. *Annals of the Royal College of Surgeons of England*, *13*(1), 17.
41. Estill, J. (1988). Belting and classic voice quality: Some physiological differences. *Medical Problems of Performing Artists*, *3*(1), 37–43.
42. Esling, J. H. (1999). Voice quality settings of the pharynx. In *Proceedings of the 14th International Congress of Phonetic Sciences* (Vol. 3, pp. 2449–2452). Berkeley: University of California.
43. Sundberg, J., & Askenfelt, A. (1983). Larynx height and voice source: A relationship? In D. M. Bless & J. H. Abbs (Eds.), *Vocal fold physiology: Contemporary research and clinical issues*. San Diego, California: College-Hill Press.
44. Welch, G. F., & Sundberg, J. (2002). Solo voice. In R. Parncutt & G. McPherson (Eds.), *The science and psychology of music performance: Creative strategies for teaching and learning* (pp. 253–268). Oxford, UK: Oxford University Press.
45. Gatys, L. A., Ecker, A. S., & Bethge, M. (2016). Image style transfer using convolutional neural networks. In *Proceedings of the IEEE Conference on Computer Vision and Pattern Recognition (CVPR), Las Vegas, USA* (pp. 2414–2423).
46. Li, Y., Wang, N., Liu, J., & Hou, X. (2017). Demystifying neural style transfer. arXiv:1701.01036.
47. Jain, A. K., Mao, J., & Mohiuddin, K. M. (1996). Artificial neural networks: A tutorial. *Computer*, *3*, 31–44.
48. Wythoff, B. J. (1993). Backpropagation neural networks: A tutorial. *Chemometrics and Intelligent Laboratory Systems*, *18*(2), 115–155.
49. Lippmann, R. P. (1987). An introduction to computing with neural nets. *IEEE ASSP Magazine*, *4*(2), 4–22.
50. Smith, H. M., Dunn, A. K., Baguley, T., & Stacey, P. C. (2016). Concordant cues in faces and voices: Testing the backup signal hypothesis. *Evolutionary Psychology*, *14*(1), 1–10.
51. Peelen, M. V., & Downing, P. E. (2007). The neural basis of visual body perception. *Nature Reviews Neuroscience*, *8*(8), 636.
52. Cunningham, M. R., Roberts, A. R., Barbee, A. P., Druen, P. B., & Wu, C. H. (1995). Their ideas of beauty are, on the whole, the same as ours: Consistency and variability in the cross-cultural perception of female physical attractiveness. *Journal of Personality and Social Psychology*, *68*(2), 261.
53. Bruce, V., & Young, A. (1998). *In the eye of the beholder: The science of face perception*. Oxford, UK: Oxford University Press.
54. Mermelstein, P. (1967). Determination of the vocal-tract shape from measured formant frequencies. *The Journal of the Acoustical Society of America*, *41*(5), 1283–1294.
55. Yang, C. S., & Kasuya, H. (1994). Accurate measurement of vocal tract shapes from magnetic resonance images of child, female and male subjects. In *Proceedings of the Third International Conference on Spoken Language Processing (ICSLP), Yokohama, Japan* (pp. 623–626).
56. Edler, R., Agarwal, P., Wertheim, D., & Greenhill, D. (2006). The use of anthropometric proportion indices in the measurement of facial attractiveness. *The European Journal of Orthodontics*, *28*(3), 274–281.
57. Farkas, L. G., & Munro, I. R. (1987). *Anthropometric facial proportions in medicine*. Springfield, Illinois: Charles C. Thomas Publisher.
58. Rakosi, T., Jonas, I., & Graber, T. (1993). *Orthodontic diagnosis. Color atlas of dental medicine*. Stuttgart, Germany: Thieme Medical Publishers.
59. Sassouni, V. (1969). A classification of skeletal facial types. *American Journal of Orthodontics*, *55*(2), 109–123.
60. Allanson, J. E., Cunniff, C., Hoyme, H. E., McGaughran, J., Muenke, M., & Neri, G. (2009). Elements of morphology: Standard terminology for the head and face. *American Journal of Medical Genetics Part A*, *149*(1), 6–28.

61. Collett, A. R., & West, V. C. (1993). Terminology of facial morphology in the vertical dimension. *Australian Dental Journal, 38*(3), 204–209.
62. Nanda, S. K. (1988). Patterns of vertical growth in the face. *American Journal of Orthodontics and Dentofacial Orthopedics, 93*(2), 103–116.
63. Byers, S. N. (2016). *Introduction to forensic anthropology*. Abingdon, UK: Taylor & Francis.
64. Enlow, D. H., & McNamara, J. A, Jr. (1973). The neurocranial basis for facial form and pattern. *The Angle Orthodontist, 43*(3), 256–270.
65. Enlow, D. H., Kuroda, T., & Lewis, A. B. (1971). The morphological and morphogenetic basis for craniofacial form and pattern. *The Angle Orthodontist, 41*(3), 161–188.
66. Farkas, L. G., Katic, M. J., & Forrest, C. R. (2005). International anthropometric study of facial morphology in various ethnic groups/races. *Journal of Craniofacial Surgery, 16*(4), 615–646.
67. Nei, M., & Roychoudhury, A. K. (1974). Genic variation within and between the three major races of man, Caucasoids, Negroids, and Mongoloids. *American Journal of Human Genetics, 26*(4), 421.
68. Goedde, H. W., Agarwal, D. P., Fritze, G., Meier-Tackmann, D., Singh, S., Beckmann, G., et al. (1992). Distribution of ADH 2 and ALDH2 genotypes in different populations. *Human Genetics, 88*(3), 344–346.
69. Hauser, G., & De Stefano, G. F. (1989). *Epigenetic variants of the human skull*. Stuttgart, Germany: Schweizerbart Science Publishers.
70. Bass, W. M. (1987). *Human osteology: A laboratory and field manual* (3rd ed.). Columbia: Missouri Archaeological Society.
71. France, D. L. (2003). *Lab manual and workbook for physical anthropology* (5th ed.). Belmont, California: West/Wadsworth Publishing Company.
72. Blanton, P. L., & Biggs, N. L. (1969). Eighteen hundred years of controversy: The paranasal sinuses. *American Journal of Anatomy, 124*(2), 135–147.
73. Chen, F. C., Ma, E. P. M., & Yiu, E. M. L. (2014). Facial bone vibration in resonant voice production. *Journal of Voice, 28*(5), 596–602.
74. Titze, I. R. (2001). Acoustic interpretation of resonant voice. *Journal of Voice, 15*(4), 519–528.
75. Katzmarzyk, P. T., & Leonard, W. R. (1998). Climatic influences on human body size and proportions: Ecological adaptations and secular trends. *American Journal of Physical Anthropology: The Official Publication of the American Association of Physical Anthropologists, 106*(4), 483–503.
76. Livshits, G., Roset, A., Yakovenko, K., Trofimov, S., & Kobyliansky, E. (2002). Genetics of human body size and shape: Body proportions and indices. *Annals of Human Biology, 29*(3), 271–289.
77. Martin, R., & Saller, K. (1957). *Textbook of anthropology*. Stuttgart, Germany: Fischer Publications.
78. Katzenberg, M. A., & Grauer, A. L. (Eds.). (2018). *Biological anthropology of the human skeleton*. New York City, New York: Wiley.
79. Wilkinson, C. (2004). *Forensic facial reconstruction*. Cambridge, UK: Cambridge University Press.
80. Iscan, M. Y., & Steyn, M. (2013). *The human skeleton in forensic medicine*. Springfield, Illinois: Charles C. Thomas Publisher.
81. Wright, S. (1918). On the nature of size factors. *Genetics, 3*(4), 367.
82. Antón, S. C., & Leigh, S. R. (2003). Growth and life history in Homo erectus. *Cambridge Studies in Biological and Evolutionary Anthropology*, 219–245.
83. Simoneau, J. A., & Bouchard, C. (1989). Human variation in skeletal muscle fiber-type proportion and enzyme activities. *American Journal of Physiology-Endocrinology And Metabolism, 257*(4), E567–E572.

Chapter 10
Applied Profiling: Uses, Reliability and Ethics

There are many uses of profiling. Currently, as represented by this book, the science of profiling is in its nascent stages. As it becomes more accurate, more uses for it will emerge. However, there is a dichotomy associated with this progression. While its increasing accuracy is likely to give rise to more applications, its potential to severely infringe on a person's privacy through them will also rise. In the context of practical applications, two issues therefore become extremely important: whether the information generated through profiling is accurate or not, and whether it is relevant and ethical or not.

The reliability of automated profiling techniques can be monitored to ensure their net benefit: for the deductions we make algorithmically, quantitative measures of confidence in their accuracy and applicability can be calculated. For their potential to breach privacy, however, we do *not* have ready solutions at this stage. Breach of privacy is a topic that has become extremely controversial in multiple contexts in today's world. Our societies today struggle to strike a difficult balance between prioritizing security in a world of growing technology-facilitated crime, and the transgressions of privacy that must be allowed to support security measures.

This chapter is divided into five sections that are relevant in different ways to both the issues mentioned above. The first is a brief overview of some possible uses of this technology. The second discusses some performance metrics that must be considered in the practical deployment of profiling systems. The third section discusses reliability and confidence measures in general, which can be applied to practical profile deductions. The fourth section discusses the subject of predictions (and their reliability) based on experimentally found correlations between entities. The final section presents the author's viewpoint on ethical issues associated with the use of this technology in the world as it stands today—a more technologically endowed but angrier world, with growing unrest, deteriorating climate, burgeoning population, diminishing resources, and a technologically brilliant but societally uncertain future.

R. Singh, *Profiling Humans from their Voice*,
https://doi.org/10.1007/978-981-13-8403-5_10

10.1 Application Areas and Uses of Profiling

10.1.1 Law Enforcement

Profiling can be used to provide additional and supporting evidence in crimes that have voice as evidence. Some crimes are entirely committed through voice, and in these the criminal's voice is often the only tangible evidence of their identity. In these cases especially, profile information may potentially help track and identify the perpetrator. The technology itself is applicable to the analysis of voiceprints from a wide variety of sources. These include radio-transmitted voice, wiretaps, telephone conversations, and voice evidence from crimes recorded on a multitude of security and surveillance devices, including wearable smart devices. Its use in the law enforcement context, however, has more challenges than in other application areas. The key challenges are listed below:

1. **Quality degradation**: Deducing accurate profiling information from voice transmissions that are made in noisy conditions over lossy channels and non-ideal acoustic environments, or from degraded recordings. It requires the accurate identification of those aspects of the voice signal that retain micro-signatures in spite of the specific degradations it may have undergone. The search for relevant micro-features can be extremely subjective to the hardware and transmission media used from the time of voice capture to storage.
2. **Presence of overlapping human voices**: Profiling people in continuous speech-on-speech scenarios, such as casually recorded speech in crowded places, crowds and multi-speaker conversations.
3. **Profusion of new signatures**: Deducing information about the speaker's immediate physical environment, including the type of equipment or device used by the speaker when the market continues to be flooded with new devices whose acoustic signatures may not be available for study. The search space however can be narrowed using some domain knowledge, e.g. in the case of hoax calls made from marine environments, the search can be limited to equipment used in maritime environments.
4. **The criminal as a voice artist**: Profiling accurately, and accomplishing all of the above in the presence of deliberate deception—in situations where the perpetrator may have attempted to disguise or mask his/her voice to evade identification etc. In the case of repeat crimes wherein voice has been disguised in each instance, an additional challenge is that of finding instances of repeat crimes in the first place. Once ascribed to a particular speaker, a combination of profiling and speaker matching technologies may be required for this purpose.

   In specific categories of voice-based crimes that include conducting fraudulent transactions over phone, uploading damaging statements in other people's voices on public websites, and making hoax emergency calls to various medical, law enforcement and citizen distress response agencies, perpetrators often engage in voice disguise by impersonation. This is largely driven by the desire to convince the victims that the call is genuine so that a response is evoked, and the concurrent desire to evade identification. While in some relevant cases the perpetrators impersonate actual people, in most cases the perpetrators are observed to modify their voice to sound like any fictional person—*other than themselves*, faking accents, voice quality, style and demeanor in many complex ways. Profiling accurately, while in the presence of such deliberate deception, remains a challenge.

Techniques to detect the presence of impersonation in the absence of comparators or references (*in-vacuo* detection of impersonation) can help in directing profiling efforts better. Detection of the presence of disguise can form the basis for counter-spoofing measures that can be used to filter out situations where a response from the victims would have been needless, saving money, lives, and resources.

There is also a counter-problem associated with such crimes: when perpetrators of such voice-based crimes are caught and confronted with their voices as evidence, they often deny that they were the originators of the voices. It then becomes necessary to address the problem of *disproving* their denial of ownership of the voice in question. Some solutions have been suggested for this [1], but more need to be devised.

5. **The vanishing nativity**: Profiling for geographical origins and nativity of the speaker is rife with a modern challenge—that of globalization of human presence. Widespread communication, exposure to media, and increased ease and necessity for movement across geographical regions, have all served to diminish the effects of origin and nativity on speech. Dialects and accents show a downward trend of prominence—tending to become less pronounced with passage of time. Although they cannot be completely obscured since they are deeply ingrained during early childhood and growing years [2], they may be increasingly harder to detect, and even harder to correlate to specific locations, due to mixed influences. This challenge remains a serious moving target for some aspects of profiling.
6. **Vintage of evidence**: An additional problem is posed by the time elapsed between the analysis of evidence and the crime itself. If found, the perpetrator's voice must be compared to prior recordings. If a long time has elapsed since those were obtained, many of the speaker's parameters may have changed and the changes in turn may have affected their voice. In this case, profiling must be done in a manner that the changes are accounted for. This comes under the purview of *predictive profiling*. This topic has not been addressed in this book, but is now a sub-area of profiling that is beginning to form.

Profiling that is done for law enforcement purposes must be done in a legally acceptable and consistent manner. In most countries, lawmakers have instituted specific criteria for the legal acceptability of scientific evidence. In the United States, these are exemplified by the Daubert Standard criteria [3–5]. While most such criteria are generic and apply to all types of scientific evidence and analysis, they are followed by specific guidelines devised for each type of evidence that are enforced in courts on a per-case basis. In the case of voice profiling, such specific guidelines remain to be devised (in any country) at the time of writing this book.

10.1.2 Security

As we have discussed in Chap. 2, due to the enormous number of parameters that play a role in creating the human voice signal, no two voices in the world are alike. Unlike fingerprints and DNA, voice carries signatures of the speaker's state and surroundings at the time of speaking. This makes voice a good biometric to use for security purposes, provided we can exploit its biometric properties appropriately.

In most current security applications, a speaker's voice is compared to those in a database. When the goal is to identify the speaker from a set of possible speakers, the system is said to perform *speaker identification*. In this case the voiceprint is matched with all the recordings in the database and the closest match is used to determine the identity of the speaker. On the other hand, when the goal is to verify whether a given speaker is who they say they are, the voiceprint is only matched to the specific speaker's voiceprint, if present, in the database. In this case, the system is said to perform *speaker verification*, or *speaker authentication*. In both cases, the underlying technology used for voiceprint comparison is the same.

However, voice comparison is not the same as profiling from voice, as must be amply evident from the content of this book so far.

Traditionally speaker matching based applications use macro level features (explained in Chap. 7) for voice comparison. These can be spoofed by adversarial algorithms designed for this purpose [6–8]. To strengthen the verification process, profile deductions can be used to match additional details about the speaker, e.g. their height, nativity, weight, facial structure etc. This would involve the use of predictive profiling. We do not discuss adversarial algorithms in this book.

10.1.3 Health Services

Profiling can be used in multiple ways in healthcare. When profiling systems are accurate enough, the technology can be used to map the physical and mental parameters of a person and provide a **diagnostic aid** to healthcare professionals, much like an X-ray or a CT scan, without the harmful effects associated with exposure to ionizing radiation. Voice is non-invasive and can be easily transmitted to remote locations for analysis. It can be used for **assessment**, **tracking** or **monitoring** the physical and mental health status of persons over time. It can be used to monitor for medical drug compliance, intoxication, substance abuse, suicidal tendencies, and myriad other factors. The technology is likely to have its greatest impact in the telemonitoring or remote monitoring of underprivileged populations, veterans, elderly people, disabled people, children, etc. For all populations, it can be applied to generate early warning signs of many diseases, which could increase the efficacy and success of available curative measures.

10.1.4 Social and Commercial Services

Harnessing the power of human speech to understand humans at deeper levels can have a tremendous transformative effect on society. The technology can drive the automated choices made by application servers, such as voice interactive systems or voice-driven human-machine interfaces to be more appropriate for the human at any given time. Responses can be tailored to their status.

If a machine is tasked with administering an oral academic skills test to a student, for example, as time progresses the questions could be made more suited to the mindset of the student. Not all students in the one-size-fits-all educational settings that are a norm today, respond to a test correctly if the question is posed in just one specific way, even if they all know the answer equally well. The results of academic tests today are greatly tied to a student's linguistic alacrity and other peripheral factors that have nothing to do with their innate abilities to grasp the subject at hand, and only influence their *expression* of their understanding.

Social monitoring is a potentially large area of application for profiling. Businesses can benefit from a deeper and more alert monitoring of customer calls for potential problems that could arise from an ongoing conversation between a customer and service agent. Profiling technologies could monitor and flag the calls for anger, rude behavior, dissatisfaction, frustration, attempts to defraud, impersonation, intoxication, and many other human-centric issues that could lead to loss of business, failed transactions, or even criminal activities. In the field of education, these technologies can be used to monitor students as they learn. It could be used to monitor (and help improve) the skills of persuasion and leadership in many settings. There are in fact many professions, including those of voice artists, where different aspects of this technology could help. Monitoring different professionals, such as drivers, doctors and nurses, machinery operators etc. for fatigue alone could help save lives.

There are other potential social uses of the technology, but their ethical status is dubious—such as profiling candidates for admission in school, or for employment, or for health insurance purposes etc. We will discuss ethical issues briefly towards the end of this chapter.

10.1.5 Gaming and Entertainment

Recreation of the human in 2D and 3D from voice, even in its nascent stages where reconstructions are not quite accurate in all respects, can open up new gaming and entertainment possibilities. Profiling-centric attractions, such as holographic reconstructions, themed creations of characters in accordance with deduced personality traits of a speaker etc. can be set up for entertainment. Depending on the context, gaming servers can tailor the content of games to the traits of the player (e.g. age-appropriate and height-appropriate characters within animated children's games).

In public exhibitions of this technology [9], people have been found to enjoy learning more about themselves and experiencing how machines perceive them. Public tolerance of the current inaccuracies of this technology has been found to be remarkably high. In public areas such as malls and theme parks, it is clear that even in its current nascent stages of development, this technology could provide a good diversion for people waiting around for other reasons. Quite obviously, also, it can lend more depth to currently existing voice-driven games.

10.2 Metrics for Operational Settings

The *operating point* of a profiling system is the (parameter) setting at which it gives optimal performance for the specific task at hand. The choice of an operating point is made by evaluating the system's performance based on some performance metrics, and tuning it so that its performance is at a desired level with reference to these. Usually, this involves the computation of *tradeoffs* between opposing metrics.

In practical applications, the errors made by a profiling system may be offset by its advantages. Setting an operating point involves weighing the cost of making errors against the advantages, and choosing what tradeoffs are acceptable in the context of the application. For example, an application for diagnosing a serious disease from voice may make an error and say that a perfectly healthy person has the disease. If this sends the person into psychological trauma, then the system should avoid making such errors. On the other hand, if a person *does* have the disease, and the system makes an error by calling the person healthy, the cost of this error is even higher. In this case, the disease would be undiagnosed and the person may die.

In this case, the operating point must be set to not make any errors when it says that the person *does not* have the disease, but may make errors when it says the person *does* have the disease (thereby calling a healthy person sick). In other words, *missed detections* must be zero, at the cost of *false alarms*. If the missed detections cannot be made zero due to inherent issues with the application or the technology, then such a system may be better off being used as diagnostic tool (such as an X-ray or CT scan machine), rather than a mechanism for making actual diagnoses.

There are many ways to set the operating points of practical systems. These trade off different metrics of performance (e.g. missed detections versus false alarms in the example above). The paragraphs below describe some key metrics that are traded off in setting the operating points of applied systems.

Let us assume that a profiling application predicts some profile parameter, say a health condition. Let us assume that voice samples from a test (reference) dataset that includes subjects who are both "positive" and "negative" for the condition are processed by this application. The results (hypotheses) given by this application for the test (either "positive" or a "negative" for each subject tested) are collated in Fig. 10.1 as the labels A, B, C and D. The following metrics can be computed from these:

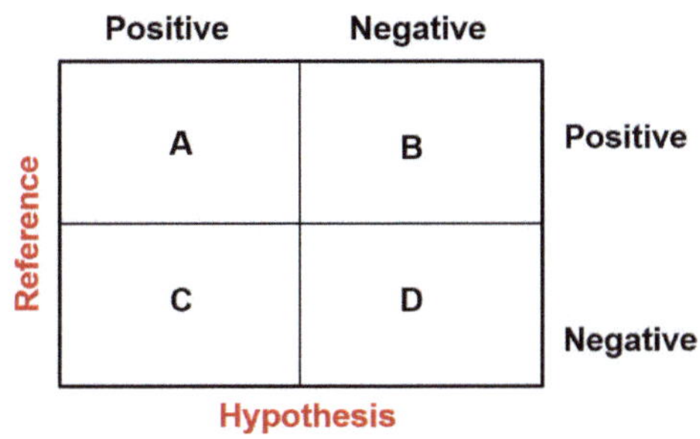

Fig. 10.1 Performance table for an application that makes binary decisions

1. **Precision** [A/(A+C)]: Of all the times it was predicted (or detected), how many times was it positive?
2. **Recall** (also called **True positive** or **Sensitivity**) [A/(A+B)]: Of all the times it was positive, how many times was it correctly detected?
3. **True negative rate** (also called **Specificity** [D/(C+D)]: Of all the times it was negative, how many times was it correctly detected as negative?
4. **False positive rate** (also called **False alarm**) [1-Specificity or 100-Specificity or C/(C+D)]: Of all the times it was negative, how many times was it detected as positive?
5. **False negative rate** (also called **Missed detection**) [B/(A+B)]: Of all the times it was positive, how many times was it detected as negative?
6. **Accuracy** [(A+D)/(A+B+C+D)]: Of all the times a hypothesis was made, how many times was it correct?
7. **Error** [(B+C)/(A+B+C+D)]: Of all the times a hypothesis was made, how many times was it wrong?

For practical deployment, operating points (or parameter settings) for any application can be chosen based on tradeoff charts (or graphs, or curves) that evaluate pairs of these metrics. Some such widely used charts are described below. and illustrated in Fig. 10.2.

1. **Receiver Operating Characteristic (ROC) curve**: Trades off true positive rate (Sensitivity) and false positive rate (1-Specificity) for different settings of the application.
2. **Precision-Recall (PR) curve**: mirrors the ROC curve.
3. **Detection Error Tradeoff (DET) curve**: Trades off false alarm (false positive) rate and missed detection (false negative) rate. The curve is drawn on a linear or log-log scale plot. The latter is more widely used.

Once an application is deployed, presumably with a fixed operating point, the reliability of the predictions/deductions made by it must be assessed. This involves computing confidence measures for the deductions made. In the following section, we will consider this very important final aspect of profiling that is essential to close the process well enough for decision-making (automated or otherwise). All of the

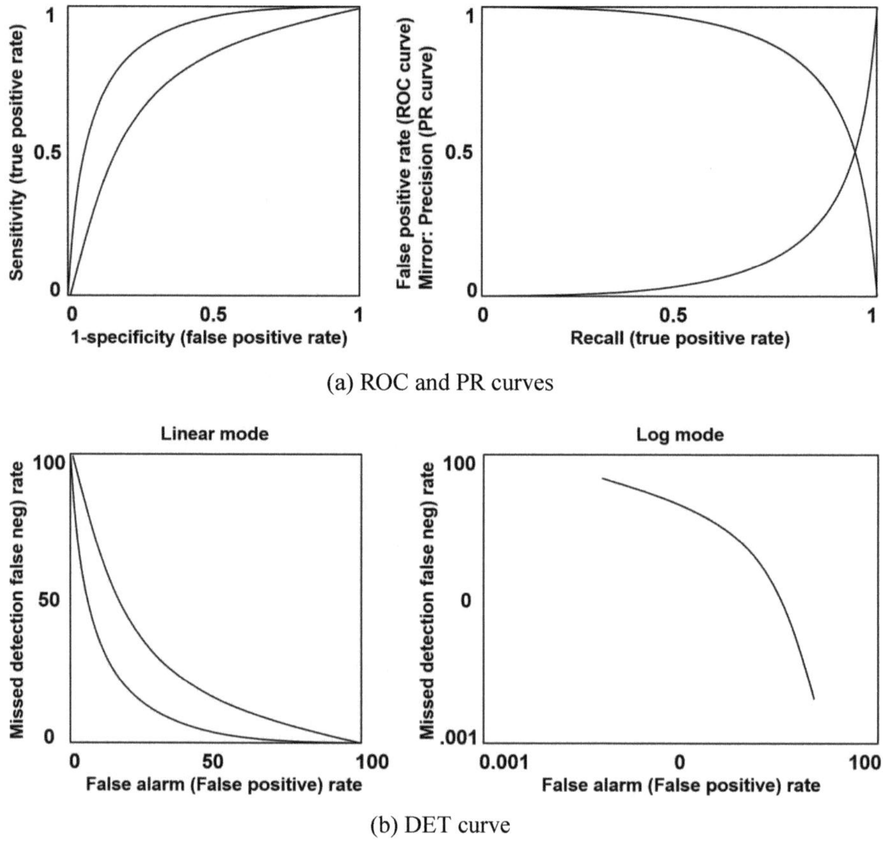

Fig. 10.2 Operating curves for performance optimization

concepts discussed in the next section fall under the general area of **Hypothesis testing** in statistics.

10.3 Establishing the Reliability of Profile Deductions

For applications that have tangible negative consequences attached to errors in profiling, it is *extremely* important to qualify the deductions made with corresponding estimates of their reliability. This is however a non-trivial task in many cases, and requires a clear understanding of the relevant statistical tests and procedures that must be done.

A profile parameter that is deduced is essentially a *hypothesis*. Its reliability must be established. Depending on the type of hypothesis, and the evidence that is considered in making it, the reliability measures differ. Not all measures are capable of supporting all aspects of a hypothesis, and must be carefully chosen to avoid meaningless results. For example, if the hypothesis is that "all cats are blue," then the test for the reliability of that hypothesis must consider *how many* cats were included in the study—obviously the higher this number, the more reliable this hypothesis would be. The reliability measure must also consider whether (a) the test represents the entire *relevant population* of cats (from which the "test" cats were chosen) adequately, and (b) whether the hypothesis itself is complete (i.e, considers the complete set of evidence or not). On the contrary, if the test set included live cats and cats drawn on paper, it would be considered incomplete for the example hypothesis above.

10.3.1 Biased and Unbiased Point Estimates of Profile Parameters

This section presents some basic concepts that are relevant in the context of profile deductions (methods and outcomes). Only some basic concepts are presented in this section. They are all well-known, and for detailed explanations we may refer to any of the numerous textbooks on statistics, e.g. [10–12].

Let X be a random variable with mean μ and variance σ^2. Given N independent random draws from its underlying distributions, the estimate of the **mean** of X is:

$$\hat{\mu} = \frac{1}{N}\sum_{i=1}^{N} X_i \tag{10.1}$$

The **sample variance** or the **biased sample variance** estimate of X is given by:

$$\hat{S}^2 = \frac{1}{N}\sum_{i=1}^{N}(X_i - \hat{\mu})^2 \tag{10.2}$$

The **unbiased sample variance** estimate of X is given by:

$$\hat{\sigma}^2 = \frac{1}{N-1}\sum_{i=1}^{N}(X_i - \hat{\mu})^2 \tag{10.3}$$

The **variance of the mean estimator** $\hat{\mu}$ is given by:

$$Var(\hat{\mu}) = \frac{\sigma^2}{N} \tag{10.4}$$

The variance of the estimator of $\hat{\mu}$ characterizes how much sample estimates of the mean, $\hat{\mu}$, would vary from one another across different experiments in which we drew N samples.

The **variance of the sample variance** S^2 is given by

$$Var(S^2) = \frac{2\sigma^4}{N-1} \tag{10.5}$$

The **variance of the unbiased estimator** $\hat{\sigma}^2$ is given by

$$Var(\hat{\sigma}^2) = \frac{2N^2\sigma^4}{(N-1)^3} \tag{10.6}$$

The variance of the estimator of S^2 or $\hat{\sigma}^2$ characterizes how much sample estimates of the respective variances would vary from one another across different experiments in which we drew N samples.

The **probability density function** (PDF) of $\hat{\mu}$ is computed equivalently through the probability distribution of its normalized value z, given by

$$z = \frac{\hat{\mu} - \mu}{\frac{\sigma}{\sqrt{N}}}. \tag{10.7}$$

The PDF of z is given by the Central Limit Theorem (CLT) as:

$$P_z(Z) = \mathcal{N}(Z) \tag{10.8}$$

where $\mathcal{N}()$ represents the standard normal distribution with mean 0 and variance 1.

We may only know the true mean μ, but not the true variance σ^2. In this case, we can use the *estimated* variance $\hat{\sigma}^2$ instead to compute $\tilde{z}$ as

$$\tilde{z} = \frac{\hat{\mu} - \mu}{\frac{\hat{\sigma}}{\sqrt{N}}}. \tag{10.9}$$

$P_{\tilde{z}}(Z)$, the PDF of $\tilde{z}$, is given by a *Student's t-distribution with* $N-1$ *degrees of freedom*, rather than a Gaussian.

$$P_{\tilde{z}}(Z) = T(Z; N-1) \tag{10.10}$$

PDF and CDF (cumulative density function) values for Student's t-distributions with different degrees of freedom can be obtained from standard tables. For small N ($N < 20$, or so), and consequently smaller degrees of freedom, t-distribution has a lower peak and heavier tail than a normal distribution. However, for large enough N ($N > 20$ or so), the Student's t is very similar to the standard normal, and we can use the Gaussian instead.

10.3.2 Confidence Intervals and Bounds

For each profile deduction that we make (e.g an estimate of the height, or weight or age of a person), there is a possibility that our estimate may be wrong in the absolute, but correct within a range of values around the estimate. This range is called the **confidence interval**, and its extremes are the **error bounds**. For example, if an algorithm for age estimation is only accurate to within 3 years of the person's true age, and it estimates a person's age to be 40 years, then the confidence interval around it would be 40 ± 3 years.

The key problems involved in computing a confidence interval are:

- How to establish error bounds on the parameter estimate, with some confidence.
- How to determine how many samples to draw, in order to establish a desired error bound on an estimator.

A **confidence bound** on any point estimator $\hat{\mu}$ of a variable μ is a range $[\hat{\mu} - \varepsilon, \hat{\mu} + \varepsilon]$ that the true value of μ is expected to lie in with high probability. In probability terms, this can be stated as:

$$P(\mu \in [\hat{\mu} - \varepsilon, \hat{\mu} + \varepsilon]) > C, \quad C \in [0, 1] \tag{10.11}$$

C is the confidence limit. The typical value of C used to specify confidence bounds is 0.95.

Alternately, we may state it in terms of $\delta = 1 - C$:

$$P(|\mu - \hat{\mu}| > \varepsilon) \leq \delta \tag{10.12}$$

How do we choose a sufficient number of samples N to estimate all of the above reliably? Given that we cannot make any assumptions about the distribution of the estimate of $\hat{\mu}$, but do know either the true variance σ^2 of X, or have an upper bound on it, we can do the following:

Let us represent this upper bound as σ^2_{max}. We know now that the expected value of $\hat{\mu}$ is μ and the variance of $\hat{\mu}$ is (bounded by) $\frac{\sigma^2_{max}}{N}$. We can now use Chebyshev's inequality to determine N [12]:

$$N \geq \frac{\sigma^2_{max}}{\delta\varepsilon^2} \tag{10.13}$$

If we expect N to be sufficiently large, then we can use the Central limit theorem to determine a tighter bound for N. Let $z_{\frac{\delta}{2}}$ be the value of z where, assuming a standard normal PDF for z, $P(|z| > z_{\frac{\delta}{2}}) = \delta$. For any δ, $z_{\frac{\delta}{2}}$ can be found from CDF tables for the standard normal distribution. We can now estimate the appropriate value of N as

$$N \geq \frac{z^2_{\frac{\delta}{2}}\sigma^2}{\varepsilon^2} \tag{10.14}$$

The above equation assumes that we know σ^2. If we only have an upper bound σ^2_{max} on the true variance, then

$$N \geq \frac{z^2_{\frac{\delta}{2}} \sigma^2_{max}}{\varepsilon^2} \tag{10.15}$$

satisfies our requirements.

10.3.3 Simple Hypothesis Testing

Hypothesis testing attempts to prove or disprove a hypothesis/claim about a data from samples of the data. It can also be used to compare classifiers. Sampling the data for this purpose must be done carefully, so that the results of hypothesis testing are correct.

The value of any variable x estimated by a profiling system comprises a hypothesis. The alternative is a **null hypothesis** H_0 which proposes a default value for the variable, instead. The purpose of hypothesis testing is to determine if we can reject the null hypothesis and accept the proposed hypothesis.

To do this, samples are chosen from the data, and used to verify if we can reject the null hypothesis. The test itself consists of computing a *statistic* Y from the sample set and checking if its value falls within a *rejection region* R (which is typically defined as one of $Y \leq \theta$, $Y \geq \theta$, or $|Y| \geq \theta$ for some threshold θ). If the value of the computed statistic Y falls within this rejection region, we decide that the null hypothesis H_0 can be rejected, if not, we cannot reject H_0.

In rejecting or accepting the null hypothesis H_0, we do not want to make errors (i.e. rejecting H_0 when it is actually correct, or accepting H_0 when it is actually wrong). If the rejection region R (or threshold θ) is improperly chosen, the likelihood of these errors can become unacceptably high. So, when specifying the rejection region R we choose it carefully, so that the probability of error is small. In practice, we first specify the largest probability of error that is acceptable (typically a value such as 0.05), and then choose R such that the error probability meets our specification.

In the following, we look at this in more detail.

10.3.3.1 Basic Formalism

Formally, a hypothesis testing procedure must specify the following four terms

1. A *null* hypothesis H_0,
2. An *alternate* hypothesis H_a which corresponds to the claim mentioned above,
3. A statistic Y that we will compute from a set of data samples,
4. A rejection region R for Y, typically specified by a threshold θ.

If the statistic Y falls within the rejection region R, we can reject the null hypothesis and accept our alternate hypothesis. Otherwise, we have not established sufficient reason to reject the null hypothesis.

Note that this does not necessarily mean that the null hypothesis is correct; it only means that we do not have the evidence needed to reject it.

10.3.3.2 Hypothesis Testing Errors

We are likely to make two kinds of errors when we run a hypothesis test:

1. **Type-1 Error**: We *reject* the null hypothesis when it is in fact true, i.e. the statistic Y falls within rejection region R even though H_0 is true.
2. **Type-2 Error**: We *accept* the null hypothesis when it is in fact false. Type-2 errors happen if Y falls *outside* the rejection region R and we accept H_0, even though H_0 is *false*.

We represent the probability of a Type-1 error by α. Ideally this must be low.

$$P(Type - 1\ Error|H_0) = \alpha \tag{10.16}$$

The converse of the probability of Type-1 error is also called the **confidence** in the test. Thus, if $\alpha = 0.05$, then the confidence is 0.95, i.e. we are 95% sure that the test wont reject the null hypothesis by mistake.

The probability of a *Type-2 error* may be interpreted as the fraction of times we would accept the null hypothesis, even though the *alternate hypothesis is true*, if we were to repeat our test using our rejection criterion a very large number of times.

We refer to the probability of a Type-2 error as β.

$$P(Type - 2\ Error|H_a) = \beta \tag{10.17}$$

To compute β, the *true* value of the parameter that the null hypothesis defines must be known. At the least, the alternate hypothesis must hypothesize a value of this parameter, instead of merely rejecting the null hypothesis (so that we can use it to compute β). In most cases this is not possible, and β cannot be known. In such case we make estimates of the *trends* in β.

The probabilities of Type-1 and Type-2 errors are primarily controlled through the rejection region R and the number of samples N in the test.

The rejection region R is the range of values that the statistic must take in order to reject the null hypothesis H_0. If the statistic falls outside the rejection region, the null hypothesis must be accepted. We call this region the *acceptance* region, and represent it by $\bar{R}$. α, the probability of Type-1 error, and β, the probability of Type-2 error, can both be modified by changing the rejection region. However, this involves a tradeoff between Type-1 and Type-2 errors. Type-1 errors occur only when the test statistic falls within R. If R is reduced, the probability of Type-1 errors (α)decreases. On the other hand, Type-2 errors only occur when the test statistic falls *outside* the rejection region, i.e. within the acceptance region $\bar{R}$. To reduce the probability of Type-2 error (β), $\bar{R}$ must be reduced. Modifying R to decrease α increases β, and vice versa. Thus, if the number of samples N is fixed, we cannot decrease both α and β.

Both α and β can however be decreased simultaneously by increasing N. *In practice* one does not usually have the information necessary to quantify β, so we attempt to obtain as many samples as we can, and choose the rejection region R such that the probability of Type-1 error falls below some acceptable threshold. This is the format of most hypothesis tests.

10.3.4 Hypothesis Testing in Practice: Introduction

The generic formulation for hypothesis testing, once we've specified our null and alternate hypothesis H_0 and H_a, is as follows:

1. We start with an α probability (probability of Type-1 error) that we think is acceptable, or alternately, the desired confidence in the test (where confidence is 1 - α). Let us call this α_z. Typical α_z values are 0.05 or 0.01.
2. We first derive the probability mass function (or probability density function) function of the statistic that we are evaluating.
3. From this distribution, we compute the probability of Type-1 error for different rejection regions. Since the rejection region is typically parameterized by a threshold θ, this equates to computing the α value for different thresholds (different θ values), until we find a threshold θ_r where the computed α value is the desired α_z value.
4. We then determine if the value of the statistic derived from our sample falls in the rejection region defined by the threshold θ_r to accept or reject the null hypothesis.

In addition to the rejection region, we may also be able to control the number of samples N. Typically, however, the focus is on the confidence.

The rejection region R is generally defined by comparing the statistic Y against a threshold θ. If, in order to reject the null hypothesis in favor of the alternate hypothesis the statistic Y must lie *below* θ, we have a **left-sided test**; if it must lie *above* θ, we have a **right-sided test**.

Often, we do not propose a *direction* for our alternate hypothesis H_a; we only propose that it contradicts the null hypothesis H_0. This is a **two-sided test**. Left–and right-sided tests are called *one-sided* tests.

In this case, in order to reject the null hypothesis H_0 we generally require the test statistic to be distant from a mean, without mentioning the direction, i.e. we reject H_0 if $|Y - \mu| \geq \theta_r$. We now choose θ_r such that the total probability of either $Y \geq \mu + \theta_r$ or $Y \leq \mu - \theta_r$ is no more that α_z, i.e., we choose θ_r such that $P(Y \geq \mu + \theta_r) + P(Y \leq \mu - \theta_r) \leq \alpha_z$. When the probability distribution of Y is *symmetric* about its mean, this simply becomes $P(Y - \mu \geq \theta_r) > \frac{\alpha_z}{2}$.

10.3.5 Hypothesis Testing in Practice: One Sample Tests

One-sample tests are tests where we test only one sample, without any comparisons to any other.

Consider, for example, a one-sample test for a Bernoulli variable X with parameter P. Bernoulli variables take one of two values, and P represents the probability of one of these values. For instance, if X represents the gender of a subject, P may be the probability that the subject is female. Similarly, X may represent the event that a binary classifier correctly classifies a test instance; in this case P would represent the probability of correct classification. P must be estimated. The hypothesis testing problem for Bernoulli variables is as follows. The null hypothesis H_0 is that P has some (constant) value P_0. The alternate hypothesis H_a is that the true value of the Bernoulli parameter is *not* P_0.

10.3.5.1 One Sided Test

One-sided tests may either be left-sided or right-sided.

In the *right-sided* test the alternate hypothesis claims specifically that the true P is greater than the value P_0 claimed by the null hypothesis.

To test the hypothesis, N samples $X_1, X_2, \ldots, X_N$ are obtained from the process. We compute the statistic $Y = \sum_{i=1}^{N} X_i$, where for instance, Y may represent the number of samples (from $X_1, \ldots, X_N$) that were correctly classified. For Bernoulli X_i, Y has a Binomial distribution: $P_Y(K) = \binom{N}{K} P^K (1-P)^{N-K}$ [11].

We know that if the true Bernoulli parameter is greater than that proposed by the null hypothesis, i.e if $P > P_0$, then the number of positive outcomes will be greater than what we would obtain under the null hypothesis. Under the null hypothesis $(P = P_0)$, the expected number of positive outcomes in N trials is NP_0. We *reject* H_0 (and accept H_a) if $Y \geq \theta_r$, i.e. Y is greater than or equal to a threshold θ_r that is sufficiently larger than NP_0.

Since we do not want to make a Type-1 error (rejecting H_0 when it is true), we specify the acceptable probability of Type-1 error, i.e. α_z. Typical values (as mentioned earlier) are 0.01 and 0.05.

The probability of Type-1 error using a threshold θ is the probability of obtaining $Y \geq \theta$ under the null hypothesis, and is given by

$$\alpha(\theta) = \sum_{K=\theta}^{N} \binom{N}{K} P_0^K (1 - P_0)^{N-K} \tag{10.18}$$

For the right-sided test, we choose a threshold θ_r as the smallest θ such that $\alpha(\theta) \leq \alpha_z$.

A left sided test would be similarly derived. Now we must find the threshold θ_r as the largest θ such that $\alpha(\theta) = \sum_{i=K}^{\theta} \binom{N}{K} P_0^K (1 - P_0)^{N-K}$ is less than α_z.

10.3.5.2 Two Sided Test

In the two-sided test, we do not specify whether the true P value is greater or less than the value P_0 proposed by the null hypothesis. Our alternate hypothesis is only that $P \neq P_0$.

So now, we try to find a threshold θ_r such that $P(|Y - NP_0| \geq \theta_r) \leq \alpha_z$, i.e. such that

$$P(|Y - NP_0| \geq \theta_r) = P(Y \leq NP_0 - \theta_r) + P(Y \geq NP_0 + \theta_r) \leq \alpha_z \quad (10.19)$$

We find the *smallest* θ value such that $\alpha(\theta) \leq \alpha_z$, and set that θ to be θ_r.

10.3.5.3 Large Sample Tests

In large sample approximations to Bernoulli variables, for large N working out the Binomial probability distribution values is hard. In these cases we can obtain an excellent approximation using the Central Limit Theorem.

We define the variable Z as:

$$Z = \frac{Y - NP_0}{\sqrt{NP_0(1 - P_0)}}. \quad (10.20)$$

The CLT gives us a standard normal PDF for Z. We now get the following relation:

$$P(Z \leq Z_\theta) = P(Y \leq NP_0 + Z_\theta\sqrt{NP_0(1 - P_0)}) \quad (10.21)$$

where, as before Z_θ is the value of Z where $P(|Z| > Z_\theta) = \alpha_z$.

Confidence intervals are found on the basis of two-sided tests.

10.3.5.4 Computing Confidence Intervals

Two-sided tests (in both the large sample and small sample cases) for hypothesis testing population means can be converted to compute **confidence intervals**: i.e., given a probability bound δ, we can find an error bound ε such that $P(|E[X] - Y| \geq \varepsilon) \leq \delta$.

Let $t_{\frac{\delta}{2}}$ be the value where, for a variable t distributed according to the Student's t-distribution, $P(|t| > t_{\frac{\delta}{2}}) = \delta$. For any δ the corresponding $t_{\frac{\delta}{2}}$ can be obtained from standard tables for the Student's t-distribution.

Given our probability bound δ, ε can now be found as $\varepsilon = t_{\frac{\delta}{2}} \frac{\hat{\sigma}_N}{\sqrt{N}}$.

10.3.5.5 Comparing and Hypothesis Testing of Population Means

In many problems, we are required to test a hypothesis about a population mean. Here the null hypothesis is that the population mean of the random variable X is μ_0. Our alternate hypothesis contradicts this null hypothesis. We may make a one-sided counter claim, e.g. that the actual population mean is greater than μ_0, or it may be two-sided, where our counter claim is merely that the real population mean μ is not equal to μ_0, but may be greater or lesser.

The testing procedure samples N instances of data. From these N samples, we compute a *sample mean* $Y = \frac{1}{N}\sum_{i=1}^{N} X_i$.

Based on N, we conduct either a large-sample test or a small-sample test. Each of these can be left-sided, right-sided or two-sided.

10.3.6 Hypothesis Testing in Practice: Two Sample Tests

10.3.6.1 Comparing and Hypothesis Testing of Unrelated Populations

We now consider the case of comparing two populations, X and U. As an example, one may want to compare the performance of two classifiers C_1 and C_2, without actually specifying their accuracies. The default hypothesis is that C_1 and C_2 have similar accuracies. The alternate hypothesis may either claim that $C_1 > C_2$, or simply that the two classifiers are not identical, i.e. $C_1 \neq C_2$.

In this case, X may represent the outcomes of classifier C_1 while U may represent the outcomes of C_2.

The null hypothesis H_0 is that $E[X] - E[U] = \mu_0$. μ_0 is 0 for this example.

The alternate hypothesis H_a either proposes that $E[X] - E[U] > \mu_0$ (for a right-sided test), or that $E[X] - E[U] \neq \mu_0$ for a two-sided test.

The approach here is to obtain a set of N_X samples $X_1, X_2, \ldots, X_{N_X}$ from X, and a set of N_U samples $U_1, U_2, \ldots, U_{N_U}$ of U. Here $X_1, \ldots, X_{N_X}$ are the outcomes of classifier C_1 on a test set of N_X samples, while $U_1, \ldots, U_{N_U}$ are the outcomes of C_2 on an *unrelated* set of N_U samples. If the samples are *related*, e.g. if they are outcomes computed on the same data, then we must use a *paired* test. This is described in the next section.

The difference in the sample means is

$$D = \hat{\mu}_X - \hat{\mu}_U = \frac{1}{N_X}\sum_{i=1}^{N_X} X_i - \frac{1}{N_U}\sum_{i=1}^{N_U} U_i \tag{10.22}$$

D can be computed from the samples $X_1, \ldots, X_{N_X}$ and $U_1, \ldots, U_{N_U}$.

The expected value of D is $E[D] = E[\hat{\mu}_X] - E[\hat{\mu}_U]$. Since $E[\hat{\mu}_X] = E[X]$ and $E[\hat{\mu}_U] = E[U]$,

$$E[D] = E[X] - E[U] \tag{10.23}$$

Under the null hypothesis, $E[D] = \mu_0$.
The sample variance of D is given by

$$\hat{\sigma}_D^2 = \frac{1}{N_X(N_X - 1)} \sum_{i=1}^{N_X} (X_i - \hat{\mu}_X)^2 + \frac{1}{N_U(N_U - 1)} \sum_{i=1}^{N_U} (U_i - \hat{\mu}_U)^2 \tag{10.24}$$

Note that like D, $\hat{\sigma}_D^2$ can also be computed from the samples $X_1, \ldots, X_{N_X}$ and $U_1, \ldots, U_{N_U}$.

D is, in fact, a linear combination of $N_X + N_U$ independent samples. If we define

$$Z = \frac{D - E[D]}{\sqrt{var(D)}} \tag{10.25}$$

which, under the null hypothesis can be computed from the samples as

$$Z = \frac{D - \mu_0}{\hat{\sigma}_D^2} \tag{10.26}$$

then for sufficiently large N_X and N_U, Z has a standard normal distribution. More generally, Z has a Student's t-distribution with $\min(N_X - 1, N_U - 1)$ degrees of freedom (the smaller of $N_X - 1$ and $N_U - 1$). Note that some books may give you a more complex formula for the true degrees of freedom; $\min(N_X - 1, N_U - 1)$ is close enough for most practical purposes.

Since Z can be computed entirely from the samples, in order to test our alternate hypothesis we only have to compare Z to a threshold.

Note that in general when the test statistic has a t-distribution, the test is called a **'t-test'**, or a **Student's t-test**.

Other topics that are relevant in this context are tests of normality and Wilcoxon-Mann-Whitney tests. We do not discuss these here.

10.3.6.2 Comparing and Hypothesis Testing of Related Populations

In the section on comparing populations above, we discussed how we can evaluate hypotheses about the difference in population means using two *unrelated* samples. This is an unpaired test, as opposed to the case where we must compare two populations that are related, for example to compare two classifiers that work on the same input data set. This is a situation where a **paired test** must be performed, since the outputs of the classifiers are not independent of each other.

Let $I_1, \ldots, I_N$ be a set of test instances. Let $X_1, \ldots, X_N$ be the outputs of classifier C_1 on this set. Let $U_1, \ldots, U_N$ be the outputs of classifier C_2. In a paired setting, we expect X to be *correlated* to Y.

The appropriate tests for this setting are explained below.

Paired Test

Let X and U represent a corresponding *pair* of measurements or characterizations of some process. For instance, X may represent "before" readings for a drug treatment, and U may be the "after" readings.

The null hypothesis claims that $E[X - U] = \mu_0$. For instance, we may claim that the average cholesterol level decreased by μ_0 after administering a drug. In the most common setting, $\mu_0 = 0$, i.e. that X and U are statistically equivalent (the drug has no effect, in our example).

The alternate hypothesis claims that $E[X - U]$ is not μ_0 and may be one-sided (claim that $E[X - U] > \mu_0$ for instance) or two-sided ($E[X - U] \neq \mu_0$). For example if $\mu_0 = 0$, the alternate hypothesis may try to specifically establish the one-sided hypothesis that $E[X - U] > 0$, i.e. the drug helps, or it may simply make the two-sided hypothesis that $E[X - U] \neq 0$ (the drug has a significant, but unknown effect).

Consider two corresponding sets of measurements $X_1, \ldots, X_N$ and $U_1, \ldots, U_N$ obtained from N test samples. Let $D_i = X_i - U_i, \quad i = 1, \ldots N$. The sample mean of D_i is

$$Y = \frac{1}{N}\sum_{i=1}^{N} D_i = \frac{1}{N}\sum_{i=1}^{N}(X_i - U_i) \tag{10.27}$$

The expected value of Y is $E[Y] = \mu_0$.

The sample variance of Y is given by

$$\hat{\sigma}_N^2 = \frac{1}{N-1}\sum_{i=1}^{N}(D_i - Y)^2 \tag{10.28}$$

Under the null hypothesis the variance of Y is $var(Y) = \frac{\hat{\sigma}_N^2}{N}$.

The rest of the test is exactly like the comparison of population means. Define

$$Z = \frac{Y - \mu_0}{\frac{\hat{\sigma}_N}{\sqrt{N}}} \tag{10.29}$$

Under the null hypothesis Z has a Student's t-distribution with $N - 1$ degrees of freedom.

For the *right-sided* test, given α_z we can now use the Student's t-tables to find the t_α such that $T(t_\alpha, N - 1) = 1 - \alpha_z$. The null hypothesis can be rejected if $Z \geq t_\alpha$. For the *two-sided* test, given α_z we can now use the Student's t-tables to find the $t_{\frac{\alpha}{2}}$ such that $T(t_{\frac{\alpha}{2}}, N - 1) = 1 - \frac{\alpha_z}{2}$. The null hypothesis can be rejected if $|Z| \geq t_{\frac{\alpha}{2}}$.

The key difference between the *paired* and the conventional tests for comparing population means is that we now directly work with $X - Y$ before performing the test, rather than computing the sample means and variances of X and Y separately. This gives us a better estimate of $\hat{\sigma}_N^2$, which automatically also factors in the correlation between X and Y.

When the Data Are Not Normally Distributed:

The paired test described above assumes that X and Y have a normal distribution. When the distribution is not normal (or bell shaped), a different test, called the **Wilcoxon signed-rank test**, is used. We will not go into the details of this test here. However, we do discuss below the special case when X and Y are Bernoulli, e.g. when we compare two classifiers.

Comparing Two Classifiers: McNemar's Test

Consider the problem of comparing two classifiers C_1 and C_2. In *McNemar's test*, the null hypothesis H_0 is that the two classifiers perform similarly on average, i.e. their accuracies are identical. The alternate hypothesis is typically that one of them (say C_2) performs better than the other. In other words, if P_1 is the probability of correct classification by C_1 and P_2 is the probability of correct classification for C_2, then the null hypothesis is that $P_1 = P_2$, while the alternate hypothesis is that $P_2 > P_1$.

To test the hypothesis we classify N test instances. Let $I_1, \ldots, I_N$. Let $X_1, \ldots, X_N$ be the outputs of classifier C_1 on this set. Let $Y_1, \ldots, Y_N$ be the outputs of classifier 2 on the same set.

For the purpose of this test, we ignore all instances on which the two classifiers C_1 and C_2 agree, and only consider the instances on which they *disagree* (i.e. one of them is correct and the other is wrong). Under these conditions, on the remaining set, under the null hypothesis the fraction of times C_1 is correct and C_2 is wrong is the same as the fraction of times C_2 is correct and C_1 is wrong. In other words, the fraction of instances on which C_1 is right is roughly 0.5, and McNemar's test becomes identical to a test of a Bernoulli variable where the null hypothesis is that the Bernoulli parameter $P = 0.5$. This is easily tested using the procedure explained in Sect. 10.3.5.

10.3.7 P-Values

The P-value of a test is the *smallest α_z value at which the test would reject the null hypothesis.*

Let us take the example of tosses of a (fair) coin. Our null hypothesis is that the coin is fair. The alternate hypothesis is that $P(heads) < P(tails)$. We test this by tossing a coin 100 times. We set $\alpha_z = 0.05$, and find from our tables that this corresponds to a threshold θ of 41 heads: If we obtain 41 or fewer heads, we can reject the null hypothesis. If we obtain 40 heads in the test comprising 100 tosses, this falls in the rejection region $(40 < 41)$, and we can reject the null hypothesis. Note that we could have used a lower α_z value and still have rejected H_0. In fact, we could have reduced α_z all the way down to 0.025. For $\alpha_z = 0.025$ the threshold is 40. In other words, if we get only 40 heads out of 100 tosses, we can actually be 97.5% confident that the coin is not fair. This then becomes the value of P.

In general, the approach to computing a P-value is as follows:

1. Compute our statistic Y from the sample. If we intend to use a *derived* statistic Z (e.g. when we use the central limit theorem, or if we employ the Student's t-distribution), then compute Z from Y.
2. Set $P = 0.001$. This is a test with 99.9% confidence. (0.001 is an arbitrary choice. We could begin with a higher or lower value).
3. Test if the null hypothesis H_0 can be rejected with $\alpha_z = P$.
4. If not, increment P slightly (e.g. $P = P + 0.001$). Return to step 3.
5. If yes, we have obtained the desired P value.

10.3.8 Likelihood Ratio Test

For decades, the most ubiquitously used reliability measure for forensic voice evidence has been the **likelihood ratio** [13]. The likelihood ratio test is, in theory, the test that provides the lowest probability of Type-2 error for any specified value for the Type-1 error [14–16], and has served well for voice-matching based forensic analysis.

The likelihood ratio test is used to compare two hypotheses based on the probability assigned to the data under each of the hypotheses. Let $P(X|H_0)$ be the probability assigned to a data instance X under the null hypothesis H_0, and let $P(X|H_a)$ be the probability assigned to it by the alternate hypothesis H_a. E.g. for a speech recording X, if the null hypothesis is that the speaker is in a state of normal health and the alternate hypothesis is that the speaker has a particular health problem, then $P(X|H_0)$ would represent the generic distribution of normal speech, while $P(X|H_a)$ would represent the distribution of speech by people who suffer from the problem.

The *likelihood* of a hypothesis is defined simply as the probability assigned to the data under that hypothesis, i.e. $L(H_0|X) = P(X|H_0)$ and $L(H_a|X) = P(X|H_a)$. The *simple* likelihood-ratio test computes the statistic

$$\Lambda = \frac{L(H_0|X)}{L(H_a|X)} \tag{10.30}$$

If $\Lambda < \Lambda_\alpha$ for some appropriately chosen threshold $\Lambda_\alpha < 1$, then it means that the alternate hypothesis is more likely than the null hypothesis, and the null hypothesis is rejected. This test is applicable where $P(X|H_0)$ and $P(X|H_a)$ have known (or pre-estimated) form and can be computed for any X. Λ_α is usually chosen to achieve a specified Type-I error rate α. For some classes of distributions for $P(X|H_0)$ and $P(X|H_a)$ such as when both are Gaussian, Λ_α can be formally derived from the formula of Eq. 10.30 [17]. In other situations such as the one described later in this section, Λ_α may be obtained from the assumed distribution of Λ. More generally, Λ_α may be obtained from empirically measured ROC curves.

In the more complex setting, $P(X|H_0)$ and $P(X|H_a)$ are parametric distributions with unknown parameters θ_0 and θ_a and have the form $P(X|H_0) = P(X; \theta_0)$ and $P(X|H_a) = P(X; \theta_a)$. The corresponding likelihood functions can be represented as $L(\theta_0|X)$ and $L(\theta_a|X)$. While θ_0 and θ_a are unknown, they are assumed to belong to known *families* of distribution parameters, i.e. it is known that $\theta_0 \in \Theta_0$ and $\theta_a \in \Theta_a$. Furthermore, the alternate hypothesis parameter set Θ_a is assumed to be a more detailed version of the null-hypothesis set Θ_0, such that the latter is nested within the former. E.g. Θ_0 may represent parameters of a Gaussian distribution, while Θ_a represents the set of parameters of a mixture of Gaussians.

In this case the likelihood ratio test is computed using the *maximum likelihood* estimates for the parameters under the two hypotheses:

$$\Lambda = \frac{\max_{\theta_0 \in \Theta_0} L(\theta_0|X)}{\max_{\theta_a \in \Theta_a} L(\theta_a|X)} \tag{10.31}$$

The actual statistic used for hypothesis testing is $Y = -2\log\Lambda$. This generally has a χ^2 distribution. The cutoff value $\Lambda = \Lambda_\alpha$ below which the null hypothesis is rejected is computed using $P(Y > -2\log\Lambda_\alpha) = \alpha$, where $P(Y)$ is the χ^2 distribution.

In the context of profiling the simple ratio test of Eq. 10.30 is generally more commonly used than the more detailed version of Eq. 10.31.

10.3.9 *Power of a Test*

The **power** of a hypothesis test is the probability of correctly rejecting the null hypothesis. Recall that the Type-2 error β is the probability of falsely accepting the null hypothesis. Thus, the power of the test is $1 - \beta$, or the probability of *not making* a Type-2 error.

α quantifies the probability of obtaining the test statistic Y (computed from the test sample) by chance when the null hypothesis is true. β is the probability of obtaining it randomly when the *alternate* hypothesis is actually true. The cutoff value of Y used to reject H_0 is based on the desired α. The distribution of Y under the alternate hypothesis is dependent on the expected value of Y under H_a and the size of the test sample used to compute Y. As a result, the power of the test is dependent the significance value α, the size of the test sample, and the expected value of the statistic under H_a.

In this setting, the **power goal** is the minimum power to be achieved, and the sample size required to achieve it can be computed from α and the known (or assumed) distribution of Y under H_a. An acceptable sample size must yield a power greater than or equal to the power goal [18–20].

10.4 Finding Correlations

The term correlation was first used by Galton in 1888 [58], interestingly in the context of anthropometry. Correlation is a measure of the relationship between two or more variables. The relationship is measured in terms of a **correlation coefficient**. Correlation coefficients have values that range from -1.00 to $+1.00$. The value -1.00 represents a perfect negative correlation, the value $+1.00$ represents a perfect positive correlation, and the value 0.00 represents a lack of correlation.

The profiling mechanisms used in practical systems attempt to map information in voice to profile parameters. This prediction is fundamentally possible only if the features extracted from voice are *correlated* with the profile parameter. How do we test whether this is true, perhaps even before we begin to use the mechanisms for profiling mentioned in this book?

For this, we must perform correlation analysis. Finding correlations among entities or populations (such as between the random variables X and Y) is different in premise from testing hypotheses.

When a profiling system generates a deduction for a profile parameter, that is a hypothesis and must be tested or assigned a confidence to measure its reliability. Hypothesis testing methods must be used for this.

On the other hand, **correlation analysis** serves to find, discover and support relationships between X and Y. It is largely useful in experimental design and procedures, such as those that attempt to discover the relationship between two entities. When a relationship is found, one entity can conceivably be used to predict the other, and the reliability of the results can be supported by the underlying correlations. This leads to the subject of hypothesis testing in predictions made on the basis of correlations, and a different set of confidence measures that apply to these. In this section we briefly outline some key ones.

10.4.1 Measurement of Correlation

10.4.1.1 Pearson-r Correlation

The most widely used measure of correlation is the Pearson-r value, or Pearson correlation coefficient. Pearson presented it in the context of heredity and evolution [21]. Pearson-r is also called the *linear* or *product-moment* correlation. It is important to note that a correlation coefficient is independent of the measurement units used for the data it is calculated on. It is simply a numerical value within the range -1.00 and $+1.00$. A linear correlation can be represented by a straight line (sloped upwards or downwards) on a graph that plots the values of the variables being correlated. This line is called the regression line or least squares line, since it is computed such that the sum of the squared distances of all the data points from the line is minimum.

The Pearson correlation applies to variables that are measured on interval scales. Given a collection of pairs of variables (X_1, Y_1), (X_2, Y_2), ... pairs, the Pearson product moment correlation coefficient between X and Y is given by:

$$r_{XY} = \frac{\sum[(X_i - \bar{X})(Y_i - \bar{Y})]}{\sqrt{(\sum(X_i - \bar{X})^2 \sum(Y_i - \bar{Y})^2)}} \tag{10.32}$$

where $\bar{X}$ is the average of the given X_i values and $\bar{Y}$ is the average of the corresponding Y_is.

The Pearson correlation coefficient assumes an affine relationship between X and Y, and measures how closely Y tracks X. For instance, if X is the true age of a subject and Y is the age predicted by forensic analysis, the Pearson correlation between the two quantifies how well the predicted age tracks the true age. A correlation of 1.0 indicates that the predicted value tracks the true value exactly. A correlation of -1.0 indicates that the trends in X and Y are exactly reversed: when X increases, Y decreases in proportion to the change in X. A correlation of 0 indicates that X has no linear correspondence to Y (although it may still have non-linear dependence).

Non-linear correspondences between X and Y can be captured through a variant of the standard Pearson correlation coefficient called the **rank correlation coefficient**. A rank correlation coefficient between two random variables is based on the ranks of the measurements of data, and not their actual values. Some examples are Spearman-r, Kendall tau, Gamma coefficient etc. [22–24].

In particular, the Spearman-r correlation coefficient is the Pearson correlation between the *rank values* of two variables. It measures the monotonic nature of the relationship between the two variables (in contrast to Pearson-r which quantifies the linear relationship the two). In the absence of repeated data values, Spearman-r can take the values -1 or $+1$ in the case of a perfect monotonic relationship between the two variables considered.

The computed correlation value is itself only an estimate of the *true* correlation between X and Y based on a finite set of samples. Thus, it is customary to establish confidence intervals on the true correlation, or alternately to compute its statistical significance. For the Pearson correlation this is performed by computing the *Fisher transformation* of the computed correlation to obtain the value

$$F(r_{XY}) = arctanh(r_{XY}) \tag{10.33}$$

For a sufficiently large number of samples, $F(r_{XY})$ has a normal distribution with variance $\frac{1}{n-3}$, where n is the size of the number of (X, Y) pairs in the sample. The mean of the distribution is $F(r_0)$, where r_0 is the value of r_{XY} predicted by the null hypothesis. The z-score

$$z = (F(r_{XY}) - F(r_0))\sqrt{n-3} \tag{10.34}$$

has a standard normal distribution, and can be used to establish both statistical significance values and confidence intervals on r_{XY}. In the case of the Spearman-r, the formula for the z-score changes to the approximation

$$z = (F(r_{XY}) - F(r_0))\sqrt{\frac{n-3}{1.06}} \tag{10.35}$$

As before, z has a standard normal distribution and can be utilized for statistical significance testing.

10.4.1.2 The Chi-Square Test

The distribution of the sum of the squares of k independent standard normal random variables is the *chi-squared* distribution with k degrees of freedom. It has the form [25–27]

$$P(x) = \frac{1}{\Gamma\left(\frac{k}{2}\right)} x^{\frac{k}{2}-1} e^{-\frac{x}{2}} \tag{10.36}$$

A χ^2 distribution with k degrees of freedom is generally represented as $\chi^2(k)$ (when the degrees of freedom are not specified, it may simply be denoted as χ^2). The distribution is only defined for positive x and assigns 0 probability to negative values of x.

The χ^2 distribution features in several statistical testing problems, of which the following two are most common.

10.4.1.3 Goodness of Fit Test

Consider a multinomial variable that can take one of k values $v_1, \ldots, v_k$. The multinomial may have also been derived by quantization or binning of a continuous-valued variable. The null hypothesis H_0 assigns the probability distribution $P = [p_1, \ldots, p_k]$ to these outcomes. In case the multinomial is derived by binning a continuous valued variable, P may be derived by integrating the probability density function of the variable over the bins.

A test sample of size N includes $n_1, \ldots, n_k$ instances respectively of the k values $v_1, \ldots, v_k$.[1] The *goodness of fit* test determines if the test sample was indeed drawn from P, with the null hypothesis being that it was, and the alternate hypothesis H_a being that it was not.

Under the null hypothesis, the *expected* number of instances of the ith outcome v_i in N trials is $e_i = Np_i$. We define the following statistic

$$z = \sum_{i=1}^{k} \frac{(n_i - e_i)^2}{e_i} \tag{10.37}$$

[1]For instance, a sample of 100 people may have 20, 30 and 50 instances of short, medium height and tall people respectively.

For large N, each n_i is approximately Gaussian, and the statistic has a χ^2 distribution. The statistic sums k terms, but since the e_i terms are constrained to sum to N, which eliminates one degree of freedom, z has a χ^2 distribution with $k-1$ degrees of freedom. Even if N is not large, z can nevertheless still be assumed to be approximately $\chi^2(k-1)$.

The goodness of fit can now be quantified by comparing z against the $\chi^2(k-1)$ distribution. To determine if the fit achieves some significance level α, the value z_α such that $P(z > z_\alpha) < \alpha$ must be identified from standard χ^2 tables, and z compared to it. Alternately, a P-value for the computed z may be derived from the table. χ^2 tables are abundantly available in the literature and on the internet, e.g. [28–31].

In practice, the null hypothesis will generally not have the *true* probabilities, but $p_1, \ldots, p_k$ must themselves be estimated. If the p_is have a parametric form with l independent parameters that have been estimated using a maximum-likelihood procedure, the statistic of Eq. 10.37 can still be employed to determine goodness of fit. However, now the corresponding χ^2 distribution will only have $k-l-1$ degrees of freedom.

10.4.1.4 Statistical Independence Test

An extension of the goodness of fit test can be used to test the statistical independence of two discrete (or discretized) variables X and Y. Such a test may be required, for instance, to establish if the height of a person is dependent on their ethnicity—here X would represent the (discretized) height, while Y would represent the ethnicity.

The null hypothesis is that the two distributions are independent. Let $x_1, \ldots, x_L$ and $y_1, \ldots, y_M$ be the sets of values that X and Y can take. Let $P_X(x_i)$ and $P_Y(y_j)$ represent the marginal probabilities that X and Y take values x_i and y_j respectively. The null hypothesis claims that the probability of any combination (x_i, y_j) is just the product of the separately computed probabilities of x_i and y_j, i.e.

$$P(x_i, y_j) = P_X(x_i)P_Y(y_j) \tag{10.38}$$

For instance, the null hypothesis may claim that the probability that a person of Chinese ethnicity is tall is no different from the probability of tallness among the general populace. So here x_i would take the value "Chinese", y_j would represent "tall," and Eq. 10.38 states that under the null hypothesis the probability that a random subject is both Chinese and tall is merely the product of the probability of randomly choosing a Chinese subject, and that of choosing a tall one. Under the *alternate* hypothesis, the distribution of heights of Chinese people would be different from that of other groups of people (and, as a consequence, from that of general populace at large). Thus Eq. 10.38 would not hold.

Consider a test sample of size N that is used to test the independence of X and Y. Let n_{ij} be the number of instances from the sample for which X takes the value x_i and Y takes the value y_j. E.g., under our example, if x_1 represented "Chinese" and y_2 represented "tall," n_{12} would be the number of tall Chinese people in our sample.

Under the null hypothesis, the expected number of N instances for which $X = x_i$ and $Y = y_j$ is given by

$$e_{ij} = NP_X(x_i)P_Y(y_j) \tag{10.39}$$

Subsequently, the problem of verifying independence becomes analogous to that of determining goodness of fit. We define the statistic.

$$z = \sum_{i,j} \frac{(n_{ij} - e_{ij})^2}{e_{ij}} \tag{10.40}$$

If the null-hypothesis distributions $P_X(x_i)$ and $P_Y(y_i)$ are specified, there are LM possible combinations of x_i and y_j, and z has a χ^2 distribution with $(LM - 1)$ degrees of freedom.

If $P_X(x_i)$ and $P_Y(y_j)$ are not specified, and the test for independence must nevertheless be performed, they can be computed using the following maximum likelihood estimates

$$P_X(x_i) = \frac{\sum_j n_{ij}}{N} \tag{10.41}$$

$$P_Y(y_j) = \frac{\sum_i n_{ij}}{N} \tag{10.42}$$

and substituted into Eq. 10.39. Since these distributions have $(L - 1)$ and $(M - 1)$ independent parameters respectively, the degrees of freedom of the χ^2 distribution must be reduced by $(L - 1) + (M - 1)$. Consequently, the statistic of Eq. 10.40 now has a χ^2 distribution with $(L - 1)(M - 1)$ degrees of freedom, and the corresponding tables must be used against z to establish the significance level of the deviation of n_{ij} from e_{ij}. If the deviation is large enough (i.e. if z is large enough), the null hypothesis can be rejected and it can be concluded that X and Y are indeed not independent.

10.4.1.5 The R^2 Value

The R^2 value is a statistical measure that quantifies how much of the inherent variation in a variable is explained by its linear dependence on another.

Consider the following exact relationship between two variables x and y, where x is a random variable with variance σ_x^2. In statistical terms, y is a dependent variable and x is the explanatory (or independent) variable that it depends on. β_0 and β_1 are constants.

$$y = \beta_0 + \beta_1 x \tag{10.43}$$

The variance of y is simply given by $\sigma_y^2 = \beta_1^2\sigma_x^2$. The variance of y is *fully* explained by the variance of x (although it is not equal to it), or alternately stated, x fully explains the variance of y.

Now consider the variant

$$y = \beta_0 + \beta_1 x + \varepsilon \tag{10.44}$$

where ε is a zero-mean noise variable that is independent of x, and has variance σ_ε^2. The variance y is now given by $\sigma_y^2 = \beta_1^2\sigma_x^2 + \sigma_\varepsilon^2$. Now x only explains a *part* of the variance of y. Specifically, the fraction of the variance of y that is *not* explained by x is $\frac{\sigma_\varepsilon^2}{\sigma_y^2}$, and the fraction that *is* explained by x is

$$1 - \frac{\sigma_\varepsilon^2}{\sigma_y^2} \tag{10.45}$$

The R^2 value, also known as the *coefficient of determination* captures the above relationship in a statistical setting.

Consider the following setting: we are given a test sample comprising N instances of the kind $(y, x_1, x_2, \ldots, x_K)$, where y is a *dependent* variable that must be explained, and $x_1, x_2, \ldots, x_K$ are the *explanatory* variables explaining it. For instance, y might be a person's blood pressure, and $x_1, \ldots, x_K$ may include variables like the subject's age, weight, height and the quantities of medications he may be taking.

We assume the following affine relation between the dependent and explanatory variables:

$$y = \beta_0 + \sum_{i=1}^{K} \beta_i x_i + \varepsilon \tag{10.46}$$

Here $\hat{y} = \beta_0 + \sum_{i=1}^{K} \beta_i x_i$ is the value predicted for y by the explanatory variables and $\varepsilon = y - \hat{y}$ is the error made in predicting it. The optimal values of $\beta_0, \beta_1, \ldots, \beta_K$ are estimated through the following least squares procedure

$$Error = E[\varepsilon^2] = E[(y - \hat{y})^2] \tag{10.47}$$

$$\{\hat{\beta}_i,\ i = 0, \ldots K\} = \arg\min_{\{\beta_i,\ i=0,\ldots K\}} Error \tag{10.48}$$

where $E[.]$ represents the expectation operator that is computed in practice through averaging, e.g. $E[\varepsilon^2] \simeq \frac{1}{N}(\varepsilon^l)^2 = \frac{1}{N}(y^l - (\beta_0 + \sum_i \beta_i x_i^l))^2$, where the superscript l represents that these are the values for y and the x_is in the lth instance in our test sample, and N is the number of instances in the sample.

Since $Error$ is quadratic in the βs, the above minimization is trivially performed by setting the derivative of $Error$ w.r.t. the β_is to zero. It can be shown that when the error is minimized using a least-squares procedure, $E[y] = E[\hat{y}]$. As a consequence $E[\varepsilon] = 0$ and the variance of ε is $\sigma_\varepsilon^2 = Error$. The variance of y is obtained as $\sigma_y^2 = E[(y - E[y])^2]$.

The R^2 value is given by

$$R^2 = 1 - \frac{\sigma_\varepsilon^2}{\sigma_y^2}. \tag{10.49}$$

Comparison with Eq. 10.45 shows that the R^2 value provides the fraction of the variance of the dependent variable y that is explained through affine dependence on the explanatory variables $x_1, \ldots, x_K$.

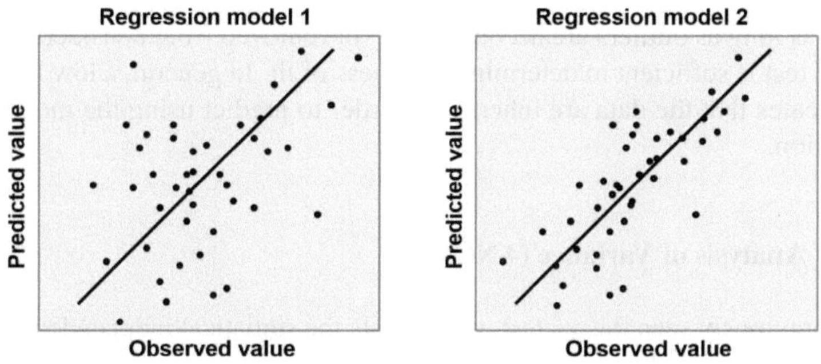

R-squared test: the goodness of fit is poor for the panel on the left. The coefficient of determination is 34%, i.e., only 34% of the variation is explained by the model (the regression line). It is better (64%) for the panel on the right.

Fig. 10.3 The R-squared test in concept

Thus, the R^2 value provides two measures.

It gives us a goodness of fit test for the affine relation assumed between the explanatory and dependent variables. R^2 values lie within $[0, 1]$. An R^2 of 1.0 gives us a perfect fit, while an R^2 of 0 indicates that the explanatory variables have no linear relation to y. They may also be expressed in terms of percentages, in which case $R^2 \in [0, 100]$. Figure 10.3 illustrates the concept behind an R-squared test.

The R^2 also tells us what fraction (or percent) of the variance of the dependent variable is explained by the explanatory variables.

When the explanatory variables $x_1, \ldots, x_K$ are also statistically independent of one another, then the contributions of the *individual* x_is to the variance of y can be computed as

$$r_i^2 = \frac{\hat{\beta}_i^2 \sigma_i^2}{\sigma_y^2}$$

where $\hat{\beta}_i$ is the optimal coefficient assigned to x_i by Eq. 10.48 and σ_i^2 is the variance of x_i. This is seen to be simply the square of the correlation of x_i to y.

The simple R-squared tests has some notable drawbacks. It is only valid when the parameters β_i are estimated using a simple least squares procedure. Otherwise the assumption that $E[\varepsilon] = 0$ becomes invalid as does the corresponding R^2 value. Thus, any form of regularization of the parameters will invalidate the test. It is also only strictly valid for affine or linear predictors for y such as the one in Eq. 10.46. Otherwise, once again, several of the assumptions (e.g. additivity of variances, zero mean error) become invalid. Even under the affine model, the estimation does not account for any consistent bias in the sample. The least-squares procedure for estimation is known to be highly sensitive to outliers in the data, since they have disproportionate representation in the error. For this reason, it is possible for a good model to yield outcomes with low R-squared value, and a bad one to yield high R-squared values.

However, as long as outliers are accounted for (or removed from consideration), the R-squared test is sufficient to determine goodness of fit. In general, a low R-squared value indicates that the data are inherently harder to predict using the model under consideration.

10.4.1.6 Analysis of Variance (ANOVA)

We have previously seen the χ^2 test, which tests the statistical independence of two discrete variables. Analysis of Variance, often called *one-way analysis of variance*, or ANOVA, is a test for the statistical independence between a discrete variable and a continuous-valued variable.

Consider a discrete variable X and a continuous valued variable Y that is potentially statistically dependent on X (i.e. X is the independent variable and Y is the dependent variable). For instance, X may be a drug ingested by a person and Y may represent a continuous-valued symptom such as their blood pressure. The test is to confirm if Y is indeed dependent on X. In our example, the objective is to determine if the drugs influence blood pressure.

The null hypothesis is that Y is not influenced by X. Let X take K discrete values, $x_1, \ldots, x_k$. Consider any test sample $T = \{(X_i, Y_i),\ i = 1 \ldots N\}$ comprising N instances of (X, Y) pairs. We segregate T into K mutually-exclusive subsets, or *groups*, according to the value of X, i.e. we segregate T into the groups $T_1, \ldots, T_K$ such that $T_k = \{(X_i, Y_i) : X_i = x_k\}$. According to the null hypothesis, the distribution of Y values in each of the groups would be statistically indistinguishable from the distribution of Y values in T itself. Figure 10.4 illustrates the situation. In our example, each group would consist of data from a set of subjects who took a particular drug. The null hypothesis would be that the drug has no influence on blood pressure, and that the distributions of blood pressure readings in the individual groups are indistinguishable from that of the populace at large (or from each other).

The alternate hypothesis is that Y is indeed influenced by X. In this case, the distribution of Y in at least one of the groups would be distinctly separate from the rest.

In order to test the hypotheses, an appropriate test statistic that can be used to evaluate the hypotheses is required. ANOVA provides a statistic for such a test under a specific set of assumptions:

- Y has a Gaussian distribution within each group T_i, i.e. the X-conditioned distribution of Y is normal. In practice, one-way ANOVA is relatively robust to non-normal (skewed or kurtotic) distributions, if the groups are large enough. If not, Y must be transformed to have a normal distribution before running the test.
- The variance $P(Y|X)$ for every value of X is equal (and as a consequence, the variance of Y within each group is statistically indistinguishable from the variance of Y in other groups).
- The test instances in each group are independent of other subsets (i.e. the Y values within any T_i must not depend on the Y values in any other T_j).

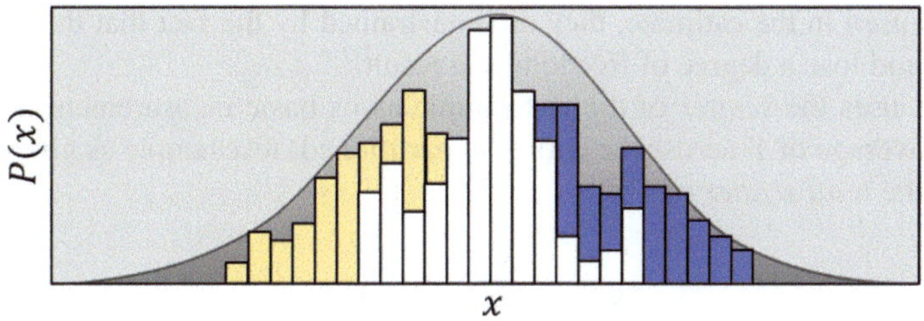

The histograms of different color represent the distributions of the variable Y for different values of X (which is assumed to take three discrete values in this example). The histograms overlap greatly and are statistically indistinguishable from each other and the overall distribution of Y (shown as the Gaussian in the background).

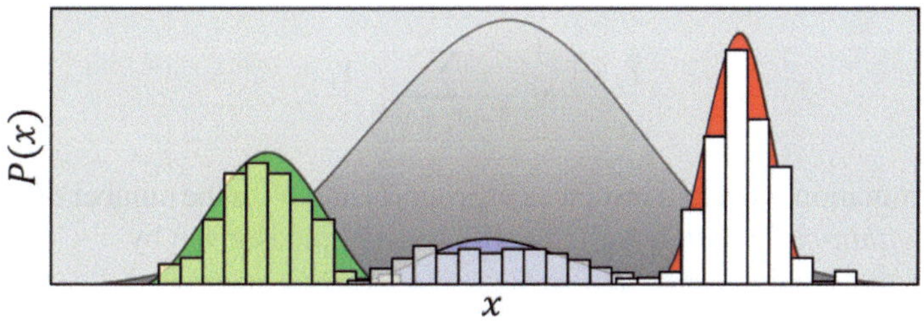

The histograms of Y segregated by X value are distinctly different. Although in this figure the variances of each subgroup of Y are different, ANOVA further assumes them to be identical.

Fig. 10.4 ANOVA: Null hypothesis versus alternate hypothesis

Under the above conditions, ANOVA posits the null hypothesis $H_0 : \mu_1(Y) = \mu_2(Y) \cdots = \mu_K(Y)$, where μ_k is the mean $P(Y|X = x_k)$. Plainly stated, the null hypothesis states that the probability distributions for Y values segregated by X value have identical means, for every value of X. Thus, the Y values in every group T_i are effectively obtained from the same distribution.

The alternate hypothesis H_a is that there are *at least* two group means that are statistically significantly different from each other.

The ANOVA statistic is based on the fact that the unbiased estimate for the variance of a distribution $P(Y)$ based on a set of N samples $Y_1, \ldots, Y_N$ drawn from it is given by

$$var(Y) = \frac{1}{N-1} \sum_{i=1}^{N} (Y_i - \bar{Y})^2 \tag{10.50}$$

where $\bar{Y} = \frac{1}{N} \sum_i Y_i$ is the estimated mean of $P(Y)$. The denominator $N - 1$ is called the *degrees of freedom* for $var(Y)$, and represents the actual number of variables that influence the estimate. The explanation for the denominator is that while there are

N samples used in the estimate, they are constrained by the fact that their average must be $\bar{Y}$ and lose a degree of freedom as a result.[2]

ANOVA uses the *scatter* of the test sample as its basic measurement. Let $\bar{Y}$ be the *global* average of Y across the entire (unpartitioned) test sample as given above. We define the *total scatter* of Y as

$$S_{tot} = \sum_{i=1}^{N} (Y_i - \bar{Y})^2 \tag{10.51}$$

This scatter can be written in terms of two components—the scatter of instances within the individual groups, and the scatter of the groups about the global average as follows.

Let $\bar{Y}_j$ represent the average Y within the jth group:

$$\bar{Y}_j = \frac{1}{N_j} \sum_{Y_i:(X_i,Y_i)\in T_j} Y_i \tag{10.52}$$

Here the summation is over all instances in group T_j and N_j is the number of instances in T_j. The *within-class scatter* for the entire test sample is given by

$$S_W = \sum_{j=1}^{K} \sum_{Y_i:(X_i,Y_i)\in T_j} (Y_i - \bar{Y}_j)^2 \tag{10.53}$$

S_W quantifies how the samples within each group are scattered around the average value for the group. If the null hypothesis holds true, the within-group average $\bar{Y}_j$ for each group is very similar to the global average $\bar{Y}$, and S_W is very similar to S_{tot}. The within-class variance, which quantifies the variances of the individual groups about their corresponding means, is defined as

$$var_W(Y) = \frac{1}{N-K} S_W \tag{10.54}$$

The denominator $N - K$ represents the fact that there are $N - K$ degrees of freedom in the estimate of $var_W(Y)$, because while S_W sums over N instances, it is constrained by the fact that within each of the K groups the average is constrained to be the group average $\bar{Y}_k$ for that group, thereby losing K degrees of freedom.

The *between-class scatter* is defined as

$$S_B = \sum_{j=1}^{K} N_j (\bar{Y}_j - \bar{Y})^2 \tag{10.55}$$

[2]If Eq. 10.50 μ, the *true* mean of $P(Y)$, instead of $\bar{Y}$, there would be N degrees of freedom and the denominator would be N.

S_B quantifies how the means of the individual groups are scattered about the global mean. Note that $S_{tot} = S_B + S_W$.

The between-class variance is given by

$$var_B(Y) = \frac{1}{K-1} S_B \tag{10.56}$$

Here the denominator $K - 1$ represents $K - 1$ degrees of freedom, arising from the fact that K separate terms are being added in S_B, however these are constrained to average to $\bar{Y}$, thereby losing one degree of freedom. Under the null hypothesis the group averages of the individual groups must be close to the global average $\bar{Y}$ and both S_B and $var_B(Y)$ are small.

ANOVA utilizes the F-statistic, defined as

$$F_{test} = \frac{var_B(Y)}{var_W(Y)} \tag{10.57}$$

Note that F_{test} is always positive. Also, if the null hypothesis holds true, the numerator of Eq. 10.57 is small, while the denominator is large, and the F_{test} is close to 0.

F_{test} is known to be distributed according to the two-parameter F-distribution $F(N - K, K - 1)$ [32], where the two parameters are, respectively, the degrees of freedom in the numerator and the denominator of Eq. 10.57. Given N and K, the F_α value for any given confidence level α can be obtained from standard F-distribution tables. If $F_{test} > F_\alpha$, the null hypothesis can be rejected, thereby concluding that X and Y are indeed dependent. The F-distribution tables [33] can also be used to identify the p-value of F_{test}.

Note that ANOVA does not actually identify which of the various groups are different from one another, only that at least one of the groups is statistically significantly separated from the rest. Also, since F-distribution tables are not defined for fewer than 2 degrees of freedom, there must be at least three groups, i.e. X must take at least three values for ANOVA to be applicable.

ANOVA is easily extended to consider more than one discrete independent or explanatory variable. Specifically, when there are two explanatory variables, the resulting test is known as *two-way* ANOVA. ANOVA is also extendable to consider *vector* valued dependent variables, a test that is known as MANOVA or multi-variate ANOVA. MANOVA is very analogous to one-way ANOVA, except that now $var_B(Y)$ and $var_W(Y)$ are matrices, and the computation used to derive the F statistic now results in a matrix $A = var_B(Y)var_W(Y)^{-1}$. The actual test must be performed on scalar statistics derived from the matrix A and a variety of solutions have been proposed for this purpose [34, 35].

10.4.1.7 Repeated Measurements Correlation (R-ANOVA, RM-ANOVA)

In some situations, the data to be correlated may not be the output of a predictor that maps one variable to another (such as age to health) and the true value of the output (true health), but rather the output of a predictor and the input variable (e.g. predicted health to age).

A **repeated measures** design uses the same input to either predict the output under different conditions, or predict different outputs under the same conditions.

The former is a *longitudinal study*, e.g. using the same subjects again and again to measure the relationship between health and age, which is expected to change each time such a measurement is taken. The condition that is varied here is time.

The latter is a *crossover study*, e.g. using the same subjects to assess the relationship between speaker's voices and environment, and speaker's voices and health.

Repeated measurement (univariate or multivariate) analysis of variance (R-ANOVA or RM-ANOVA) is used to measure correlation in such cases. We do not discuss the details here, there are many excellent explanations of these measures in the literature, e.g. [36–38].

An alternative to repeated measurements design of experiments is **multilevel modeling**, where the function (linear or any other kind) fitted to the data has a slope and intercept that is allowed to vary depending on the type of experiment. In a longitudinal study, for example, this would allow for the prediction of differences in trends over varying conditions of an independent variable. A Multilevel model can be viewed as single consolidated predictor and does not require repeated measurements analysis of variance.

This completes our brief and selective discussion of statistical tests for reliability.

10.5 Computational Ethics in Profiling

There are two facets to computational ethics: one is the **ethical use of algorithms** by people when the algorithms themselves can be used in an unethical fashion, and the other is the **design of algorithms that can solve ethical problems** that arise from the unethical use of algorithms. In the case of profiling, both facets are important.

The issue of ethical use of algorithms has been under furious debate lately with transgressions of privacy and mass manipulation of users by commercial enterprises, e.g. [39]. In the context of profiling, the solutions are in the realm of societies and lawmakers, and may be difficult to resolve because of the pervasiveness of the use of voice in our daily lives.

The latter—algorithms that are specifically designed to solve ethical (or ethical-use) problems—are the panacea we need for addressing the ethical problems related to profiling. These are more in the realm of the scientist. In the following two sections we discuss each facet very briefly.

10.5.1 Ethical Use of Profiling Algorithms

Let us assume that an individual keeps a record of their DNA sequence in a locker. A health insurance company would like to see it prior to giving the individual insurance coverage, and asks the individual for their DNA record. The individual, who has a legal right to deny such a request, refuses to show the record to the insurance company. The company then steals the key to the locker, analyzes the record, and denies insurance to the currently perfectly healthy individual, based on a genetic propensity to some disease. Is this ethical? Clearly not. In fact, in modern society this would not only be considered unethical, but a crime, and discrimination at the deepest levels. Societies institute laws to prevent such acts.

Voice carries a wealth of information about the speaker. The individual must use it—expose it to others—on a daily basis, and has no control over the information it gives away. Using that information to deny the person insurance would be the equivalent of stealing the key to the locker in the case above. Just because the locker *cannot be locked*, the claim that access was therefore voluntarily provided cannot be made. The only way, in fact, for the individual to deny others access to the information present in their voice is to not speak—a measure that would penalize the individual seriously in many ways for no fault of theirs.

Ethical issues in profiling may not always be complicated. Some aspects of it may be clear. For example, consider a speech translation application that asks the user for consent to record their voice in order to translate its content. In the process it also extracts profile information (such as ethnicity, age and native language of the user) from the user's voice. It does not specify whether this information will be stored or deleted, or how exactly it may be used. Clearly the waiver must include this information, especially if the information is stored for further use by the provider of the application. Implying otherwise, or obtaining consent by misrepresenting facts is unethical. Voice represents information. Neither voice nor its information must be coerced by misrepresentation in any fashion whatsoever.

When it comes to ethical use of voice profiling algorithms, the speaker's awareness is key. Profiling must not be done if the speaker is not aware of and in agreement with the potential use of the information given away. The information extracted from profiling must not be distributed without reason and permission, and must be obliterated when the need for it is spent.

The question of *what supersedes awareness* is what makes this issue complicated. It may be debated that in cases where the speaker is unable to give consent, profiling may be permissible if doing so would prolong the life of the speaker without compromising their freedom, opportunities, health status or quality of life, or alternatively, help correct a deliberate wrongdoing by the speaker and thereby protect the life, freedom, opportunities or health or quality of life of others. It may be debated whether awareness is replaceable by the speaker's well-being in other settings (such as automated driving, safe driving, medical diagnostics etc.), or replaceable by societal well-being when the speaker is clearly and deliberately a perpetrator of a crime against society. It may be debated whether the suspicion of propensity to crime is enough to waive the requirement for consent. There may be many other aspects of the use of profiling that may be debated. These debates, as mentioned earlier, are in the realm of societies and lawmakers.

10.5.1.1 When Unethical Use Turns into Crime

Misuse of the results of profiling is one thing. Deliberate disruption of the applicability and accuracy of profiling systems is quite another. In situations where a profiling system may be deployed for critical use (e.g. as a medical diagnostic aid or tele-monitoring system for health related problems), the deliberate use of technology to maliciously cause the system to report false results is an example of an *adversarial attack* on a profiling system. These cases no longer belong in a debate on ethics. They are clear violations of ethical-use norms, and fall under the umbrella of crime.

Unfortunately, such adversarial algorithms are being developed in the context of **all** voice-processing systems, and in their wake, adversarial attacks on profiling systems (when deployed for public use) are expected to follow. These may be both *black-box attacks* (where the attacker is only exposed to the results of profiling, and not to the details of the profiling systems or algorithms), or white-box attacks, where the attacker is aware of the algorithms used, and their implementation details. In most cases, adversarial attacks are expected to be carried out using tampered voice signals, wherein the attacker may insert or modify the information in the voice signal, that when extracted and processed, may cause the system to generate erroneous outcomes.

The scientific community continues to work towards generating such algorithms for the purpose of pre-empting true malicious attacks on all audio-processing systems. Some examples are [6, 40]. One effective way to counter such attacks is to deploy audio-tampering detection technologies to screen data before they are processed by any profiling system. Other techniques involve fortifying the internal algorithms used, such as neural-network based AI algorithms [41]. It may be argued that voice disguise for the purpose of misleading profiling systems also falls within the category of adversarial attacks. However, voice disguise by humans is not algorithmic, and can be detected more easily than malicious and targeted micro-alterations of the voice signal. Again, since this subject is outside the purview of a discussion on ethics. We will not discuss it further here.

10.5.2 Algorithms to Solve Ethical Problems

In general, ethical issues can be confounding. Even the simplest rules in ethics can have recursive exceptions. Problems of this nature are best solved using the method of science. We must define the problems as modularly as possible, and then hypothesize their solutions. Then we must prove or disprove the hypotheses using proven methods.

The science of computational ethics builds upon this approach [42]. It builds upon existing solutions to ethical problems that have *quantifiable outcomes*. In ethics, exactness in posing a problem—i.e. the problem statement itself is extremely important. In many cases, the principles of logic can then be used to solve them, if they are exact enough. Based on their scope and approach, algorithms that can address ethical problems may be divided into many categories. Some of those are listed below (in no particular order):

1. **Statistical preference algorithms**: Many problems can be solved by casting the notion of "the greater good" in statistical terms, and using statistical methods to devise solutions with quantifiable outcomes. For example when it is not clear whether a car that is newly on the market may be allowed to sell or not on the suspicion of an unproven fault, and if taking it off the market is likely to cost billions to the car company, then statistical algorithms that can take the predictions and outcomes of many real and hypothetical situations into account, may be used to decide which course of action might be best.
2. **Fairness algorithms**: Algorithms for "fairness," seek to resolve ethical issues related to the outcome of applications, such as those based on game theory [43], by enforcing the principles of fairness [44] on the algorithms used in them.
3. **Bias reducing algorithms**: Bias in an algorithm translates to bias in its use, making it unethical by default, by our standards. Unfortunately, machine learning and artificial intelligence algorithms have inherent biases in them [45]. Examples include those that are used in the personalization of online content, search results [46, 47], computer vision applications [48] etc. Statistical algorithms have explicitly recognized biases, such as the sampling bias. Algorithms can be devised to overcome these biases. For example some algorithms that overcome sampling bias are [49–51].
4. **Privacy preserving algorithms**: An algorithmic way to *prevent* most of the ethical problems that are likely to be raised in the context of profiling , is to use *privacy preserving voice processing algorithms*. Privacy preserving technologies are based on mathematical foundations that provide solutions to voice processing problems without requiring any decipherable exposure to the voice signal itself. These algorithms are based on principles of cryptography, and currently represent a science that is also in its infancy. Some of the state of art is represented by [52–54].
5. **Policing algorithms**: These are algorithms that constantly screen the outcomes of algorithms in an application, and alter the mode of their operation automatically so that transgressions of ethically-motivated rules are minimized. The screening may involve regression analysis of their effects on users, reception by users, estimated trends and outcomes for the future, and many other factors.

Many other categories of algorithms exist or can be defined for use in computational ethics. The field of computational ethics builds upon the solutions reached for ethical problems within other technologies that deal with information, e.g. as in [55–57]. Where ambiguities arise, human judgment may always be the final word.

10.6 On the Misinterpretation of Visualized Results

Conclusions based on plots generated on computers can be misleading, especially when they are used to interpret the nature and trends of the finer characteristics of data. Graphing is done by connecting points on a fine grid of pixels. This grid is obviously discretized, and can only represent a sampled view of a continuous valued

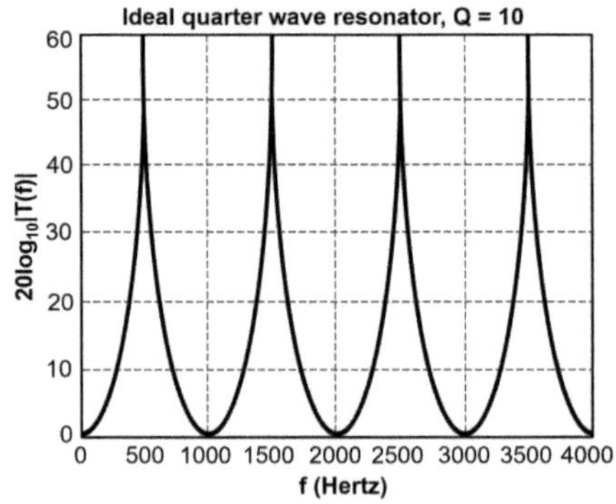

(a) The original plot of Figure 2.10

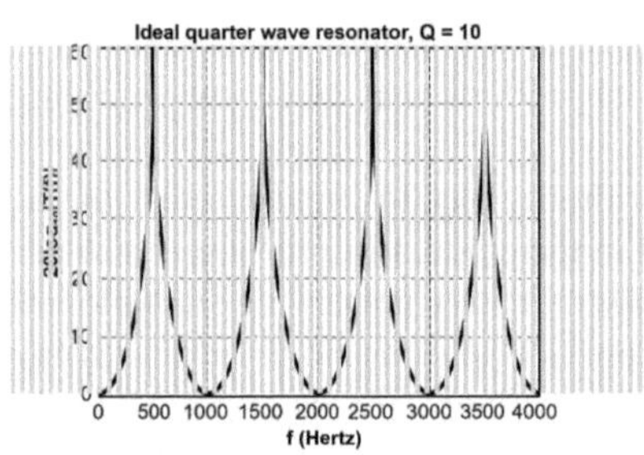

(b) Sampling on a grid. A coarse grid is shown to illustrate the concept of misinterpretation of such plots due to discrete sampling as on a computer.

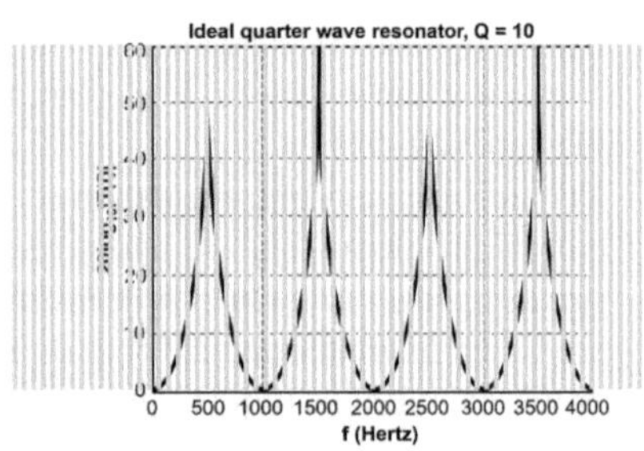

(c) The grid above shifted slightly with reference to the underlying plot.

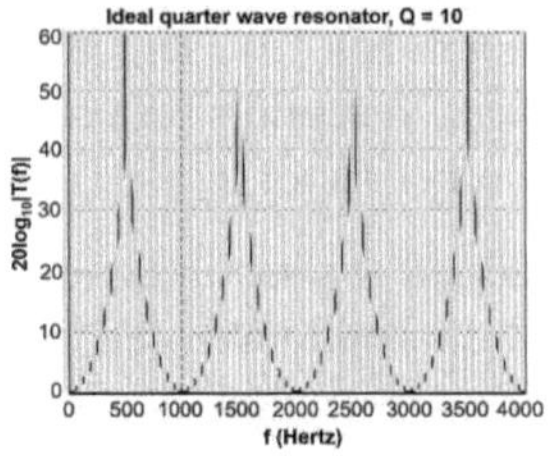

(d) Fine-grid sampling

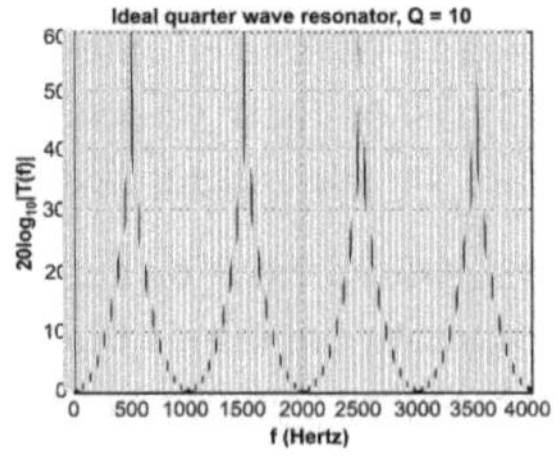

(e) Shift 1

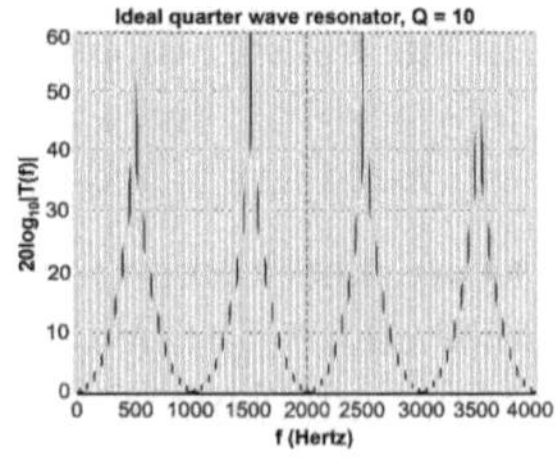

(f) Shift 2

Fig. 10.5 Misinterpretations in the visualization of fine variations: an illustration

co-ordinate system. Consider a spectral plot against a frequency axis that goes from 0 to π, for a spectral component at a frequency that is a rational fraction of π. A plotting routine that samples the axis in increments that are not also rational functions of π will never represent that frequency. Artifacts such as this can result in visual displays that are likely to be highly misleading when the features on the plot are very fine in nature—such as trends and patterns in micro-features.

As a specific example, let us consider the plot of the transfer function of a lossless uniform tube closed at one end, reproduced from Chap. 2. This is shown in Fig. 10.5a. The system response is infinite at pole frequencies, which are uniformly spaced on the frequency axis. Assuming that the plotting routine places a ceiling on the displayed height of the system response, a plot of the system response must show uniformly spaced peaks of uniform height as (supposedly) shown. To convey the notion of discretization errors, we place a grid on this plot, as shown in Fig. 10.5b. The grid represents the pixel-based sampling grid on a computer-displayed plot. The effect of this can be clearly seen in the Moiré patterns (Moiré effect) on the plot. Moiré patterns result from optical illusions created by the interference of patterns in visual images. These are due to visual undersampling of objects that happens when parts of an image are systematically obscured (as in a grid pattern, a concentric circular pattern etc). As we see, the transfer function as interpreted through the grid no longer represents the correct response of the system. Shifting the grid slightly, as in Fig. 10.5c, results in a transfer function that is significantly different from the one in Fig. 10.5b. Both are misinterpretations of the transfer function of Fig. 10.5a. This effect persists for finer sampling for plots such as these, as shown in Fig. 10.5d, e and f. These effects of course also come under the purview of the sampling theorem.

In profiling, wherein the interpretation of functions of micro-features is a central theme, the possibility of such misinterpretations in reading results always exists, since digital systems must read and process them in a sampled fashion.

References

1. Singh, R., Jimenéz, A., & Öland, A. (2017). Voice disguise by mimicry: Deriving statistical articulometric evidence to evaluate claimed impersonation. *IET Biometrics*, *6*(4), 282–289.
2. Menn, L. (1983). Development of articulatory, phonetic, and phonological capabilities. *Language Production*, *2*, 3–50.
3. Giannelli, P. C. (1993). Daubert: Interpreting the federal rules of evidence. *Cardozo Law Review*, *15*, 1999.
4. Lally, S. J. (2003). What tests are acceptable for use in forensic evaluations? A survey of experts. *Professional Psychology: Research and Practice*, *34*(5), 491.
5. Hamilton, H. G. (1997). The movement from Frye to Daubert: Where do the states stand. *Jurimetrics*, *38*, 201.
6. Cisse, M. M., Adi, Y., Neverova, N., & Keshet, J. (2017). Houdini: Fooling deep structured visual and speech recognition models with adversarial examples. In *Proceedings of the Thirty-First Annual Conference on Neural Information Processing Systems (NIPS)* (pp. 6977–6987). Long Beach, California.

7. Kreuk, F., Adi, Y., Cisse, M., & Keshet, J. (2018). Fooling end-to-end speaker verification with adversarial examples. In *Proceedings of the International Conference on Acoustics, Speech and Signal Processing (ICASSP)* (pp. 1962–1966). Calgary, Canada: IEEE.
8. Brendel, W., Rauber, J., & Bethge, M. (2017). Decision-based adversarial attacks: Reliable attacks against black-box machine learning models. arXiv:1712.04248.
9. Xia, L. (2018). China focus: Technologies at summer Davos offer a glimpse into future, News Article, XinhuaNet, China, 20 Sept 2018.
10. Wonnacott, T. H., & Wonnacott, R. J. (1990). *Introductory statistics* (Vol. 5). New York: Wiley.
11. Spiegel, M. R., Schiller, J. J., & Srinivasan, R. (2013). *Probability and statistics*. New York: McGraw-Hill.
12. Wasserman, L. (2013). *All of statistics: A concise course in statistical inference*. Berlin: Springer Science & Business Media.
13. Vuong, Q. H. (1989). Likelihood ratio tests for model selection and non-nested hypotheses. *Econometrica: Journal of the Econometric Society, 57*, 307–333.
14. Garside, G. R., & Mack, C. (1976). Actual type 1 error probabilities for various tests in the homogeneity case of the 2×2 contingency table. *The American Statistician, 30*(1), 18–21.
15. Chen, P. N. (1996). General formulas for the Neyman-Pearson type-II error exponent subject to fixed and exponential type-I error bounds. *IEEE Transactions on Information Theory, 42*(1), 316–323.
16. Lehmann, E. L. (1993). The Fisher, Neyman-Pearson theories of testing hypotheses: One theory or two? *Journal of the American Statistical Association, 88*(424), 1242–1249.
17. Self, S. G., & Liang, K. Y. (1987). Asymptotic properties of maximum likelihood estimators and likelihood ratio tests under nonstandard conditions. *Journal of the American Statistical Association, 82*(398), 605–610.
18. Cohen, J. (1992). A power primer. *Psychological Bulletin, 112*(1), 155.
19. Donner, A., & Eliasziw, M. (1987). Sample size requirements for reliability studies. *Statistics in Medicine, 6*(4), 441–448.
20. Olson, C. L. (1976). On choosing a test statistic in multivariate analysis of variance. *Psychological Bulletin, 83*(4), 579.
21. Pearson, K. (1896). Mathematical contributions to the theory of evolution III. Regression, heredity, and panmixia. *Philosophical Transactions of the Royal Society of London. Series A, Containing Papers of a Mathematical or Physical Character, 187*, 253–318.
22. Kendall, M. G. (1955). *Rank correlation methods*. Oxford, England: Griffin Publishers.
23. Siegal, S. (1956). *Nonparametric statistics for the behavioral sciences*. New York: McGraw-Hill.
24. Siegel, S. C., & Castellan, N. J., Jr. (1988). *Nonparametric statistics for the behavioural sciences*. New York: McGraw-Hill.
25. Abramowitz, M., & Stegun, I. A. (1965). *Handbook of mathematical functions: With formulas, graphs, and mathematical tables* (Vol. 55). USA: Courier Corporation.
26. Johnson, N. L., Kotz, S., & Balakrishnan, N. (1994). Chi-squared distributions including chi and Rayleigh. *Continuous univariate distributions* (2nd ed., Vol. 1, pp. 415–493). New York: Wiley.
27. Mood, A. M., Graybill, F. A., & Boes, D. (1974). *Introduction to the theory of statistics* (3rd ed., pp. 241–246). USA: McGraw-Hill.
28. Beyer, W. H. (2018). *Handbook of mathematical science*. USA: CRC Press.
29. Kres, H. (2012). *Statistical tables for multivariate analysis: A handbook with references to applications*. Berlin: Springer Science & Business Media.
30. Krishnamoorthy, K. (2006). *Handbook of statistical distributions with applications*. USA: Chapman and Hall/CRC Press.
31. Simon, M. K. (2007). *Probability distributions involving Gaussian random variables: A handbook for engineers and scientists*. Berlin: Springer Science & Business Media.
32. Patnaik, P. B. (1949). The non-central χ^2 and F-distribution and their applications. *Biometrika, 36*(1/2), 202–232.
33. Rohlf, F. J., & Sokal, R. R. (1995). *Statistical tables*. USA: Macmillan.

34. O'brien, R. G., & Kaiser, M. K. (1985). MANOVA method for analyzing repeated measures designs: An extensive primer. *Psychological Bulletin*, *97*(2), 316.
35. Huberty, C. J., & Olejnik, S. (2006). *Applied MANOVA and discriminant analysis* (Vol. 498). Hoboken, New Jersey: Wiley.
36. Davis, C. S. (2002). *Statistical methods for the analysis of repeated measurements*. Berlin: Springer Science & Business Media.
37. Vonesh, E., & Chinchilli, V. M. (1996). *Linear and nonlinear models for the analysis of repeated measurements*. Boca Raton, Florida: CRC Press.
38. Jones, B., & Kenward, M. G. (2014). *Design and analysis of cross-over trials*. Monographs on statistics and applied probability. USA: Chapman & Hall, CRC Press.
39. Tufekci, Z. (2015). Algorithmic harms beyond Facebook and Google: Emergent challenges of computational agency. *Colorado Technology Law Journal*, *13*, 203.
40. Kreuk, F., Adi, Y., Cisse, M., & Keshet, J. (2018). Fooling end-to-end speaker verification by adversarial examples. arXiv:1801.03339.
41. Adi, Y., Baum, C., Cisse, M., Pinkas, B., & Keshet, J. (2018). Turning your weakness into a strength: Watermarking deep neural networks by backdooring. In *Proceedings of the Twenty-Seventh Security Symposium (USENIX)* (pp. 1615–1631). Baltimore, USA.
42. Bostrom, N., & Yudkowsky, E. (2014). The ethics of artificial intelligence. *The Cambridge handbook of artificial intelligence*, *1*, 316–334.
43. Rabin, M. (1993). Incorporating fairness into game theory and economics. *The American Economic Review*, *83*, 1281–1302.
44. Fehr, E., & Schmidt, K. M. (1999). A theory of fairness, competition, and cooperation. *The Quarterly Journal of Economics*, *114*(3), 817–868.
45. Haussler, D. (1988). Quantifying inductive bias: AI learning algorithms and Valiant's learning framework. *Artificial Intelligence*, *36*(2), 177–221.
46. Bozdag, E. (2013). Bias in algorithmic filtering and personalization. *Ethics and Information Technology*, *15*(3), 209–227.
47. Edelman, B. (2011). Bias in search results: Diagnosis and response. *Indian Journal of Technology*, *7*, 16.
48. Tommasi, T., Patricia, N., Caputo, B., & Tuytelaars, T. (2017). A deeper look at dataset bias. *Domain adaptation in computer vision applications* (pp. 37–55). Cham: Springer.
49. Ananny, M. (2016). Toward an ethics of algorithms: Convening, observation, probability, and timeliness. *Science, Technology, & Human Values*, *41*(1), 93–117.
50. Kraemer, F., Van Overveld, K., & Peterson, M. (2011). Is there an ethics of algorithms? *Ethics and Information Technology*, *13*(3), 251–260.
51. Huang, J., Gretton, A., Borgwardt, K., Schölkopf, B., & Smola, A. J. (2007). Correcting sample selection bias by unlabeled data. In *Advances in Neural Information Processing Systems* (pp. 601–608).
52. Jiménez, A., & Raj, B. (2017). Privacy preserving distance computation using somewhat-trusted third parties. In *Proceedings of the IEEE International Conference on Acoustics, Speech and Signal Processing (ICASSP)* (pp. 6399–6403). New Orleans, Louisiana, USA: IEEE.
53. Pathak, M. A., Raj, B., Rane, S. D., & Smaragdis, P. (2013). Privacy-preserving speech processing: Cryptographic and string-matching frameworks show promise. *IEEE Signal Processing Magazine*, *30*(2), 62–74.
54. Pathak, M. A., & Raj, B. (2011). Privacy preserving speaker verification using adapted GMMs. In *Proceedings of the Twelfth Annual Conference of the International Speech Communication Association*, Florence, Italy.
55. Buchanan, E. A. (1999). An overview of information ethics issues in a world-wide context. *Ethics and Information Technology*, *1*(3), 193–201.
56. Chen, H., Chiang, R. H., & Storey, V. C. (2012). Business intelligence and analytics: From big data to big impact. *Management Information Systems (MIS) Quarterly*, *36*, 1165–1188.
57. Lipinski, T. A., & Britz, J. (2000). Rethinking the ownership of information in the 21st century: Ethical implications. *Ethics and Information Technology*, *2*(1), 49–71.
58. Galton, F. (1888). Co-relations and their measurement, chiefly from anthropometric data. *Proceedings of the Royal Society of London*, *45*(273–279), 135–145.

Index

R. Singh, *Profiling Humans from their Voice*,
https://doi.org/10.1007/978-981-13-8403-5

MIX
Papier aus verantwortungsvollen Quellen
Paper from responsible sources
FSC® C105338

If you have any concerns about our products,
you can contact us on
ProductSafety@springernature.com

In case Publisher is established outside the EU,
the EU authorized representative is:
Springer Nature Customer Service Center GmbH
Europaplatz 3, 69115 Heidelberg, Germany

Printed by Libri Plureos GmbH
in Hamburg, Germany